# Neurochemical Aspects of Neurotraumatic and Neurodegenerative Diseases

Akhlaq A. Farooqui

# Neurochemical Aspects of Neurotraumatic and Neurodegenerative Diseases

Springer

Akhlaq A. Farooqui
Department of Molecular and Cellular Biochemistry
Ohio State University
1645 Neil Avenue
Columbus, Ohio 43210, USA
farooqui.1@osu.edu

ISBN 978-1-4419-6651-3 e-ISBN 978-1-4419-6652-0
DOI 10.1007/978-1-4419-6652-0
Springer New York Dordrecht Heidelberg London

Library of Congress Control Number: 2010931168



Printed on acid-free paper

Springer is part of Springer Science+Business Media (www.springer.com)

*This monograph is dedicated to my wife (Tahira), daughter (Soofia), and son (Seraj). Thank you for sharing your lives with me. You all are always in my heart.*

*Akhlaq A. Farooqui*

# Preface

American population is aging and an increasing number of Americans are afflicted with stroke, spinal cord trauma, traumatic brain injury, and neurodegenerative diseases. These neurological conditions result in the acute as well as gradual and progressive neurodegeneration, which leads to brain dysfunction. Known risk factors for stroke and neurodegenerative diseases include increasing age, genetic polymorphisms, endocrine dysfunction, oxidative stress, neuroinflammation, excitotoxicity, hypertension, infection, and exposure to neurotoxins. In contrast, spinal cord trauma and traumatic brain injury due to motor cycle and car accidences are major causes of death and disability among young people below the mid-thirties in the USA. According to the NINDS approximately 30–40 million Americans are affected by stroke and neurodegenerative diseases each year. The number of people affected with neurological disorders will double every 20 years and will cost the US economy billions of dollars each year in direct health-care costs and lost opportunities. As the baby boomer's generation ages and the prevalence of neurotraumatic and neurodegenerative diseases increases in the American society, the need to confront and solve the present day health-care crisis becomes more critical than ever before. In fact, there is now an urgent need to expand significantly the national and international efforts to solve the problem of neurotraumatic and neurodegenerative diseases, with special emphasis on prevention. It is estimated that $100 billion/year will be spent on Alzheimer disease alone. In addition to the financial cost, there is an immense emotional burden on patients, their relatives, and caregivers.

Although molecular mechanisms associated with the pathogenesis of neurotraumatic and neurodegenerative diseases remain unknown, oxidative stress, excitotoxicity, inflammation, misfolding, aggregation, and accumulation of proteins, perturbed $Ca^{2+}$ homeostasis, and apoptosis have been implicated as possible causes of neurodegeneration in the above neurological disorders. There have been remarkable developments not only on neurochemical aspects but also on target-based pharmacological therapeutic intervention in neurotraumatic and neurodegenerative diseases in a variety of animal and cell culture models in past 20 years. In the clinical setting, however, these treatments have failed not only due to the heterogeneity (occurrence of neurons, astrocytes, oligodendrocytes, and microglial cells) of brain and spinal cord tissues but also because degenerating neurons and injured axons within brain and spinal cord are unable to regenerate spontaneously. The therapeutic strategies to re-establish lost neuronal connections in neurotraumatic and neurodegenerative diseases are currently unavailable. The main objective of this monograph is to present readers with cutting edge and comprehensive overview on neurochemical aspects of neurotraumatic (stroke, spinal cord trauma, and traumatic head injury) and neurodegenerative diseases (Alzheimer disease, Parkinson disease, Amyotrophic Lateral Sclerosis, Huntington disease, and prion disease) in a manner that is useful not only to students and teachers but also to researcher scientists and clinicians. This monograph has 10 chapters. Chapter 1 deals with molecular mechanisms associated with neurodegenerative processes in the brain and spinal cord. Chapters 2 and 3 describe molecular mechanism of neurodegeneration in stroke and potential therapeutic approaches for the treatment of ischemic injury in the brain. Chapters 4 and 5 describe cutting-edge information on neurochemical mechanisms of secondary injury in spinal cord trauma and potential therapeutic strategies for spinal cord injury. Chapters 6 and 7 describe molecular mechanism and treatment strategies for traumatic brain injury. Chapters 8 and 9 describe potential molecular mechanisms associated with the pathogenesis of neurodegenerative diseases and progress on pharmacological approaches that can be used for the treatment of neurodegenerative diseases. Finally, Chapter 10 provides readers and researchers with perspective that will be important for the future research work on neurotraumatic and neurodegenerative diseases in brain and spinal cord.

This monograph can be used as supplemental text for a range of neuroscience and neurochemistry courses. Clinicians (neurologists, pathologists, and psychiatrists) will find this book useful for understanding molecular aspects of neurotraumatic and neurodegenerative diseases. These topics fall in a fast-paced research area related to neurodegeneration that provides opportunities for target-based therapeutic intervention. Although many edited books are separately available on molecular mechanism of stroke, spinal cord trauma, traumatic brain injury, and neurodegenerative diseases but, to the best of my knowledge no one has written a monograph on the neurochemical aspects of neurotraumatic and neurodegenerative diseases. The present monograph is the first to provide a comprehensive and comparative description of neurochemical changes in stroke, spinal cord trauma, traumatic brain injury, and various neurodegenerative diseases along with progress on their pharmacological therapy. This monograph not only provides background and refresher information on neurotraumatic and neurodegenerative diseases in the brain and spinal cord to readers not working in this field but also presents a thorough and unique overview on progress that has been made on the neurochemistry and treatment of stroke, spinal cord trauma, traumatic brain injury, and various neurodegenerative diseases for researcher scientists, who are actively working in the field of neurodegeneration.

The choices of topics presented in this monograph are personal. They are based on my interest not only in the neurochemistry of stroke, spinal cord injury, traumatic brain injury, and various neurodegenerative diseases but also in areas where major progress has been made. I have tried to ensure uniformity and mode of presentation as well as a logical progression of subject from one topic to another and have provided extensive bibliography. For the sake of simplicity and uniformity a large number of figures with chemical structures of drugs used for the treatment of above neurological disorders and line diagrams of colored signal transduction pathways are also included. I hope that my attempt to integrate and consolidate the knowledge on the neurochemistry of neurotraumatic and neurodegenerative diseases will provide the basis of more dramatic advances and developments not only on molecular mechanisms but also on causes and treatment of neurotraumatic and neurodegenerative diseases.

Columbus, Ohio Akhlaq A. Farooqui

# Acknowledgments

I thank late Professor Lloyd A. Horrocks for introducing and mentoring me to studies on neurodegeneration in acute neural trauma and neurodegenerative diseases. I also express my gratitude to Ann H. Avouris and Melissa Higgs of Springer, New York, for their cooperation, rapid responses to my queries, and professional and able manuscript handling. It has been a pleasure working with them for many years.

Columbus, Ohio Akhlaq A. Farooqui

# Contents

# About the Author

Dr. Akhlaq A. Farooqui is a leader in the field of brain phospholipases $A_2$, bioactive ether lipid metabolism, polyunsaturated fatty acid metabolism, glycerophospholipid-, sphingolipid-, and cholesterol-derived lipid mediators, glutamate-induced neurotoxicity, and neurological disorders. He has discovered the stimulation of plasmalogen-selective phospholipase $A_2$ (PlsEtn-$PLA_2$) in brains from patients with Alzheimer disease. Stimulation of PlsEtn-$PLA_2$ produces plasmalogen deficiency and increases levels of eicosanoids that may be related to the loss of synapses, induction of neuroinflammation, and oxidative stress in brains of patients with Alzheimer disease. Dr. Farooqui has published cutting-edge research on the generation and identification of glycerophospholipid-, sphingolipid-, and cholesterol-derived lipid mediators in kainic acid neurotoxicity by lipidomics. He has previously authored five monographs: *Glycerophospholipids in Brain: Phospholipase $A_2$ in Neurological Disorders* (2007); *Neurochemical Aspects of Excitotoxicity* (2008); *Metabolism and Functions of Bioactive Ether Lipids in Brain* (2008); *Hot Topics in Neural Membrane Lipidology* (2009); *and Beneficial Effects of Fish Oil on Human Brain* (2009). All monographs are published by Springer. Dr. Farooqui has also edited two books: *Biogenic Amines: Pharmacological, Neurochemical and Molecular Aspects in the CNS* Nova Science Publisher, Hauppauge, NY (2010) *and Molecular Aspects of Neurodegeneration and Neuroprotection, Bentham Science Publishers Ltd* (2010).

# List of Abbreviations

| | |
|---|---|
| AD | Alzheimer disease |
| ALS | Amyotrophic lateral sclerosis |
| ARA | Arachidonic acid |
| BDNF | Brain-derived neurotrophic factor |
| Cer | Ceramide |
| PlsCho | Choline plasmalogen |
| COX | Cyclooxygenase |
| DHA | Docosahexaenoic acid |
| EPOX | Epoxygenase |
| PlsEtn | Ethanolamine plasmalogen |
| HD | Huntington disease |
| Ins-1,4,5-$P_3$ | Inositol-1,4,5-trisphosphate |
| LOX | Lipoxygenase |
| PD | Parkinson disease |
| PtdIns4$P$ | Phosphatidylinositol 4-phosphate |
| PtdH | Phosphatidic acid |
| PtdCho | Phosphatidylcholine |
| PtdEtn | Phosphatidylethanolamine |
| PtdIns | Phosphatidylinositol |
| PtdIns(4,5)$P_2$ | Phosphatidylinositol 4,5-bisphosphate |
| PtdSer | Phosphatidylserine |
| $PLA_2$ | Phospholipase $A_2$ |
| PLC | Phospholipase C |
| PLD | Phospholipase D |
| PKC | Protein kinase C |
| ROS | Reactive oxygen species |
| Sph | Sphingosine |

# Chapter 1
# Neurodegeneration in Neural Trauma, Neurodegenerative Diseases, and Neuropsychiatric Disorders

## 1.1 Introduction

Neurodegeneration is a complex, progressive, and multifaceted process that results in neural cell dysfunction and death in brain and spinal cord. Adult brain and spinal cord contain terminally differentiated postmitotic neurons with downregulated cell division controlling mechanisms (silencing of cyclin-dependent kinases) and upregulated anti-apoptotic mechanisms such as neurotrophic factor signaling, antioxidant enzymes, protein chaperones, anti-apoptotic proteins, and ionostatic systems (Nguyen et al., 2002). Under pathological conditions these adaptations are lost, resulting neuronal re-entry into the cell cycle before death (Becker and Bonni, 2005; Krantic et al., 2005). Like other tissues, in brain neural cell death occurs either through (a) apoptosis or (b) necrosis. The necrosis is characterized by the passive cell swelling, intense mitochondrial damage with rapid loss of ATP, alterations in neural membrane permeability, high calcium influx, and disruption of ion homeostasis. This type of cell death leads to membrane lysis and release of intracellular components that induce inflammatory reactions. In contrast, apoptosis is an active process in which caspases (a group of endoproteases with specificity for aspartate residues in protein) are stimulated. Apoptotic cell death is accompanied by cell shrinkage, dynamic membrane blebbing, chromatin condensation, DNA laddering, loss of phospholipids asymmetry, low ATP levels, and mild calcium overload (Sastry and Subba Rao, 2000; Farooqui et al., 2004; Farooqui, 2009). Thus, apoptosis and necrosis are two extremes of a wide spectrum of cell death processes with different mechanistic and morphological features. However, they may share some common mediators and signal transduction processes that are often inseparable. Neurodegeneration occurs at many different levels of neuronal circuitry. It is often accompanied by atrophy of the affected central or peripheral nervous system structures. Neurodegeneration is regulated by many different factors, including, but not limited to, inherited genetic abnormalities, problems in the immune system, and metabolic or mechanical insults to the brain or spinal cord tissues. Neurodegeneration occurs not only in acute neural trauma (ischemia and traumatic injury to brain and spinal cord) but also in neurodegenerative diseases (Alzheimer disease, AD; Parkinson disease, PD; Huntington disease, HD; and

A.A. Farooqui, *Neurochemical Aspects of Neurotraumatic and Neurodegenerative Diseases*, DOI 10.1007/978-1-4419-6652-0_1,

amyotrophic lateral sclerosis, ALS) and neuropsychiatric disorders (schizophrenia and depression) (Farooqui and Horrocks, 2007; Farooqui, 2009). Neurodegeneration in many of above conditions is accompanied with dementia, a multi-faceted cognitive, memory, and functional progressive impairments, which advance with age (Wehr et al., 2006). Thus, dementia is a behavioral syndrome that is closely associated with cerebrovascular dysfunction in neurodegenerative diseases and stroke (Schaller, 2008). It should be noted that vascular dementia literature lacks a clear consensus regarding the neuropsychological and other constituent characteristics associated with various cerebrovascular changes. The rate of neurodegeneration and dementia varies considerably from one disease to another (Fig. 1.1). Dementia is a syndrome due to a chronic or progressive neural disease, with alterations in multiple cortical functions, such as memory, orientation, comprehension, learning, language, and judgment. Demented subjects are unable to perform spoken and written communication, preparing meals, driving, and leisure activities with the same level of independence as they had enjoyed earlier in life (Schaller, 2008). In addition, they also show deterioration in emotions, personal care, and social behavior.

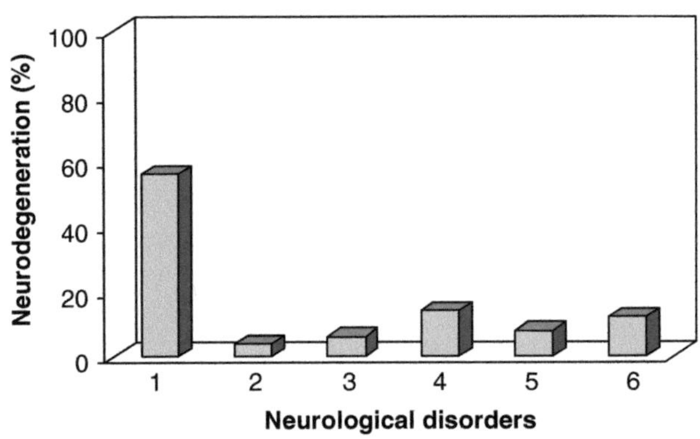


**Fig. 1.1** Rate of neurodegeneration in neurodegenerative conditions. Alzheimer disease (1); head injury (2); other causes (3); multifactorial dementia (4); Parkinson disease (5); and multiple cause dementia (6)

Neurodegeneration in acute neural trauma and neurodegenerative diseases is also associated with disturbed glycerophospholipid metabolism in neural membranes, activation of phospholipases $A_2$, and generation of glycerophospholipid degradation products, which include the production of reactive oxygen species (ROS) and lipid hydroperoxides. Both these metabolites induce oxidative stress (Farooqui and Horrocks, 2007; Farooqui, 2009). A major source for vascular and neuronal ROS is a family of non-phagocytic NADPH oxidases, including the prototypic Nox2 homolog-based NADPH oxidase, as well as other NADPH oxidases, such as Nox1 and Nox4 (Sun et al., 2007). Other possible sources include mitochondrial electron transport enzymes, xanthine oxidase, cyclooxygenase, lipoxygenase, and uncoupled nitric oxide synthase. NADPH oxidase-derived ROS plays a physiological role in the regulation of neural and endothelial function. At present, pathophysiological importance of neural membrane glycerophospholipid breakdown in acute neural trauma and neurodegenerative diseases is not fully understood. However, it is proposed that glycerophospholipid degradation in acute neural trauma may be an earliest event (Farooqui and Horrocks, 2007). In contrast, in neurodegenerative diseases (AD) alterations in neural membrane glycerophospholipids precede the clinical manifestations of the disease (dementia) (Pettegrew et al., 1995).

Neurodegenerative diseases and neuropsychiatric disorders fall in a large group of neurological disorders with heterogeneous clinical and pathological expressions affecting specific subsets of neurons in specific functional anatomic regions of brain and spinal cord. Although the exact cause and molecular mechanism of acute neural trauma, neurodegenerative diseases, and neuropsychiatric disorders are not fully understood, it is becoming increasingly evident that multiple factors and mechanisms may contribute to the pathogenesis of above neurological disorders (Bossy-Wetzel et al., 2004; Farooqui and Horrocks, 2007; Farooqui, 2009). For ischemic injury, the most important factor is lack of oxygen and blood flow resulting from blocked blood vessels (stroke), traumatic injury which is caused by shear force of trauma (head and spinal cord injuries), and familial form of neurodegenerative diseases which involve genetic mutations. The most important risk factors for sporadic neurodegenerative diseases are old age, positive family history, unhealthy lifestyle, endogenous factors, and exposure to toxic environment (Fig. 1.2) (Farooqui and Horrocks, 2007). In the brain tissue, aging process is associated with elevated mutation load in mitochondrial DNA, defects in mitochondrial

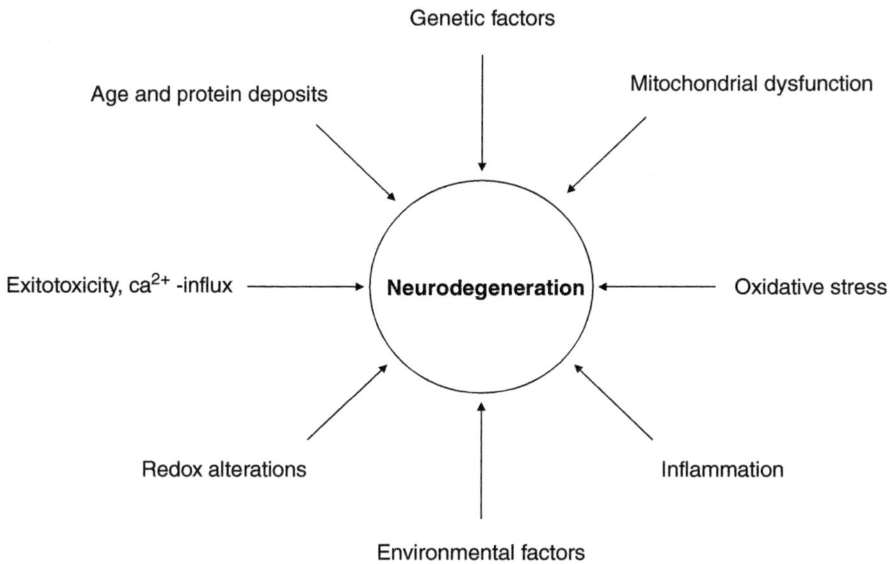


**Fig. 1.2** Factor effecting neurodegeneration in neurological disorders

respiration, and increased oxidative damage (Farooqui and Farooqui, 2009). In aging brain, decline in respiratory function not only results in production of less ATP but also causes elevation in the generation of ROS as by-products of aerobic metabolism. Aging also induces alterations in activities of free radical-scavenging enzymes. In addition, the accumulation of mitochondrial DNA mutations accelerates normal aging, promotes oxidative damage to nuclear DNA, and impairs gene transcription. Thus, normal aging process is accompanied by some level of neurodegeneration, which falls below the threshold of a clinical pathology (Graeber et al., 1998; Farooqui and Farooqui, 2009).

Based on epidemiological and molecular biological studies, it is suggested that in vast majority of sporadic neurodegenerative subjects, genetic contribution to the neurodegenerative process is minimal. Instead, toxic environmental factors and unhealthy lifestyle may contribute to the initiation of neurodegenerative processes (BenMoyal-Segal and Soreq, 2006; Farooqui and Farooqui, 2009). This view is based on the observation that some neurodegenerative diseases arise in geographic or temporal clusters. For example, Guam-type amyotrophic lateral sclerosis/parkinsonism dementia (ALS/PDC) is caused by the presence of β-methylaminoalanine (BMAA) in *Cycas circinalis*, an indigenous plant commonly ingested as a food or medicine by the Chamorros of Guam (Murck et al., 2004; Ince and Codd, 2005). Intoxication with 1-methyl-4-phenyl-1,2,3,6-tetrahydropyridine (MPTP) causes a severe and irreversible parkinsonian syndrome that is similar, but not identical, to PD in pathology and progression. In addition, exposure to certain insecticides and herbicides, such as paraquat and rotenone, also produces a Parkinson-like syndrome (Brown et al., 2006; Kamel and Hoppin, 2004; Keifer and Firestone, 2007). However, evidence for the involvement of environmental factors in pathogenesis of neurodegenerative diseases is weak and contradicted by several large-scale epidemiological studies. These studies have failed to show any definitive association between environmental factors and occurrence of neurodegenerative diseases such as AD, PD, HD, and ALS (Brown et al., 2006; Kamel and Hoppin, 2004; Keifer and Firestone, 2007).

Protein folding is a normal biological process associated with the conversion of newly synthesized proteins into physiologically functional molecules. This process is regulated by molecular chaperones that facilitate normal folding, prevent inappropriate interaction between non-native polypeptides, and promote the refolding of proteins that have become misfolded as a result of cellular stress (Muchowski and Wacker, 2005). Cell death in neurodegenerative diseases is accompanied by the accumulation of abnormal extracellular and intracellular deposits caused by misfolding and aggregation of some proteins in some neurons in specific area of the brain (Ross and Poirier, 2004; Farooqui and Farooqui, 2009). Accumulating evidence indicates that at least two pathways modulate protein folding: the ubiquitin-proteasome system (UPS) and molecular chaperone pathway. Downregulation of UPS results in misfolding and aggregation of specific proteins that are often trapped in misfolded conformations in neurodegenerative diseases (Bossy-Wetzel et al., 2004; Ross and Poirier, 2004). To handle a buildup of abnormal misfolded proteins, cells employ a complicated machinery of molecular chaperones and various proteolytic systems associated with endoplasmic reticulum (Scheper and Hoozemans, 2009). Chaperones promote refolding of misfolded polypeptides, inhibit protein aggregation, and mediate the formation of aggresome, a centrosome-associated body to which small cytoplasmic aggregates are transported (Merlin and Sherman, 2005). The ubiquitin-proteasome proteolytic system is critical for downregulating the levels of soluble abnormal proteins, while autophagy (a lysosomal pathway) plays the major role in clearing of cells from protein aggregates. The accumulation of prone protein aggregates modulates signal transduction pathways that control cell death, including JNK pathway that regulates viability of a cell in various models of PD and HD (Merlin and Sherman, 2005). Most molecular chaperones passively prevent protein aggregation by interacting with misfolding protein intermediates. Some molecular chaperones and chaperone-related proteases, such as those in proteasome, perform their function by hydrolyzing ATP and forcefully converting stable harmful protein aggregates into harmless natively refoldable, or protease-degradable, polypeptides (Hinault et al., 2006). Collective evidence suggests that molecular chaperones and chaperone-related proteases modulate the delicate balance between natively folded functional proteins and aggregation-prone misfolded proteins, which may accumulate during the lifetime leading to neurodegeneration (Hinault et al., 2006). The major chaperone protein, Hsp72, interferes with this signaling pathway and thus promotes neural cell survival. Other molecular chaperones include protein disulfide isomerase and glucose-regulated protein 78. These proteins also provide neuroprotection from aberrant proteins by facilitating proper folding and thus preventing their aggregation. Molecular chaperones are first line of defense against misfolded, aggregation-prone proteins and are among the most potent suppressors of neurodegeneration. In neurodegenerative diseases, consequences of aggregation and deposition of misfolded proteins are impairment of the ubiquitin-proteasome degradation system and suppression of the heat shock response (Merlin and Sherman, 2005). A common feature of neurodegenerative diseases is a long course in period until sufficient protein accumulates, followed by a cascade of symptoms over many years with increasing disability leading to death. Although normal aging is accompanied by the ability of the brain to modify its own structural organization and functioning that result in loss of some cognitive function, neurodegenerative diseases are accompanied by dramatic impairment in ability to modulate structural organization and functioning of the brain tissue causing a progressive loss of complete cognitive function (Farooqui, 2009).

Recent studies also indicate that generation of excessive nitric oxide (NO) and reactive oxygen species (ROS), in part, due to overactivity of the NMDA subtype of glutamate receptor, can mediate protein misfolding in the absence of genetic predisposition. S-Nitrosylation, or covalent reaction of NO with specific protein thiol groups, represents one mechanism contributing to NO-mediated protein misfolding and neurotoxicity (Uehara, 2007; Nakamura and Lipton, 2009). In addition, a functional relationship between inhibitory S-nitrosylation of the redox enzyme protein disulfide isomerase defects in regulation of protein folding within the endoplasmic reticulum and neurodegeneration. Examination of brains from PD and AD patients supports a causal role for the S-nitrosylation of protein disulfide isomerase and consequent endoplasmic reticulum stress in these prevalent neurodegenerative disorders (Benhar et al., 2006). Furthermore, increase in levels of S-nitrosylation of dynamin-related protein 1 (SNO-Drp1) triggers neurodegeneration in AD (Cho et al., 2009), and the blockade of nitrosylation of Drp1 by cysteine mutation prevents cell death in AD. Nitrosylation modifies function of many proteins by altering the hydrophobicity, hydrogen bonding, and electrostatic properties within the targeted protein. Nitrosylation in general and S-nitrosylation in particular are regarded as important redox signaling mechanisms in the regulation of many neural cell functions. However, deregulation of S-nitrosylation has been linked to neurodegenerative disorders. Although nitrosative stress has long been considered as a major mediator of neurodegeneration, the molecular mechanism of how NO can contribute to neurodegeneration is not fully established. It is recently suggested that nitration and nitrosylation of proteins contribute to the neurodegenerative process by inducing protein aggregation (Benhar et al., 2006; He et al., 2007; Nakamura and Lipton, 2009).

In addition, under pathophysiological conditions, the excessive generation of NO due to the overactivation of NMDA receptor in neurons or by inducible NO synthase from neighboring glia (microglial cells and astrocytes) results in the interaction between NO and superoxide anion, generated by the mitochondria (2% of the $O_2$ consumed by healthy mitochondria is converted to superoxide) or by other mechanisms, leading to the formation of the powerful oxidant species, peroxynitrite. Furthermore, the activation of $NAD^+$-consuming enzyme poly(ADP-ribose) polymerase-1 (PARP-1) is another likely mechanism for NO-mediated energy failure and neurotoxicity. Although under mild oxidative stress the activation of PARP-1 is a repair process for neuronal protection, under high oxidative stress it causes neuronal energy compromise leading to neurodegeneration (Moncada and Bolanos, 2006; Farooqui, 2009). Nitric oxide also binds to cytochrome *c* oxidase and is able to inhibit cell respiration in a process that is reversible and in competition with oxygen. This action leads to the release of more superoxide anion from the mitochondrial respiratory chain. Collective evidence suggests that brain aging is accompanied by a higher degree of ROS and NO production, and by diminished functions of mitochondria, endoplasmic reticulum, and the proteasome system, which are responsible for the maintenance of the normal protein homeostasis of the cell. In the event of mitochondrial and endoplasmic reticulum dysfunction, unfolded proteins aggregate forming potentially toxic deposits, which tend to be resistant to degradation. As stated above, neural cells possess adaptive mechanisms, molecular chaperone, and the ubiquitin proteasome system to avoid the accumulation of incorrectly folded proteins to fulfill cellular protein quality control functions (Moncada and Bolanos, 2006; Farooqui, 2009).

Thus, the diversity of neurodegenerative diseases can be explained through the combination of the above pathogenic events: one specific and associated with the aggregation of a particular protein in the nervous system and the other non-specific and associated with aging and with the production and harmful actions of ROS and RNS. This interpretation indicates that the development of drugs capable either of inhibiting the production or aggregation of proteins specifically implicated in neurodegenerative diseases or blocking the generation or action of ROS and RNS in the brain (Christen, 2002) may be useful for the treatment of neurodegenerative diseases.

Accumulating evidences also support the view that endogenous "biometals," such as copper, iron, zinc, and exogenous metal ion, aluminum, may also be involved in the etiopathogenesis of a variety of neurodegenerative diseases. Among above metal ions, iron plays a role in oxygen transportation, myelin synthesis, neurotransmitter production, and transfer of electrons (Campbell et al., 2001; Ong and Farooqui, 2005; Valko et al., 2005). Although iron is a crucial cofactor in normal brain metabolism, increased levels of brain iron may promote neurotoxicity due to free radical formation, lipid peroxidation, and ultimately, cellular death. Advanced neuroimaging studies indicate that elevated levels of iron have been observed in patients with neurological diseases, including AD, PD, and stroke. It is also proposed that alterations in the homeostasis of above metal ions may not only contribute to misfolding of accumulating proteins but also promote initiation of plaque aggregation (Zatta et al., 2009).

Neuropsychiatric disorders include both neurodevelopmental disorders and behavioral or psychological difficulties associated with some neurological disorders. An important characteristic of neuropsychiatric disorders is the impairment of cognitive processing. This includes not only ability to learn and store the memory but also to retrieve stored memory for further use and to apply the stored memory to efficiently solve problems (Gallagher, 2004). The impairment of cognitive process may be caused by overexpression or underexpression of certain genes or other unknown factors that result in behavioral symptoms, such as thoughts or actions, delusions, and hallucinations, which are the hallmarks of many neuropsychiatric disorders including schizophrenia, depression, and bipolar disorders. Metabolic defects of the brain, involving myelin sheath (multiple sclerosis) and brain infections (meningitis), do not fall under neurodegenerative disorders.

## 1.2 Neurodegeneration in Ischemic Injury

Normal functioning of brain needs an uninterrupted supply of both glucose and oxygen. Glucose and oxygen are needed by brain for the synthesis of ATP, which is required not only for maintaining the appropriate ionic gradients across neural membranes (low intracellular $Na^+$, high $K^+$, and very low cytosolic $Ca^{2+}$) but also for creating optimal cellular redox potentials (Farooqui and Horrocks, 2007). Stroke is a metabolic insult induced by severe reduction or blockade in cerebral blood flow. This blockade not only causes deficiency of oxygen and reduction in glucose metabolism but also results in ATP depletion and accumulation of toxic products. Reduction in ATP is accompanied by impairment in ion homeostasis, glutamate release, and ROS and RNS generation, resulting in neuronal injury and cell death (Farooqui et al., 1994). Within minutes of ischemic insult, proinflammatory genes are upregulated, and adhesion molecules are expressed on the vascular endothelium.

This is accompanied by the migration of neutrophils from the blood into the brain parenchyma within hours after reperfusion (Emerich et al., 2002), followed by the entry of macrophages and monocytes within a few days. Activated microglial cells contribute to vast majority of macrophages in the infarct area before macrophage infiltration from the blood (Schilling et al., 2003). Animal studies indicate that microglial activation also extends beyond the core and can contribute to peri-infarct neuronal death (Mabuchi et al., 2000; Block and Hong, 2005). Microglial activation is accompanied by inflammation, a neuroprotective process (Danton and Dietrich, 2003) associated with promotion of plasticity, modulation of neurotrophic factors, and removal of dead cells (Lalancette-Hebert et al., 2007; Farooqui, 2010).

Few studies have been performed on human stroke due to the inability to collect biopsy and postmortem tissues at time points after the onset of stroke where neuronal death occurs. Information on stroke has been obtained from global or focal animal models of ischemic injury in rodents. In both cases, blood flow disruptions limit the delivery of oxygen and glucose to neurons, causing symptoms and neurochemical changes similar to human stroke. Following stroke, the released glutamate accumulates in the extracellular space and mediates prolonged stimulation of glutamate receptors and a sustained increase in intracellular calcium concentration not only through NMDA receptor channels but also through calcium channels and glutamate transporters operating in the reverse mode. These processes also contribute to the cerebral edema, which is the primary cause of patient mortality after stroke (Farooqui et al., 2008). Neurons are particularly vulnerable to ROS- and RNS-mediated damage not only because of alterations in mitochondrial membrane potential and generation of ROS and RNS but also due to inactivation of glutamine synthetase (Atlante et al., 2000). It decreases glutamate uptake by glial cells and increases glutamate availability at the synapse, producing excitotoxicity (Farooqui et al., 2008). Morphologically glutamate-mediated neurodegeneration (excitotoxicity) is characterized by somatodendritic swelling, chromatin condensation into irregular clumps, and organelle damage. In addition, glutamate also produces neural cell demise by a transporter-related mechanism involving the inhibition of cystine uptake, which decreases glutathione in neural cells and makes them vulnerable to toxic-free radicals (Matute et al., 2006). Major proportions of free radicals originate from glutamate-mediated enhancement of calcium influx, stimulation of phospholipase $A_2$, and oxidation of released arachidonic acid through arachidonic acid cascade, activation of NADPH oxidase, and mitochondrial dysfunction. This increase in intracellular $Ca^{2+}$ also mediates the uncoupling of mitochondrial electron transport and stimulates $Ca^{2+}$-dependent enzymes including calpains, nitric oxide synthase, protein phosphatases, and various protein kinases (Farooqui et al., 2008). Neurons undergoing severe ischemic injury die rapidly (minutes–hours) by necrotic cell death at the core of injury site, whereas neurons in penumbral region display delayed vulnerability and die through apoptosis (Farooqui et al., 2004, 2008). Which neurons degenerate in ischemic injury depends on which blood vessel is blocked, but often neurons in the cerebral cortex, hippocampus, and striatum are affected. The extent of stroke injury varies according to the age of animals. Thus, 10- and 21-day-old rats develop greater damage from stroke-mediated insult than

6-week, 9-week, and 6-month-old rats (Yager et al., 1996; Yager and Thronhill, 1997). Younger rats may be more susceptible to stroke because of an unbalanced maturation of excitatory versus inhibitory neurotransmitter systems (Hattori and Wasterlain, 1990).

## 1.3 Neurodegeneration in Traumatic Brain Injury and Spinal Cord Trauma

Few studies have been performed on human brain and spinal cord tissues due to the inability to collect biopsy or postmortem tissue at time points after the onset of traumatic injury. Information on traumatic brain and spinal cord injury has been obtained from global or focal animal models in rodents. Traumatic injury to brain and spinal cord is defined by two broad components: a primary component, attributable to the mechanical insult itself, and a secondary component that consists of series of systemic and local neurochemical changes that occur in the brain and spinal cord after the initial traumatic insult (Klussmann and Martin-Villalba, 2005). The primary injury causes a rapid deformation of brain and spinal cord tissues, leading to the rupture of neural cell membranes and the release of intracellular contents. In contrast, secondary injury to brain and spinal cord includes glial cell reactions involving both activated microglia and astroglia and demyelination involving oligodendroglia (Beattie et al., 2000). Neurochemically, secondary injury is characterized by the release of glutamate from intracellular stores (Panter et al., 1990; Sundstrom and Mo, 2002) and overstimulation of glutamate receptors (excitotoxicity) resulting in a large $Ca^{2+}$ influx into neurons (Katayama et al., 1990), which not only uncouples of the mitochondrial electron transport but also stimulates $Ca^{2+}$-dependent phospholipases $A_2$ ($PLA_2$), phospholipase C (PLC), calpains, nitric oxide synthase, protein phosphatases, matrix metalloproteinases, and various protein kinases (Bazan et al., 1995; Ray et al., 2003; Ellis et al., 2004; Arundine and Tymianski, 2004; Xu et al., 2006). The stimulation of these enzymes not only generates a variety of lipid mediators (Table 1.1) but also rapidly decreases in ATP level, changes ion homeostasis, and alters cellular redox, resulting in the neurodegeneration in the traumatic brain injury and spinal cord trauma. Following brain and spinal cord injury, necrotic cell death normally occurs at the core of injury site whereas apoptotic cell death occurs several hours or days after injury in the surrounding area. Accumulating evidence suggests that excitotoxicity and oxidative stress are major components of brain injury and spinal cord trauma (Farooqui et al., 2004).

## 1.4 Neurodegeneration in Neurodegenerative Diseases

In general, neurodegeneration in neurodegenerative diseases is accompanied by site-specific premature and slow death of certain neuronal populations in central and peripheral nervous systems (Graeber and Moran, 2002). For example in AD,

**Table 1.1** Neurochemical events that are common to acute neural trauma, neurodegenerative diseases, and neuropsychiatric disorders

| Parameter | Acute neural trauma | Neurodegenerative diseases | Neuropsychiatric diseases |
|---|---|---|---|
| Glutamate levels | Increased | Alterations in glutamate receptors | Altered |
| Calcium | Increased | Altered | Altered |
| Cytokines | Increased | Increased | Increased |
| Neuroinflammation | Increased | Increased | Increased |
| Oxidative stress | Increased | Increased | Increased |
| Accumulation of aggregated proteins | None | Yes | None |
| Mitochondrial dysfunction | Yes | Yes | Yes |
| 4-Hydroxynonenal levels | Increased | Increased | – |
| Isoprostanes | Increased | Increased | |
| Apoptotic cell death | Yes | Yes | Yes |
| Blood–brain barrier permeability | Abnormal | Abnormal | Abnormal |

Summarized from Farooqui and Horrocks (1994, 2007), Farooqui et al. (2007), McIntosh et al. (1998), Beattie et al. (2000), Block and Hong (2005), and Farooqui (2009).

neurodegeneration mainly occurs in the nucleus basalis and hippocampal area, whereas in PD, dopaminergic neurons in the substantia nigra undergo neurodegeneration. In HD, neurodegeneration occurs in striatal medium spiny neurons and motor neurons located in the anterior part of spinal cord degenerate in ALS and spinal muscular atrophy (SMA). In Friedreich ataxia (FA), motor neurons found in the posterior part of the spinal cord undergo neurodegeneration (Table 1.2). Some neurodegenerative diseases produce neurodegeneration in cerebellum and cortical atrophy lesions are confined to the Purkinje cells and the inferior olive cells, while in pontocerebellar atrophy neurodegeneration occurs in several cerebellar structures. Despite the important differences in neurochemistry and clinical manifestation, neurodegenerative diseases share some common characteristics such as their commencement late in life, the extensive neuronal death, and loss of synapses, and the presence of cerebral deposits of misfolded protein aggregates (Soto, 2003; Ross and Poirier, 2004). These deposits are a typical disease signature, and although the main protein component of deposits is different in each disease, many accumulated proteins have similar morphological, structural, and staining characteristics. Deposits may be found either outside or inside the dead or dying cells and are generated by abnormal interactions between proteins. Examples of extracellular aggregates are amyloid plaques in AD and prion protein aggregates in bovine spongiform encephalopathy (mad cow disease). Examples of intracellular inclusions are the neurofibrillary tangles in AD and Lewy bodies in PD and the polyglutamine expanded protein aggregates in HD. It should be noted that protein misfolding and deposition in neurodegenerative diseases is the result of an altered balance between protein synthesis, aggregation rate, and clearance. Loss of synapse may also cause protein

**Table 1.2** Neurodegeneration sites in neurodegenerative diseases

| Disease | Neurodegeneration site | References |
|---|---|---|
| AD | Nucleus basalis and hippo-campus | Bossy-Wetzel et al. (2004), Ross and Poirier (2004) |
| PD | Substantia nigra | Bossy-Wetzel et al. (2004), Ross and Poirier (2004) |
| HD | Striatum | Bossy-Wetzel et al. (2004), Ross and Poirier (2004) |
| ALS | Anterior spinal cord | Bossy-Wetzel et al. (2004), Ross and Poirier (2004) |
| SMA | Anterior spinal cord | Bossy-Wetzel et al. (2004), Ross and Poirier (2004) |
| FA | Posterior spinal cord | Lodi et al. (2006) |
| CCA | Cerebellum | Bossy-Wetzel et al. (2004), Ross and Poirier (2004) |
| PCA | Cerebellum | Bossy-Wetzel et al. (2004), Ross and Poirier (2004) |

Alzheimer disease (AD); Parkinson disease (PD); Huntington disease (HD); amyotrophic lateral sclerosis (ALS); spinal muscular atrophy (SMA); Friedreich ataxia (FA); cerebellum cortical atrophy (CCA); and pontocerebellar atrophy (PCA).

accumulation, which may be correlated with cognitive impairment in normal aging and different types of dementia in neurodegenerative diseases. Numerous studies indicate the disruption of microtubule-based transport mechanisms as a contributor to synaptic degeneration (Butler et al., 2007). Reported reductions in a microtubule stability marker, acetylated α-tubulin, indicate that disruption transport occurs in AD neurons, and such a reduction is known to be associated with transport failure and synaptic compromise in a hippocampal slice model of protein accumulation (Butler et al., 2007). Collective evidence suggests that degeneration of synapse and disruption of microtubule-based transport may be correlated with cognitive impairment.

Most neurodegenerative diseases are accompanied by elevation in energy demands and reduction in energy production and supply. In neurodegenerative diseases the energy demands of brain are increased due to (a) partial depolarization (Blanchard et al., 2002); (b) impairment in $Ca^{2+}$ homeostasis (Farooqui and Horrocks, 2007); (c) glutamate-mediated increase in neuronal activity (Farooqui et al., 2008); (d) increase in oxidative stress (Farooqui and Horrocks, 2007); and (e) decrease in $Na^+/K^+$-ATPases and $Ca^{2+}$-dependent ATPases (Dickey et al., 2005). At the same time energy production and supply of brain are significantly decreased because of (a) mitochondrial dysfunctions (Kwong et al., 2006; Farooqui et al., 2008); (b) changes in blood flow; and (c) decrease in glucose metabolism/supply (Farooqui and Horrocks, 2007; Farooqui et al., 2008). There is considerable overlapping among above processes and many are coupled by positive feedback mechanisms, as is the energy balance (Kwong et al., 2006). Increased energy deficit promotes increased energy demand and slow neurodegeneration in neurodegenerative diseases.

The neuronal population, which degenerates in neurodegenerative diseases, modulates movements, learning and memory, processing sensory information, and decision-making processes (Rao and Balachandran, 2002). Other risk factors for neurodegenerative diseases include neuroinflammation, autoimmunity, cerebral blood flow, and blood–brain barrier dysfunction (Farooqui et al., 2007; Farooqui and Horrocks, 2007; Farooqui, 2009). For the most part, the nature, time course, and molecular causes of neuronal cell death in neurodegenerative diseases remain unknown, but age-mediated decrease in cellular antioxidant defenses and resultant accumulation of lipid, protein, and DNA damage in central nervous system has been proposed to play an important role in the etiology and pathogenesis of neurodegenerative diseases (Farooqui, 2009) (Fig. 1.3).

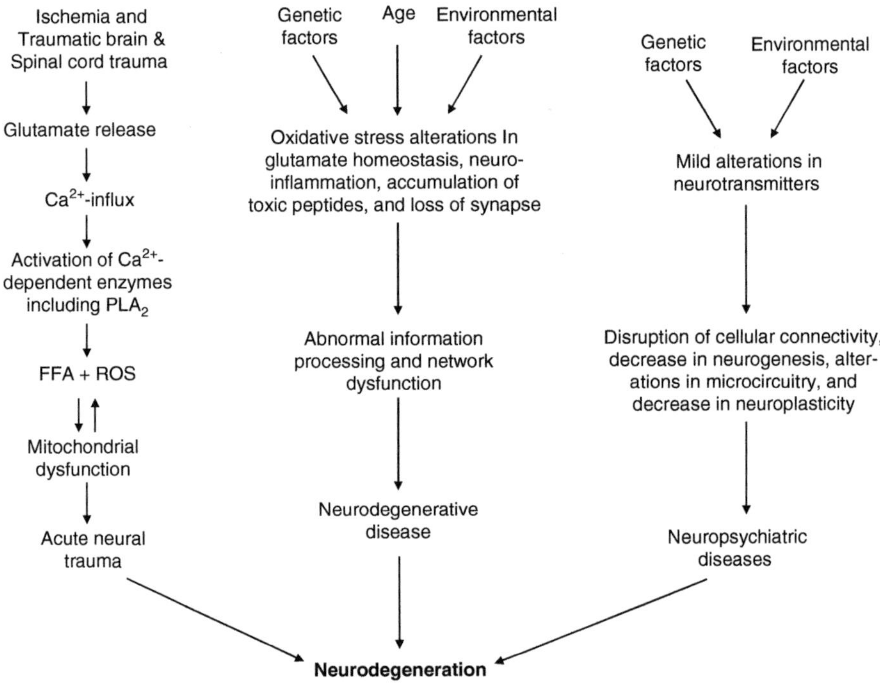


**Fig. 1.3** Neurochemical events associated with ischemia and traumatic injuries, and neurodegenerative diseases and neuropsychiatric disorders

In many neurodegenerative diseases, neurodegeneration shortens the life expectancy of patients, but other neurodegenerative diseases are fatal per se. Only those diseases in which neurological structures impair ability to control or execute such vital functions as respiration, heart rate, or blood pressure are deadly (Przedborski et al., 2003). Thus, in ALS, loss of lower motor neurons innervating respiratory muscles leads the patient to succumb to respiratory failure. Alternatively, in diseases like Friedreich ataxia, the association of neurodegeneration with heart disease can also cause the death of the patient although, in this case, death is not due to any neuronal loss but due to serious cardiac problems, such as congestive heart failure (Przedborski et al., 2003). In other neurodegenerative diseases, death is attributed neither to the disease of the nervous system nor to associated extra-nervous system degeneration, but caused by motor and cognitive impairments that increase the risk of fatal accidental falling, aspiration pneumonia, pressure skin ulcers, malnutrition, and dehydration (Przedborski et al., 2003).

Although some progress has been made on neurochemical alterations and in understanding factors that may trigger neurodegenerative diseases, the precise molecular pathways that lead to neurodegeneration are not fully understood (Farooqui and Horrocks, 2007; Farooqui, 2009). It is proposed that complex interplay between inflammatory mediators, aging, genetic background, oxidative stress, and environmental factors may regulate the progression of chronic neurodegeneration. It should be noted that for every neurodegenerative disease, multiple hypotheses have been proposed to explain the cause of neurodegeneration and neural dysfunction. In many cases, common pathways have been proposed for multiple neurodegenerative diseases (Bossy-Wetzel et al., 2004). Most common hypotheses include interactions among neuroinflammation, oxidative stress, and excitotoxicity; mitochondrial dysfunction; alterations in calcium homeostasis; proteasomal dysfunction; protein aggregation; decrease in blood flow; alterations in blood–brain barrier, and neuronal cell cycle induction (Farooqui and Horrocks, 2007; Golde, 2009). However, placing these pathways in the proper relationship to the onset, time course, and progress of neurodegeneration and its relationship to cytoskeletal pathology are challenging issues that are not fully understood (Golde, 2009).

As stated above, the molecular mechanism of neurodegeneration in neurodegenerative diseases is very complex. These diseases progress slowly over time, often taking several years to reach the end stage. Does this observation mean that degenerating neurons yield to the disease only after a prolong agony or neurodegeneration occurs suddenly? Histochemical studies indicate that neurodegeneration corresponds to an asynchronous death, in that neurons within a neuronal population die at very different times with different rates. Thus, in a neurodegenerative disease at any given time, only a small number of neurons actually degenerate, while others are at various stages along the neuronal death pathway (Bossy-Wetzel et al., 2004; Ross and Poirier, 2004). This situation complicates clinical and biochemical measurements, which provide information on the entire population of cells in a particular brain region. Therefore, the rate of neurochemical alterations essentially reflects the changes in the entire population of affected cells in a particular brain region and provides very little insight into the pace at which the death of an individual neuronal cell occurs (Przedborski et al., 2003). Still, large body of in vitro data indicates that once a neuron becomes sick, the entire process of neurodegeneration proceeds control and prolonged clinical progression of neurodegenerative disease may reflect a small number of neurons dying rapidly at any given point in time (Przedborski et al., 2003).

## 1.5 Neurodegeneration in Neuropsychiatric Diseases

Neuropsychiatric disorders are closely associated with the abnormalities in cerebral cortex and limbic system (thalamus, hypothalamus, hippocampus, and amygdale). An important characteristic of neuropsychiatric disorders is the abnormality in cognitive processing, which is mediated by signal transduction processes associated with everyday problem-solving behavior. This includes the ability to learn and store the memory, to retrieve stored memory for further use, and to apply the stored memory to efficiently solve problems (Gallagher, 2004). At the cellular level, abnormalities in cognitive process may be regulated by overexpression or underexpression of genes and molecular mechanisms involved in modulation of behavioral symptoms, such as thoughts or actions, delusions, and hallucinations. These behavioral abnormalities are the hallmarks of many neuropsychiatric diseases, including schizophrenia, depression, and compulsive and bipolar disorders. In addition to abnormalities in signal transduction processes, neuropsychiatric disorders are also linked to gray matter atrophy caused by decreased neuronal and glial size, increased cellular packing density suggesting a disruption in neuronal connectivity, particularly in the dorsolateral prefrontal cortex, and distortions in neuronal orientation (Arnold and Trojanowski, 1996; Blitzer et al., 2005). These observations are supported by neuroimaging studies that indicate a number of anatomical and neurochemical abnormalities in neurocircuits in specific brain area of neuropsychiatric patients. In addition, neurochemical studies also indicate that in neuropsychiatric diseases several neurotransmitter systems are simultaneously altered within a single microcircuit and each transmitter system shows circuitry changes in more than one region. Changes in microcircuits and neurotransmitters (synthesis and transport) may not only vary on a region-by-region basis but also from one neuropsychiatric disease to another. Both macro- and microcircuitry within the specific brain system (such as limbic system) may serve as "triggers" for the onset of neuropsychiatric condition (Benes, 2000; Harrisson, 1999). Neurochemical and neuroimaging studies also indicate alterations in cerebral blood flow and glucose utilization in the limbic system and prefrontal cortex of patients with major depression and other neuropsychiatric diseases (Ito et al., 1996; Kimbrell et al., 2002). Collective evidence suggests that genetic factors, alterations in blood flow, disruption of cellular connectivity, decrease in neurogenesis, alterations in microcircuitry, decrease in neuroplasticity along with mild oxidative stress, and mild neuroinflammation are major risk factors for neuropsychiatric diseases. In addition, both AD and PD are accompanied by neuropsychiatric symptoms due to age-related changes in neurotransmission, neuroplasticity, and signal transduction processes (Hornykiewicz, 1987; Becker et al., 1997; Blitzer et al., 2005), supporting the view that there is an overlap among some neurochemical mechanisms associated with neurodegenerative and neuropsychiatric diseases.

## 1.6 Similarities and Differences Between Ischemic, Neurotraumatic Injuries, Neurodegenerative Diseases, and Neuropsychiatric Disorders

Ischemic and traumatic injuries to brain and spinal cord arise from very different kinds of initial insults to brain and spinal cord tissues. Ischemic injury is a metabolic insult caused by severe reduction in cerebral blood flow due to blocked blood vessel. This blockade not only decreases oxygen and glucose delivery to brain tissue but also results in the buildup of potentially toxic products such as ROS and RNS in brain. Because neurons lack the ability to store glycogen, oxygen deficiency results in a rapid reduction in ATP production causing not only a marked impairment in ion homeostasis and release of glutamate from neurons but also a decrease in glutamate uptake ability of glial cells. These processes not only potentiate excitotoxicity but also upregulate the production of ROS and RNS compounding the neuronal injury and cell death (Siesjö et al., 2000; Liu et al., 2004; Farooqui and Horrocks, 1994). Ischemic injury also involves the loss of synapses and damage to presynaptic nerve terminal. Neurodegeneration in severe ischemic injury occurs rapidly (minutes–hours) through necrotic cell death at the core of ischemic injury site, whereas in the injury surrounding area (penumbral region) neurodegeneration takes place through apoptosis (Farooqui et al., 2004; Farooqui, 2009). As stated above, neurochemically apoptosis is characterized by alterations in mitochondrial membrane permeability, unaltered levels of ATP, release of cytochrome *c*, activation of caspases, induction of p53, Bax, and Par-4 (Beattie et al., 2000; Mattson, 2003; Farooqui et al., 2004; Farooqui, 2009). Other subcellular organelles, such as plasma membrane, mitochondria, and endoplasmic reticulum, remain active during apoptosis. In contrast, necrosis is accompanied by the permeabilization of plasma membrane, deficiency of ATP, loss of ion homeostasis, glutathione depletion, and activation of lysosomal enzymes resulting in a passive cell death through lysis (Nicotera and Lipton, 1999; Farooqui et al., 2004). Major participants in necrotic cell death irrespective of the stimulus are calcium, ROS and RNS, and lysosomal enzymes. During necrosis, elevated cytosolic calcium levels produce not only mitochondrial calcium overload and abnormalities in bioenergetics but also activation of proteases and $PLA_2$ and PLC. As stated earlier, ROS and RNS initiate damage to lipids, proteins, and DNA that consequently result in loss of ion homeostasis due to compromised neural membrane integrity. Furthermore, necrosis results in the release of cellular contents with immunomodulatory factors that lead to recognition and engulfment by phagocytes and the subsequent immunological response (Farooqui et al., 2007). Accumulating evidence suggests that apoptosis and necrosis are interrelated mechanisms with some overlap between biochemical events. Apoptosis and necrosis not only require well-organized signaling cascade but also involve extensive cross talk between several biochemical and molecular events at different cellular levels (Farooqui, 2009).

Autophagy is a cell survival mechanism that involves degradation and recycling of cytoplasmic long-lived proteins and organelles. In addition, autophagy mediates cell death in ischemic injury under specific circumstances (Rami and Kogel, 2008). Increasing evidence suggests that the effects of autophagy are highly contextual. An insufficient autophagic response may make neural cells more susceptible to stress whereas prolonged overactivation of autophagy may lead to a complete self-digestion of the cell. The extent of autophagy may represent a master switch between cell survival and cell death. Although autophagy and apoptosis are remarkably distinct processes, several pathways regulate both autophagic and apoptotic machinery. It remains to be seen whether autophagy is primarily a strategy for survival or whether autophagy can also be a part of a cell death program and thus contribute to cell death after cerebral ischemia (Rami and Kogel, 2008). Recent studies also indicate that ischemic injury also involves the upregulation of autophagy regulator called Beclin 1 (Bcl2 interacting protein) and subcellular redistribution of the autophagic marker LC3 (microtubule-associated protein 1 light chain 3) to vacuolic structures in injured neurons (Rami and Kogel, 2008; Rami et al., 2008). Neuronal cells that overexpress Beclin 1 show damaged DNA but without changes in nuclear morphology indicating that not all the Beclin 1-upregulating cells are predestined to die. The upregulation of Beclin 1 and related changes of LC3 in the ischemic penumbra may represent enhanced autophagy either as a mechanism to recycle injured cells and reduce damage or a process leading to cell demise.

Glial cells respond to ischemic injury in a complex manner. On one hand, astrocytes protect neurons from excitotoxicity through the intake of glutamate, and on the other hand, they may also contribute to the extracellular glutamate increase during severe ischemic insult (Dronne et al., 2007). Thus, under conditions of mild ischemic insult, astrocytes take up glutamate via the glutamate transporter, and potassium via the $Na^+/K^+/Cl^-$ cotransporter that limit glutamate levels and increase potassium in the extracellular space. In contrast, under severe ischemic insult, astrocytes are unable to maintain potassium homeostasis and contribute to the excitotoxicity by expelling glutamate out of the cells via the reversed glutamate transporter (Dronne et al., 2007). Oligodendroglial cells are highly vulnerable to glutamate-mediated ischemic injury. Competitive inhibition of cystine uptake and accumulation of intracellular peroxides along with chromatin fragmentation and condensation are also associated with ischemia–reperfusion injury-mediated oligodendroglial cell death (Farooqui and Horrocks, 2007).

Microglial cells respond to ischemic injury by transforming themselves into activated form. They not only change their shape into “ameboid” morphology but also release matrix metalloproteinases, ROS, RNS, and other proinflammatory cytokines (Farooqui and Horrocks, 2007), followed by neutrophil entry after the onset and monocyte infiltration later at the injury site. Microglial cells contain a wide range of receptors that allow them to identify and internalize numerous pathogens. In the brain tissue, NF-κB, and mitogen-activated protein kinase (MAPK), p38 are associated with proinflammatory cytokine production, generation of ROS, production of eicosanoids, and neurodegeneration following acute metabolic injury (Sun et al., 2007).

In contrast to metabolic injury in ischemia, traumatic injury to brain and spinal cord is caused by the mechanical impact and shear forces (McIntosh et al., 1998;

Fiskum et al., 1999; Bramlett and Dietrich, 2004). Thus, the traumatic injury to head and spinal cord consists of mechanical insult, which is followed by a series of systemic and local neurochemical and pathophysiological changes that occur in brain and spinal cord (Bramlett and Dietrich, 2004; Klussmann and Martin-Villalba, 2005). The primary injury produces a rapid deformation of brain and spinal cord tissues, leading to rupture of neural cell membranes, release of intracellular contents, and disruption of blood flow and breakdown of the blood–brain barrier. In contrast, morphological changes include activation of microglia and astroglia and demyelination, involving oligodendroglial cells (Beattie et al., 2000; Farooqui et al., 2004). Neurochemical and pathophysiological changes in brain and spinal cord tissues involve release of high levels of glutamate inducing excitotoxicity, generation of oxygen free radicals producing oxidative stress, and generation of cytokines inducing neuroinflammation (Farooqui et al., 2004, 2008; Farooqui, 2009). Several enzymes including $PLA_2$, cyclooxygenases, and p38 MARK mediate signal transduction processes associated with propagation and maintenance of the excitotoxicity, oxidative stress, and neuroinflammation. In addition, the complement system also participates and contributes to ischemic and traumatic injuries. The complement system is a crucial mediator of neuroinflammation and cell lysis after ischemic injury. Complement components C1q, C3c, and C4d have been detected in all ischemic lesions, suggesting activation via the classical pathway. C9, C-reactive protein, and IgM can be detected in necrotic zones. Marked CD59 and weak CD55 expression are found in normal brains, but these complement regulators have been virtually absent in ischemic lesions (Pedersen et al., 2009). Modest amounts of mannose-binding lectin (MBL), MBL-associated serine protease-2, and factor B are found in both ischemic lesions and controls. Increased deposition of complement components combined with decreased expression of complement regulators may be closely associated with brain damage following ischemic injury to human brain (Pedersen et al., 2009).

Like ischemic injury, neurodegeneration in head and spinal cord injuries occurs rapidly (hours–days) at the injury site. Thus, at the core of traumatic injury site neurons die through necrosis, whereas in the surrounding area neurons undergo apoptotic cell death (several days–months) (McIntosh et al., 1998; Farooqui et al., 2004). Like ischemic injury (Kogel, 2008; Rami et al., 2008), head injury and spinal cord trauma are accompanied by dramatic increase in the expression of Beclin 1, a Bcl2 interacting protein, at the injury site suggesting the participation of autophagic cell death during traumatic injuries (Kanno et al., 2009a). In hemisection model of mice spinal cord elevation in expression of Beclin 1 starts from 4 h, peaks at 3 days, and lasts for at least 21 days after hemisection (Kanno et al., 2009b). The Beclin 1 expression occurs in neurons, astrocytes, and oligodendrocytes. In Beclin 1 expressing cells, nuclei have round shape, which is a characteristic feature of cells undergoing autophagic cell death. This is in contrast to apoptotic cell death, which is characterized by either shrunken or fragmented nuclei (Kanno et al., 2009b). In head injury, the overexpression of Beclin 1 occurs only in neurons without any change in nuclear morphology. It is suggested that elevation of Beclin 1 at the site of injury may represent enhanced autophagy as a mechanism to discard injured cells and reduce damage to cells by disposing of injured components (Diskin et al., 2005; Erlich et al., 2006).

The fact that spinal cord trauma and head injury result from shear force that produces diverse cellular vulnerability pattern, which damages neuronal cell bodies, white matter structures, and vascular beds along with many signal transduction pathway alterations that are similar but not identical to ischemic insult (Bramlett and Dietrich, 2004; Farooqui and Horrocks, 2007). Severe cerebral ischemic injury induces metabolic stress, ionic perturbations, and a complex cascade of biochemical and molecular events that are similar, but not identical to traumatic injuries to brain and spinal cord in terms of rate of generation of lipid mediators and extent of endogenous neuroprotecting mechanisms (Farooqui and Horrock, 2007). This is temping to suggest that similar therapeutic strategies can be utilized for the treatment of ischemic and traumatic brain injuries.

In contrast to ischemic and traumatic brain injuries, neurodegenerative diseases are characterized by neural injury or death caused by many different factors, including, but not limited to genetic abnormalities, alterations in neural membrane composition, changes in neurotransmitters and their receptors, alterations in cerebral blood flow and blood–brain barrier, and problems in the immune system. Thus, neurodegenerative diseases represent a large group of neurological disorders with heterogeneous clinical and pathological expressions affecting specific subsets of neurons in specific functional anatomical and progress slowly in a relentless manner. They do not involve edema, hemorrhage, and trauma of the brain tissue. As stated above, the most important risk factors for sporadic neurodegenerative diseases are old age, positive family history, unhealthy lifestyle, endogenous factors, and exposure to toxic environment (Farooqui and Horrocks, 2007; Farooqui, 2009). Very little information is available on the rate of neurodegeneration and clinical expression of neurodegenerative diseases with age. These diseases commence late in life and are accompanied by the loss of synapses and accumulation of misfolded protein aggregates (Soto, 2003; Farooqui, 2009). The chemical nature of misfolded protein aggregate is different in each neurodegenerative disease. For example, β-amyloid peptide and τ-protein aggregate and accumulate in plaques and tangle of AD patients, α-synuclein and perkin accumulate in Lewy bodies of PD patients, huntingtin accumulates as nuclear inclusion in HD patients, and mutation in Cu/Zn superoxide dismutase occurs in some inherited form of ALS.

Although each neurodegenerative disease has a separate etiology with distinct morphological and pathophysiological characteristics, they may also share the similar terminal neurochemical common processes such as excitotoxicity, oxidative stress, and inflammation with ischemic and neurotraumatic injuries (Farooqui and Horrocks, 1994, 2007). It remains controversial whether excitotoxicity, oxidative stress, and inflammation are the cause or consequence of neurodegeneration in ischemic and traumatic injuries and neurodegenerative diseases (Andersen, 2004; Juranek and Bezek, 2005). In addition, there are several similarities as well as differences in neurochemistry of acute neural trauma and neurodegenerative diseases. Similarities include decrease in activity of cytochrome oxidase (Fiskum et al., 1999; Schinder et al., 1996) and overexpression of endogenous mitochondrial uncoupling proteins (UCP). These proteins are known to decrease the mitochondrial membrane potential and increase neuronal cell death following oxidative stress (Sullivan et al., 2004). The overexpression of UCP activity promotes the excitotoxicity-mediated ROS generation (Sullivan et al., 2004). Furthermore, activation of microglia and astrocytes in acute neural trauma and neurodegenerative diseases induce expression of cytokines and chemokines (Bramlett and Dietrich, 2004; Farooqui and Horrocks, 2007). In addition, ischemic injury and neurodegenerative diseases may have a cerebrovascular pathogenic component often in the form of reduced cerebral blood flow (Farkas et al., 2002; de la Torre and Stefano, 2000; de la Torre, 2008). Chronic cerebral hypoperfusion has been shown to adversely affect metabolic, anatomic, and cognitive function. In aged animals, chronic brain hypoperfusion results in regional pre- and postsynaptic changes, protein synthesis abnormalities, energy metabolic dysregulation, reduced glucose utilization, cholinergic receptor loss, and visuospatial memory deficits. Furthermore, keeping old animals for prolonged periods of time after chronic brain hypoperfusion causes brain capillary degeneration in CA1 hippocampus and neuronal damage extending from the hippocampal region to the temporo-parietal cortex where neurodegenerative tissue atrophy eventually forms (de la Torre, 2000). Similarly in humans, vascular risk factors in old age may create a critically threshold for cerebral hypoperfusion that triggers regional brain microcirculatory disturbances and impairs optimal energy production (reduced ATP synthesis) needed for normal brain cell function. Neuronal energy compromise enhances oxidative stress through the production of ROS, induction of aberrant protein synthesis, alterations in ionic membrane pump function, impairment in signal transduction, changes in neurotransmitter release, and abnormal processing of accumulating protein. Thus, the outcome of this defect may generate a chain of events that result in progressive evolution of brain metabolic, cognitive, and tissue pathology that characterizes ischemic injury and many neurodegenerative diseases (de la Torre, 2000, 2008). It is not known whether the reduced blood flow is a primary cause or secondary symptom in the neuropathological progression of ischemia and neurodegenerative diseases.

Differences between neurotraumatic diseases (ischemic and traumatic injuries) and neurodegenerative diseases include sudden lack of oxygen, quick depletion in ATP, rapid release of glutamate, and sustained increase in calcium influx resulting in rapid neurodegeneration (minutes–hours) in ischemic and traumatic injuries. In contrast, in neurodegenerative diseases, oxygen, nutrients, and ATP are available to neurons and ion homeostasis is maintained to a limited extent, neurons may take longer time period (years) to degenerate. Low levels of ATP and limited ion homeostasis may be related to diminished supply of growth factors (NGF and BDNF). Thus, reduced ATP synthesis, alterations ion homeostasis, diminished NGF and BDNF in brains may lead a molecular cascade that initiates the activation of region specific neuroglial death pathway in neurodegenerative diseases. Thus, many neurodegenerative diseases occur later in life and their onset is consistent with prolonged exposure to low excitotoxicity, oxidative stress, and neuroinflammation. Importantly, neurogenesis, a process associated with birth and maturation of functional new hippocampal neurons, is impaired by interplay among excitotoxicity, oxidative stress, and neuroinflammation accounting for brain atrophy in patients with neurodegenerative diseases. Another important point is that in neurodegenerative diseases, neurons increase their defenses by developing compensatory responses (oxidative strength) (Moreira et al., 2005; Numazawa et al., 2003) aimed to avoid or at least reduce cellular damage caused by the interplay among excitotoxicity, oxidative stress, and neuroinflammation. This hypothesis is supported by the view that Aβ deposition may not be the initiator of AD pathogenesis, but rather a downstream protective adaptation mechanism developed by cells in response to coordinated and upregulated interplay among excitotoxicity, oxidative stress, and neuroinflammation (Numazawa et al., 2003; Lee et al., 2004; Moreira et al., 2005). This observation supports the neuroprotective role of Aβ and explains why many aged individuals, despite having a high number of senile plaques in their brain, show little or no alterations in cognitive function.

## 1.7 Conclusion

Neurodegeneration is defined as a pathological process that results in the death of neural cells. Neurodegeneration occurs in ischemic and traumatic injuries to brain and spinal cord. Neurodegeneration also occurs in neurodegenerative diseases and neuropsychiatric disorders. In recent years, a remarkable progress has been made on molecular mechanisms underlying the pathogenesis of acute neural trauma, neurodegenerative diseases, and neuropsychiatric disorders. Although, growing evidence indicates an overlap in molecular mechanisms of neurodegeneration, there are remarkable differences in molecular, clinical, and neurophysiological aspects of acute neural trauma, neurodegenerative diseases, and neuropsychiatric disorders. Three basic mechanisms of neurodegeneration include autophagy, apoptosis, and necrosis. AD, PD, HD, and ALS are the most debilitating neurodegenerative diseases that induce alterations in skilled movements, cognition, and memory. Although neurodegeneration in acute neural trauma and neurodegenerative diseases is accompanied by the abnormal accumulation of extracellular and intracellular filamentous deposits in neurons, the precise molecular mechanisms that lead to neurodegeneration are not fully understood. However, it is proposed that interactions among neuroinflammation, oxidative stress, and excitotoxicity, mitochondrial dysfunction, alterations in calcium homeostasis, proteasomal dysfunction, protein aggregation, and neuronal cell cycle induction may play important roles in neurodegenerative process. In ischemic and traumatic brain and spinal cord injuries, neurons degenerate rapidly (in minutes–hours) because of the sudden lack of oxygen and a quick drop in ATP and alteration in ion homeostasis. In contrast, in neurodegenerative diseases oxygen and nutrients and ATP are available to the neurons and ion homeostasis is maintained to a limited extent, neuronal cell may take a longer time period (years) to die. Although common basis of many neurodegenerative dementias is found in increased production, misfolding and pathological aggregation of proteins, such as β-amyloid, τ-protein, huntingtin, and α-synuclein, the exact mechanism of different pathologies with regard to the development of neurodegenerative diseases is not fully understood.

However, it is becoming increasingly evident that neuronal mechanisms known to be disrupted at early stages in multiple neurodegenerative disorders include gene expression, protein interactions (manifesting as pathological protein aggregation and disrupted signaling), synaptic function, and neuroplasticity.

## References


Andersen JK (2004) Oxidative stress in neurodegeneration: cause or consequence? Nat Med 10 Suppl:S18–S25

Arnold SE, Trojanowski JQ (1996) Recent advances in defining the neuropathology of schizophrenia. Acta Neuropathol (Berl) 92:217–231

Arundine M, Tymianski M (2004) Molecular mechanisms of glutamate-dependent neurodegeneration in ischemia and traumatic brain injury. Cell Mol Life Sci 61:657–668

Atlante A, Calissano P, Bobba A, Azzariti A, Marra E, Passarella S (2000) Cytochrome c is released from mitochondria in a reactive oxygen species (ROS)-dependent fashion and can operate as a ROS scavenger and as a respiratory substrate in cerebellar neurons undergoing excitotoxic death. J Biol Chem 275:37159–37166

Bazan NG, Rodriguez de Turco EB, Allan G (1995) Mediators of injury in neurotrauma: intracellular signal transduction and gene expression. J Neurotrauma 12:791–814

Beattie MS, Farooqui AA, Bresnahan JC (2000) Review of current evidence for apoptosis after spinal cord injury. J Neurotrauma 17:915–925

Becker T, Becker G, Seufert J, Hofmann E, Lange KW, Naumann M (1997) Parkinson's disease and depression: evidence for an alteration of the basal limbic system detected by transcranial sonography. J Neurol Neurosurg Psych 63:590–596

Becker EB, Bonni A (2005) Beyond proliferation–cell cycle control of neuronal survival and differentiation in the developing mammalian brain. Semin Cell Dev Biol 16:439–448

Benes FM (2000) Emerging principles of altered neural circuitry in schizophrenia. Brain Res Brain Res Rev 31:251–269

Benhar M, Forrester MT, Stamler JJ (2006) Nitrosative stress in the ER: a new role for S-nitrosylation in neurodegenerative diseases. ACS Chem Biol 1:355–358

BenMoyal-Segal L, Soreq H (2006) Gene-environment interactions in sporadic Parkinson's disease. J Neurochem 97:1740–1755

Blanchard BJ, Thomas VL, Ingram VM (2002) Mechanism of membrane depolarization caused by the Alzheimer Abeta1-42 peptide. Biochem Biophys Res Commun 293:1197–1203

Blitzer RD, Iyengar R, Landau EM (2005) Postsynaptic signaling networks: cellular cogwheels underlying long-term plasticity. Biol Psychiatry 57:113–119

Block ML, Hong J-S (2005) Microglia and inflammation-mediated neurodegeneration: multiple triggers with a common mechanism. Prog Neurobiol 76:77–98

Bossy-Wetzel E, Schwarzenbacher R, Lipton SA (2004) Molecular pathways to neurodegeneration. Nat Med 10 Suppl:S2–S9

Bramlett HM, Dietrich WD (2004) Pathophysiology of cerebral ischemia and brain trauma: similarities and differences. J Cereb Blood Flow Metab 24:133–150

Brown TP, Rumsby PC, Capleton AC, Rushton L, Levy LS (2006) Pesticides and Parkinson's disease – is there a link? Environ Health Perspect 114:156–164

Butler D, Bendiske J, Michaelis ML, Karanian DA, Bahr BA (2007) Microtubule-stabilizing agent prevents protein accumulation-induced loss of synaptic markers. Eur J Pharmacol 562:20–27

Campbell A, Smith MA, Sayre LM, Bondy SC, Perry G (2001) Mechanisms by which metals promote events connected to neurodegenerative diseases. Brain Res Bull 55: 125–132

Cho DH, Nakamura T, Fang J, Cieplak P, Godzik A, Gu Z, Lipton SA (2009) S-nitrosylation of Drp1 mediates beta-amyloid-related mitochondrial fission and neuronal injury. Science 324: 102–105

Christen Y (2002) Proteins and mutations: a new vision (molecular) of neurodegenerative diseases. J Soc Biol 196:85–94

Danton GH, Dietrich WD (2003) Inflammatory mechanisms after ischemia and stroke. J Neuropathol Exp Neurol 62:127–136

de la Torre JC (2000) Critically attained threshold of cerebral hypoperfusion: the CATCH hypothesis of Alzheimer's pathogenesis. Neurobiol Aging 21:331–342

de la Torre JC, Stefano GB (2000) Evidence that Alzheimer's disease is a microvascular disorder: the role of constitutive nitric oxide. Brain Res Rev 34:119–136

de la Torre JC (2008) Pathophysiology of neuronal energy crisis in Alzheimer's disease. Neurodegener Dis 5:126–132

Dickey CA, Gordon MN, Wilcock DM, Herber DL, Freeman MJ, Morgan D (2005) Dysregulation of $Na^+/K^+$ ATPase by amyloid in $APP^+PS1$ transgenic mice. BMC Neurosci 6:7

Diskin T, Tal-Or P, Erlich S, Mizrachy L, Alexandrovich A, Shohami E, Pinkas-Kramarski R (2005) Closed head injury induces upregulation of Beclin 1 at the cortical site of injury. J Neurotrauma 22:750–762

Dronne MA, Grenier E, Dumont T, Hommel M, Boissel JP (2007) Role of astrocytes in grey matter during stroke: a modelling approach. Brain Res 1138:231–242

Ellis RC, Earnhardt JN, Hayes RL, Wang KKW, Anderson DK (2004) Cathepsin B mRNA and protein expression following contusion spinal cord injury in rats. J Neurochem 88:689–697

Emerich DF, Dean IIIRL, Bartus RT (2002) The role of leukocytes following cerebral ischemia: pathogenic variable or bystander reaction to emerging infarct? Exp Neurol 173:168–181

Erlich S, Shohami E, Pinkas-kramarski R (2006) Neurodegeneration induces upregulation of Beclin 1. Autophagy 2:49–51

Farkas E, de Wilde MC, Kiliaan AJ, Luiten PG (2002) Chronic cerebral hypoperfusion-related neuropathologic changes and compromised cognitive status: window of treatment. Drugs Today (Barc) 38:365–376

Farooqui AA, Haun S, Horrocks LA (1994) Ischemia and hypoxia. In: Siegel GJ, Agranoff BW, Albers RW, Molinoff PB (eds) Basic Neurochemistry. Raven Press, New York, NY, pp 867–883

Farooqui AA, Horrocks LA (1994) Excitotoxicity and neurological disorders: involvement of membrane phospholipids. Int Rev Neurobiol 36:267–323

Farooqui AA (2009) Hot topics in neural membrane lipidology. Springer, New York, NY

Farooqui T, Farooqui AA (2009) Aging: an important factor for the pathogenesis of neurodegenerative diseases. Mech Aging Dev 130:203–215

Farooqui AA, Horrocks LA (2007) Glycerophospholipids in the brain: phospholipases $A_2$ in neurological disorders. Springer, New York, NY, pp 1–394

Farooqui AA, Horrocks LA, Farooqui T (2007) Modulation of inflammation in brain: a matter of fat. J Neurochem 101:577–599

Farooqui AA, Ong WY, Horrocks LA (2004) Biochemical aspects of neurodegeneration in human brain: involvement of neural membrane phospholipids and phospholipases $A_2$. Neurochem Res 29:1961–1977

Farooqui AA, Ong WY, Horrocks LA (2008) Neurochemical aspects of excitotoxicity. Springer, New York, NY

Farooqui AA (2010) Neurochemical aspects in inflammation in brain. In: Farooqui AA, Farooqui T (eds) Molecular aspects of neurodegeneration and neuroprotection. Bentham Science Publishers Ltd, Sharjah (E. book) in press

Fiskum G, Murphy AN, Beal MF (1999) Mitochondria in neurodegeneration: acute ischemia and chronic neurodegenerative diseases. J Cereb Blood Flow Metab 19:351–369

Gallagher S (2004) Neurocognitive models of schizophrenia: a neurophenomenological critique. Psychopathology 37:8–19

Golde TE (2009) The therapeutic importance of understanding mechanisms of neuronal cell death in neurodegenerative disease. Mol Neurodegener 4:8

Graeber MB, Grasbon-Frodl E, Eitzen UV, Kösel S (1998) Neurodegeneration and aging: role of the second genome. J Neurosci Res 52:1–6
Graeber MB, Moran LB (2002) Mechanisms of cell death in neurodegenerative diseases: fashion, fiction, and facts. Brain Pathol 12:385–390
Harrison PJ (1999) Neurochemical alterations in schizophrenia affecting the putative receptor targets of atypical antipsychotics.Focus on dopamine ($D_1$, $D_3$, $D_4$) and 5-$HT_2$a receptors. Brain 122:593–624
Hattori H, Wasterlain CG (1990) Excitatory amino acids in the developing brain: ontogeny, plasticity, and excitotoxicity. Pediatr Neurol 6:219–228
He J, Wang T, Wang P, Han P, Chen C (2007) A novel mechanism underlying the susceptibility of neuronal cells to nitric oxide: the occurrence and regulation of protein S-nitrosylation is the checkpoint. J Neurochem 102:1863–1874
Hinault MP, Ben-Zvi A, Goloubinoff P (2006) Chaperones and proteases: cellular fold-controlling factors of proteins in neurodegenerative diseases and aging. J Mol Neurosci 30:249–265
Hornykiewicz O (1987) Neurotransmitters changes in human brain during aging. In: Govoni S, Battaini F (eds) Modification of cell to cell signals during normal and pathological aging. NATO ASI series, Heidelberg, Springer, pp 169–182
Ince PG, Codd GA (2005) Return of the cycad hypothesis – does the amyotrophic lateral sclerosis/parkinsonism dementia complex (ALS/PDC) of Guam have new implications for global health? NeuropatholAppl Neurobiol 31:345–353
Ito H, Kawashima R, Awata S, Ono S, Sato K, Goto R, Koyama M, Sato M, Fukuda H (1996) Hypoperfusion in the limbic system and prefrontal cortex in depression: SPECT with anatomic standardization technique. J Nucl Med 37:410–414
Juranek I, Bezek S (2005) Controversy of free radical hypothesis: reactive oxygen species – cause or consequence of tissue injury? Gen Physiol Biophys 24:263–278
Kamel F, Hoppin JA (2004) Association of pesticide exposure with neurologic dysfunction and disease. Environ Health Perspect 112:950–958
Kanno H, Ozawa H, Sekiguchi A, Itoi E (2009a) The role of autophagy in spinal cord injury. Autophagy 5:390–392
Kanno H, Ozawa H, Sekiguchi A, Itoi E (2009b) Spinal cord injury induces upregulation of Beclin 1 and promotes autophagic cell death. Neurobiol Dis 33:143–148
Katayama Y, Shimizu J, Suzuki S, Memezawa H, Kashiwagi F, Kamiya T, Terashi A (1990) Role of arachidonic acid metabolism on ischemic brain edema and metabolism. Adv Neurol 52: 105–108
Keifer MC, Firestone J (2007) Neurotoxicity of pesticides. J Agromedicine 12:17–25
Kimbrell TA, Ketter TA, George MS, Little JT, Benson BE, Willis MW, Herscovitch P, Post RM (2002) Regional cerebral glucose utilization in patients with a range of severities of unipolar depression. Biol Psychiatry 51:237–252
Klussmann S, Martin-Villalba A (2005) Molecular targets in spinal cord injury. J Mol Med 83: 657–671
Krantic S, Mechawar N, Reix S, Quirion R (2005) Molecular basis of programmed cell death involved in neurodegeneration. Trends Neurosci 28:670–676
Kwong JQ, Beal MF, Manfredi G (2006) The role of mitochondria in inherited neurodegenerative diseases. J Neurochem 97:1659–1675
Lalancette-Hebert M, Gowing G, Simard A, Weng YC, Kriz J (2007) Selective ablation of proliferating microglial cells exacerbates ischemic injury in the brain. J Neurosci 27:2596–2605
Lee HG, Casadesus G, Zhu XW, Takeda A, Perry G, Smith MA (2004) Challenging the amyloid cascade hypothesis – senile plaques and amyloid-beta as protective adaptations to Alzheimer disease. In: DeGrey ADN (ed) Strategies for engineered negligible senescence: why genuine control of aging may be foreseeable. Ann NY Acad Sci, New York, NY, pp 1–4
Liu CL, Siesjo BK, Hu BR (2004) Pathogenesis of hippocampal neuronal death after hypoxia-ischemia changes during brain development. Neuroscience 127:113–123
Lodi R, Tonon C, Calabrese V, Schapira AH (2006) Friedreich's ataxia: from disease mechanisms to therapeutic interventions. Antioxid Redox Signal 8:438–443

Mabuchi T, Kitagawa K, Ohtsuki T, Kuwabara K, Yagita Y, Yanagihara T (2000) Contribution of microglia/macrophages to expansion of infarction and response of oligodendrocytes after focal cerebral ischemia in rats. Stroke 31:1735–1743

Mattson MP (2003) Excitotoxic and excitoprotective mechanisms: abundant targets for the prevention and treatment of neurodegenerative disorders. Neuromolecular Med 3:65–94

Matute C, Domercq M, Sánchez-Gómez MV (2006) Glutamate-mediated glial injury: mechanisms and clinical importance. Glia 53:212–224

McIntosh TK, Saatman KE, Raghupathi R, Graham DI, Smith DH, Lee VM, Trojanowski JQ (1998) The Dorothy Russell Memorial Lecture. The molecular and cellular sequelae of experimental traumatic brain injury: pathogenetic mechanisms. Neuropathol Appl Neurobiol 24:251–267

Merlin AB, Sherman MY (2005) Role of molecular chaperones in neurodegenerative disorders. Int J Hyperthermia 21:403–419

Moncada S, Bolanos JP (2006) Nitric oxide, cell bioenergetics and neurodegeneration. J Neurochem 97:1676–1689

Moreira PI, Oliveira CR, Santos MS, Nunomura A, Honda K, Zhu XW, Smith MA, Perry G (2005) A second look into the oxidant mechanisms in Alzheimer's disease. Curr Neurovasc Res 2: 179–184

Muchowski PJ, Wacker JL (2005) Modulation of neurodegeneration by molecular chaperones. Nat Rev Neurosci 6:11–22

Murck H, Song C, Horrobin DF, Uhr M (2004) Ethyl-eicosapentaenoate and dexamethasone resistance in therapy-refractory depression. Int J Neuropsychopharmacol 7:341–349

Nakamura T, Lipton SA (2009) Cell death: protein misfolding and neurodegenerative diseases. Apoptosis 14:455–468

Nguyen MD, Mushynski WE, Julien JP (2002) Cycling at the interface between neurodevelopment and neurodegeneration. Cell Death Differ 9:1294–1306

Nicotera P, Lipton SA (1999) Excitotoxins in neuronal apoptosis and necrosis. J Cereb Blood Flow Metab 19:583–591

Numazawa S, Ishikawa M, Yoshida A, Tanaka S, Yoshida T (2003) Atypical protein kinase C mediates activation of NF-E2-related factor 2 in response to oxidative stress. Am J Physiol Cell Physiol 285:C334–C342

Ong WY, Farooqui AA (2005) Iron, neuroinflammation, and Alzheimer's disease. J Alzheimer Dis 8:183–200

Panter SS, Yum SW, Faden AI (1990) Alteration in extracellular amino acids after traumatic spinal cord injury. Ann Neurol 27:96–99

Pedersen ED, Løberg EM, Vege E, Daha MR, Maehlen J, Mollnes TE (2009) In situ deposition of complement in human acute brain ischaemia. Scand J Immunol 69:555–562

Pettegrew JW, Klunk WE, Kanal E, Panchalingam K, McClure RJ (1995) Changes in brain membrane phospholipid and high-energy phosphate metabolism precede dementia. Neurobiol Aging 16:973–975

Przedborski S, Vila M, Jackson-Lewis V (2003) Neurodegeneration: what is it and where are we? J Clin Invest 111:3–10

Rami A, Bechmann I, Stehle JH (2008) Exploiting endogenous anti-apoptotic proteins for novel therapeutic strategies in cerebral ischemia. Prog Neurobiol 85:273–296

Rami A, Kögel D (2008) Apoptosis meets autophagy-like cell death in the ischemic penumbra: two sides of the same coin? Autophagy 4:422–426

Rao AV, Balachandran B (2002) Role of oxidative stress and antioxidants in neurodegenerative diseases. Nutr Neurosci 5:291–309

Ray SK, Hogan EL, Banik NL (2003) Calpain in the pathophysiology of spinal cord injury: neuroprotection with calpain inhibitors. Brain Res Rev 42:169–185

Ross CA, Poirier MA (2004) Protein aggregation and neurodegenerative disease. Nat Med 10 Suppl:S10–S17

Sastry PS, Rao KS (2000) Apoptosis and the nervous system. J Neurochem 74:1–20

Schaller BJ (2008) Strategies for molecular imaging dementia and neurodegenerative diseases. Neuropsychiatr Dis Treat 4:585–612

Scheper W, Hoozemans JJ (2009) Endoplasmic reticulum protein quality control in neurodegenerative disease: the good, the bad and the therapy. Curr Med Chem 16:615–626

Schilling M, Besselmann M, Leonhard C, Mueller M, Ringelstein EB, Kiefer R (2003) Microglial activation precedes and predominates over macrophage infiltration in transient focal cerebral ischemia: a study in green fluorescent protein transgenic bone marrow chimeric mice. Exp Neurol 183:25–33

Schinder AF, Olson EC, Spitzer NC, Montal M (1996) Mitochondrial dysfunction is a primary event in glutamate neurotoxicity. J Neurosci 16:6125–6133

Siesjo BK, Kristian T, Shibasaki F, Uchino H (2000) The role of mitochondrial dysfunction in reperfusion damage in the brain. In: Kriegistein J, Klumpp S (eds) Pharmacology of cerebral ischemia. Wissenschaftliche Verlagsgeselischaft Mbh, Stuttgart, pp 163–175

Soto C (2003) Unfolding the role of protein misfolding in neurodegenerative diseases. Nat Rev Neurosci 4:49–60

Sullivan PG, Springer JE, Hall ED, Scheff SW (2004) Mitochondrial uncoupling as a therapeutic target following neuronal injury. J Bioenerg Biomembr 36:353–356

Sun GY, Horrocks LA, Farooqui AA (2007) The roles of NADPH oxidase and phospholipases $A_2$ in oxidative and inflammatory responses in neurodegenerative diseases. J Neurochem 103:1–16

Sundström E, Mo LL (2002) Mechanisms of glutamate release in the rat spinal cord slices during metabolic inhibition. J Neurotrauma 19:257–266

Uehara T (2007) Accumulation of misfolded protein through nitrosative stress linked to neurodegenerative disorders.Antioxid Redox Signal 9:597–601

Valko M, Morris H, Cronin MT (2005) Metals, toxicity and oxidative stress. Curr Med Chem 12:1161–1208

Wehr H, Bednarska-Makaruk M, Łojkowska W, Graban A, Hoffman-Zacharska D, Kuczyńska-Zardzewiały A, Mrugała J, Rodo M, Bochyńska A, Sułek A, Ryglewicz D (2006) Differences in risk factors for dementia with neurodegenerative traits and for vascular dementia. Dement Geriatr Cogn Disord 22:1–7

Xu Z, Wang BR, Wang X, Kuang F, Duan XL, Jiao XY, Ju G (2006) ERK1/2 and p38 mitogen-activated protein kinase mediate iNOS-induced spinal neuron degeneration after acute traumatic spinal cord injury. Life Sci 79:1895–1905

Yager JY, Shuaib A, Thornhill J (1996) The effect of age on susceptibility to brain damage in a model of global hemispheric hypoxia-ischemia. Brain Res Dev Brain Res 93:143–154

Yager JY, Thornhill JA (1997) The effect of age on susceptibility to hypoxic-ischemic brain damage. Neurosci Biobehav Rev 21:167–174

Zatta P, Drago D, Bolognin S, Sensi SL (2009) Alzheimer's disease, metal ions and metal homeostatic therapy. Trends Pharmacol Sci 30:346–355

# Chapter 2
# Neurochemical Aspects of Ischemic Injury

## 2.1 Introduction

The brain has the highest metabolic rate of all organs and depends predominantly on oxidative metabolism as a source of energy. Thus, it utilizes about 20% of respired oxygen for normal function, even though it represents only 5% of the body weight. Much of oxygen taken up by neurons is utilized for producing ATP, which is needed not only for maintaining the appropriate ionic gradients across the neural membranes but also creating the proper cellular redox potentials. Full and transient deficits in glucose and oxygen can rapidly compromise ATP production and threaten cellular integrity by either not maintaining or abnormally modulating ion homeostasis and cellular redox. The initial response to a transient insufficiency of energy is depolarization resulting in $Na^+$ influx into axons. Prolonged energy insufficiency results in a massive influx of $Ca^{2+}$ that facilitates neural cell death resulting in irreversible loss of neurologic function (Farooqui and Horrocks, 1994). All subcelluar organelles participate and contribute to neuronal cell death. Thus, $Ca^{2+}$-entry through plasma membrane exposes cytoplasm to increased levels of $Ca^{2+}$. Many phospholipases, kinases, and proteases are localized in cytosol and are activated directly or indirectly by the ischemic insult. Some enzymes generate proinflammatory and pro-apoptotic lipid metabolites while others produce anti-inflammatory and anti-apoptotic metabolites. Those neurons, which degenerate due to ischemic insult, synthesize proinflammatory and pro-apoptotic lipid metabolites, but ones that survive possess anti-inflammatory and anti-apoptotic metabolites. Mitochondria play the central role in apoptosis. The release of cytochrome *c* from mitochondria is the key step in apoptotic cascade in neurons injured by ischemia. In neural cell, endoplasmic reticulum (ER) not only mediates proteins processing but also modulates intracellular calcium homeostasis and cell death signal activation. ER dysfunction occurs at an early stage after ischemic injury and may be the initial step in apoptotic cascades in neurons (Lipton, 1999; Hayashi and Abe, 2004). Golgi apparatus and lysosomes also contribute to apoptotic cell death in some situations. Nucleus is the organelle that contains genomic DNA. Many studies have demonstrated that ischemic injury causes nitric oxide-mediated DNA fragmentation in neurons that would die later, but whether this is the cause or merely the result of the ischemic insult remains uncertain (Lipton, 1999; Hayashi and Abe, 2004).

A.A. Farooqui, *Neurochemical Aspects of Neurotraumatic and Neurodegenerative Diseases*, DOI 10.1007/978-1-4419-6652-0_2,

As stated in Chapter 1, stroke (ischemia) is a metabolic insult induced by severe reduction or blockade in cerebral blood flow due to cerebrovascular disease. This blockade not only decreases oxygen and glucose delivery to brain tissue but also results in the breakdown of blood–brain barrier (BBB) and buildup of potentially toxic products in brain. Breakdown of BBB integrity in ischemic injury not only results in transmigration of numerous immune system cells including monocytes and lymphocytes but also causes hyperpermeability induced by enhanced transcytosis and gap formation between endothelial cells. According to American Stroke Association, stroke is an emergency with its characteristic signs (Fig. 2.1). It initiates a complex cascade of events at genomic, molecular, cellular, subcellular levels producing heterogeneous changes in brain oxygenation (Fig. 2.2). There are two major types of strokes: ischemic and hemorrhagic. Ischemic strokes are brought about by critical decrease in blood flow to various brain regions causing neuronal cell death. Ischemic stroke is the most common type of stroke, constituting around 80% of all strokes, of which 60% are attributable to large-artery ischemia (Feigin et al., 2003). Hemorrhagic strokes are caused by a break in the wall of the artery resulting in spillage of blood inside the brain or around the brain. Age is a prominent risk factor for stroke. Thus, at the age of 55–64 years the prevalence of stroke is 11%. The risk increases to 43% in subjects that are older than 85 years. The reason for age-mediated vulnerability for stroke is not fully understood. However, potential mechanisms of age-mediated vulnerability include changes in brain plasticity-promoting factors, unregulated expression of neurotoxic factors, or differences in the generation of scar tissue that impedes the formation of new axons and blood vessels in the infarcted region (Popa-Wagner et al., 2007). In addition,

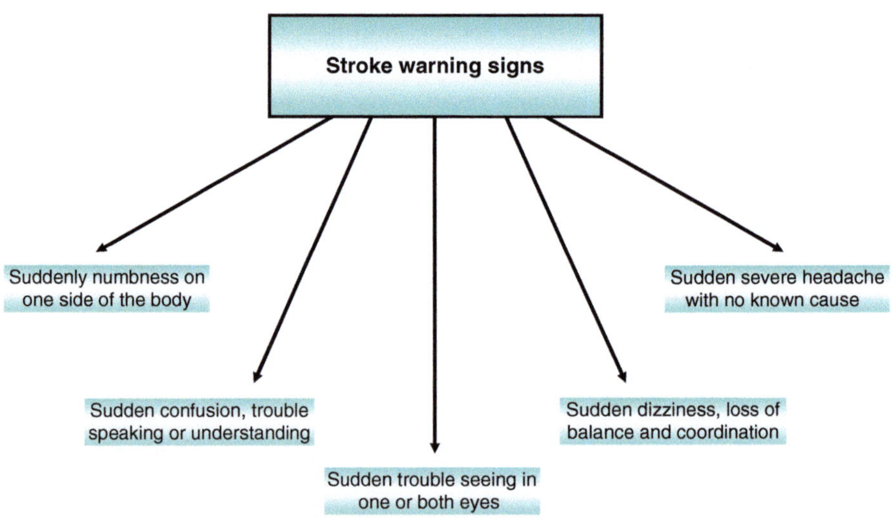


**Fig. 2.1** Stroke warning signs as stated by American Stroke Association, a division of American Heart Association

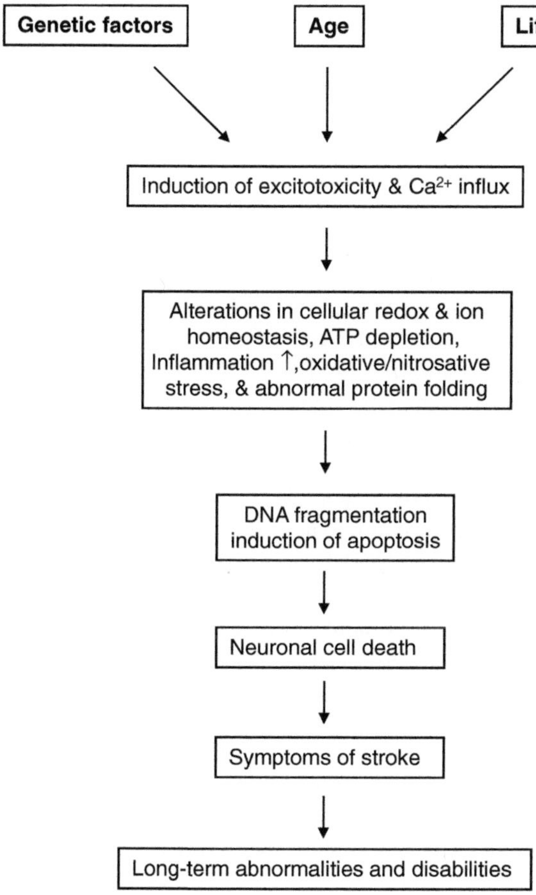


**Fig. 2.2** Risk factors and neurochemical processes associated with the pathogenesis of ischemic injury

vascular factors may also partially contribute to this vulnerability. It is also shown that white matter is inherently more vulnerable to ischemic injury in older mice, and the mechanisms of white matter injury change as a function of age (Baltan, 2006, 2009). Ischemic injury in white matter of older mice is predominantly caused by a $Ca^{2+}$-independent excitotoxicity involving overactivation of AMPA/kainate receptors (Baltan, 2009). It is suggested that increased vulnerability of aging white matter to ischemic injury is a consequence of age-related alterations in white matter molecular architecture (Baltan, 2006; Hinman et al., 2006; Baltan, 2009). Thus, older patients have less chance of surviving a stroke: 37% of patients 45–64 may die after a hemorrhagic stroke, whereas that number increases to 44% of patients over 65 years of age (Rosamond et al., 2007; Salaycik et al., 2007). Animal studies have shown that the aged brain has the ability to mount a cytoproliferative response to ischemic injury, but the timing of the cellular and genetic response to cerebral insult are dysregulated in aged animals, thereby compromising functional recovery (Popa-Wagner et al., 2007).

Unlike neurodegenerative diseases where neuronal damage occurs in a relatively homogenous population of neurons in a specific area (Farooqui, 2009), stroke affects multiple different neuronal phenotypes. For example, an infarct might involve the thalamus, hippocampus, and striate visual cortex, affecting three or more very different neuronal populations including neurons, oligodendrocytes, astrocytes, and endothelial cells (Savitz et al., 2003, 2004). Other risk factors for stroke include hypertension, diabetes mellitus, abnormal apolipoprotein E metabolism, high alcohol consumption, cigarette smoke, oral contraceptive, and underlying clotting disorders. According to American stroke Association, hypertension contributes to 30–40% stroke risk, cigarette smoking 12–18%, and diabetes between 5 and 27%. Some of the above risk factors can be mitigated. For example, the use of antihypertensive drugs to lower blood pressure and statins to treat hyperlipidemia has proven effective. Furthermore, changing lifestyle (stopping smoking and decreasing body weight), healthy diet (fruits, vegetable, legumes, and fish), and physical activity undoubtedly lower the risk of suffering a stroke.

Ischemia can be focal (regional) or global (forebrain). The two principal models for human stroke are produced in animals either by global or focal ischemia. In both cases, blood flow disruptions limit the delivery of oxygen and glucose to neurons by not only producing ATP depletion but also impairing ion homeostasis, inducing glutamate release, and initiating excitotoxic cascades that are deleterious for neurons (Fig. 2.2). An important difference between humans and controlled animal model studies is the physiological variability with frequent elevations and variability in blood pressure, glucose, temperature, and oxygenation in contrast to experimental models where animals are anesthetized and physiological parameters controlled. Furthermore, Stroke patients often have other conditions such as heart disease or pre-existing neurodegenerative disorders.

Stroke initiates excitotoxic insult, which involves the hyperactivation of glutamate receptors and release of excess glutamate in the extracellular space inducing neuron depolarization and dramatic increase of intracellular calcium that in turn activates multiple intracellular death pathways (Farooqui and Horrocks, 1994). Thus, stroke triggers a complex series of biochemical and molecular mechanisms that impairs the neurologic functions through the breakdown of cellular and subcellular integrity mediated by excitotoxic glutamatergic signaling, $Ca^{2+}$-influx, alterations in ionic balance and redox, and free-radical generation. These processes also lead to the activation of signaling mechanisms involving phospholipases $A_2$, C, and D ($PLA_2$, PLC, and PLD); calcium/calmodulin-dependent kinases (CaMKs); mitogen-activated protein kinases (MAPKs) such as extracellular signal-regulated kinase (ERK), p38, and c-jun N-terminal kinase (JNK); nitric oxide synthases (NOS); calpains; calcinurin; and endonucleases. Stimulation of these enzymes (Fig. 2.3) bring them in contact with appropriate substrates and modulates cell survival/degeneration mechanisms (Hou and MacManus, 2002; Farooqui and Horrocks, 2007). Degenerative mechanisms include apoptosis, necrosis, and autophagy in traumatized neurons in vitro ischemia models.

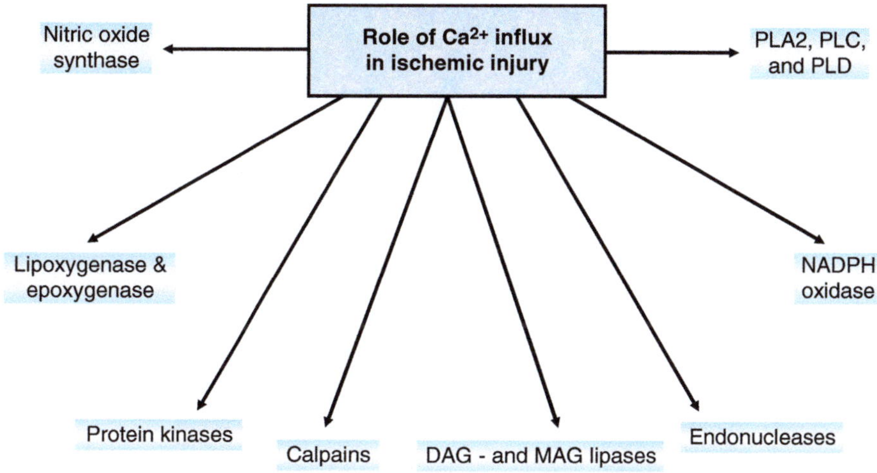


**Fig. 2.3** Stimulatory effect of $Ca^{2+}$ influx on enzymic activities following ischemic injury to the brain

## 2.2 Ischemic Injury-Mediated Alterations in Glycerophospholipid Metabolism

During ischemic injury interruption in oxygen supply, depletion in ATP generation, and mitochondrial dysfunction result in production of reactive oxygen species (ROS), such as superoxide, hydroxyl anion, and reactive nitrogen species (RNS), such as NO and $ONOO^-$. The initial response to ATP depletion in ischemic injury is depolarization, which causes $Na^+$ influx into axons. Prolonged depletion of ATP produces a massive $Ca^{2+}$ influx and accumulation that facilitates neurodegeneration (Dienel 1984; Farooqui and Horrocks, 2007) (Fig. 2.4). At the injury site, all vascular cells (endothelial cells, vascular smooth muscle cells, and adventitial fibroblasts) produce ROS primarily via cell membrane-bound NADPH oxidase (Sun et al., 2007). Other sources of ROS include oxygenases and mitochondria, which generate significant levels of ROS during normal respiration as well as cell death. Oxidative stress occurs either from an excessive generation or decrease in clearance of ROS. In addition, oxidation of biogenic amines by monoamine oxidases generates hydrogen peroxide ($H_2O_2$), which in the presence of copper generates hydroxyl radicals ($^{\cdot}OH$). Neurons are particularly vulnerable to oxidative damage not only because of alterations in mitochondrial membrane potential (Atlante et al., 2000) but also due to inactivation of glutamine synthetase. This decreases glutamate uptake by glial cells and increases glutamate availability at the synapse, producing excitotoxicity, a process by which high levels of glutamate and its analogs excite neurons and bring about their demise (Olney et al., 1979; Choi, 1988; Farooqui et al., 2008). Glutamate exerts its effect by interacting with excitatory amino acid receptors. These receptors include *N*-methyl-D-aspartate (NMDA), α-amino-3-hydroxy-5-methyl-4-isoxazole

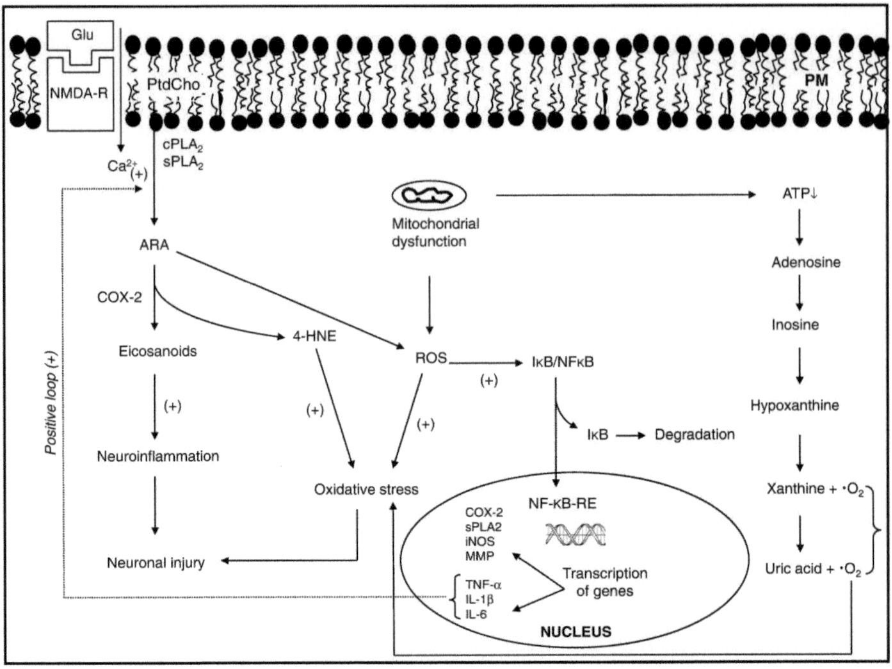


**Fig. 2.4** Diagram showing the effect of ischemic injury on glycerophospholipid-derived lipid mediators in brain. Plasma membrane (PM); *N*-methyl-D-aspartate receptor (NMDA-R); glutamate (Glu); phosphatidylcholine (PtdCho); lyso-phosphatidylcholine (lyso-PtdCho); cytosolic phospholipase $A_2$ ($cPLA_2$); secretory phospholipase $A_2$ ($sPLA_2$); cyclooxygenase (COX-2); arachidonic acid (ARA); platelet-activating factor (PAF); 4-hydroxynonenal (4-HNE); reactive oxygen species (ROS); nuclear factor kappaB (NF-κB); nuclear factor kappaB response element (NF-κB-RE); inhibitory subunit of NFκB (IκB); tumor necrosis factor-α (TNF-α); interleukin-1β (IL-1β); interleukin-6 (IL-6); matrix metalloproteinases (MMPs); positive sign (+) represents upregulation

propionate (AMPA), kainate (KA), and metabotropic glutamate receptors (Farooqui et al., 2008). Excitotoxicity-mediated calcium influx initiates a cascade of events that result in mitochondrial dysfunction, ROS production, and activation of many $Ca^{2+}$-dependent enzymes (Table 2.1) including $PLA_2$, nitric oxide synthases, protein kinases, cyclooxygenase-2 (COX-2), lipoxygenases (LOX), and epoxygenases (EPOX) (Fig. 2.3) (Phillis et al., 2006). Activation of $PLA_2$ results in the release of arachidonic acid (ARA), which is then oxidized by cyclooxygenases, lipoxygenases, and epoxygenases resulting in the generation of oxygenated metabolites of ARA. Non-enzymic oxidation of ARA (arachidonic acid cascade) generates reactive oxygen species (ROS), which includes oxygen-free radicals (superoxide radicals, hydroxyl and alkoxyl radicals, lipid peroxy radicals), and peroxides (hydrogen peroxide and lipid hydroperoxide). At higher concentrations, ROS contribute to neural membrane damage when the balance between reducing and oxidizing (redox) forces shifts toward oxidative stress. Thus, glutamate-mediated uncontrolled "arachidonic

**Table 2.1** NF-κB-mediated stimulation of enzymes associated with ischemic injury

| Enzyme | Effect | References |
|---|---|---|
| Cytosolic phospholipase $A_2$ | Stimulated | Edgar et al. (1982), Farooqui and Horrocks (2007) |
| Cyclooxygenase | Stimulated | Phillis et al. (2006) |
| Inducible nitric oxide synthase | Stimulated | Li et al. (2007) |
| NADPH oxidase | Stimulated | Sun et al. (2007), Farooqui and Horrocks (2007) |
| Superoxide dismutase | Stimulated | Block and Hong (2005) |
| Matrix metalloproteinase | Stimulated | Block and Hong (2005) |
| PKC-δ | Stimulated | Farooqui and Horrocks (2007) |

acid cascade" produces in an irreversible neural cell injury (Farooqui and Horrocks, 1994, 2006; Farooqui and Horrocks, 2009). Other sources of ROS are the mitochondrial respiratory chain and NADPH oxidase (Fig. 2.3). This enzyme catalyzes the production of superoxide radical by the one-electron reduction of oxygen, using NADPH as the electron donor. NADPH oxidase plays a pivotal role in glutamate-mediated inflammatory response. A downstream target of NADPH oxidase-derived superoxide radicals is the transcription factor NF-κB, which controls the expression of a large array of genes involved in immune function, inflammation, and cell survival. NF-κB itself is a key factor in controlling NADPH oxidase expression and function (Anrather et al., 2006). Glutamate-mediated increase in ROS leads to chemical cross-linking between ROS and unsaturated fatty acids. This causes peroxidative injury to neuronal membrane. This depletion of unsaturated fatty acids in neuronal membranes is associated with an alteration in membrane fluidity changing in the activity of membrane-bound enzymes, ion channels, and receptors (Farooqui and Horrocks, 2007). The presence of peroxidized glycerophospholipids in neural membranes induces a membrane-packing defect, making the *sn*-2 ester bond at glycerol moiety more accessible to the action of calcium-independent $PLA_2$. In fact, glycerophospholipid hydroperoxides are a better substrate for $PLA_2$ than native glycerophospholipids (Farooqui and Horrocks, 2007). Glycerophospholipid hydroperoxides inhibit the reacylation of lyso-glycerophospholipids in neuronal membranes (Zaleska and Wilson, 1989). This inhibition may constitute another important mechanism whereby peroxidative processes contribute to irreversible neuronal injury and death.

ARA is also metabolized to 4-hydroxynonenal (4-HNE). This metabolite impairs the activities of $Na^+$, $K^+$-ATPase, glucose 6-phosphate dehydrogenase, and several kinases, including c-jun amino-terminal kinase (JNK) and p38 mitogen-activated protein kinase (Mark et al., 1997; Camandola et al., 2000). The impairment of $Na^+$, $K^+$-ATPase depolarizes neuronal membranes leading to the opening of NMDA receptor channels and influx of additional $Ca^{2+}$ into neurons.

Lysophospholipid is the other product of $PLA_2$ catalyzed reaction. Lysophospholipids regulate a broad range of cellular processes including signal transduction. Its focal injections produce demyelination (Farooqui and

Horrocks, 2007). Under certain conditions, lyso-PtdCho also causes cell fusion. The accumulation of lyso-PtdCho induces neural cell demyelination and injury under pathological situations. In addition, lysophospholipids can also be converted to platelet-activating factor (PAF) through acetylation. This lipid mediator that induces neuroinflammation (Fig. 2.4) and modulates a variety of neural cell functions, including upregulation in activities of mitogen-activated protein (MAP) kinases and extracellular signal-regulated kinases, c-jun N-terminal kinase, and p38 kinases in primary hippocampal neurons in vitro (Mukherjee et al., 1999; DeCoster et al., 1998), suggesting MAP kinase and PAF may regulate pathways promoting neural cell survival or death, depending on the cellular context in which they are activated. The PAF receptor antagonist, hetrazepine BN 50730 can prevent MAP-kinase activation.

Pathophysiologically, PAF is associated with neuroinflammation, allergic reactions, and immune responses. High levels of PAF induce the release of cytokines and expression of cell adhesion molecules (Maclennan et al., 1996; Ishii et al., 2002; Honda et al., 2002). Glutamate-mediated elevation in PAF has been implicated in the mitochondrial swelling, membrane permeability transition (mPT), and release of cytochrome *c* (Parker et al., 2002) in rat brain mitochondrial preparations. The PAF antagonist BN50730 can block this process supporting the view that glutamate-mediated neural cell injury is associated with PAF elevation.

Glutamate also mediates damage to glial cells through alterations in glutamate uptake (Oka et al., 1993; Matute et al., 2006). It is well known that glutamate uptake from the extracellular space by specific glutamate transporters is essential for maintaining excitatory postsynaptic currents (Auger and Attwell, 2000) and for blocking excitotoxic death due to overstimulation of glutamate receptors (Farooqui et al., 2008). Out of 5 glutamate transporters, at least two glutamate transporters, namely excitatory amino acid transporter E1 (EAAT1) and excitatory amino acid transporter E2 (EAAT2), are expressed in astrocytes, oligodendrocytes, and microglial cells (Matute et al., 2006). Exposure of astroglial, oligodendroglial, and microglial cell cultures to glutamate induces glial cell death through the inhibition of cystine uptake and reduction in glutathione making glial cells vulnerable to ROS (Oka et al., 1993; Matute et al., 2006). The addition of cystine or cysteine totally blocks the glutamate-induced toxicity to oligodendroglia. A decreased glutathione level, through inhibition of glutathione synthesis, is accompanied by increased excitotoxic response to NMDA, degeneration of mitochondria, and larger infarct areas in stroke models (Janaky et al., 1999).

In brain, glutamate stimulates the synthesis of nitric oxide (NO) from L-arginine by $Ca^{2+}$/calmodulin-dependent nitric oxide synthase (NOS) (Bolanos et al., 1997) (Table 2.1). Low levels of NO are associated with signal transduction, but glutamate-induced excessive NO generation contributes to neurotoxicity. Nitric oxide synthase (NOS) inhibitor, *N*-ω-nitro-L-arginine methyl ester (NAME) or the NMDA receptor antagonist 2-amino-5-phosphonopentanoate (APV) blocks the neurotoxic effects of NO (Almeida et al., 1998). Excitotoxicity-induced neurodegeneration occurs through a mechanism involving NO and superoxide formation and the generation of peroxynitrite ($ONOO^-$) (Fig. 2.5). $ONOO^-$ not only reacts with SH groups of

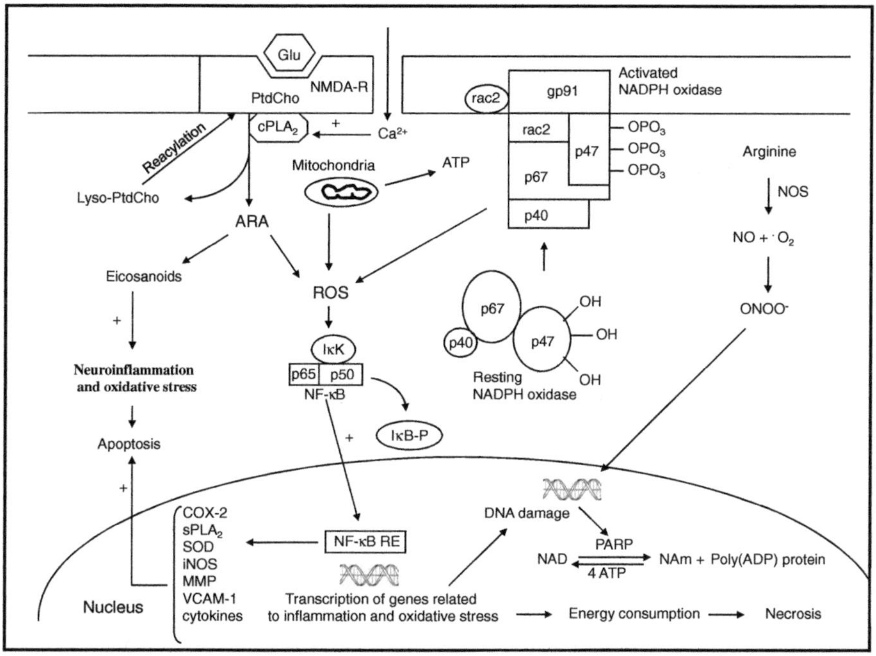


**Fig. 2.5** Diagram showing effect of oxidative and nitrosative stress on neuronal injury. Plasma membrane (PM); *N*-methyl-D-aspartate receptor (NMDA-R); glutamate (Glu); phosphatidylcholine (PtdCho); lyso-phosphatidylcholine (lyso-PtdCho); cytosolic phospholipase $A_2$ ($cPLA_2$); secretory phospholipase $A_2$ ($sPLA_2$); cyclooxygenase (COX-2); arachidonic acid (ARA); reactive oxygen species (ROS); nuclear factor kappaB (NF-κB); nuclear factor kappaB response element (NF-κB-RE); inhibitory subunit of NFκB (IκB); inducible nitric oxide synthase (iNOS); peroxynitrite ($ONOO^-$); Superoxide ($^{\bullet}O_2$); matrix metalloproteinases (MMPs); vascular cell adhesion molecule-1 (VCAM-1); poly(ADP-ribose) polymerase (PARP); nicotinamide (Nam); nicotinamide adenine dinucleotide (NAD); positive sign (+) represents upregulation

enzymes but also S-nitrosylates (transfer of NO to a critical thiol group) a number of proteins. Recently, S-nitrosylation-mediated post-translational protein misfolding has also been implicated in excitotoxicity (Lipton, 2007; Lipton et al., 2007). Protein disulfide isomerase (PDI), the enzyme responsible for normal protein folding is located at the endoplasmic reticulum (ER). S-Nitrosylation of PDI during excitotoxicity compromises the function of this enzyme and leads protein misfolding that may cause neurodegeneration in brain tissue. Another enzyme, whose S-nitrosylation may cause abnormal protein misfolding is the E3 ubiquitin ligase, a protein that covalently attaches ubiquitin to a lysine on a target protein via an isopeptide bond (Lipton, 2007; Lipton et al., 2007). E3 ubiquitin ligases contain cysteine residues in their RING domains. This cysteine thiol reacts with NO to form an S-nitrosylated derivative and thus alters ubiquitin-proteasome system degradative pathway and contribute to protein aggregation. In addition, $ONOO^-$ inhibits mitochondrial respiration, disturbs membrane pumps, decreases cellular glutathione, and damages DNA through the activation of poly (ADP-ribose) synthase, an enzyme that leads to cellular energy depletion (Pryor and Squadrito, 1995; Radi et al., 1991; Qi et al., 2000). All these processes are associated with neuronal energy deficiency and glutamate-mediated neurotoxicity.

In addition to the above-mentioned oxidation of neuronal molecules by ROS and RNS, the occurrence of novel pathways for molecular modifications has been reported (Perez-Pinzon et al., 2005). Two examples of these pathways explain why lethal ischemic insults lead to the translocation of protein kinase Cδ (PKCδ), which plays a role in apoptosis after cerebral ischemia, or why sublethal ischemic insults, such as in ischemic preconditioning, lead to the translocation of PKCζ, which plays a pivotal role in neuroprotection. A better understanding of the mechanisms by which ROS and/or RNS modulate key protein kinases may also play an important role in cell death and survival after cerebral ischemia (Perez-Pinzon et al., 2005).

## 2.3 Ischemic Injury-Mediated Alterations in Protein Metabolism

It is well known that protein synthesis is very sensitive to ATP, which is depleted following ischemic injury. Translational step of protein synthesis is more vulnerable to ischemic injury than transcriptional step. Following brief ischemia, protein synthesis is markedly decreased in all neurons but recovers during reperfusion, except in vulnerable neurons, such as those in CA1 region of hippocampus. Ischemic injury disaggregates polyribosomes, where proteins are synthesized into monosomes after reperfusion (Abe et al., 1995). Under normal conditions, protein synthesis requires a functional translation initiation complex, a key element of which is eukaryotic initiation factor 2 (eIF2), which in a complex with GTP introduces the met-tRNAi. Under ischemic conditions, phosphorylation of $Ser^{51}$ on the α-subunit of eIF2 [eIF2α(P)] generates a competitive inhibitor of eIF2B, thereby preventing the replenishment of GTP onto eIF2, thus blocking translation initiation. The mechanisms leading to cellular damage from ischemic/reperfusion injury are complex and multifactorial. Accumulating evidence suggests that oxidative stress plays a major role in brain damage. Ischemic/reperfusion injury facilitates $Ca^{2+}$ influx to activate many protein-degrading enzymes, including μ-calpain, calcineurin, and caspases, which mediate the progressive proteolysis of structural proteins such as spectrin, tubulin, eIF2, and eIF4 (DeGracia et al., 2002; DeGracia and Montie, 2004; DeGracia, 2004). In selectively vulnerable neurons, calpain-mediated proteolytic degradation of eIF4G and cytoskeletal proteins alter translation initiation mechanisms that substantially reduce total protein synthesis and impose major alterations in message selection, downregulate survival signal transduction, and caspase activation. Thus, ischemic/reperfusion injury causes inhibition of protein synthesis in neurons. In all eukaryotic cells, the endoplasmic reticulum is the site where folding and assembly occurs for proteins destined to the extracellular space, plasma membrane, and the exo/endocytic compartments. Following ischemic/reperfusion injury, phosphorylation of the α-subunit of eIF2 [eIF2(αP)] by the endoplasmic reticulum transmembrane eIF2α kinase (PERK) leads to inhibition of translation initiation. PERK activation, depletion of endoplasmic reticulum $Ca^{2+}$, inhibition of the endoplasmic reticulum $Ca^{2+}$-ATPase suggest that an endoplasmic reticulum unfolded protein response (UPR) is induced as a result of brain ischemic/reperfusion injury (DeGracia et al., 2002; DeGracia and Montie, 2004; DeGracia, 2004). It is shown that in mammalian brain, the upstream unfolded protein response components PERK, inositol requiring enzyme 1 (IRE1), and activating transcription factor 6 (ATF6) not only upregulate prosurvival mechanisms (e.g., transcription of GRP78, PDI, SERCA2b) but also promote pro-apoptotic mechanisms (i.e., activation of Jun N-terminal kinases, caspase-12, and CHOP transcription). Sustained activation of eIF2(αP) is achieved by inducing the synthesis of ATF4, the CHOP transcription factor, through "bypass scanning" of 5′ upstream open-reading frames in ATF4 messenger RNA; these upstream open-reading frames normally inhibit access to the ATF4 coding sequence (DeGracia et al., 2002; DeGracia and Montie, 2004; DeGracia, 2004). Detailed studies have shown that following ischemic/reperfusion injury, several transcription factors (XBP1, ATF4, and ATF6f) are produced and they collaborate with each other to activate unfolded protein response (UPR), a neural cell stress program activated by misfolded proteins accumulation in the endoplasmic reticulum lumen (Haze et al., 1999; Lin et al., 2007). UPR activation not only causes a PERK-mediated phosphorylation of eIF2α, inhibition of protein synthesis, and prevention of further accumulation of unfolded proteins in the endoplasmic reticulum but also upregulation of genes coding for endoplasmic reticulum-resident enzymes and chaperone proteins via eIF2α(p) and ATF6 and IRE1 activation (DeGracia et al., 2002; DeGracia and Montie, 2004; DeGracia, 2004) suggesting that UPR-mediated transcription increases capacity of the endoplasmic reticulum to process misfolded proteins. Prolonged endoplasmic reticulum stress and the UPR accumulation lead to apoptotic cell death (DeGracia et al., 2002; DeGracia and Montie, 2004; DeGracia, 2004). Accumulating evidence suggests that ischemic/reperfusion injury is accompanied by multiple forms of endoplasmic reticulum stress. The UPR following brain ischemic/reperfusion injury is not an isomorphic process. Although PERK and IRE1 are activated in the initial hours of reperfusion, the total PERK is decreased, ATF6 is not activated, and there is delayed appearance of UPR-induced mRNAs. In addition, ischemic/reperfusion injury also facilitates caspase-3-mediated proteolysis of eIF4G, which shifts message selection to $m^7$G-cap-independent translation initiation of messenger RNAs containing internal ribosome entry sites. This internal ribosome entry site-mediated translation initiation promotes apoptosis. Thus, alterations in eIF2 and eIF4 have major implications for which messenger RNAs are translated by residual protein synthesis in neurons during brain reperfusion, in turn constraining protein expression of changes in gene transcription induced by ischemia and reperfusion (DeGracia et al., 2002; DeGracia and Montie, 2004; DeGracia, 2004). In addition, brain ischemic/reperfusion injury activates the expression of a number of genes involved in pro-survival pathways (Truettner et al., 2009). As stated above, the pro-survival pathways involve the sequestration and elimination of misfolded and aggregated proteins. Recent studies suggest that the endoplasmic reticulum, mitochondria, and cytoplasm respond individually to the accumulation of unfolded proteins by induction of organelle-specific molecular chaperones and folding enzymes (Ma and Hendershot, 2004; Truettner et al., 2009). These chaperones and folding enzymes not only prevent protein unfolding and block aggregation, but also promote the proper folding and assembly of proteins in the endoplasmic reticulum (Ma and Hendershot, 2004). Some endoplasmic reticulum chaperones are also involved in signaling the endoplasmic reticulum stress response, targeting misfolded proteins for degradation, and perhaps even shutting down the UPR when the stress subsides. Chaperones and folding enzymes include heat shock protein 70 (Hsp70 cytoplasmic), Hsp60 (mitochondrial), endoplasmic reticulum luminal proteins glucose response proteins GRP78 and GRP94, protein disulphide isomerase (PDI), homocysteine-inducible, endoplasmic reticulum stress-inducible protein (HERP), and calnexin. Thus, in hippocampus induction of mRNA and expression of Hsp70 is observed at 4 h while those of Hsp60, GRP78, GRP94 is seen after 24 h following reperfusion. This suggests that subcellular responses to ischemic/reperfusion insult vary among various subcellular compartments and are most prevalent in the cytoplasm and, to a lesser degree, in the mitochondrial matrix and endoplasmic reticulum lumen (Truettner et al., 2009). Collective evidence suggests that molecular chaperones and chaperone-related proteases thus control the delicate balance between natively folded functional proteins and aggregation-prone misfolded proteins, which may form during ischemic/reperfusion injury. Beside chaperones and folding enzymes, neurons also express neuroglobin (Ngb), a recently discovered protein that is distantly related to hemoglobin and myoglobin. This protein is predominantly expressed in the brain following hypoxic or ischemic injury. It provides protection against hypoxic or ischemic neuronal injury (Khan et al., 2006). In transgenic mice with overexpression of Ngb, the occlusion of the middle cerebral artery produces 30% reduction in volume of cerebral infarcts compared with wild-type littermates. Mice overexpressing Ngb also show enhanced expression of NOS in vascular endothelial cells. The molecular mechanism associated with the action of Ngb is not fully understood. However, it is proposed that neuroprotective actions of Ngb may involve inhibition of Pak1 kinase activity and Rac1-GDP-dissociation inhibitor disassociation (Khan et al., 2006; Greenberg et al., 2008).

Neural cells respond to ischemic/reperfusion injury differently. Thus, glial cell are more resistant to short ischemic/reperfusion injury than neurons. Astrocytes express many proteins that provide resistance to ischemic injury. These proteins include selenoprotein-S, CHOP, endothelin, and oxygen-regulated protein 150 (ORP150) (Ho et al., 2001; Kuwabara et al., 1996; Fradejas et al., 2008; Benavides et al., 2005). Localized in endoplasmic reticulum, selenoprotein-S not only protects astrocytes against oxygen, and glucose deprivation (OGD), but also prevents the deleterious consequences of accumulation of misfolded proteins oxidative damage, inflammation, and apoptosis (Fradejas et al., 2008). Astrocytes also contain CEBP homologous protein CHOP and (CHOP)-coding gene (Benavides et al., 2005). CHOP is also localized in endoplasmic reticulum and like selenoprotein-S, it protects astrocytes from OGD. Astrocytes undergo apoptosis only when CHOP is permanently upregulated and not when CHOP increases are transient (Benavides et al., 2005). Endothelin, a 21-amino-acid peptide, is found in astrocytes and has been reported to protect astrocytes against ischemic stress through the upregulation of endothelin and increase in levels of the endocannabinoid (anandamide), which participates in paracrine signaling toward neurons and microglia. Thus, astrocytes either repair their neighboring damaged neurons or participate in forming a protective boundary of the injured cells of the brain after ischemic injury (Ho et al., 2001). ORP150, a 150 kDa protein, is localized in endoplasmic reticulum of astrocytes. It protects astrocyte from hypoxic injury (Kuwabara et al., 1996). Collective evidence suggests that astrocytes respond to oxygen deprivation through the expression of several proteins that not only protect them from oxidative stress but also initiate adaptive responses that promote enhancement for the survival of neurons in penumbra.

## 2.4 Ischemic Injury-Mediated Alterations in Nucleic Acid Metabolism

Ischemic/reperfusion injury alters nucleic acid metabolism and damages neuronal nucleic acids through two mechanisms. First mechanism involves non-specific endonucleases and nitric oxide synthase (Gavrieli et al., 1992; Liu et al., 1997). Endonucleases are key enzymes that mediate regulated DNA fragmentation and chromatin condensation in response to ischemic/perfusion injury signal. This type of nucleic acid damage is irreversible and is referred to as DNA fragmentation (Chen et al., 1997). It occurs at sites between nucleosomes, protein-containing structures that occur in chromatin at ~200-BP intervals. DNA fragmentation is initiated by proteases (caspases) (Enari et al., 1998; Liu et al., 1997; Cao et al., 2001) or by neuronal NOS (Yoshida et al., 1994; Kamii et al., 1996; Huang et al., 2000). This type of nucleic acid damage becomes apparent between few hours to few days after cerebral ischemia. It depends on the duration of ischemic insult. A 40 kDa nuclear enzyme that is activated by caspase-3 and promotes apoptotic DNA degradation (CAD/DFF40) has been cloned from rat brain (Cao et al., 2001). Studies in the involvement of CAD/DFF40 in the induction of internucleosomal DNA fragmentation in the hippocampus of rat model of transient global ischemia indicate that after 8–72 h of ischemia, there occurs an induction of CAD/DFF40 mRNA and protein in the degenerating hippocampal CA1 neurons. CAD/DFF40 forms a heterodimeric complex in the nucleus with its natural inhibitor CAD (ICAD) and is activated after ischemia in a delayed manner (>24 h) by caspase-3, which is translocated into the nucleus and cleaves ICAD (Cao et al., 2001). Furthermore, an induction of CAD/DFF40 activity can be also detected in nuclear extracts, and the DNA degradation activity of CAD/DFF40 can be blocked by purified ICAD protein. These results support the view that CAD/DFF40 is the endogenous endonuclease that mediates caspase-3-dependent internucleosomal DNA degradation and related nuclear alterations in ischemic neurons (Fig. 2.5) (Cao et al., 2001; Widlak, 2000; Woo et al., 2004). DNA fragmentation and chromatin condensation are hallmark of apoptotic neuronal cell death.

The second mechanism is oxidative DNA damage that occurs early after ischemia (within the first 30 min of reperfusion) (Liu et al., 1996; Cui et al., 1999; Huang et al., 2000). In addition to DNA strand breaks (11, 15), this type of DNA damage involves base modifications (Liu et al., 1996; Cui et al., 1999, 2000) and DNA lacking a base (Huang et al., 2000). Evidence suggests that ROS (most likely NO, superoxide ions, and hydroxyl radicals) mediate this type of nucleic acid damage, which is often referred to as oxidative DNA damage (Epe et al., 1996; Liu et al., 1996; Cui et al., 1999; Huang et al., 2000; Beckman and Ames, 1997). Thus, oxidative DNA damage is closely associated with the delayed neuronal death in ischemic injury. These ischemic DNA lesions are similar to those found after ionizing radiation (Epe et al., 1996) and are generally reversible by DNA repair mechanisms (Beckman and Ames, 1997), with the exception of those in RNA (Kamath-Loeb et al., 1997).

Immunocytochemical studies indicate that 8-hydroxy-2′-deoxyguanosine (8-OHdG) immunoreactivity is present in the nucleus of neurons, glia, and endothelial cells in the hippocampus. The level of 8-OHdG is increased significantly in CA1 area at the end of 30 min after ischemia, and there is no increase within CA2 and CA3 areas. The increase in 8-OHdG immunoreactivity coincides with neuronal death in CA1 area (Won et al., 1999). It is not clear how the brain repairs oxidative DNA lesions in both the mitochondria and nuclei (Hanawalt, 1994; Lin et al., 2000; Sobol et al., 1996). However, it is becoming increasingly evident that base-excision repair (BER) pathway is the main mechanism employed by neurons to repair various types of oxidative DNA damage. BER involves the concerted effort of several repair proteins that recognize and excise specific DNA damages, eventually replacing the damaged moiety with a normal nucleotide. BER has two sub-pathways, both of which are initiated by the action of a DNA glycosylase. This enzyme interacts specifically with a target base and hydrolyzes the N-glycosylic bond, liberating the inappropriate or damaged base while keeping the sugar phosphate backbone of the DNA intact. This cleavage generates an AP (apyrimidinic/apurinic) or abasic site (i.e., the site of base loss) in the DNA. The AP site is processed by APE1 (AP endonuclease-1, also called HAP1/REF1/APEX), which cleaves the phosphodiester backbone immediately 5′ to the AP site, resulting in a 3′-hydroxyl group and a transient 5′-dRP (abasic deoxyribose phosphate) (Demple and Sung, 2005). Removal of the dRP is followed by the action of DNA Pol β (Polymerase β), which adds one nucleotide to the 3′-end of the nick, and removes the dRP moiety via its associated AP lyase activity. A DNA ligase seals the strand nick, thus restoring the integrity of the DNA. Replacement of the damaged base with a single new nucleotide is referred to as short-patch repair and represents approximately 80–90% of all BER. Among repair enzymes, 8-oxoguanine glycosylase/apyrimidinic/apurinic lyase (OGG) removes 8-OHdG from damaged DNA. Studies on 8-OHdG-removing activity in the cell nuclei of male C57BL/6 mouse brains following ischemic injuries indicate that OGG removes 8-OHdG with the greatest efficiency on the oligodeoxynucleotide duplex containing 8-OHdG/dC and with less efficiency on the heteroduplex containing 8-OHdG/dT, 8-OHdG/dG, or 8-OHdG/dA suggesting that the OGG1 protein may excise 8-OHdG in the mouse brain and that the activity of OGG1 may have a functional role in reducing oxidative gene damage in the brain after forebrain ischemia–reperfusion injury (Lin et al., 2000). It is also shown that the cellular BER activity is highly controlled (up- or downregulated) after ischemic brain injury, and this regulation may contribute to the outcome of cell injury. Although the molecular mechanism through which cellular BER is regulated in response to neuronal injury is not fully understood, it has been suggested that the functional impairment of the BER pathway after severe focal cerebral ischemia may be due to the loss-of-function post-translational modifications of repair enzymes (Luo et al., 2007). In addition, a major base modification is induced by the reaction between peroxynitrite and guanine, guanosine, and 2′- deoxyguanosine, either free or in DNA or RNA. These reactions involve myeloperoxidase–$H_2O_2$–nitrite system and results in conversion of guanine to 8-nitroguanine, 8-hydroxyadenine, 5-hydroxycytosine, and the deamination guanine to form xanthine (Love, 1999; Cui et al., 2000). 8-Nitroguanine acts as a specific marker for peroxynitrite-mediated DNA damage in ischemic and cancer tissues. Peroxynitrite-mediated damage results in breaking of DNA strand and in turn activating poly(ADP-ribose) polymerase (PARP).

PARP is a family of enzymes, which catalyzes poly(ADP-ribosyl)ation of DNA-binding proteins. To date, seven isoforms namely PARP-1, PARP-2, PARP-3, PARP-4, PARP-5, PARP-7, and PARP-10 have been identified. PARP-1, the best characterized member of PARP family is enriched in the nucleus. Upon activation, PARP-1 hydrolyzes $NAD^+$ to nicotinamide and transfers ADP ribose units to a variety of nuclear proteins, including histones and PARP-1 itself (Fig. 2.5). This process is important in facilitating DNA repair. Thus, under normal conditions, PARP plays an important role in maintaining genomic stability. However, under ischemic conditions, massive DNA injury is accompanied by excessive activation of PARP that may not only deplete stores of $NAD^+$ (the PARP substrate) but also cause marked reduction in ATP (Skaper, 2003a). PARP activation also enhances the expression of proinflammatory molecules and adhesion molecules in ischemic brain. These processes may lead to cell death. Accumulating evidence suggests that PARP activation plays a major role in neuronal death induced by cerebral ischemia (Park et al., 2004; Cui et al., 2000).

The secondary damage to surviving neurons in stroke accounts for the infarct volume and the subsequent loss of brain function. Microglial migration is strongly controlled in brain tissue through the expression of integrin CD11a, which is regulated in turn by PARP-1. This suggests that downregulation of PARP-1 may be a promising strategy in protecting neurons from secondary injury. PARP-1 has emerged as a major enzyme that plays an important role in the regulation of gene transcription (Skaper, 2003a, b). This observation further increases the importance and intricacy of poly(ADP-ribosyl)ation in the control of cell homeostasis and challenges the notion that ATP depletion is the sole mechanism by which poly(ADP-ribose) formation contributes to cell death. It is proposed that PARP(s) may regulate cell fate as essential modulators of death and survival transcriptional programs through its interactions with NF-κB and inhibitors of poly(ADP-ribosyl)ation may therefore retard the deleterious consequences of neuroinflammation by suppressing NF-κB activity (Skaper, 2003b).

## 2.5 Ischemic Injury-Mediated Alterations in Enzymic Activities

As stated above, ischemic/reperfusion injury is accompanied with the activation of many enzymes including $cPLA_2$, PLC, NOS, protein kinases, calpains, calcinurin, and endonucleases (Fig. 2.3). Many of these enzymes are activated by $Ca^{2+}$, which enters neurons through NMDA receptor and voltage-dependent $Ca^{2+}$ channels at the plasma membrane level, and mobilization of $Ca^{2+}$ from intracellular stores through PLC-mediated generation of InsP3 is indispensable for neural injury. As stated above, $cPLA_2$ is a major enzyme that releases ARA and induces global and focal cerebral ischemia-induced oxidative injury, BBB dysfunction, edema, and inflammation (Clemens et al., 1996; Nito et al., 2008). Immunocytochemical studies indicate that both reactive astrocytes and microglia contain elevated levels of $cPLA_2$ following ischemia/reperfusion injury (Clemens et al., 1996). Following focal cerebral ischemia/reperfusion injury, $cPLA_2$ is activated through phosphorylation by p38 mitogen-activated protein kinase (MAPK). In transient focal cerebral ischemia (tFCI) model in rats, determination of MARK and $cPLA_2$ activities along with western blot analysis indicates a significant increase in activities and expression of phospho-p38 MAPK and phospho-$cPLA_2$ in rat brain cortex after tFCI. Intraventricular administration of SB203580 not only significantly suppresses activation and phosphorylation of $cPLA_2$ but also attenuates BBB extravasation and subsequent edema (Nito et al., 2008). Moreover, overexpression of copper/zinc-superoxide dismutase remarkably decreases the activation and phosphorylation of both p38 MAPK and $cPLA_2$ after reperfusion. These results suggest that the p38 MAPK/$cPLA_2$ pathway plays a key role in inducing oxidative stress, promoting BBB disruption with initiating secondary vasogenic edema following ischemia–reperfusion injury (Nito et al., 2008).

Three cytosolic $Ca^{2+}$ sensors, calmodulin, protein kinases C (PKCs), and p21(ras)/phosphatidylinositol 3-kinase (PtdIns3K)/Akt pathways, are simultaneously involved in the steps linking the $Ca^{2+}$ to NF-κB-mediated neuronal injury (Lilienbaum and Israel, 2003; Marchetti et al., 2004). It is suggested that the duration of NF-κB activation is a critical determinant for excitotoxic stress-mediated neuronal injury and is dependent on a differential upstream and downstream signaling associated with various kinases. Extracellular signal-regulated kinase 1/2 (ERK1/2) is a member of the mitogen-activated protein kinase (MAPK) family. It mediates several processes including metabolism, motility, inflammation, neural cell death, and survival. It is phosphorylated and activated through a three-tiered MEK mode via cell surface receptors stimulated by growth factors or cytokines (Sawe et al., 2008). Levels of phosphorylated ERK1/2 are increased after cerebral ischemia/reperfusion. It is proposed that ROS and RNS contribute to ERK1/2 activation. It remains to be seen whether an increase in ERK1/2 phosphorylation is protective or detrimental to neural cells (Sawe et al., 2008). Contribution of NOS and endonucleases to DNA damage has been mentioned above. ROS activates many signaling pathways (Fig. 2.5) including ataxia-telangectasia mutated pathway (ATM), heat shock transcription factor 1(HSF1), PtdIns3K, and Janus protein kinase (JAK) pathway. Magnitude and duration of the oxidative stress along with cell type determine the involvement of above pathways. Low oxidative stress results in cell survival, whereas high oxidative stress and high levels of $Ca^{2+}$ lead to neurodegeneration (Farooqui and Horrocks, 2007).

## 2.6 Ischemic Injury-Mediated Alterations in Nuclear Transcription Factor-κB (NF-κB)

NF-κB (nuclear factor-κB) is a collective name for inducible dimeric family of transcription factors composed of five DNA binding proteins sharing the N-terminal Rel-homology domain (RHD): NF-κB1 (p50/p105), NF-κB2 (p52/p100), RelA (p65), cRel, and RelB that recognize a common sequence motif. NF-κB is found in neuronal and glial cells, and is involved in activation and modulation of a large number of genes in response to ischemic injury, immune responses, neuroinflammation, macrophage infiltration factors, cell adhesion molecules, cell survival, and other stressful situations requiring rapid reprogramming of gene expression. Five different proteins of NF-κB factor, namely p50, RelA/p65, c-Rel, RelB, and p52, can combine differently to form active dimers in response to external stimuli. RelA is activated by neurotoxic agents while c-Rel produces neuroprotective effects (Sarnico et al., 2009). In brain ischemia, RelA and p50 factors rapidly activate, but how they associate with c-Rel to form active dimers and contribute to the changes in diverse dimer activation for neuron susceptibility is unknown. Ischemic injury causes persistently activation of RelA and p50 factors of NF-κB in neurons that are destined to die. There are several potential routes through which NF-κB can act to induce neuronal death, including induction of death proteins and an aborted attempt to reenter the cell cycle. Under normal conditions, p50 and p65 protein subunits of NF-κB reside in the cytoplasm as an inactive complex bound by inhibitor proteins, Iκ-Bα and Iκ-Bβ. In response to ischemic injury, Iκ-B is phosphorylated by Iκ-B kinase and ubiquitinated and degraded by the proteasome; simultaneously, the active heterodimer translocates to the nucleus where it initiates gene transcription (Stephenson et al., 2000) (Fig. 2.5). The mechanism by which NF-κB mediates cell death remains unknown. It is proposed that translocation of NF-κB from cytoplasm to the nucleus results in its binding with target sequences in the genome and facilitates the expression of a number of proteins including many enzymes ($sPLA_2$, COX-2, NADPH oxidase and inducible nitric oxide synthase, superoxide dismutase) and cytokines (TNF-α, IL-1β, and IL-6) (Fig. 2.4). Activation of p50/RelA complex in the nucleus also induces the pro-apoptotic Bim and Noxa genes. Upregulation of $sPLA_2$, COX-2, NADPH oxidase and inducible nitric oxide synthase, and cytokines is closely associated with neuronal cell death in ischemic/reperfusion injury. Thus, NF-κB activation represents a paradigm for controlling the function of a regulatory protein via ubiquitination-dependent proteolysis, as an integral part of a phosphorylation-based signaling cascade.

In addition to ischemic injury, many agonists, viral and bacterial infections, and LPS stimulate NF-κB. Interactions of ROS with NF-κB also promote the translocation of NF-κB to the nucleus. Antioxidants prevent NF-κB translocation to the nucleus (Stephenson et al., 2000). A variety of other signaling events, including phosphorylation of NF-κB, hyperphosphorylation of I-κK, induction of I-κB synthesis, and the processing of NF-κB precursors, provide additional mechanisms that modulate the level and duration of NF-κB activity. Hypothermia decreases NF-κB translocation and binding activity by affecting NF-κB regulatory proteins. Mild hypothermia suppresses phosphorylation of NF-κB′s inhibitory protein (I-κB-α) by decreasing expression and activity of I-κB kinase-γ (IKK). As a consequence, hypothermia suppresses gene expression of two NF-κB target genes, inducible NOS and TNF-α. Accumulating evidence suggests that the protective effect of hypothermia on cerebral injury is, in part, related to NF-κB inhibition due to decreased activity of IKK (Yenari and Han, 2006). Protein-energy malnutrition (PEM) worsens functional outcome following global ischemia and is clinically relevant since 16% of elderly are nutritionally compromised at the time of hospitalization for stroke. It is proposed that this worsening is correlated with increasing activation of NF-κB and reactive gliosis, which involves increase in inflammatory response (Ji et al., 2008). Studies on transgenic mice expressing the I-κBα superrepressor (I-κBα mutated at serine-32 and serine-36, κ-Bα-SR) under transcriptional control of the neuron-specific enolase (NSE) and the glial fibrillary acidic protein (GFAP) promoter suggest that induction of c-myc and transforming growth factor-β2 in permanent middle cerebral artery occlusion (MCAO) model of cerebral ischemia is downregulated by neuronal expression of κ-Bα-SR, whereas induction of GFAP by MCAO is decreased by astrocytic expression of κ-Bα-SR. Neuronal, but not astrocytic, expression of the NF-κB inhibitor reduce both infarct size and cell death 48 h after permanent MCAO. In summary, these studies show that NF-κB is activated in neurons and astrocytes during cerebral ischemia and that NF-κB activation in neurons contributes to the ischemic damage (Zhang et al., 2005).

NF-κB also plays an important role in neuronal survival. Although the molecular mechanism of NF-κB-mediated neuroprotection is not fully understood, recent studies have indicated that c-Rel-containing dimers, p50/c-Rel and RelA/c-Rel, but not p50/RelA, promotes Bcl-xL transcription (Sarnico et al., 2009). Thus, the oxygen glucose deprivation (OGD) of cortical neurons not only results in Bim induction but also downregulation of Bcl-xL promoter activity and reduction in endogenous Bcl-xL protein content. These findings indicate that within the same neuronal cell, the balance between activation of p50/RelA and c-Rel-containing complexes fine-tunes the threshold of neuron vulnerability to the ischemic insult (Sarnico et al., 2009). NF-κB dimer (p50/p65) participates in the pathogenesis of post-ischemic injury by inducing pro-apoptotic gene expression, while c-Rel-containing dimers increase neuron resistance to ischemia by inducing anti-apoptotic gene transcription (Pizzi et al., 2009). In addition, NF-κB activation may prevent neuronal cell death through the induction of inhibitor of apoptosis proteins (IAPs) and manganese superoxide dismutase (Mn-SOD). NF-κB-mediated neuroprotective signaling produces changes in the structure and function of neuronal circuits (Mattson and Meffert, 2006). Collective evidence suggests that the ultimate survival or death of neurons depends on which, where, and when the NF-κB factors are activated.

In addition to NF-κB, ischemic injury is also associated with the activation of other transcription factors; for example, activator protein 1 (AP-1) [97], cAMP response element-binding protein (CREB), and hypoxia inducible factors (HIFs) (Miao et al., 2005; Walton et al., 1996; Bergeron et al., 2000). AP-1 is involved in the control of cell proliferation, differentiation, and death via the regulation of multiple gene families. Members of the AP-1 transcription factors include c-fos, fra-1, fra-2, fosB, c-jun, junB, and junD. Ischemic injury is accompanied by significant changes in their expression. For example, marked increases are observed in c-fos and c-jun, junB, junD Krox-24 mRNAs in a rat model of ischemia (Kiessling et al., 1993; An et al., 1993). It is reported that ischemic tolerance is associated with short increases in AP-1 binding activity, which peaks at 3 h. Similar changes occur in cells that are destined to survive in the hippocampal CA1 areas. Ischemic injury also involves phosphorylation of CREB and increases in the expression of CREB-dependent genes in the brain (Walton et al., 1996). CREB participates in cellular proliferation, survival, and differentiation (Carlezon et al., 2005). In the brain, CREB-mediated gene expression is caused by stimulation of glutamate receptor and increase in cytosolic calcium, which facilitates learning and memory, as well as in neuron survival and differentiation. Activation of CREB is associated with preconditioning (Lee et al., 2004). Hypoxia-inducible factor-1 (HIF-1) is another transcription factor that regulates the adaptive response to hypoxia in mammalian cells. HIF-1 consists of $O_2$-regulated subunit, HIF-1α, and the constitutively expressed aryl hydrocarbon receptor nuclear translocator, HIF-1β. Under hypoxic conditions, HIF-1α is stable, accumulates, and migrates to the nucleus where it binds to HIF-1β to form the complex (HIF-1α + HIF-1β). Transcription is initiated by the binding of the complex (HIF-1α + HIF-1β) to hypoxia responsive elements (HREs). The complex [(HIF-1α + HIF-1β) + HREs] stimulates the expression of target genes involved in angiogenesis, anaerobic metabolism, vascular permeability, and inflammation (Zaman et al., 1999; Hamrick et al., 2005).

## 2.7 Ischemic Injury-Mediated Alterations in Genes

Cerebral ischemia is one of the strongest stimuli for gene induction in the brain (Koistinaho and Hökfelt, 1997; Millán and Arenillas, 2006). Hundreds of genes have been found to be induced by brain ischemia. Genes modulating excitotoxicity, inflammatory response, and neuronal apoptosis are involved in neurodegeneration (Table 2.2). Beside above genes, cerebral ischemic injury also modulates neuroprotective gene expression, which is associated with reformatting and reprogramming

**Table 2.2** Induction of genes in ischemic brain

| Gene | Location | References |
|---|---|---|
| Immediate early genes | | |
| c-fos | Cortex, CA1, CA3 | Koistinaho and Hökfelt (1997), Akin et al. (1996) |
| Fos-B | Cortex, CA1, CA3 | Koistinaho and Hökfelt (1997), Akin et al. (1996) |
| c-jun | Cortex, CA1, CA3 | Koistinaho and Hökfelt (1997), Akin et al. (1996) |
| Jun B | Cortex, CA1, CA3 | Koistinaho and Hökfelt (1997), Akin et al. (1996) |
| Jun D | Cortex, CA1, CA3 | Koistinaho and Hökfelt (1997), Akin et al. (1996) |
| Zif268 | Cortex, CA1, CA3 | Koistinaho and Hökfelt (1997), Akin et al. (1996) |
| Krox 20 | Cortex | Koistinaho and Hökfelt (1997), Akin et al. (1996) |
| Nurr-1 | Cortex | Koistinaho and Hökfelt (1997) |
| Nurr-77 | Forebrain | Koistinaho and Hökfelt (1997) |
| Apoptotic genes | | |
| bcl-2 | CA1, CA3 | Koistinaho and Hökfelt (1997) |
| bcl-x | CA1, CA3 | Koistinaho and Hökfelt (1997) |
| bax | CA1, CA3 | Koistinaho and Hökfelt (1997) |
| P53 | Cortex | Koistinaho and Hökfelt (1997) |
| Fas | CA1, astrocyte | Koistinaho and Hökfelt (1997) |
| SGP-2 | CA1, astrocyte | Koistinaho and Hökfelt (1997) |
| BDNF | Contralateral side | Koistinaho and Hökfelt (1997) |
| bFGF | Cortex | Koistinaho and Hökfelt (1997) |
| TGF | Cortex | Koistinaho and Hökfelt (1997) |
| Calbindin | Cortex | Koistinaho and Hökfelt (1997) |

processes in the injured brain. These genes include immediate early genes, anti-apoptotic genes, Hsp genes, and genes encoding growth factors (BDNF). Many of above genes encode protein products that are associated directly or indirectly in neuronal survival. For example, enhanced expression of Hsps, growth factors, and anti-apoptosis genes promotes recovery. Neurodegeneration is promoted by induction of apoptotic and inflammatory genes, such as genes for iNOS, COX-2, and $sPLA_2$. Although so many ischemic injury inducible genes have been identified, there is a general reduction in gene transcription and inhibition of protein translation following ischemic injury. In fact modulation of genes for excitotoxicity, inflammatory response, apoptosis, anti-apoptotic genes, heat shock protein genes, and genes encoding for BDNF determines the clinical outcome after stroke.

The development of microarray techniques for gene expression profiling has facilitated the screening of large numbers of genes, following ischemic insult (Jin et al., 2001; Yakubov et al., 2004; Büttner et al., 2009). Oligonucleotide microarrays studies in complete global ischemia model indicate that levels of 576 transcripts are significantly altered in response to ischemic injury. Four hundred and nineteen transcripts are upregulated and 157 are downregulated. Reperfusion-induced transcript changes occur in a time-dependent manner. Thus, 1 h of reperfusion alters 39 transcripts, while 6 h of reperfusion produces changes in 174 transcripts, and 24 h of reperfusion causes changes in 462 transcripts. Quantitative real-time reverse transcription PCR studies of 18 selected genes show excellent agreement with the microarray results. Analyses of gene ontology patterns and the most strongly regulated transcripts show that the immediate response to an ischemia/reperfusion is mediated by the induction of specific transcription factors and stress genes. Delayed gene expression response is characterized by inflammation and immune-related genes. These results support the view that the response of brain tissue to ischemia is an active, specific, and coordinated process (Büttner et al., 2009). Similarly, quantitative reverse transcription polymerase chain reaction of 20 selected genes at 2, 4, and 24 h after ischemic injury following permanent cerebral occlusions shows early upregulated genes at 2 h including Narp, Rad, G33A, HYCP2, Pim-3, Cpg21, JAK2, CELF, Tenascin, and DAF. Late upregulated genes at 24 h include cathepsin C, Cip-26, cystatin B, PHAS-I, TBFII, Spr, PRG1, and LPS-binding protein (Lu et al., 2003). Glycerol 3-phosphate dehydrogenase, which is involved in mitochondrial reoxidation of glycolysis-derived NADH, is upregulated more than 60-fold. In addition, transcripts for plasticity-related genes such as Narp, agrin, and Cpg21 are also upregulated (Lu et al., 2003). Other genes that are upregulated in ischemic brain include C/EBP induction of Egr-1 (NGFI-A) with downstream induction of PAI-1, VEGF, ICAM, IL1, and MIP1. Genes regulated acutely after stroke may modulate cell survival and death; also, late regulated genes may be related to tissue repair and functional recovery (Lu et al., 2003).

Collectively, these studies suggest that ischemic injury induces the expression of selective gene in the brain. In the acute phase, the ischemic injury induces immediate early gene, followed by genes responsible for the induction of Hsps, proinflammatory genes (cytokines and chemokines), and apoptosis-related genes. Many immediate early genes code for transcription factors. Additional genes, including those encoding for neurotrophic factors and neurotransmitter systems, are induced in a delayed fashion after cerebral ischemia (Akin et al., 1996). As stated above, some of these genes are associated with neuronal death while other genes are related to neuronal survival (Yagita et al., 2008). In the later phase of ischemic injury, genes related to neurogenesis and tissue remodeling are expressed in the brain. These genes are associated with the recovery of neurological function. Many of these genes are expressed mainly in the glial cells in this phase (Yagita et al., 2008).

Ischemic tolerance is powerful protective mechanism against ischemic injury established by preconditioning with a mild insult of short duration. Tolerance evoked by brief ischemic injury is similar to transient ischemic attack that often precedes full-blown ischemic stroke in a clinical setting. Ischemic tolerance is commenced 24–48 h following sublethal ischemia. Since gene expression is altered during this period, it is proposed that gene expression may be involved in ischemic tolerance (Yagita et al., 2008). The induction of Hsp genes is closely associated with some part in ischemic tolerance. Thus, induction of Hsp27 has been reported to occur in gerbil brain with a 2-min period of sublethal ischemia (Kato et al., 1995a). In contrast, DNA microarray analysis indicates that gene suppression, rather than expression, may contribute to the molecular mechanism of ischemic tolerance. These observations suggest that gene expression profiles in ischemic brain injury and ischemic tolerance may involve different gene expression profiles (Yagita et al., 2008).

## 2.8 Ischemic Injury-Mediated Alterations in Cytokines and Chemokines

Ischemic/perfusion injury causes the expression of three major cytokines, namely, tumor necrosis factor (TNF-α), interleukin (IL)-1, IL-8, and IL-6 in different regions of rat brain as well as in cell culture experiments (Table 2.3) (Al-Bahrani et al., 2007; Tuttolomondo et al., 2008). All neural cells (neurons, astrocytes, microglia, and oligodendrocytes) produce inflammatory cytokine, which mediate cellular intercommunication through autocrine, paracrine, or endocrine mechanisms. Their actions involve a complex network linked to feedback loops and cascades. Cytokines produce their effects by binding to specific membrane-associated receptors that are composed of an extracellular ligand-binding region, a membrane-spanning region, and an intracellular region that is activated by binding of cytokines, and hence delivering a signal to the nucleus (Rothwell and Relton, 1993). Cytokine receptors are expressed constitutionally throughout the brain tissue at low levels. Cytokines play an important role not only in neuronal development, maturation, survival, regeneration but also in recovery process following neural insult (Rothwell and Relton, 1993). In addition, cytokines contribute to interconnection between brain and the immune system through hormonal cascades and cell-to-cell interactions. Indeed, the balance between pro- and anti-inflammatory cytokines not only determines the prowess of the immunological response but also influences the fate of the injured neurons following ischemic insult. In addition to cytokines, microglia and macrophages express adhesion molecules including selectin, immunoglobulin

**Table 2.3** NF-κB-mediated stimulation of $Na^+/Ca^{2+}$ exchangers, cytokines, chemokines, and adhesion molecules following ischemic injury

| Target | Effect | References |
|---|---|---|
| NCX1 | Upregulation | Formisano et al. (2008), Sirabella et al. (2009) |
| NCX3 | Downregulated | Formisano et al. (2008), Sirabella et al. (2009) |
| TNF-1α | Upregulation | Al-Bahrani et al. (2007) |
| IL-1, IL-8, IL-6 | Upregulation | Al-Bahrani et al. (2007), Tuttolomondo et al. (2008) |
| MCP-1 | Upregulation | Terao et al. (2009) |
| MIP-1α (protein-3α) | Upregulation | Terao et al. (2009) |
| Adhesion molecules | Upregulation | Wen et al. (2006) |

superfamily, integrins, and matrix metalloproteinases, which intensify and support neuroinflammation (Farooqui et al., 2007).

The molecular mechanism associated with cytokine mRNA expression in neurons is not fully understood. However, it is becoming increasingly evident that cytokines play important role in neuroinflammation following ischemic injury (Farooqui and Horrocks, 2007). In brain, inflammation is promoted by eicosanoids, which are generated through $PLA_2$/cyclooxygenase cascade reactions. Several mechanisms of $cPLA_2$ stimulation by cytokines are possible. One molecular mechanism of $cPLA_2$ stimulation by TNF-α and IL-1β involves the phosphorylation of $cPLA_2$ by mitogen-activated protein kinase in the presence of agents that mobilize intracellular $Ca^{2+}$ (Clark et al., 1995). Another mechanism involves TNF-α-mediated activation of caspase-3 and the proteolytic cleavage of $cPLA_2$ by caspase-3 (Wissing et al., 1997). Acetyl-Asp-Glu-Val-Asp-aldehyde, a tetrapeptide inhibitor of caspase-3 prevents the proteolytic cleavage and activation of $cPLA_2$ indicating that caspase-3-mediated $cPLA_2$ proteolysis retards cell death. Arachidonoyl trifluoromethyl ketone, a potent inhibitor of $cPLA_2$ activity, also blocks neural cell death (Wissing et al., 1997). Thus, the stimulation of $cPLA_2$ and caspase-3 along with induction of cyclooxygenase results in the oxidative stress, mitochondrial dysfunction, and calcium ion overload along with the release of cytochrome *c* and the activation of downstream caspase-9 and caspase-3 resulting in cell death (Farooqui, 2009). In addition, cytokines also facilitate the expression of other enzymes, such as cyclooxygenase-2, inducible nitric oxide synthase, and myeloperoxidase. These enzymes promote neuroinflammation following ischemic injury. The release of glutamate in activated microglia may also induce the expression of cytokines (Phillis et al., 2006).

Chemokines are a large family of structurally related small cytokines (8–10 kDa) originally identified as factors regulating the migration of leukocytes in inflammatory and immune responses (Minami and Satoh, 2000, 2003). Examples of chemokines are MCP-1, MIP-1α, and CINC. Some chemokines such as SDF-1 and fractalkine are constitutively produced in the brain and are associated with maintenance of brain homeostasis or determination of the patterning of neurons and/or glial cells in the developing brain and normal adult brain (Minami and Satoh, 2003). Ischemic injury not only stimulates the generation and release of chemokines but also increases the number of chemokine receptors in the brain. It is shown that mRNA expression for monocyte chemoattractant protein-1 (MCP-1) and macrophage inflammatory protein-1α (MIP-1α) is induced in the rat brain after focal cerebral ischemia. Chemokine signaling is associated with the post-ischemic inflammatory response. Overlapping pathways involving ROS, Toll-like receptor (TLR) activation, and the nuclear factor NF-κB system mediate both CXC and CC chemokines in ischemic tissues. Reperfusion accentuates chemokine expression promoting an intense inflammatory reaction (Frangogiannis, 2007). ELR-containing CXC chemokines regulate neutrophil infiltration in the ischemic area, whereas CXCR3 ligands may mediate recruitment of Th1 cells. CC chemokines, on the other hand, mediate mononuclear cell infiltration and macrophage activation. Accumulating evidence demonstrates that chemokine signaling mediates actions beyond leukocyte chemotaxis and activation, regulating angiogenesis and fibrous tissue deposition (Frangogiannis, 2007).

Intracerebroventricular injection of viral macrophage inflammatory protein-II (vMIP-II), a broad-spectrum chemokine receptor antagonist, reduces infarct volume in a dose-dependent manner (Minami and Satoh, 2000, 2003). These observations suggest that brain chemokines are involved in ischemic injury, and that chemokine receptors are potential targets for therapeutic intervention in stroke. Another potential target to suppress the harmful effect of chemokines is the signal transmission system(s) regulating the chemokine production. It is reported that induction of MCP-1 occurs through the activation of NMDA receptors in the cortico-striatal slice cultures. Almost all of the MCP-1 immunoreactivity is located on astrocytes, but NMDA treatment does not increase the MCP-1 synthesis in the enriched astrocyte cultures due to the absence of NMDA receptors on astrocytes (Minami and Satoh, 2003).

It is well established that cytokines modulate inflammation through proinflammatory cytokines, such as TNF-$\alpha$, IL-1$\beta$, and IL-6. These cytokines alter blood flow and increase vascular permeability, thus leading to secondary ischemia and accumulation of immune cells in the brain. The generation of cytokines is initiated by signaling through Toll-like receptors (TLRs) that recognize host-derived molecules released from injured tissues and cells. Recently, advances have been made in understanding of regulation of the innate immune system, particularly the signaling mechanisms of TLRs, which contribute to inflammatory response. This response is required to remove cell debris and to start regenerative process. However, inflammatory response can exacerbate cerebral damage and is associated with intensification in brain damage. Therefore, mammals have developed different mechanisms to regulate inflammatory response. An accurate balance between inflammation and anti-inflammation is necessary to assure the removal of cell debris, avoid secondary cell damage, and survival of neural cells around the injury site (Brea et al., 2009).

## 2.9 Ischemic Injury-Mediated Alterations in Heat Shock Proteins

Brain respond to ischemic injury by inducing the expression of a family of heat shock proteins, which include Hsp27, Hsp32, Hsp47, and Hsp70 (Kato et al., 1995b; Giffard and Yenari, 2004; Nishino and Nowak, 2004). Hsp27 has a potent ability to increase cell survival in response to oxidative stress. Detailed investigations indicate that ischemic injury induces the expression of Hsp27. Using transgenic and viral overexpression of Hsp27, it is shown that the overexpression of Hsp27 confers long-lasting tissue preservation and neurobehavioral recovery, as measured by infarct volume, sensorimotor function, and cognitive tasks up to 3 weeks following focal cerebral ischemia (Stetler et al., 2008). Similarly, the addition of Hsp27 to the culture medium of astrocyte and primary neuronal cells results in rapid entry into cells and protection from the oxidative stress (An et al., 2008).

In addition, intraperitoneal injections of Hsp27 into gerbils retard neuronal cell death in the CA1 region of the hippocampus in response to transient forebrain ischemia. Thus, Hsp27 protects neurons against ischemic injury-mediated cell death in vitro and in vivo settings. The signal transduction mechanism associated with Hsp27-mediated neuroprotection is not fully understood. However, neuropharmacological studies indicate that Hsp27 overexpression causes the suppression of the MKK4/JNK kinase cascade (Stetler et al., 2008). Although, Hsp27 overexpression has no effect on the activation of an upstream regulatory kinase of the MKK/JNK cascade and ASK1, but Hsp27 effectively blocks ASK1 activity via a physical association through its N-terminal domain and the kinase domain of ASK1 (Stetler et al., 2008). The N-terminal region of Hsp27 is necessary for neuroprotective effect against in vitro ischemic injury. Moreover, knockdown of ASK1 or inhibition of the ASK1/MKK4 cascade effectively prevents cell death following ischemia. These observations underscore the importance of this kinase cascade in the progression of ischemic neuronal death. Inhibition of PtdIns3K has no effect on Hsp27-mediated neuroprotection, suggesting that Hsp27 does not promote cell survival via activation of PtdInsK3/Akt. Based on these observations, it is suggested that the overexpression of Hsp27 results in long-lasting neuroprotective effect against ischemic brain injury through the inhibition of ASK1 kinase signaling (Stetler et al., 2008). Based on excitotoxic brain damage, it is shown that the expression of Hsp32, Hsp27, and Hsp47 in glial cells provides neuroprotection through different mechanisms. Thus, Hsp32 may promote antioxidant protective mechanisms to microglia/macrophages, whereas Hsp47 is associated with extracellular matrix remodeling, and Hsp27 may stabilize the astroglial cytoskeleton and participate in astroglial antioxidant mechanisms (Acarin et al., 2002). It is also suggested that Hsps reciprocally modulate cytokine production in response to the ischemic injury and other stressful stimuli.

Hsp70, in combination with cochaperones Hip and Hop, promotes the refolding of partially denatured proteins. Hsp70, in combination with cochaperones BAG-1 and CHIP, targets proteins for degradation in the proteasome (Ran et al., 2007). The mitochondrial Hsp70 (mtHsp70), together with cochaperones, facilitates protein/peptide entry into mitochondria. Hsp70 also modulates caspase-3-dependent and caspase-independent apoptotic cell death by binding apoptotic protease activation factor-1 (Apaf-1) and disrupting the formation of the Apaf-1/cytochrome *c*/caspase-9 apoptosome, which blocks activation of caspase-3. Collective evidence suggests that Hsps promote optimal protein folding and protein refolding, disaggregate proteins, and chaperone proteins across membranes; or they can target proteins for degradation and even stimulate cell apoptosis (Ran et al., 2007).

## 2.10 Ischemic Injury-Mediated Alterations in Adehesion Molecules

Cell adhesion molecules (CAMs) are transmembrane proteins located on the cell surface. They interact with other cells or with the extracellular matrix (ECM) in the process called cell adhesion. In addition to cytokines, CAMs (vascular cell adhesion molecule type 1, intercellular adhesion molecule type 1), endothelial leukocyte adhesion molecule 1 (ELAM-1), CD11/CD18 integrins, P-selectin, and metalloproteinases are also induced and participate in the early and delayed phases of ischemic damage (Koistinaho and Hökfelt, 1997). Matrix metalloproteinases are essential for the breakdown of the extracellular matrix around cerebral blood vessels and neurons, and their action leads to opening of the blood–brain barrier, brain edema, and hemorrhage (Jian Liu and Rosenberg, 2005). Matrix metalloproteinases act as cell surface sheddases and can affect cell signaling initiated by growth factors or death receptors. These enzymes also participate in tissue repair by promoting angiogenesis and neurogenesis. Metalloproteinases and vascular cell adhesion molecule levels are useful in the diagnosis of ischemic stroke. Inflammatory cytokines and adhesion cell molecules play important roles in early neurological deterioration and infarct volume. It is becoming increasingly evident that following ischemic injury the vasculature endothelium promotes inflammation through upregulation of adhesion molecules such as intercellular adhesion molecule (ICAM), vascular cell adhesion molecule-1 (VCAM-1), ELAM, E-selectin, and P-selectin. These molecules interact and bind to circulating leukocytes and facilitate migration of leukocytes into the central nervous system (CNS). Once being in the CNS, leukocytes produce cytotoxic molecules that promote cell death (Kim, 1996; Wen et al., 2006). The induction of adhession molecules in ischemic brain is time-locked process and is controlled in a highly regulated manner during the ischemic cascade. The functional role, interrelationship, and basic mechanism of action of adhesion molecules are being increasingly recognized, while trials such as anti-adhesion antibody molecules, growth factors, and anti-cytokine antibodies have shown some encouraging results in terms of reduction in neuronal damage in animals subjected to ischemic injury (Kim, 1996). However, clinical trials using immune blockade of adhesion molecules by antibodies have failed due to immune reactions of the host (Yilmaz and Granger, 2008). It is recently shown that mice lacking key adhesion molecules are more resistance to ischemic insult than wild-type mice. This suggests that further clinical trials are needed to judge the efficacy of humanized antibodies or non-immunogenic agents that interfere with cell adhesion mechanisms.

## 2.11 Ischemic Injury-Mediated Alterations in Apoptosis-Inducing Factor

Apoptosis-inducing factor (AIF) is a mitochondrial protein that upon translocation to nucleus produces large-scale DNA fragmentation. Although some studies indicate that mitochondrial dysfunction triggers the translocation of AIF to the nucleus, the molecular mechanism and stimulus for the cytosolic release and nuclear translocation AIF remains unknown (Chaitanya and Babu, 2008; Li et al., 2007). The time course of nuclear translocation of AIF after experimental stroke varies with the severity of injury and is increased by oxidative stress associated with reperfusion and nitric oxide (NO) production. AIF translocation to the nucleus triggers chromatin condensation, DNA fragmentation through the activation of poly(ADP-ribose) polymerase-1 (PARP-1) causing nuclear shrinkage. In addition to its apoptogenic activity on nuclei, AIF also participates in the regulation of apoptotic mitochondrial membrane permeabilization and exhibits an NADH oxidase activity (Cande et al., 2002). AIF-mediated neuronal cell death is caspase independent (Dawson and Dawson, 2004; van Wijk and Haegeman, 2005). Inhibition of neuronal NO synthase reduces formation of poly(ADP-ribose) polymer and nuclear AIF accumulation. Gene deletion of neuronal NO synthase also prevents nuclear AIF accumulation. Although reperfusion increases AIF translocation to the nucleus after 60 min of focal ischemia, translocation of AIF is markedly delayed when ischemia duration is decreased to 30 min. Prolonged focal ischemia with or without reperfusion induces translocation of AIF to the nucleus (Li et al., 2007). Based on detailed investigations, it is proposed that neuronally derived NO is a major factor responsible for the nuclear AIF accumulation after stroke. Hsp70, a heat shock protein, neutralizes AIF in a reaction that is independent of ATP or the ATP-binding domain (ABD) of Hsp70 and thus differs from the previously described Apaf-1/Hsp70 interaction (which requires ATP and the Hsp70 ABD). Intriguingly, Hsp70 lacking ABD (Hsp70 δ ABD) prevents apoptotic cell death mediated by serum withdrawal, staurosporin, and menadione. Microinjections of anti-AIF antibody or genetic ablation of AIF blocks AIF-mediated apoptosis. This observation suggests that AIF-mediated cell death is caspase independent (Ravagnan et al., 2001; Cande et al., 2002). Hepatocyte growth factor (HGF) protects hippocampal cornu ammonis (CA) subregion 1 neurons from apoptotic cell death after transient forebrain ischemia. It is proposed that HGF not only attenuates the increase in the expression of AIF protein in the nucleus after transient forebrain ischemia but also prevents the primary oxidative DNA damage as judged by using anti-8-OHdG (8-hydroxy-2′-deoxyguanosine) antibody (Niimura et al., 2006). Collectively, these studies suggest that AIF mediates neuronal cell death after focal cerebral ischemia and that caspase-independent signaling pathways downstream of mitochondria play an important role in the regulation of caspase-independent cell death after experimental stroke.

## 2.12 Ischemic Injury-Mediated Alterations in $Na^+/Ca^{2+}$ Exchanger

The $Na^+/Ca^{2+}$ exchanger (NCX), an ion transport protein, is expressed in the neural cell plasma membrane (PM). It extrudes $Ca^{2+}$ in parallel with the PM ATP-driven $Ca^{2+}$ pump. As a reversible transporter, it also mediates $Ca^{2+}$ entry in parallel with various ion channels. The energy for net $Ca^{2+}$ transport by the $Na^+/Ca^{2+}$ exchanger and its direction depend on the $Na^+$, $Ca^{2+}$, and $K^+$ gradients across the PM, the membrane potential, and the transport stoichiometry. Under normal conditions, the $Na^+/Ca^{2+}$ exchanger is known to transport one calcium ion out of the cell and three sodium ions into the cell. This is known as the calcium exit, or "forward" mode.

Under certain conditions, however, the exchanger reverses and transports calcium ions into the cell (calcium entry mode). Because dysregulation of sodium and calcium homeostasis is an integral part of ischemic brain injury, the role of the NCX in neurons has been studied in in vivo and in vitro models of ischemia (Blaustein and Lederer 1999; Pignataro et al., 2004; Farooqui et al., 2008). Five genes that code for the exchangers have been identified in mammalian tissues including brain: three in the $Na^+/Ca^{2+}$ exchanger family (NCX1, NCX2, and NCX3) and two in the $Na^+/Ca^{2+}$ plus $K^+$ family (NCKX1 and NCKX2) (Blaustein and Lederer 1999; Formisano et al., 2008). Exposure of cortical neurons to 3 h of oxygen and glucose deprivation (OGD) produces dissimilar effects on the NCX1, NCX2, and NCX3. First, OGD induces an upregulation in NCX1 transcript and protein expression (Table 2.3). This change is exerted at the transcriptional level because the inhibition of NF-κB translocation by small interfering RNA against p65 and SN-50 blocks oxygen and glucose deprivation-induced NCX1 upregulation. Second, OGD elicits a downregulation of NCX3 protein expression. This change, unlike NCX1, occurs at the post-transcriptional level because it is inhibited by the proteasome inhibitor MG-132 (Formisano et al., 2008). Finally, it is shown that OGD significantly increases NCX1 both in the forward and reverse modes of operation and facilitates an increase in endoplasmic reticulum $Ca^{2+}$ accumulation. Interestingly, such accumulation is blocked by the silencing of NCX1 or by NCX inhibitor CB-DMB treatment that triggers caspase-12 activation. NF-κB-dependent NCX1 upregulation may play a fundamental role in $Ca^{2+}$ refilling in the endoplasmic reticulum, thus helping neurons to retard OGD-mediated endoplasmic reticulum stress (Pignataro et al., 2004; Formisano et al., 2008).

Studies on NCX1, NCX2, and NCX3 protein levels in the rat hippocampus at 3, 6, 12, 18, 24, and 48 h following a 3 and 8 min durations of global cerebral ischemic injury indicate that NCX1 protein levels are significantly increased by 22.3 and 20.6% at the 6 and 12 h respective time points following a 3 min duration of global ischemia, while NCX2 and NCX3 protein levels remain unchanged (Bojarski et al., 2008). Following 8 min global ischemic injury, NCX1 protein levels remain relatively constant, while NCX2 protein levels are downregulated by 6.9, 10.8, 14.4, and 10.3% at the 6, 18, 24, and 48 h time points, respectively, and NCX3 protein levels are upregulated by 22.1% at the 18 h time point (Bojarski et al., 2008). In a permanent middle cerebral artery occlusion model of ischemia, all three NCX proteins have been reported to be downregulated in ischemic core; NCX3 is decreased in periinfarctual area, whereas NCX1 and NCX2 remain unchanged (Pignataro et al., 2004). Collectively, these results show that NCX subtype protein expression is sensitive to cerebral ischemic injury and indicate that alterations in NCX activity may play an important role in calcium maintenance and neuronal outcome following ischemia. Studies on dysregulation of NCX in cerebral ischemia have been controversial. The effects of KB-R7943, a specific inhibitor of the reverse mode of NCX, indicate that this drug significantly inhibits effluxes of phosphoethanolamine, but has no effect on glutamate, aspartate, taurine, or GABA levels (Pilitsis et al., 2001). KB-R7943 also produces significant reductions in levels of myristic, docosahexaenoic, and arachidonic acid during ischemia and in reperfusion levels of arachidonic and docosahexaenoic acids. These data indicate that inhibition of $Na^+/Ca^{2+}$ exchange likely blocks the activation of phospholipases that usually occurs following an ischemic insult as evidenced by its attenuation of phosphoethanolamine and free fatty acid efflux. The inhibition of phospholipases may be an essential component of the neuroprotective benefits of $Na^+/Ca^{2+}$ exchange inhibitors in ischemia–reperfusion injury and may provide a basis for their possible use in therapeutic strategies for stroke (Pilitsis et al., 2001). The majority of in vivo studies in focal cerebral ischemia model indicate that blocking NCX activity is neurodamaging, while increasing NCX activity is neuroprotective (Annunziato et al., 2004). However, others have failed to reproduce these results. Thus, more studies are needed on the role of NCX in ischemic injury.

## 2.13 Mechanism of Neurodegeneration in Ischemia/Reperfusion Injury

At the molecular level, ischemic injury is accompanied by the release of neurotoxic concentrations of glutamate, which interacts with its receptors and mediates $Ca^{2+}$ influx through NMDA receptor $Ca^{2+}$ channel. A sustained increase in $Ca^{2+}$ levels is harmful for the survival of neurons. $Ca^{2+}$ influx stimulates neural membrane glycerophospholipid degradation through the activation of isoforms of $PLA_2$, cyclooxygenases (COX), and lipoxygenases (LOX). Stimulation of these enzymes releases ARA and lyso-glycerophospholipids. Lyso-glycerophospholipids are either reacylated to the native glycerophospholipids or acetylated to proinflammatory platelet-activating factor (PAF) (Fig. 2.6). ARA is metabolized to eicosanoids. The non-enzymic oxidation of ARA produces ROS (Phillis et al., 2006). Mitochondrial dysfunction following ischemia/reperfusion generates ROS, which stimulates NF-κB. As stated above in cytoplasm, NF-κB is present in an inhibitory form attached to its inhibitory protein, I-κB (Yamamoto and Gaynor, 2004). Ischemia/reperfusion results in dissociation of I-κB, which is ubiquinated, and then degraded by proteasomes. Translocation of active NF-κB to the nucleus produces the transcription of numerous genes that not only influence the survival of neural cells and maintenance of normal functional integrity but also induces many genes associated with neuroinflammation, oxidative stress, and immune responses. These genes code for $sPLA_2$, iNOS, COX-2, intracellular adhesion molecule-1 (ICAM-1), vascular adhesion molecule-1 (VCAM-1), E-selectin, cytokines, and matrix metalloproteinases (MMP). Activation of NF-κB also leads to the local generation of more cytokines and chemokines, which in turn promulgate glutamate-mediated signals and potentiates the activation of NF-κB activity (Block and Hong, 2005).

The molecular mechanism associated with neuronal injury depends on the brain region affected by stroke. The factors that initiate, propagate, and maintain ischemic injury include many enzymes such as multiple forms of $PLA_2$, cyclooxygenases (COX), and lipoxygenases (LOX) generating lyso-glycerophospholipids, platelet-activating factor (PAF), pro-inflammatory prostaglandins (Fig. 2.6). In addition,

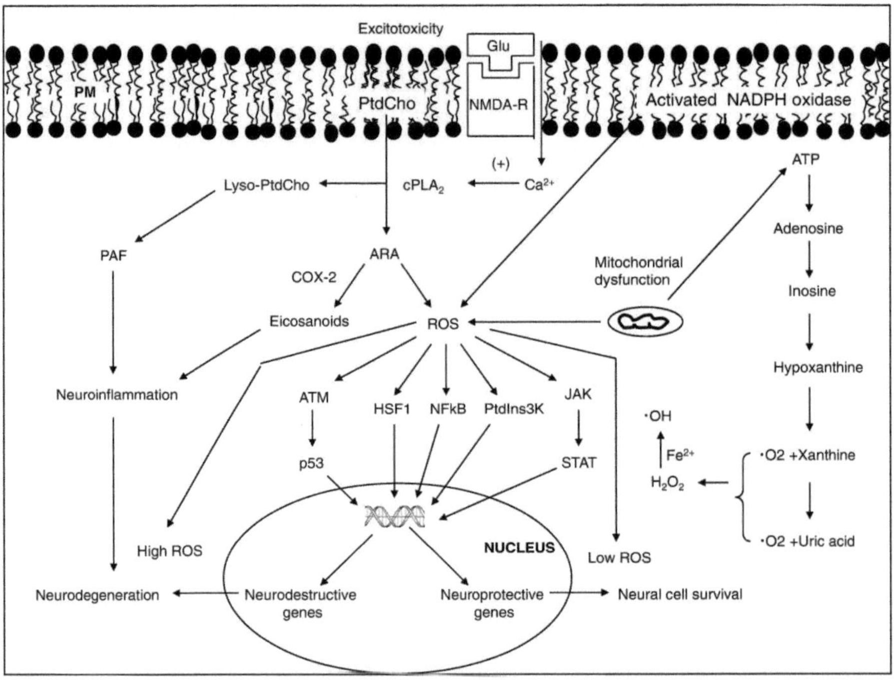


**Fig. 2.6** Activation of major signaling pathways and transcription factors by oxidative stress. Ataxia-telangectasia mutated (ATM); Heat shock transcription factor 1 (HSF1); nuclear factor-kappaB (NF-κB); and Janus protein kinase (JAK); cytosolic phospholipase $A_2$ ($cPLA_2$); cyclooxygenase-2 (COX-2); phosphatidylcholine (PtdCho); lyso-phosphatidylcholine (lyso-PtdCho); arachidonic acid (ARA); platelet-activating factor (PAF); superoxide ($^{\bullet}O_2$); and hydroxyl radical ($^{\bullet}OH$)

ischemic injury stimulates inducible nitric oxide synthases, endonucleases, proteases, protein kinases, and protein phosphatases (Phillis et al., 2006). Following ischemic injury, the stimulation of nitric oxide synthase by $Ca^{2+}$ generates nitric oxide, which reacts with superoxide to form peroxynitrite. Peroxynitrite produces single-stranded breaks in DNA, which activate poly(adenosine diphosphate ribose) polymerase leading to NAD and ATP depletion. This may be another mechanism that may contribute to ischemia/reperfusion-mediated neural cell death. In addition, proinflammatory eicosanoids and platelet-activating factor interact with their receptors and modulate signaling in brain (Chabot et al., 1998; Phillis et al., 2006) through cross talk among glutamate, eicosanoids, platelet-activating factor, and thromboxane receptors. Under physiological conditions, this cross talk refines their communication among neurons, macroglial cells, microglial cells, and vascular cells, but under pathological situations, this cross talk initiates and promotes neuronal injury depending on the magnitude of $PLA_2$, COX, and LOX expression, production of arachidonic acid metabolites, synthesis of platelet-activating factor, and generation of ROS (Farooqui and Horrocks, 2007). Collective evidence suggests that mechanisms leading to cellular damage from ischemia–reperfusion injury are complex and multifactorial. Accumulating evidence suggests an important role for oxidative stress in the regulation of neuroinflammation following stroke. Gene expression studies have revealed that the increase in oxygen radicals post-ischemia triggers the expression of a number of proinflammatory genes. Although these processes are confirmed in a variety of animal models of cerebral ischemia, the exact mechanism is still uncertain (Wong and Crack, 2008).

## 2.14 Conclusion

Stroke is a rapidly developing cerebrovascular event caused by a thrombus or embolism in an extraparenchymal cerebral vessel (commonly in the middle and anterior cerebral arteries) and resulting in impairment of brain function due to the interruption of blood flow to the brain. Stroke triggers a complex and highly interconnected cascade of cellular and molecular events. Early events induced following ischemic injury, including excitotoxicity, calcium overload, and oxidative stress that rapidly result in cell death in the infarct core. Later events, such as neuroinflammation and apoptosis, are relevant to the death of the ischemic penumbra. Stroke also initiates breakdown of cellular integrity, ionic imbalance, and production of ROS and NRS. Ischemic injury is accompanied by rapid release of glutamate and sustained calcium influx at the core of injury site but not in surrounding area. $Ca^{2+}$ influx activates $PLA_2$, PLC and PLD, CaMKs, MAPKs, NOS, calpains, calcinurin, and endonucleases. These enzymes are closely associated with neuronal cell death. In addition, ischemic injury causes expression of many genes, transcription factors, adhesion molecules, heat shock proteins, and apoptosis-inducing factor. Excitotoxicity-mediated generation of superoxide and nitric oxide leads to formation of highly reactive products, including peroxynitrite and hydroxyl radical, which have potential to irreversibly damage lipids, proteins, and DNA.

Peroxynitrite and hydroxyl radical also play a critical role in the initiation of mitochondrial dysfunction, apoptotic cell death, and poly(ADP-ribose) polymerase activation. This provides additional mechanisms for oxidative damage. Coordination of all subcellular organelles is necessary for neuronal death cascade. Although all subcellular organelles participate in ischemic injury-mediated neurodegeneration, mitochondria and nucleus play a major role in delayed neurodegeneration caused by apoptotic cell death. Nucleus is the organelle that contains genomic DNA. Ischemic injury not only causes DNA breakage but also contributes to the expression of proinflammatory enzymes, cytokines, chemokines, and Bax (pro-apoptotic), which intensify proinflammatory lipid mediators through cytokine and chemokine positive loops. Whether these processes are the cause or merely the result of the ischemic insult remains an open question. Activation of anti-apoptotic signaling cascades also occurs in neurons in animal models of ischemic injury. Anti-apoptotic signaling factors and pathways are initiated and activated by the expression of neurotrophic factors (BDNF), certain cytokines, antioxidant enzymes, Bcl-2 (anti-apoptotic), and calcium-regulating proteins.

## References


Abe K, Aoki M, Kawagoe J, Yoshida T, Hattori A, Kogure K, Itoyama Y (1995) Ischemic delayed neuronal death: a mitochondrial hypothesis. Stroke 26:1478–1489

Acarin L, Paris J, González B, Castellano B (2002) Glial expression of small heat shock proteins following an excitotoxic lesion in the immature rat brain. Glia 38:1–14

Akin Sp, Liu PK, Hsu CY (1996) Immediate early gene expression in response to cerebral ischemia: friend or foe? Stroke 27:1682–1687

Al-Bahrani A, Taha S, Shaath H, Bakhiet M (2007) TNF-alpha and IL-8 in acute stroke and the modulation of these cytokines by antiplatelet agents. Curr Neurovasc Res 4:31–37

Almeida A, Heales SJ, Bolanos JP, Medina JM (1998) Glutamate neurotoxicity is associated with nitric oxide-mediated mitochondrial dysfunction and glutathione depletion. Brain Res 790:209–216

An G, Lin TN, Liu JS, Xue JJ, He YY, Hsu CY (1993) Expression of c-fos and c-jun family genes after focal cerebral ischemia. Ann Neurol 33:457–464

An JJ, Lee YP, Kim SY, Lee SH, Lee MJ, Jeong MS, Kim DW, Jang SH, Yoo KY, Won MH, Kang TC, Kwon OS, Cho SW, Lee KS, Park J, Eum WS, Choi SY (2008) Transduced human PEP-1-heat shock protein 27 efficiently protects against brain ischemic insult. FEBS J 275: 1296–1308

Annunziato L, Pignataro G, Di Renzo GF (2004) Pharmacology of brain $Na^{+}/Ca^{2+}$ exchanger: from molecular biology to therapeutic perspectives. Pharmacol Rev 56:633–654

Anrather J, Racchumi G, Iadecola C (2006) NF-kappaB regulates phagocytic NADPH oxidase by inducing the expression of gp91phox. J Biol Chem 281:5657–5667

Atlante A, Calissano P, Bobba A, Azzariti A, Marra E, Passarella S (2000) Cytochrome c is released from mitochondria in a reactive oxygen species (ROS)-dependent fashion and can operate as a ROS scavenger and as a respiratory substrate in cerebellar neurons undergoing excitotoxic death. J Biol Chem 275:37159–37166

Auger C, Attwell D (2000) Fast removal of synaptic glutamate by postsynaptic transporters. Neuron 28:547–558

Baltan S (2006) Surviving anoxia: a tale of two white matter tracts. Crit Rev Neurobiol 18:95–103

Baltan S (2009) Ischemic injury to white matter: an age-dependent process. Neuroscientist 15: 126–133

Beckman KB, Ames BN (1997) Oxidative decay of DNA. J Biol Chem 272:19633–19636

Benavides A, Pastor D, Santos P, Traque P, Calvo S (2005) CHOP plays a pivotal role in the astrocyte death induced by oxygen and glucose deprivation. Glia 52:261–275

Bergeron M, Gidday JM, Yu AY, Semenza GL, Ferriero DM, Sharp FR (2000) Role of hypoxia-inducible factor-1 in hypoxia-induced ischemic tolerance in neonatal rat brain. Ann Neurol 48:285–296

Blaustein MP, Lederer WJ (1999) Sodium/calcium exchange: its physiological implications. Physiol Rev 79:763–854

Block ML, Hong JS (2005) Microglia and inflammation-mediated neurodegeneration: multiple triggers with a common mechanism. Prog Neurobiol 76:77–98

Bojarski C, Meloni BP, Moore SR, Majda BT, Knuckey NW (2008) $Na^{+/}Ca^{2+}$ exchanger subtype (NCX1, NCX2, NCX3) protein expression in the rat hippocampus following 3 min and 8 min durations of global cerebral ischemia. Brain Res 1189:198–202

Bolanos JP, Almeida A, Stewart V, Peuchen S, Land JM, Clark JB, Heales SJ (1997) Nitric oxide-mediated mitochondrial damage in the brain: mechanisms and implications for neurodegenerative diseases. J Neurochem 68:2227–2240

Brea D, Sobrino T, Ramos-Cabrer P, Castillo J (2009) Inflammatory and neuroimmunomodulatory changes in acute cerebral ischemia. Cerebrovasc Dis 27(Suppl 1):48–64

Büttner F, Cordes C, Gerlach F, Heimann A, Alessandri B, Luxemburger U, Türeci O, Hankeln T, Kempski O, Burmester T (2009) Genomic response of the rat brain to global ischemia and reperfusion. Brain Res 1252:1–14

Camandola S, Poli G, Mattson MP (2000) The lipid peroxidation product 4-hydroxy-2,3-nonenal increases AP-1-binding activity through caspase activation in neurons. J Neurochem 74: 159–168

Cande C, Cohen I, Daugas E, Ravagnan L, Larochette N, Zamzami N, Kroemer G (2002) Apoptosis-inducing factor (AIF): a novel caspase-independent death effector released from mitochondria. Biochimie 84:215–222

Cao G, Pei W, Lan J, Stetler RA, Luo Y, Nagayama T, Graham SH, Yin XM, Simon RP, Chen J (2001) Caspase-activated DNase/DNA fragmentation factor 40 mediates apoptotic DNA fragmentation in transient cerebral ischemia and in neuronal cultures. J Neurosci 21:4678–4690

Carlezon WA Jr., Duman RS, Nestler EJ (2005) The many faces of CREB. Trends Neurosci 28:436–445

Chabot C, Gagné J, Giguère C, Bernard J, Baudry M, Massicotte G (1998) Bidirectional modulation of AMPA receptor properties by exogenous phospholipase $A_2$ in the hippocampus. Hippocampus 8:299–309

Chaitanya GV, Babu PP (2008) Multiple apoptogenic proteins are involved in the nuclear translocation of Apoptosis Inducing Factor during transient focal cerebral ischemia in rat. Brain Res 1246:178–190

Chen J, Jin K, Chen M, Pei W, Kawaguchi K, Greenberg DA, Simon RP (1997) Early detection of DNA strand breaks in the brain after transient focal ischemia: implications for the role of DNA damage in apoptosis and neuronal cell death. J Neurochem 69:232–245

Choi DW (1988) Glutamate neurotoxicity and diseases of the nervous system. Neuron 1:628–634

Clark JD, Schievella AR, Nalefski EA, Lin L-L (1995) Cytosolic phospholipase $A_2$. J Lipid Mediat Cell Signal 12:83–117

Clemens JA, Stephenson DT, Smalstig EB, Roberts EF, Johnstone EM, Sharp JD, Little SP, Kramer RM (1996) Reactive glia express cytosolic phospholipase $A_2$ after transient global forebrain ischemia in the rat. Stroke 27:527–535

Cui J, Holmes EH, Liu PK (1999) Oxidative damage to the c-fos gene and reduction of its transcription after focal cerebral ischemia. J Neurochem 73:1164–1174

Cui J, Holmes EH, Greene TG, Liu PK (2000) Oxidative DNA damage precedes DNA fragmentation after experimental stroke in rat brain. FASEB J 14:955–967

Dawson VL, Dawson TM (2004) Deadly conversations: nuclear-mitochondrial cross-talk. J Bioenerg Biomembr 36:287–294

DeCoster MA, Mukherjee PK, Davis RJ, Bazan NG (1998) Platelet-activating factor is a downstream messenger of kainate-induced activation of mitogen-activated protein kinases in primary hippocampal neurons. J Neurosci Res 53:297–303

DeGracia DJ, Kumar R, Owen CR, Krause GS, White BC (2002) Molecular pathways of protein synthesis inhibition during brain reperfusion: implications for neuronal survival or death. J Cereb Blood Flow Metab 22:127–141

DeGracia DJ, Montee HL (2004) Cerebral ischemia and the unfolded protein response. J Neurochem 91:1–8

DeGracia DJ (2004) Acute and persistent protein synthesis inhibition following cerebral reperfusion. J Neurosci Res 77:771–776

Demple B, Sung JS (2005) Molecular and biological roles of Ape1 protein in mammalian base excision repair. DNA Repair (Amst) 4:1442–1449

Dienel GA (1984) Regional accumulation of calcium in postischemic rat brain. J Neurochem 43:913–925

Edgar AD, Strosznajder J, Horrocks LA (1982) Activation of ethanolamine phospholipase $A_2$ in brain during ischemia. J Neurochem 39:1111–1116

Enari M, Sakahira H, Yokoyama H, Okawa K, Iwamatsu A, Nagata S (1998) A caspase-activated DNase that degrades DNA during apoptosis, and its inhibitor ICAD. Nature (London) 391: 43–50

Epe B, Ballmaier D, Roussyn I, Briviba K, Sies H (1996) DNA damage by peroxynitrite characterized with DNA repair enzymes. Nucleic Acids Res 24:4105–4110

Farooqui AA, Horrocks LA (1994) Excitotoxicity and neurological disorders: involvement of membrane phospholipids. Int Rev Neurobiol 36:267–323

Farooqui AA, Horrocks LA (2007) Glycerophospholipids in brain. Springer, New York, NY

Farooqui AA, Horrocks LA (2006) Phospholipase $A_2$-generated lipid mediators in the brain: the good, the bad, and the ugly. Neuroscientist 12:245–260

Farooqui AA, Horrocks LA (2009) Glutamate and cytokine-mediated alterations of phospholipids in head injury and spinal cord trauma. In: Banik N, Ray SK (eds) Brain and spinal cord trauma, Handbook of neurochemistry Lajtha, A. Springer, New York, NY, pp 71–89

Farooqui AA (2009) Hot topics in neural membrane lipidology. Springer, New York, NY

Farooqui AA, Horrocks LA, Farooqui T (2007) Modulation of inflammation in brain: a matter of fat. J Neurochem 101:577–599

Farooqui AA, Ong WY, Horrocks LA (2008) Neurochemical aspects of excitotoxicity. Springer, New York, NY

Feigin VL, Lawes CM, Bennett DA, Anderson CS (2003) Stroke epidemiology: a review of population-based studies of incidence, prevalence, and case-fatality in the late 20th century. Lancet Neurol 2:43–53

Formisano L, Saggese M, Secondo A, Sirabella R, Vito P, Valsecchi V, Molinaro P, Di Renzo G, Annunziato L (2008) The two isoforms of the Na+/Ca2+ exchanger, NCX1 and NCX3, constitute novel additional targets for the prosurvival action of Akt/protein kinase B pathway. Mol Pharmacol 73(3):727–737 13 Dec 2007 [Epub ahead of print]

Fradejas N, Pastor MD, Mora-Lee S, Tranque P, Calvo S (2008) SEPS1 gene is activated during astrocyte ischemia and shows prominent antiapoptotic effects. J Mol Neurosci 35:259–265

Frangogiannis NG (2007) Chemokines in ischemia and reperfusion. Thromb Haemost 97:738–747

Gavrieli Y, Sherman Y, Ben-Sasson SA (1992) Identification of programmed cell death in situ via specific labeling of nuclear DNA fragmentation. J Cell Biol 119:493–501

Giffard RG, Yenari MA (2004) Many mechanisms for Hsp70 protection from cerebral ischemia. J Neurosurg Anesthesiol 16:53–61

Greenberg DA, Jin K, Khan AA (2008) Neuroglobin: an endogenous neuroprotectant. Curr Opin Pharmacol 8:20–24

Hamrick SE, McQuillen PS, Jiang X, Mu D, Madan A, Ferriero DM (2005) A role for hypoxia-inducible factor-1alpha in desferoxamine neuroprotection. Neurosci Lett 379:96–100

Hanawalt PC (1994) Transcription-coupled repair and human disease. Science 266:1957–1958

Hayashi T, Abe K (2004) Ischemic neuronal cell death and organelle damage. Neurol Res 26: 827–834

Haze K, Yoshida H, Yanagi H, Yura T, Mori K (1999) Mammalian transcription factor ATF6 Is synthesized as a transmembrane protein and activated by proteolysis in response to endoplasmic reticulum stress. Mol Biol Cell 10:3787

Hinman JD, Peters A, Cabral H, Rosene DL, Hollander W, Rasband MN, Abraham CR (2006) Age-related molecular reorganization at the node of Ranvier. J Comp Neurol 495:351–362

Ho MC, Lo AC, Kurihara H, Yu AC, Chung SS, Chung SK (2001) Endothelin-1 protects astrocytes from hypoxic/ischemic injury. FASEB J 15:618–626

Honda Z, Ishii S, Shimizu T (2002) Platelet-activating factor receptor. J Biochem 131:773–779

Hou ST, MacManus JP (2002) Molecular mechanisms of cerebral ischemia-induced neuronal death. Int Rev Cytol 221:93–148

Huang D, Shenoy A, Cui JK, Huang W, Liu PK (2000) In situ detection of AP sites and DNA strand breaks with 3′-phosphate ends in ischemic mouse brain. FASEB J 14:407–417

Ishii S, Nagase T, Shimizu T (2002) Platelet-activating factor receptor. Prostaglandins Other Lipid Mediat 68–69:599–609

Janaky R, Ogita K, Pasqualotto BA, Bains JS, Oja SS, Yoneda Y, Shaw CA (1999) Glutathione and signal transduction in the mammalian CNS. J Neurochem 73:889–902

Ji L, Nazarali AJ, Paterson PG (2008) Protein-energy malnutrition increases activation of the transcription factor, nuclear factor-κB, in the gerbil hippocampus following global ischemia. J Nutr Biochem 19:770–777

Jian Liu K, Rosenberg GA (2005) Matrix metalloproteinases and free radicals in cerebral ischemia. Free Rad Biol Med 39:71–80

Jin K, Mao XO, Eshoo MW, Nagayama T, Minami M, Simon RP, Greenberg DA (2001) Microarray analysis of hippocampal gene expression in global cerebral ischemia. Ann Neurol 50:93–103

Kamath-Loeb AS, Hizi A, Tabone J, Solomon MS, Loeb LA (1997) Inefficient repair of RNA × DNA hybrids. Eur J Biochem 250:492–501

Kamii H, Mikawa S, Murakami K, Kinouchi H, Yoshimoto T, Reola L, Carlson E, Epstein CJ, Chan PH (1996) Effects of nitric oxide synthase inhibition on brain infraction in SOD-1-transgenic mice following transient focal cerebral ischemia. J Cereb Blood Flow Metab 16:1153–1157

Kato H, Araki T, Itoyama Y, Kogure K, Kato K (1995a) An immunohistochemical study of heat shock protein-27 in the hippocampus in a gerbil model of cerebral ischemia and ischemic tolerance. Neurosci 68:65–71

Kato H, Kogure K, Liu XH, Araki T, Kato K, Itoyama Y (1995b) Immunohistochemical localization of the low molecular weight stress protein Hsp27 following focal cerebral ischemia in the rat. Brain Res 679:1–7

Khan AA, Mao XO, Banwait S, DerMardirossian CM, Bokoch GM, Jin K, Greenberg DA (2006) Regulation of hypoxic neuronal death signaling by neuroglobin. FASEB J 22:1737–1747

Kiessling M, Stumm G, Xie Y, Herdegen T, Aguzzi A, Bravo R, Gass P (1993) Differential transcription and translation of immediate early genes in the gerbil hippocampus after transient global ischemia. J Cereb Blood Flow Metab 13:914–924

Kim JS (1996) Cytokines and adhesion molecules in stroke and related diseases. J Neurol Sci 137:69–78

Koistinaho J, Hökfelt T (1997) Altered gene expression in brain ischemia. Neuroreport 8:i–viii

Kuwabara K, Matsumoto M, Ikeda J, Hori O, Ogawa S, Maeda Y, Kitagawa K, Imuta N, Kinoshita T, Stern DM, Yanagi H, Kamada T (1996) Purification and characterization of a novel stress protein, the 150-kDa oxygen-regulated protein (ORP150), from cultured rat astrocytes and its expression in ischemic mouse brain. J Biol Chem 271:5025–5032

Lee HT, Chang YC, Wang LY, Wang ST, Huang CC, Ho CJ (2004) cAMP response element-binding protein activation in ligation preconditioning in neonatal brain. Ann Neurol 56: 611–623

Li X, Nemoto M, Xu Z, Yu SW, Shimoji M, Andrabi SA, Haince JF, Poirier GG, Dawson TM, Dawson VL, Koehler RC (2007) Influence of duration of focal cerebral ischemia and neuronal nitric oxide synthase on translocation of apoptosis-inducing factor to the nucleus. Neurosci 144:56–65

Lilienbaum A, Israel A (2003) From calcium to NF-kappa B signaling pathways in neurons. Mol Cell Biol 23:2680–2698

Lin L, Cao S, Yu L, Cui J, Hamilton WJ, Liu PK (2000) Up-regulation of base excision repair activity for 8–2′ deoxyhydroxyl guanosine in the mouse brain after forebrain ischemia-reperfusion. J Neurochem 74:101–108

Lin JH, Li H, Yasumura D, Cohen HR, Zhang C, Panning B, Shokat KM, Lavail MM, Walter P (2007) IRE1 signaling affects cell fate during the unfolded protein response. Science 318: 944–949

Lipton P (1999) Ischemic cell death in brain neurons. Physiol Rev 79:1431–1568

Lipton SA, Gu Z, Nakamura T (2007) Inflammatory mediators leading to protein misfolding and uncompetitive/fast off-rate drug therapy for neurodegenerative disorders. Int Rev Neurobiol 82:1–27

Lipton SA (2007) Pathologically-activated therapeutics for neuroprotection: mechanism of NMDA receptor block by memantine and S-nitrosylation. Curr Drug Targets 8:621–632

Liu PK, Hsu CY, Dizdaroglu M, Floyd RA, Kow YW, Karakaya A, Rabow LE, Cui JK (1996) Damage, repair and mutagenesis in nuclear genes after mouse forebrain ischemia-reperfusion. J Neurosci 16:6795–6806

Liu XS, Zou H, Slaughter C, Wang XD (1997) DFF: a heterodimeric protein that functions downstream of caspase-3 to trigger DNA fragmentation during apoptosis. Cell 89:175–184

Love S (1999) Oxidative stress in brain ischemia. Brain Pathol 9:119–131
Lu A, Tang Y, Ran R, Clark JF, Aronow BJ, Sharp FR (2003) Genomics of the periinfarction cortex after focal cerebral ischemia. J Cereb Blood Flow Metab 23:786–810
Luo Y, Ji X, Ling F, Li W, Zhang F, Cao G, Chen J (2007) Impaired DNA repair via the base-excision repair pathway after focal ischemic brain injury: a protein phosphorylation-dependent mechanism reversed by hypothermic neuroprotection. Front Biosci 12:1852–1862
Ma Y, Hendershot LM (2004) ER chaperone functions during normal and stress conditions. J Chem Neuroanat 28:51–65
Maclennan KM, Smith PF, Darlington CL (1996) Platelet-activating factor in the CNS. Prog Neurobiol 50:585–596
Marchetti L, Klein M, Schlett K, Pfizenmaier K, Eisel UL (2004) Tumor necrosis factor (TNF)-mediated neuroprotection against glutamate-induced excitotoxicity is enhanced by *N*-methyl-D-aspartate receptor activation. Essential role of a TNF receptor 2-mediated phosphatidylinositol 3-kinase-dependent NF-kappa B pathway. J Biol Chem 279:32869–32881
Mark RJ, Lovell MA, Markesbery WR, Uchida K, Mattson MP (1997) A role for 4-hydroxynonenal, an aldehydic product of lipid peroxidation, in disruption of ion homeostasis and neuronal death induced by amyloid β-peptide. J Neurochem 68:255–264
Mattson MP, Meffert MK (2006) Roles for NF-kappaB in nerve cell survival, plasticity, and disease. Cell Death Differ 13:852–860
Matute C, Domercq M, Sánchez-Gómez MV (2006) Glutamate-mediated glial injury: mechanisms and clinical importance. Glia 53:212–224
Miao B, Yin XH, Pei DS, Zhang QG, Zhang GY (2005) Neuroprotective effects of preconditioning ischemia on ischemic brain injury through down-regulating activation of JNK1/2 via *N*-methyl-D-aspartate receptor-mediated Akt1 activation. J Biol Chem 280:21693–21699
Millán M, Arenillas J (2006) Gene expression in cerebral ischemia: a new approach for neuroprotection. Cerebrovasc Dis 21(Suppl 2):30–37
Minami M, Satoh M (2000) Chemokines as mediators for intercellular communication in the brain. Nippon Yakurigaku Zasshi 115:193–200
Minami M, Satoh M (2003) Chemokines and their receptors in the brain: pathophysiological roles in ischemic brain injury. Life Sci 74:321–327
Mukherjee PK, DeCoster MA, Campbell FZ, Davis RJ, Bazan NG (1999) Glutamate receptor signaling interplay modulates stress-sensitive mitogen-activated protein kinases and neuronal cell death. J Biol Chem 274:6493–6498
Niimura M, Takagi N, Takagi K, Mizutani R, Ishihara N, Matsumoto K, Funakoshi H, Nakamura T, Takeo S (2006) Prevention of apoptosis-inducing factor translocation is a possible mechanism for protective effects of hepatocyte growth factor against neuronal cell death in the hippocampus after transient forebrain ischemia. J Cereb Blood Flow Metab 26:1354–1365
Nishino K, Nowak TS Jr (2004) Time course and cellular distribution of hsp27 and hsp72 stress protein expression in a quantitative gerbil model of ischemic injury and tolerance: thresholds for hsp72 induction and hilar lesioning in the context of ischemic preconditioning. J Cereb Blood Flow Metab 24:167–178
Nito C, Kamada H, Endo H, Niizuma K, Myer DJ, Chan PH (2008) Role of the p38 mitogen-activated protein kinase/cytosolic phospholipase $A_2$ signaling pathway in blood-brain barrier disruption after focal cerebral ischemia and reperfusion. J Cereb Blood Flow Metab 28: 1686–1696
Oka A, Belliveau MJ, Rosenberg PA, Volpe JJ (1993) Vulnerability of oligodendroglia to glutamate: pharmacology, mechanisms, and prevention. J Neurosci 13:1441–1453
Olney JW, Fuller T, de Gubareff T (1979) Acute dendrotoxic changes in the hippocampus of kainate treated rats. Brain Res 176:91–100
Park EM, Cho S, Frys K, Racchumi G, Zhou P, Anrather J, Iadecola C (2004) Interaction between inducible nitric oxide synthase and poly(ADP-ribose) polymerase in focal ischemic brain injury. Stroke 35:2896–2901

Parker MA, Bazan HEP, Marcheselli V, Rodriguez de Turco EB, Bazan NG (2002) Platelet-activating factor induces permeability transition and cytochrome c release in isolated brain mitochondria. J Neurosci Res 69:39–50

Perez-Pinzon MA, Dave KR, Raval AP (2005) Role of reactive oxygen species and protein kinase C in ischemic tolerance in the brain. Antioxid Redox Signal 7:1150–1157

Phillis JW, Horrocks LA, Farooqui AA (2006) Cyclooxygenases, lipoxygenases, and epoxygenases in CNS: their role and involvement in neurological disorders. Brain Res Rev 52:201–243

Pignataro G, Gala R, Cuomo O, Tortiglione A, Giaccio L, Castaldo P, Sirabella R, Matrone C, Canitano A, Amoroso S, Di Renzo G, Annunziato L (2004) Two sodium/calcium exchanger gene products, NCX1 and NCX3, play a major role in the development of permanent focal cerebral ischemia. Stroke 35:2566–2570

Pilitsis JG, Diaz FG, O'Regan MH, Phillis JW (2001) Inhibition of $Na^+/Ca^{2+}$ exchange by KB-R7943, a novel selective antagonist, attenuates phosphoethanolamine and free fatty acid efflux in rat cerebral cortex during ischemia-reperfusion injury. Brain Res 916:192–198

Pizzi M, Sarnico I, Lanzillotta A, Battistin L, Spano P (2009) Post-ischemic brain damage: NF-kappaB dimer heterogeneity as a molecular determinant of neuron vulnerability. FASEB J 276:27–35

Popa-Wagner A, Carmichael ST, Kokaja Z, Kessler C, Walker LC (2007) The response of the aged brain to stroke: too much, too soon? Curr Neurovas Res 4:216–227

Pryor WA, Squadrito GL (1995) The chemistry of peroxynitrite: a product from the reaction of nitric oxide with superoxide. Am J Physiol 268:L699–L722

Qi W, Reiter RJ, Tan DX, Manchester LC, Siu AW, Garcia JJ (2000) Increased levels of oxidatively damaged DNA induced by chromium(III) and $H_2O_2$: protection by melatonin and related molecules. J Pineal Res 29:54–61

Radi R, Beckman JS, Bush KM, Freeman BA (1991) Peroxynitrite oxidation of sulfhydryls. The cytotoxic potential of superoxide and nitric oxide. J Biol Chem 266:4244–4250

Ran A, Lu A, Xu H, Tang Y, Sharp FR (2007) Heat-shock protein regulation of protein folding, protein degradation, protein function, and apoptosis. In Handbook of neurochemistry and molecular biology. Springer, New York, NY, pp 89–107

Ravagnan L, Gurbuxani S, Susin SA, Maisse C, Daugas E, Zamzami N, Mak T, Jaattela M, Penninger JM, Garrido C, Kroemer G (2001) Heat-shock protein 70 antagonizes apoptosis-inducing factor. Nat Cell Biol 3:839–843

Rosamond W, Flegal K, Friday G, Furie K, Go A, Greenlund K, Haase N, Ho M, Howard V, Kissela B et al (2007) Heart disease and stroke statistics – 2007 update: a report from the American Heart Association Statistics Committee and Stroke Statistics Subcommittee. Circulation 115:e69–e171

Rothwell NJ, Relton JK (1993) Involvement of cytokines in acute neurodegeneration in the CNS. Neurosci Biobehav Rev 17:217–227

Salaycik KJ, Kelly-Hayes M, Beiser A, Nguyen AH, Brady SM, Kase CS, Wolf PA (2007) Depressive symptoms and risk of stroke: the Framingham Study. Stroke 38:16–21

Sarnico I, Lanzillotta A, Boroni F, Benarese M, Alghisi M, Schwaninger M, Inta I, Battistin L, Spano P, Pizzi M (2009) NF-kappaB p50/RelA and c-Rel-containing dimers: opposite regulators of neuron vulnerability to ischaemia. J Neurochem 108:475–485

Savitz SI, Malhotra S, Gupta G, Rosenbaum DM (2003) Cell transplants offer promise for stroke recovery. J Cardiovasc Nurs 18:57–61

Savitz SI, Dinsmore JH, Wechsler LR, Rosenbaum DM, Caplan LR (2004) Cell therapy for stroke. NeuroRx 1:406–414

Sawe N, Steinberg G, Zhao H (2008) Dual roles of the MAPK/ERK1/2 cell signaling pathway after stroke. J Neurosci Res 86:1659–1669

Sirabella R, Secondo A, Pannaccione A, Scorziello A, Valsecchi V, Adornetto A, Bilo L, Di Renzo G, Annunziato L (2009) Anoxia-induced NF-kappaB-dependent upregulation of NCX1 contributes to $Ca^{2+}$ refilling into endoplasmic reticulum in cortical neurons. Stroke 40(3):922–929 22 Jan 2009 [Epub ahead of print]

Skaper SD (2003a) Poly(ADP-Ribose) polymerase-1 in acute neuronal death and inflammation: a strategy for neuroprotection. Ann NY Acad Sci 993:217–228
Skaper SD (2003b) Poly(ADP-ribosyl)ation enzyme-1 as a target for neuroprotection in acute central nervous system injury. Curr Drug Targets CNS Neurol Disord 2:279–291
Sobol RW, Horton JK, Kuhn R, Gu H, Singhal RK, Prasad R, Rajewsky K, Wilson SH (1996) Requirement of mammalian DNA polymerase-β in base-excision repair. Nature (London) 379:183–186
Stephenson D, Yin T, Smalstig EB, Hsu MA, Panetta J, Little S, Clemens J (2000) Transcription factor nuclear factor-kappa B is activated in neurons after focal cerebral ischemia. J Cereb Blood Flow Metab 20:592–603
Stetler RA, Cao G, Gao Y, Zhang F, Wang S, Weng Z, Vosler P, Zhang L, Signore A, Graham SH, Chen J (2008) Hsp27 protects against ischemic brain injury via attenuation of a novel stress-response cascade upstream of mitochondrial cell death signaling. J Neurosci 28:13038–13055
Sun GY, Horrocks LA, Farooqui AA (2007) The role of NADPH oxidase and phospholipases $A_2$ in mediating oxidative and inflammatory responses in neurodegenerative diseases. J Neurochem 103:1–16
Terao Y, Ohta H, Oda A, Nakagaito Y, Kiyota Y, Shintani Y (2009) Macrophage inflammatory protein-3alpha plays a key role in the inflammatory cascade in rat focal cerebral ischemia. Neurosci Res 64:75–82
Truettner JS, Hu K, Liu CL, Dietrich WD, Hu B (2009) Subcellular stress response and induction of molecular chaperones and folding proteins after transient global ischemia in rats. Brain Res 1249:9–18
Tuttolomondo A, Di Raimondo D, di Sciacca R, Pinto A, Licata G (2008) Inflammatory cytokines in acute ischemic stroke. CurrPharm Des 14:3574–3589
van Wijk SJ, Haegeman GJ (2005) Poly(ADP-ribose) polymerase-1 mediated caspase-independent cell death after ischemia/reperfusion. Free Rad Biol Med 39:81–90
Walton M, Sirimanne E, Williams C, Gluckman P, Dragunow M (1996) The role of the cyclic AMP-responsive element binding protein (CREB) in hypoxic-ischemic brain damage and repair. Brain Res Mol Brain Res 43:21–29
Wen YD, Zhang HL, Oin ZH (2006) Inflammatory mechanism in ischemic neuronal injury. Neurosci Bull 22:171–182
Widlak P (2000) The DFF40/CAD endonuclease and its role in apoptosis. Acta Biochem Pol 47:1037–1044
Wissing D, Mouritzen H, Egeblad M, Poirier GG, Jäättelä M (1997) Involvement of caspase-dependent activation of cytosolic phospholipase $A_2$ in tumor necrosis factor-induced apoptosis. Proc Natl Acad Sci USA 94:5073–5077
Won MH, Kang TC, Jeon GS, Lee JC, Kim DY, Choi EM, Lee KH, Choi CD, Chung MH, Cho SS (1999) Immunohistochemical detection of oxidative DNA damage induced by ischemia-reperfusion insults in gerbil hippocampus in vivo. Brain Res 836:70–78
Wong CH, Crack PJ (2008) Modulation of neuro-inflammation and vascular response by oxidative stress following cerebral ischemia-reperfusion injury. Curr Med Chem 15:1–14
Woo EJ, Kim YG, Kim MS, Han WD, Shin S, Robinson H, Park SY, Oh BH (2004) Structural mechanism for inactivation and activation of CAD/DFF40 in the apoptotic pathway. Mol Cell 14:531–539
Yagita Y, Sakoda S, Kitagawa K (2008) Gene expression in brain ischemia. Brain Nerve 60: 1347–1355
Yakubov E, Gottlieb M, Gil S, Dinerman P, Fuchs P, Yavin E (2004) Overexpression of genes in the CA1 hippocampus region of adult rat following episodes of global ischemia. Mol Brain Res 127:10–25
Yamamoto Y, Gaynor RB (2004) I-κB kinases: key regulators of the NF-κB pathway. Trends Biochem Sci 29:72–79
Yenari MA, Han HS (2006) Influence of hypothermia on post-ischemic inflammation: role of nuclear factor kappa B (NFkappaB). Neurochem Int 49:164–169

Yilmaz G, Granger DN (2008) Cell adhesion molecules and ischemic stroke. Neurol Res 30: 783–793

Yoshida T, Limmroth V, Irikura K, Moskowitz MA (1994) The NOS inhibitor, 7-nitroindazole, decreases focal infract volume but not the response to topical acetylcholine in pial vessels. J Cereb Blood Flow Metab 14:924–929

Zaleska MM, Wilson DF (1989) Lipid hydroperoxides inhibit reacylation of phospholipids in neuronal membranes. J Neurochem 52:255–260

Zaman K, Ryu H, Hall D, O'Donovan K, Lin KI, Miller MP, Marquis JC, Baraban JM, Semenza GL, Ratan RR (1999) Protection from oxidative stress-induced apoptosis in cortical neuronal cultures by iron chelators is associated with enhanced DNA binding of hypoxia-inducible factor-1 and ATF-1/CREB and increased expression of glycolytic enzymes, p21(waf1/cip1), and erythropoietin. J Neurosci 19:9821–9830

Zhang W, Potrovita I, Tarabin V, Herrmann O, Beer V, Weih F, Schneider A, Schwaninger M (2005) Neuronal activation of NF-kappaB contributes to cell death in cerebral ischemia. J Cereb Blood Flow Metab 25:30–40

# Chapter 3
# Potential Neuroprotective Strategies for Ischemic Injury

## 3.1 Introduction

Stroke (ischemic injury) is the leading cause of mortality and morbidity worldwide, accounting for 5 million deaths per year. Oxygen deprivation due to stroke leads to rapid neuronal death and dysfunction of the body part controlled by the affected neurons. Thus, stroke is not only responsible for mortality and morbidity but also for serious long-term disability, including paralysis, cognitive deficits, dementia, dizziness, vertigo, impaired vision, memory loss, language deficits, emotional difficulties, pain, and depression. About 4.7 million stroke survivors currently live in the USA. The number of stroke patients is expected to increase worldwide as the population continues to age. In most cases strokes can be prevented through risk-factor modification (Fig. 3.1) and application of effective preventive therapies (Papademetriou and Doumas, 2009). The recovery of stroke patients can be enhanced by intensive rehabilitation, which probably acts through brain plasticity-mediated mechanisms. Furthermore, dyslipidemia treatment by statins and fish oil, control of hypertension, diabetes mellitus, and cessation of smoking substantially enhance chances of stroke prevention (Papademetriou and Doumas, 2009) (Fig. 3.1). Other risk factors for stroke include arrhythmia, prior heart attack, surgery on the carotid arteries, and diseases that increase the risk of blood clot (emboli) formation. About 80% of strokes are caused by an interruption of blood flow to the brain due to occlusion of a blood vessel caused by thrombus or emboli from the heart, aorta, or carotid or vertebral arteries (ischemic stroke). Remaining 20% of stroke cases are hemorrhagic and result from rupture of a blood vessel. It is important to distinguish between ischemic and hemorrhagic strokes because most initial treatment strategies are designed to reduce coagulation, which may end up in exacerbating hemorrhagic stroke. Stroke therapy to limit the progression of injury and upregulate repair process after stroke is not only complicated by the heterogeneous nature of neural cell death but also by the multiple barriers to functional recovery.

A.A. Farooqui, *Neurochemical Aspects of Neurotraumatic and Neurodegenerative Diseases*, DOI 10.1007/978-1-4419-6652-0_3,

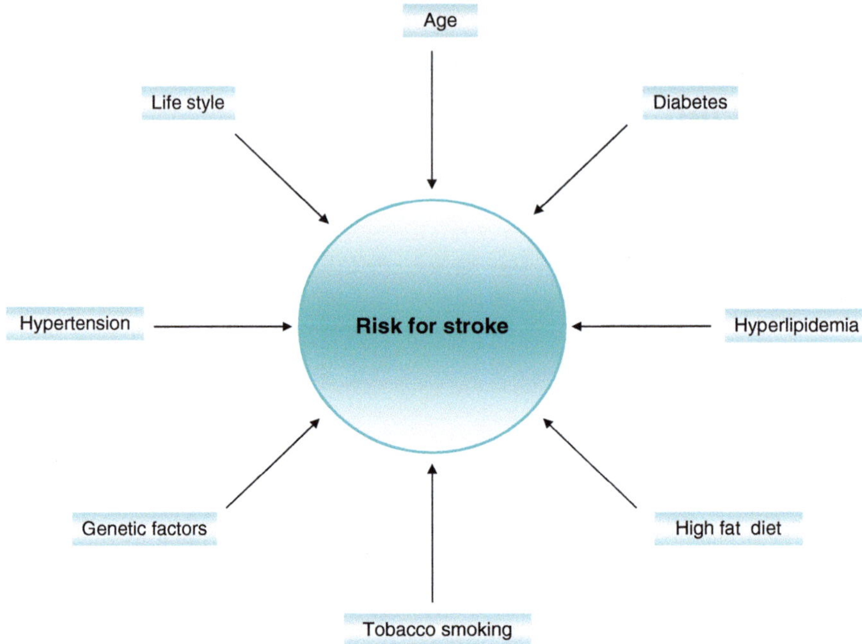


**Fig. 3.1** Risk factors for stroke injury in humans

## 3.2 Potential Treatment Strategies for Ischemic Injuries

Pharmacological mechanisms associated with stroke treatments involve thrombolysis, neuroprotection, and perfusion/reperfusion enhancers (Fagan et al., 1999). It is well known that ischemic injury results in the formation of infarct that has a core that contains irreversibly damaged cells and an area surrounding the core called penumbra. This area contains viable neurons that can be salvaged through neuroprotective strategies. The first window of opportunity for restoration of cerebral blood flow (CBF) is accomplished by the administration of tissue plasminogen activator (tPA), a thrombolytic agent that lyses the clot and restores CBF to the penumbra within 3–5 h of stroke onset (Alberts, 1999). tPA has no affect on infarct core but revitalizes the penumbra by restoring blood flow. This therapy is only used in about 4% of patients presenting after an acute ischemic stroke. tPA may also produce serious bleeding in the brain, which can be fatal (Wardlaw et al., 2003). It interacts with a protein called the platelet-derived growth factor-CC (PDGF-CC), and PDGF-CC receptor leading to usually impervious "blood–brain barrier" leakage (Su et al., 2008). Gleevec, a kinase inhibitor prevents the PDGF-CC receptor, apparently counteracting tPA's effect. Thus, the use of tPA beyond this time frame, or outside FDA approved protocols, may be hazardous.

The mechanical embolus removal in cerebral ischemia retriever (MERCI retriever) or endovascular mechanical embolectomy is a recently developed effective device to remove or break up blood clots in a small blood vessel stopping blood flow (Smith, 2006). This is performed by carefully passing a special device from a blood vessel in the leg all the way into the blood vessel in the brain where the blood clot is located. The MERCI retriever captures the clot and pulls it out of the body, thus facilitating blood flow to the affected brain area (Smith, 2006; Kim et al., 2006). First-generation MERCI retriever has achieved recanalization rates of 48%, and when coupled with intraarterial thrombolytic drugs, recanalization rates as high as 60%. It is suggested that enhancements and refinements in the embolectomy device design may improve recanalization rates in ischemic injury patients. Hyperoxia may be another powerful neuroprotective strategy to salvage acutely ischemic brain tissue and extend the time window for acute stroke treatment (Singhal, 2007). Although earlier trials have failed due to several shortcomings (delayed time to therapy, inadequate sample size, and use of excessive chamber pressures), new studies indicate that hyperbaric and even normobaric oxygen therapy can be effective if used appropriately and raises possibility of using hyperoxia to extend the narrow therapeutic time window for stroke thrombolysis (Singhal, 2006). Majority of stroke patients display a slow evolution of brain injury that occurs in penumbra over several hours. This "evolving stroke" is a realistic target for therapeutic intervention, with the goal of blocking the progression of detrimental changes that normally occur following the acute ischemic event. Preventing or reducing this delayed neural injury may improve neurological outcome and also facilitate brain recovery from ischemic injury. Thus, attempts have been made to protect injured neurons from delayed ischemic injury. Studies in animals indicate a period of at least 4 h after onset of complete ischemia in which many potentially viable neurons exist in the ischemic penumbra. In humans, the ischemic injury may be less and the time window may be longer, but human patients are older and have other pathological conditions that may limit benefit (Zivin, 1998). Since many neuroprotective drugs reduce ischemic injury in animal models of stroke, it is proposed that this approach holds great promise. Restoration of blood flow results in reperfusion and increased production of ROS and RNS, which intensifies ischemic brain damage. This increase in ROS and RNS production is due to the stimulation of phospholipase $A_2$, cyclooxygenase, and nitric oxide synthase activities (Farooqui et al., 1994). As stated in Chapter 2, nitric oxide reacts with the superoxide anion to form peroxynitrite, a highly reactive nitrogen species, which promote brain injury through DNA damage.

Intense research is underway to discover a safe agent that can limit ischemic damage in human stroke. A combination of thrombolytic therapy with a neuroprotective agent produce an additive in some ischemic models, as is the combination of a thrombolytic with an agent that facilitates reperfusion (thromboxane $A_2$ receptor antagonist and neutrophil adhesion/activation inhibition). Combinations of neuroprotective agents, such as glutamate antagonists and calcium channel antagonists may also induce additive effects, and other combinations of neuroprotective agents, such as a glutamate antagonist with a $\gamma$-aminobutyric acid (GABA) agonist, produce synergistic effects in a rat stroke model (Fagan et al., 1999). It is also suggested that lower doses of toxic drugs may be used together to yield a positive neurologic outcome. The success of additive or synergistic effects of stroke therapy in animal model depends not only on the type model, the timing of drug administration and doses of the drugs, but also on the primary neurologic endpoint (Fagan et al., 1999).

Although there have been important developments in the molecular pathophysiology and therapeutic strategies for ischemic stroke in past 25 years, no drug has been approved by FDA for the neuroprotection therapy. Neuroprotection is defined as any strategy, or combination of strategies, that antagonizes, interrupts, or slows the sequence of injurious neurochemical and molecular events that, if left unchecked, may facilitate and contribute to irreversible ischemic injury. The goal of neuroprotection strategies is to limit neuronal death after brain injury and attempt to maintain the highest possible integrity of cellular interactions in the brain resulting in an undisturbed neural function. Successful neuroprotection is limited by the short window (3–6 h) of opportunity for active intervention (Zivin, 1998). Various neuroprotective agents have reached phase III efficacy trials in focal ischemic stroke, but none has proven effective, despite successful preceding animal studies.

Ischemic injury is accompanied by the excessive activation of excitatory amino acid receptors, $Ca^{2+}$ influx, and release of other toxic products that intensify cellular injury (Fig. 3.2). By preventing excitatory neurotransmitter release, neuroprotective agents may reduce deleterious effects of ischemic injury on neural cells. In addition to NMDA receptor channel, $Ca^{2+}$ also enters through voltage gated $Ca^{2+}$ channels

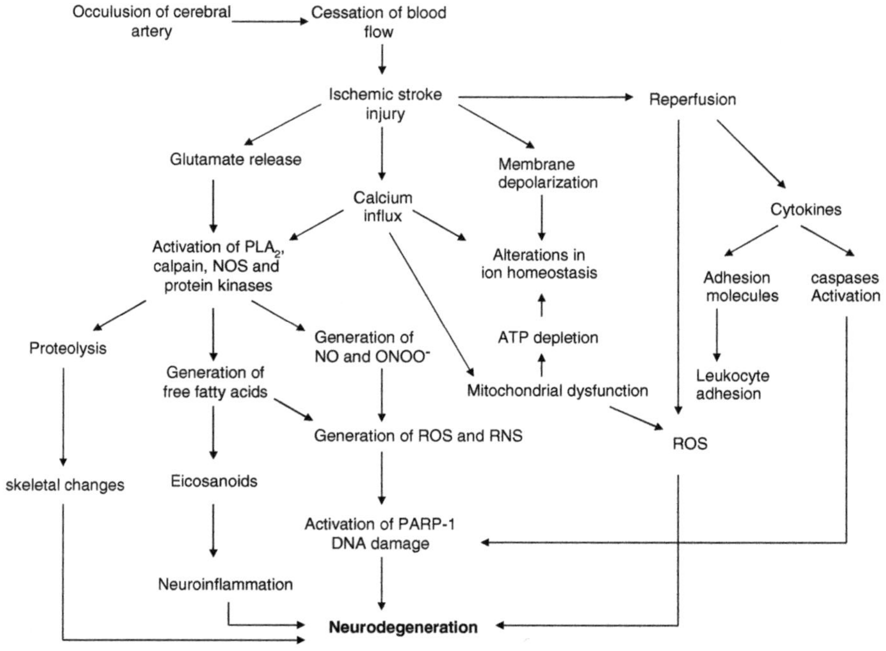


**Fig. 3.2** Diagram showing neurochemical changes in ischemic injury. Phospholipase $A_2$ ($PLA_2$); nitric oxide synthase (NOS); reactive oxygen species (ROS); reactive nitrogen species (RNS); nitric oxide (N); peroxynitrite ($ONOO^-$); and poly (ADP-ribose) polymerase-1 (PARP-1)

and through the blockade of the $Na^+/Ca^{2+}$ transporter. The increase in cytosolic $Ca^{2+}$ plays a prominent role in the development and intensification of ischemic injury through the activation of phospholipases, kinases, nitric oxide synthases, and proteases.

Generation of eicosanoids from enzymic metabolism of arachidonic acid initiates inflammatory reactions and the damaged tissue is infiltrated by leukocytes and microglia. The relatively slow pace of these processes, which occur in hours and days that follow an initial stroke, makes them an attractive target for neuroprotective therapy.

As stated in Chapter 2, the activation of these enzymes causes neuronal death by necrosis and/or apoptosis, via ROS and RNS generation, proteolysis, and DNA damage (Fig. 3.3) (Farooqui et al., 2008). ROS modulate p38/MARK, JNK/MARK, ERK/MARK pathways and RNS activate PARP-1. Under these conditions, in the infarct core neuronal death occurs within minutes to less than an hour (Lo et al., 2003). This is followed by a second wave of neuronal demise in the ischemic penumbra and neuroanatomically connected sites. This delayed cell death (secondary degeneration) occurs via apoptosis and often exceeds the initial damage of

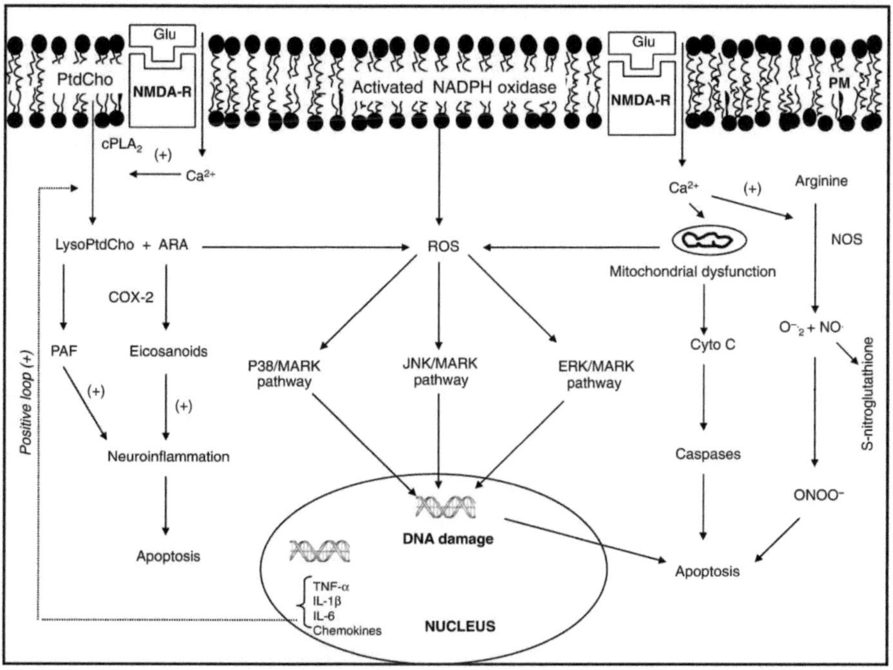


**Fig. 3.3** Pathways associated with ROS- and RNS-mediated cell death in cerebral ischemia. Phospholipase $A_2$ ($PLA_2$); nitric oxide synthase (NOS); cyclooxygenase-2 (COX-2); arachidonic acid (ARA); lyso-phosphophatidylcholine (lyso-PtdCho); platelet-activating factor (PAF); reactive oxygen species (ROS); reactive nitrogen species (RNS); superoxide ($O_2$); peroxynitrite ($ONOO^-$); transcription factor-α (TNF-α); interleukin-1β (IL-1β); and interleukin-6 (IL-6)

stroke and, thus, contributes pivotally to significant losses in neurological functions. The major contributors of the apoptotic cell death are a family of cysteine proteases (caspases). They initiate apoptosis by cleaving key components of the neuronal infrastructure and activating the factors responsible for neural cell damage.

Inflammation is another mechanism that contributes to ischemic injury. Inflammation is aided by increased production of proinflammatory cytokines including tumor necrosis factor-α (TNF-α), interleukin-1β (IL-1β, IL-6, and IL-18). These cytokines stimulate the expression of transcription factor NF-κB and endothelial adhesion molecules, intracellular adhesion molecule-1 (ICAM-1), E-selectin, and P-selectins that directly facilitate ischemic injury-mediated neuroinflammation (Farooqui et al., 2007a). Since most ischemia-mediated changes in penumbra occur simultaneously, it is difficult to design an ideal neuroprotection strategy. At the present time, neuroprotectants are designed to protect neurodegeneration that occurs in the penumbra during delayed cell death in stroke cascade. These neuroprotectants may have a longer therapeutic window than tPA.

Pathogenesis of stroke is a multifactorial process that involves multiple signal transduction pathways. An ideal ischemic injury treatment may require either a single drug that can act on multiple targets or a combination of multiple drug therapies to attain better protection in addition to tPA treatment. These drugs are not available. However, several neuroprotective agents have reached phase III efficacy trials, but have shown mixed results. They include NMDA receptor antagonists, calcium channel blockers, citicoline (CDP-choline), the free radical scavenger tirilazad, anti-intercellular adhesion molecule-1 (ICAM-1) antibody, $GM_1$ ganglioside, clomethiazole, a sedative and muscle relaxant, and fosphenytoin, an antiepileptic seizure drug, and piracetam, a nootropic drug. However, despite the development of over 1,000 compounds that have been proven effective in animal models of stroke, none has demonstrated efficacy in patients in over 100 clinical trials (Green, 2008). The failure of these clinical trials raises significant concerns about neuroprotection strategies alone as a therapeutic intervention for the treatment of ischemic injury in humans. At the present time, neuroprotection strategies are focused on injured neurons and the neurotoxic environment induced by the ischemic injury. The complex processes that are induced by post-ischemic injury events require the targeting of not only neurons but also non-neuronal cells (glial, endothelial, and inflammatory cells) along with alterations in axons and white matter. Ideal neuroprotective drugs should be chronically active and well tolerated. They should be able to cross the blood–brain barrier, have regional specificity, and should be able to reach the site where neurodegenerative process is taking place. Although neural cells in the ischemic penumbra can be protected through neuroprotective intervention strategies much later post-ischemic injury than the ischemic core, future interventions should be designed to target multiple cell types involved in maintaining the integrity of neural and non-neural cells in penumbra (Barone, 2009). In addition, steps should be taken to target ischemia-mediated white matter injury, a process that has been virtually ignored in clinical trials and needs to be monitored more systematically in the treatment of stroke (Ho et al., 2005; Wen and Sachdev, 2004). Temporal heterogeneity and complexity of ischemic events make the intervention of ischemic injury very complex, and it is not surprising that many clinical trials have failed. This is tempting to suggest that multiple cellular mechanisms should be targeted for the successful treatment of ischemic injury.

### *3.2.1* N-*Methyl-*D-*Aspartate Receptor Antagonists and Stroke Therapy*

Although the rationale behind using NMDA receptor antagonists for ischemic injury sounds perfect, many NMDA antagonists may cause behavioral and physiological side effects (Farooqui et al., 2008). Competitive and non-competitive NMDA antagonists interact with a specific site on the NMDA receptor complex to block or retard the $Ca^{2+}$ influx. Some of these antagonists cross the blood–brain barrier easily. NMDA antagonists include drugs like MK 801, selfotel, dextrorphan, dextrometorphan, aptiganel (Cerestat), eliprodil, and ifenprodil (Fig. 3.4). These drugs have been used for the treatment of stroke in animal models and human patients. In humans these drugs produce adverse clinical and behavioral effects regardless of the molecular mechanism of action. Low doses cause alterations in sensory perception, dysphoria, nystagmus, and hypotension, whereas higher doses may cause psychological adverse events such as excitement, paranoia, hallucinations, agitation, confusion, paranoia, somnolence, severe motor retardation, leading ultimately to catatonia (Grotta et al., 1990; Schäbitz et al., 2000; Labiche and Grotta, 2004). YM872 is an AMPA antagonist that reduces infarct volume in animal models (Shimizu-Sasamata et al., 1996; Kawasaki-Yatssugi et al., 2000). Two clinical trials of YM872 have been performed and terminated prematurely. In both trials YM872 caused multiple severe side effects, such as hallucination, agitation, and catatonia

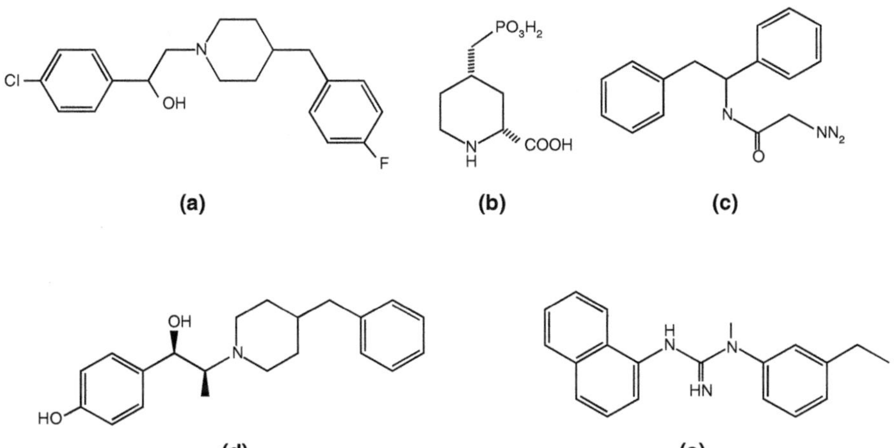


**Fig. 3.4** Chemical structures of some NMDA antagonists that have been used in clinical trials in humans. Eliprodil (**a**); selfotel (**b**); remacemide (**c**); ifenprodil (**d**); and aptiganel (**e**)

in patients. Although the reason for the failure of NMDA and AMPA antagonist for the treatment of stroke is not fully understood, their inability to protect white matter injury from ischemic damage may partly contribute to the failure.

### 3.2.2 Calcium Channel Blockers and Stroke Therapy

Calcium channel blockers prevent the entry of calcium ions into the ischemic neurons. They do not antagonize the effect of calcium ions but dilate arteries. These drugs block calcium ions from gaining access to its intracellular site of action. Calcium channel blockers have been reported to reduce the incidence of stroke in hypertensive patients. USDA approved calcium channel blockers include nisoldipine (Sular), nifedipine (Adalat, Procardia), nicardipine (Cardene), isradipine (Dynacirc), nimodipine (Nimotop), felodipine (Plendil), amlodipine (Norvasc), diltiazem (Cardizem), and verapamil (Calan, Isoptin). These drugs have been widely used for the treatment of hypertension because several clinical trials demonstrate their strong action on lowering blood pressure and their role in preventing cerebrovascular and cardiovascular events. More than 80% of acute stroke patients have high blood pressure. Several small randomized trials have assessed cerebral blood flow with calcium channel blockers in acute ischemic stroke. Overall, these studies demonstrate no change in cerebral perfusion. Calcium channel blockers do not alter outcome after ischemic stroke in 29 trials with 7,665 patients (Sare et al., 2009). Although the mechanism of calcium channel blocker's action in cerebral ischemia is still unclear, major mechanisms of their actions may include normalization of blood pressure and their antioxidative properties (Papademetriou and Doumas, 2009). Control of hypertension with calcium channel blockers in principle should reduce the risk of first and recurrent stroke. The most common side effects of calcium channel blockers are slow heart rate, constipation, nausea, edema, headache, drowsiness, and dizziness.

### 3.2.3 Free Radical Scavengers and Stroke Therapy

Free radicals play an important role in stroke by exacerbating membrane damage through peroxidation of unsaturated fatty acids of neural cell membrane, leading to neuronal death and brain edema. In the body, free radical-mediated oxidative stress is balanced by endogenous antioxidant systems. Thus, interplay between free radicals and free radical scavengers is not only important for maintaining normal health but also protection from ischemic injury (Gilgun-Sherki et al., 2006; Wang et al., 2006). The molecular mechanism involved in the neuroprotective effects of free radical scavengers not only depends on the antioxidant activity of neurons but also on the downregulation of NF-κB activity (Shen et al., 2003), suppression of genes induction by proinflammatory cytokines, and stabilization of synapses (Gilgun-Sherki et al., 2006; Wang et al., 2006). In addition, the effectiveness of free radical scavengers in protecting against stroke depends not only on their ability to cross the blood–brain barrier but their potential in terms of subcellular distribution in mitochondria, plasma membrane, and cytoplasm (Gilgun-Sherki et al., 2006; Tan et al., 2003). Examples of free radical scavengers include edaravone (3-methyl-1-phenyl-2-pyrazolin-5-one), tirilazad, ebselen, and disodium 2,4-disulfophenyl-*N*-tert-butylnitrone (NXY-059) (Fig. 3.5). These drugs are well tolerated in human stroke patients and can be administered to produce plasma concentrations exceeding those effective in animal models.

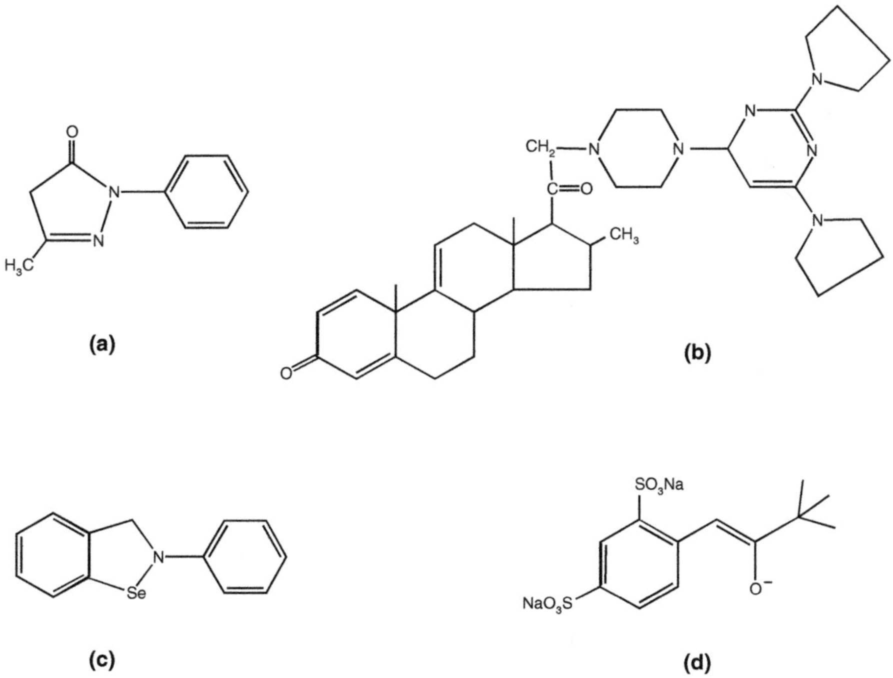


**Fig. 3.5** Chemical structures of free radical scavengers that have been used for the treatment of stroke. Edaravone (**a**); tirilazad (**b**); ebselen (**c**); and NXY-059 (**d**)

Edaravone or 3-methyl-1-phenyl-2-pyrazolin-5-one (Fig. 3.5) is a lipophilic drug with multiple mechanisms of action. It exerts neuroprotective effects by inhibiting endothelial injury and by ameliorating neuronal damage in brain ischemia. It provides the desirable features of NOS: it increases eNOS (beneficial NOS for rescuing ischemic stroke) and decreases nNOS and iNOS (detrimental NOS). Post-reperfusion brain edema and hemorrhagic events induced by thrombolytic therapy may be reduced by edaravone pretreatment (Yoshida et al., 2006; Higashi, 2009).

Clinical experience with edaravone suggests that this drug has a wide therapeutic time window. The combination therapy (a thrombolytic plus edaravone) is likely to target brain edema, reduce stroke death, and improve the recovery from neurological deficits in stroke patients. This drug improves the core neurological deficits, impaired activities of daily living, and disability, without serious safety problems (Lapchak and Zivin, 2009). Edaravone was approved in Japan for the treatment of acute brain infarction within 24 h after onset in April 2001.

Ebselen (Fig. 3.5), a selenium compound with glutathione peroxidase-like activity, is a modestly effective neuroprotectant in a rat transient middle cerebral artery occlusion model when given before the start of ischemia, but not when the insult is severe. Data from the permanent middle cerebral artery occlusion model and an embolic stroke model result in a bell-shaped dose–response curve. This weak preclinical profile explains the lack of success in clinical trials in humans (Green and Ashwood, 2005).

Tirilazad (Fig. 3.5) is a non-glucocorticoid, 21-aminosteriod that blocks lipid peroxidation. It has neuroprotective effects in experimental ischemic stroke. Tirilazad mesylate (Freedox) has been used for phase I, II, and III trials in patients with acute ischemic stroke. These trials were stopped because the drug did not improve overall functional outcome. It increases death and disability by about one-fifth when given to patients with acute ischemic stroke (No author listed, 2000). Although further trials of tirilazad are now unwarranted, analysis of individual patient data from the trials may help elucidate why tirilazad appears to worsen outcome in acute ischemic stroke. Tirilazad reduces angiographic vasospasm after experimental subarachnoid hemorrhage (SAH). Five randomized clinical trials of tirilazad have been conducted in patients with SAH and meta-analysis indicating that tirilazad has unfavorable outcome, but decreases symptomatic vasospasm in five trials of aneurysmal SAH (Jang et al., 2009).

NXY-059 (Cerovive) (Fig. 3.5) is a novel nitrone-free radical trapping agent capable of blocking the reaction of superoxide and nitric oxide, thus preventing the generation of peroxynitrite. During this process NXY-059 is hydrolyzed generating *t*-butylhydroxylamine (NtBHA), a powerful radical scavenger. NtBHA is further oxidized to 2-methyl-2-nitrosopropane (MNP), which is reduced back to NtBHA either by ascorbic acid or by mitochondria. MNP generates nitric oxide, which dilates blood vessels and facilitates cerebral blood flow, resulting into neuroprotection. In preclinical studies, NXY-059 has been found to be a very effective agent in transient and permanent transient middle cerebral artery occlusion and thromboembolic models of acute ischemic stroke (McCulloch and Dewar, 2001; Green and Ashwood, 2005). Its preclinical trials have resulted in recommendations of the Stroke Therapy Academic Industry Roundtable (STAIR) group. It has been investigated in phase III clinical trials using a therapeutic time window and plasma concentrations that are effective in rat and primate models of stroke (Green and Ashwood, 2005). It is well tolerated in patients with acute stroke at concentrations known to be associated with neuroprotection in animal models of transient cerebral ischemia; however, higher target concentrations appear necessary on the basis of animal models of permanent ischemia (McCulloch and Dewar, 2001; Green and Ashwood, 2005). Although NXY-059 showed neuroprotective effects in the Stroke-Acute Ischemic NXY Treatment I (SAINT I) trial by reducing disability in patients with acute ischemic stroke (Lees et al., 2006), SAINT II trial NXY-059 did not show any efficacy in the treatment of stroke patients (Shuaib et al., 2007). The reasons for the failure of SAINT II trial are not fully understood. However, usage of stored NXY-059 preparation and lack of sufficient preclinical studies in animal may be responsible for the failure of SAINT II trial. Ischemic injury under experimental conditions in animal model is not homologous to pathological stroke in human subject because there are substantial anatomical differences between the rodent and human brains, particularly that the rodent brain has a higher gray-to-white matter ratio. Furthermore, in animal model studies occlusion of blood vessel is performed by artificial methods, whereas during ischemia in vivo occlusion occurs through the clot formation. Animal model studies ignore the effect of clot-derived substances (such as thrombin) that may be flushed into the ischemic region by residual flow, possibly confounding the ischemic insult (Feuerstein et al., 2008). Furthermore, in ischemic stroke patients, occlusion occurs in large or small vessels and may be secondary to in situ thrombosis, artery-to-artery embolism, or cardiac embolism. This type of injury may affect very different areas of the human brain (Ford, 2008). The consequences of small-vessel occlusion may differ from large-vessel occlusion with respect to the effect of neuroprotection, and good animal models of small-vessel occlusion have not been developed (Ford, 2008). Mimicking all aspects of human stroke in one animal model is not possible because ischemic stroke is itself a very heterogeneous condition. Thus, better modeling of the human condition focusing on the embolic cause of stroke needs to be rigorously developed and implemented in stroke experimental models (Feuerstein et al., 2008). Collectively, these studies suggest that unlike the standard animal model of permanent or temporary middle cerebral artery occlusion, clinical stroke injury is a very heterogeneous process in which drug distribution and levels of biomarkers indicating recovery in various regions of brain should be monitored with sensitive neuroimaging techniques (Ford, 2008; Green, 2008; Chacon et al., 2008).

Following improvements in the experimental design have been recommended (Ford, 2008; Feuerstein and Ruffolo, 2007): (a) Experimental animal models should be more reflective of older stroke patients with physiological derangement; (b) it should be clearly demonstrated that in human, drug reaches at the injury site where neurodegeneration is taking place; (c) patients should agree for salvaging of their tissues; (d) treatment should be performed very early after the onset of stroke; and (e) refinement of measurement of neurological impairment and disability to be made before starting clinical trials (Ford, 2008; Feuerstein and Ruffolo, 2007). Another important difference between ischemic injury in patients and controlled stroke injury in animals is the presence of considerable physiological variability with frequent elevations and variability in blood pressure, glucose, temperature, and oxygenation in contrast to experimental stroke models where animals are anesthetized and all physiological parameters controlled (Ford, 2008; Feuerstein et al., 2007; Shuaib and Hussain, 2008).

### 3.2.4 $GM_1$ Ganglioside and Stroke Therapy

Gangliosides are sialic acid-containing glycosphingolipids (Fig. 3.6) that are enriched in neuronal membranes. Addition of exogenous $GM_1$ gangliosides to cell cultures and their injection in vivo results in their incorporation into neural membranes. This incorporation not only stabilizes neural membranes but also induces neuritogenesis. The molecular mechanism through which gangliosides exert their effect on neural membranes remains elusive. However, it is proposed that gangliosides not only regulate $Ca^{2+}$ influx channels and $Ca^{2+}$ exchange proteins (Ledeen and Wu, 2002) but also modulate activities of enzymes involved in signal transduction. These enzymes include adenylate cyclases, protein kinases, phospholipases $A_2$, PLC, and $Na^+$, $K^+$ ATPases (Goettl et al., 2003; Farooqui et al., 2008). Recruitment of protein kinases (MEK/ERK kinase or JNK kinase) and phospholipases ($PLA_2$ and PLC) generates lipid mediators that promote $GM_1$-mediated neurogenerative effects that facilitate neural cell survival after ischemic injury. $GM_1$ ganglioside prevents lipid peroxidation in synaptosomes and phagocytic cells (Avrova et al., 2002). $GM_1$ ganglioside may act as a membrane stabilizer, an antiexcitotoxic agent, and an antioxidant in brain tissue. Based on many studies, it is suggested that systemic administration of $GM_1$ ganglioside may reduce ischemia-evoked glutamate and aspartate release and oxidative stress. Gangliosides have been tested in many clinical trials in patients with stroke. The results of these studies were inconclusive. There is no evidence that treatment with ganglioside reduces disability after stroke

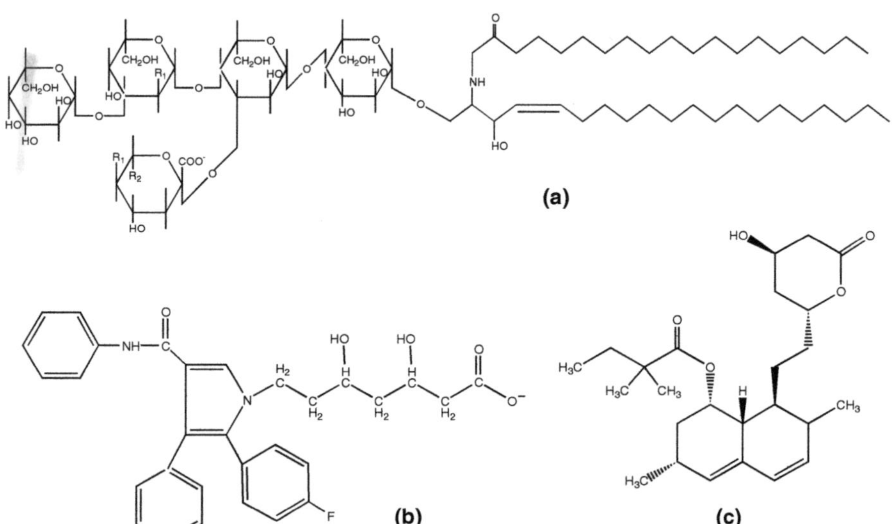


**Fig. 3.6** Chemical structures of $GM_1$ ganglioside and statins. $GM_1$ ganglioside (**a**); atorvastatin (Lipitor) (**b**); and simvastatin (Zocor) (**c**)

(Candelise and Ciccone, 2001). Caution is warranted because of reports of sporadic cases of Guillain–Barré syndrome after ganglioside therapy.

### *3.2.5 Statins and Stroke Therapy*

Statins are potent cholesterol-lowering drugs (Fig. 3.6) that act by inhibiting 3-hydroxy-3-methylglutaryl coenzyme A reductase (HMG-CoA reductase). Although the evidence for the association between hypercholesterolemia and ischemic stroke is weak, in randomized clinical trials, statins appear to consistently reduce stroke risk (Greisenegger et al., 2004). Statin trials indicate 21% relative risk reductions for stroke. The reasons for the positive statin effect on stroke end point are not understood, but positive results of statin trials have been obtained only in patients with an average or a low serum cholesterol level. Statins reduce stroke incidence in high-risk (mainly coronary heart disease, diabetics, and hypertensives) population even with a normal baseline blood cholesterol level. In patients with prior strokes, statins reduce the incidence of coronary events, but it is not yet proven if drugs of this class actually reduce the incidence of recurrent strokes in terms of secondary prevention (Parnetti et al., 2006). Statins reduce the risk of stroke by a variety of mechanisms, which are beyond cholesterol lowering effect of cholesterol (Farooqui et al., 2007b). Thus, statins interfere with platelet aggregation and have anti-inflammatory and antioxidative properties. Also statins promote stabilization of atherosclerotic plaques and improve blood flow to the ischemic brain. The protective effects of statins are also caused by their direct effect on endothelial cells leading to improved nitric oxide (NO) bioavailability (Chudzik et al., 2005; Asahi et al., 2005; Farooqui et al., 2007b). It is proposed that statins also act by upregulating endogenous tissue plasminogen activator (tPA) and enhancing clot lysis in a mouse model of embolic focal ischemia. They increase tPA mRNA levels but produce no change in mRNA levels of PAI-1 (Asahi et al., 2005). Statins attenuate the inflammatory cytokine responses that accompany stroke. Many of these effects are due to the inhibition of isoprenoid intermediates, which serve as lipid attachments for a variety of intracellular signaling molecules. It must be mentioned here that sudden discontinuation of statin treatment leads to a rebound effect with downregulation of NO production. Acute termination of statin treatment or withdrawal of statin treatment has been reported to impair vascular function, increase morbidity and mortality in patients with vascular diseases (Endres, 2005). Collective evidence suggests that the protective effects of statins on stroke are mediated through multiple mechanisms. Further studies on the usefulness of statins in stroke using neuroimaging and advanced cognitive techniques are underway to judge the efficacy of statin-mediated neuroprotection on brain.

In addition to the above drugs, nitric oxide inhibitor (Lubeluzole), opioid antagonist (Nalmefene), serotonin agonist (Repinotan), and sodium channel blocker

(Cerebyx or fosphenytoin) have been used in several clinical trials but failed to give positive results.

### 3.2.6 ω-3 Fatty Acids and Stroke

American diet is rich in ω-6 or *n*–6 fatty acids due to the consumption of vegetable oils. This diet elevates levels of ω-6 fatty acids, production of eicosanoids, and upregulates the expression of proinflammatory cytokines. These metabolites and cytokines promote hyperneuroinflammation and oxidative stress. Although inflammation is a neuroprotective mechanism, too much inflammation following stroke can be very harmful (Farooqui et al., 2007a). Under normal conditions, the resolution phase of neuroinflammation is a highly coordinated process, which is involved in restoration of original tissue homeostasis. Resolution phase of inflammation is controlled by pro-resolving lipid mediators (see below) that not only terminates leukocyte trafficking to the inflammed site and reversal of vasodilation and vascular permeability but also involved in the removal of inflammatory leukocytes, exudates, and fibrin. Under pathological conditions, retardation of resolution phase can result in scarring and fibrosis (Gilroy et al., 2004). In contrast, ω-3 or *n*–3 fatty acids consumption produces anti-inflammatory, antithrombotic, antiarrhythmic, hypolipidemic, vasodilatory, and immunosuppressive effects (Simopoulos, 2006; Farooqui et al., 2009a, b). ω-3 fatty acids produce neuroprotective effects against ischemic brain damage in rats (Okada et al., 1996; Terano et al., 1999; Cao et al., 2007; Strokin et al., 2006; Bas et al., 2007). DHA consumption and administration promote cerebral blood flow; inhibit $PLA_2$, cyclooxygenase, and lipoxygenase activities; and reduce levels of brain post-ischemic prostaglandins, thromboxanes, and leukotrienes. In addition, consumption of DHA may produce several other beneficial effects. DHA not only decreases blood–brain barrier disruption and reduces brain edema (Hossain et al., 1998) but also has antioxidant properties (Hossain et al., 1999). DHA inhibits production of inflammatory cytokines and antagonizes the metabolism of arachidonic acid and its downstream metabolites. It also downregulates NF-κB. In addition, the infusion of neuroprotectin $D_1$ ($NPD_1$), an endogenous lipid mediator derived from 15-lipoxygenase-catalyzed oxidation of DHA, following ischemic reperfusion injury downregulates neuroinflammation, oxidative stress, and blocks neurodegeneration (Fig. 3.7). $NPD_1$ also upregulates the anti-apoptotic Bcl-2 proteins (Bcl-2 and bclxL) and decreases the expression of the pro-apoptotic proteins (Bax and Bad) (Bazan, 2005). $NPD_1$ blocks reperfusion-induced leukocyte infiltration, pro-inflammatory signaling, and infarct size. $NPD_1$ not only inhibits cytokine-mediated cyclooxygenase-2 expression but also promotes homeostatic regulation of the integrity of neural cells particularly during oxidative stress, and this protective signaling may be relevant to neural cell survival following ischemic injury (Bazan, 2009). These processes strengthen the survival mechanisms through ERK-mediated and/or Bcl-2-mediated prosurvival cascade.

Resolvins are another group of proresolving and anti-inflammatory lipid mediator of *n*–3 fatty acid metabolism that have neuroprotective effects (Serhan, 2005a,

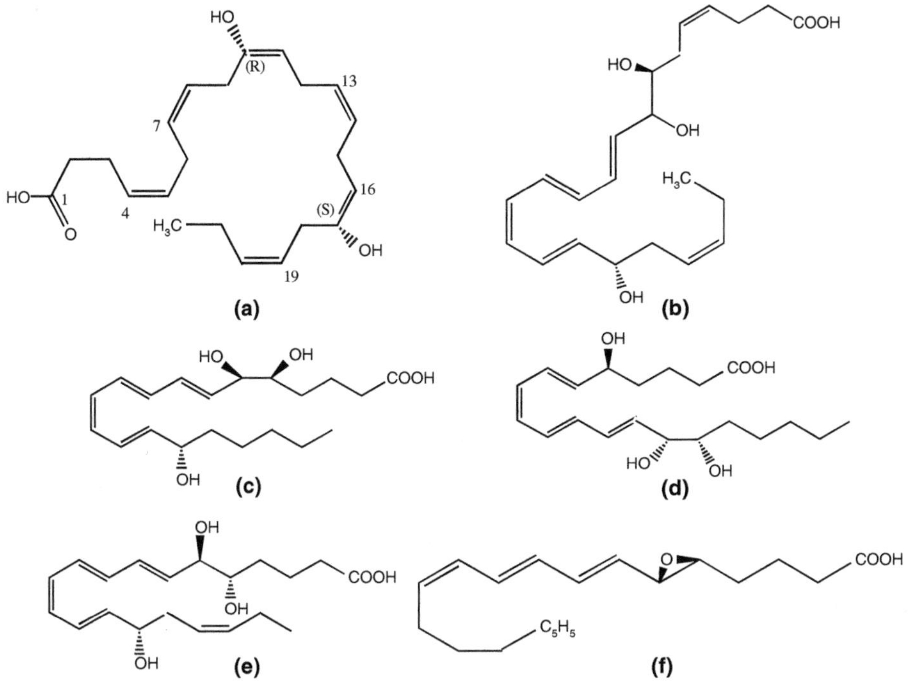


**Fig. 3.7** Chemical structures of DHA-, EPA-, and ARA-derived lipid mediators. Neuroprotectin $D_1$ (**a**); resolvin $D_1$ (**b**); lipoxin A4 (LXA4) (**c**); lipoxin (LXB4) (**d**); EPA-derived lipoxin A5 (LXA5) (**e**); and EPA-derived leukotrienes (**f**)

b; Bazan, 2005, 2009). DHA is metabolized to resolvin D series ($RvD_1$, $RvD_2$, $RvD_3$, $RvD_4$, $RvD_5$, and $RvD_6$), whereas eicosapentaenoic acid (EPA) is converted to resolvin E series (RvE and $RvE_2$) (Fig. 3.7). Like resolvin D series metabolite, RvE series blocks the activation of NF-κB by TNF-α (Arita et al., 2007). It is reported that $RvE_1$ binds to $BLT_1$ as a partial agonist and locally dampens the $BLT_1$-mediated signals on leukocytes along with other receptors (e.g., ChemR23-mediated counter-regulatory actions) to mediate the resolution of inflammation (Arita et al., 2006, 2007).

Arachidonic acid (ARA)-derived endogenous anti-inflammatory lipid mediators are called as lipoxins ($LXA_4$, $LXB_4$, 15 epi-$LXA_4$, and 15 epiLXB4) (Fig. 3.7). They are generated by the action of lipoxygenases on hydroperoxyeicosatetraenoic acid (HPETE) and hydroxyeicosatetraenoic acid (HETE). Lipoxins participate in the resolution phase of acute inflammation. Lipoxins interact with high affinity to G protein-coupled ALX and LXA receptors that transduce counter-regulatory signals in part via intracellular polyisoprenyl phosphate remodeling (Norel and Brink, 2004; Serhan et al., 2007). They inhibit neutrophil trafficking and stimulate nonphlogistic phagocytosis of apoptotic neutrophils by monocyte-derived macrophages (Serhan et al., 2004; Kantarci and Van Dyke, 2003; Chiang et al., 2006). Based on above evidence, it can be suggested that comsumption of ω-3 fatty acid enriched diet can protect animals and human from inflammatory processes following ischemic injury (Farooqui, 2009b).

### 3.2.7 Citicoline (CDP-Choline) and Stroke Therapy

Citicoline (Fig. 3.8) is an important intermediate in the biosynthetic pathway of neural membrane glycerophospholipids, particularly phosphatidylcholine. Its administration by oral and parenteral routes results in its hydrolysis generating cytidine and choline. CDP-choline is resynthesized from cytidine triphosphate and phosphocholine by CTP-phosphocholine cytidylyltransferase (CCT), the rate-limiting enzyme in PtdCho biosynthesis. CCT is regulated by sterol regulatory element binding proteins (SREBPs) at the transcriptional level. In addition to PtdCho synthesis, SREBPs also regulate lipid homeostasis by controlling the expression of a range of enzymes required for endogenous cholesterol, fatty acid, and triacylglycerol. CDP-choline is widely distributed throughout the body and serves as a choline donor in the biosynthesis of acetylcholine. It can cross the blood–brain barrier and reach into brain tissue, where it incorporates into the glycerophospholipids of plasma membrane and microsomal fractions. CDP-choline activates biosynthesis of structural glycerophospholipids of neuronal membranes, increases brain metabolism, and acts upon the levels of different neurotransmitters (Secades and Lorenzo, 2006). CDP-choline modulates several enzymic activities in brain (Table 3.1). Thus, it restores mitochondrial ATPase and membrane $Na^+/K^+$-ATPase activities, but has

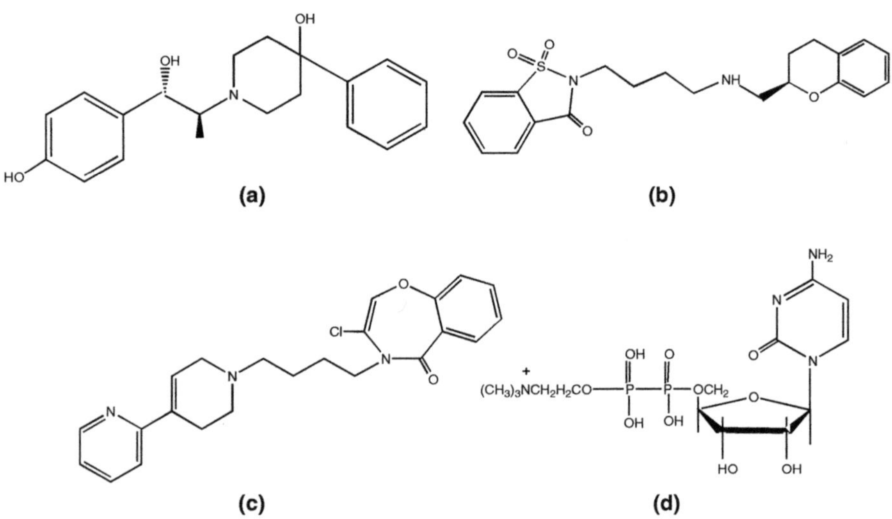


**Fig. 3.8** Chemical structures of new compounds that are in pipeline for the treatment of stroke. Traxoprodil (**a**); Branosyn (**b**); SUN-N4057 (**c**); and CDP-choline (**d**)

**Table 3.1** Enzymic activities and pathways targeted by CDP-choline

| Target | Effect | References |
|---|---|---|
| Phospholipase $A_2$ | Inhibition | Adibhatla et al. (2006) |
| $Na^+$, $K^+$-ATPase | Stimulation | Secades and Lorenzo (2006) |
| $Mg^{2+}$-ATPase | No effect | Plataras et al. (2003) |
| Procaspase | Inhibition | Krupinski et al. (2002) |
| Caspase-3 | Inhibition | Barrachina et al. (2002), Krupinski et al. (2002) |
| Excitotoxicity | Inhibition | Mir et al. (2003) |
| Bcl-2 | Stimulation | Sobrado et al. (2003) |
| Acetylcholine | Stimulation | Goldberg et al. (1985) |
| TNF-α release | Inhibition | Adibhatla et al. (2004) |
| β-Amyloid toxicity | Inhibition | Alvarez et al. (1999) |
| Homocysteine levels | Inhibition | Gimenez and Aguilar (2003) |
| 6-Hydroxydopamine toxicity | Inhibition | Barrachina et al. (2003) |

no effect on $Mg^{2+}$-ATPase activity (Plataras et al., 2003). The differential effect on various ATPases may be closely associated with modulations of cholinergic neurotransmission, neural excitability, metabolic energy production, $Mg^{2+}$ homeostasis, and protein synthesis. Pretreatment of rat cerebellar granule cells (CGCs) with CDP-choline results in a dose- and time-dependent reduction of glutamate-induced excitotoxicity (Mir et al., 2003). CGCs neurodegeneration can be retarded >50% when 100 μM CDP-choline is added 6 days before the glutamate-mediated neurotoxicity, but less than 20% when added concomitantly with glutamate. Furthermore, pretreatment of CGCs with CDP-choline protects from apoptotic cell death by >80%, indicating that CDP-choline exerts a neuroprotective effect by inhibiting the apoptotic pathway mediated by glutamate.

Transient middle cerebral artery occlusion (tMCAO) is known to increase secretory $PLA_2$ ($sPLA_2$)-IIA mRNA and protein levels, PtdCho-PLC activity, and PLD2 protein expression following reperfusion (Adibhatla et al., 2006). CDP-choline treatment attenuates $PLA_2$ activity, $sPLA_2$-IIA mRNA and protein levels, and PtdCho-PLC activity, but has no affect on PLD2 protein expression. tMCAO produces decrease in CTP:phosphocholine cytidylyltransferase (CCT) activity and CCTalpha protein and CDP-choline partially restores CCT activity (Adibhatla et al., 2006). No changes are observed in cytosolic $PLA_2$ or calcium-independent $PLA_2$ activities. Citicoline treatment also attenuates the infarction volume by 55±5% after 1 h of tMCAO and 1 day of reperfusion. Collectively, these results suggest that CDP-choline restores PtdCho levels by differentially affecting $sPLA_2$-IIA, PtdCho-PLC, and CCTalpha after transient focal cerebral ischemia (Adibhatla et al., 2006) (Fig. 3.9). CDP-choline not only blocks apoptotic cell death associated with cerebral ischemia but also potentiates neuroplasticity related mechanisms in certain neurodegeneration models (Fioravanti and Yanagi, 2005; Secades and Lorenzo, 2006). In ischemic and hemorrhagic stroke, it has been shown an excellent safety record and efficacy in several clinical trials outside of the USA. Results on the administration of CDP-choline in human stroke trials have been inconclusive. Meta-analysis

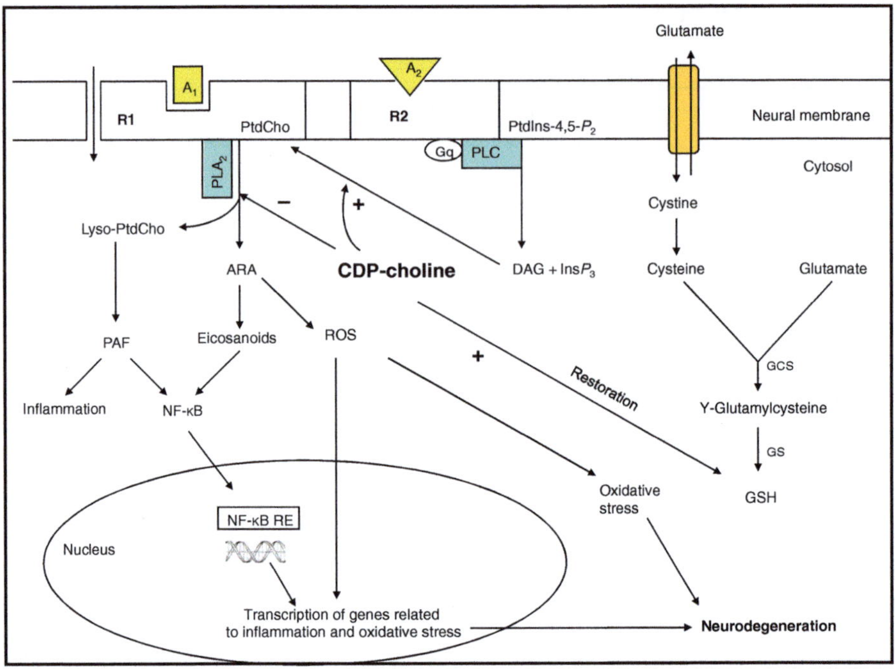


**Fig. 3.9** Neuroprotective mechanisms associated with the effects of CDP-choline following ischemic injury. Agonist ($A_1$ and $A_2$); receptors ($R_1$ and $R_2$); phosphatidylcholine (PtdCho); lyso-phosphatidylcholine (lyso-PtdCho); inositol 4,5-bisphosphate (PtdIns(4,5)$P_2$); inositol 1,4,5-trisphosphate (Ins$P_3$); diacylglycerol (DAG); platelet-activating factor (PAF); phospholipase $A_2$ ($PLA_2$); phospholipase C (PLC); cystine/glutamate antiporter (Cys-Glu-A.); γ-glutamylcysteine synthase (GCS); glutathione synthetase (GS); and glutathione (GSH). Positive sign indicates stimulation and negative sign indicates inhibition

of 10 trials enrolling 2,279 patients indicates that patients receiving CDP-choline have substantially reduced frequencies of death and disability. Reinvestigation of CDP-choline with modern neuroimaging and clinical trial methods are underway. These studies may provide more definitive information regarding the mechanistic and clinical effects of this neurotherapeutic agent (Saver, 2008; Clark, 2009).

### *3.2.8 Peroxisome Proliferator-Activated Receptor γ-Agonists and Stroke*

The peroxisome proliferator-activated receptors (PPARs) are ligand-activated transcription factors of the nuclear hormone receptor superfamily. The three PPAR isoforms (α, β/δ, and γ) are known to occur in mammalian tissues. In response to specific agonists, these receptors form dimers and translocate to the nucleus, where they act as agonist-dependent transcription factors and regulate gene expression by binding to specific promoter regions of target genes that not only regulate glucose and fat metabolism but also attenuate neurodegenerative and inflammatory processes in the brain (Kapadia et al., 2008). Although the natural ligand for PPARγ are long-chain fatty acids, 15d-prostaglandin $J_2$ (15dPG$J_2$), and thiazolidinediones (TZDs) are potent exogenous agonists. Due to their insulin-sensitizing properties, 2 TZDs, rosiglitazone and pioglitazone, are currently FDA approved for type 2 diabetes treatment. It is also shown that TZDs produce significant neuroprotection in animal models of focal ischemia by multiple mechanisms. The pleiotropic actions of TZDs have been observed through PPARγ-dependent as well as independent mechanisms involving anti-inflammatory activities of these drugs on peripheral immune cells (macrophages and lymphocytes), as well as direct effects on neural cells including cerebral vascular endothelial cells, neurons, and glial cells. The major mechanism of TZD-mediated neuroprotection involves the suppression of microglial activation and inflammatory cytokine and chemokine expression (Kapadia et al., 2008). TZDs also retard the activation of proinflammatory transcription factors at the same time promoting the antioxidant mechanisms in the injured brain (Kapadia et al., 2008). In addition, intracerebroventricular infusion of pioglitazone over a 5-day period before and 2 days after middle cerebral artery occlusion (MCAO) reduces the infarct size, the expression of TNF-α, COX-2, and the number of cells positively stained for COX-1 and COX-2 in the peri-infarct cortical regions (Zhao et al., 2006). The neuroprotective effect of pioglitazone can be reversed after cotreatment with GW 9662, a selective antagonist of the PPARγ, indicating the involvement of a PPARγ-dependent mechanism (Zhao et al., 2006; Culman et al., 2007). Pioglitazone also inhibits LPS-mediated iNOS expression and NO generation in dopaminergic neurons. In addition, inhibition of p38 MAPK, but not JNK, is also blocked by LPS-induced NO generation suggesting that PPARγ activation may differentially regulate neuroinflammation through the modulation of p38 MAPK (Xing et al., 2008). Recent studies have also shown that pioglitazone effectively reduces the number of IL-6 immunoreactive cells and IL-6 protein levels after MCAO supporting the view that PPARγ activation with pioglitazone may be a potent therapeutic option for preventing inflammation and neuronal damage following ischemic injury (Patzer et al., 2008).

### *3.2.9 Hypoxia-Inducible Factor 1 and Stroke Therapy*

Hypoxia-inducible factor (HIF) is a heterodimeric transcription factor associated with the regulation of transcriptional responses to hypoxia (Loor and Schumacker, 2008). It is composed of HIF-1α and HIF-1β protein subunits, which are constitutively expressed. In normoxia, HIF-1β is destabilized by post-translational hydroxylation of Pro-564 and Pro-402 by a family of oxygen-sensitive dioxygenases. During hypoxia, HIF-1α binds with HIF-1β to form HIF-1, which interacts with promoter elements in hypoxia-responsive target genes in the nucleus. This causes upregulation of HIF target genes, which include vascular endothelial cell growth factor, erythropoietin, iNOS, glucose transporter-1, glycolytic enzymes, and many other genes that protect the brain against ischemia 24 h later (Ran et al., 2005). In addition, non-HIF pathways including MTF-1 and Egr-1 act directly or indirectly on other target genes to also promote hypoxia-induced preconditioning. Thus, HIF target genes modulate glycolysis, glucose metabolism, mitochondrial function, cell survival, apoptosis, vasomotor control, angiogenesis, erythropoiesis, cell proliferation, and resistance to oxidative stress (Loor and Schumacker, 2008). Accumulating evidence suggests that activation of HIF-1$\alpha$ is involved in triggering cellular protection and metabolic alterations from the consequences of oxygen deprivation.

Similarly, focal ischemic injury not only increases mRNAs for HIF-1$\alpha$ at the core of infarct but also upregulates the expression of glucose transporter-1 and several glycolytic enzymes in the peri-infarct penumbra (Bergeron et al., 1999). Regional cerebral blood flow is moderately decreased at 1 and 24 h after the ischemic injury in core and peri-infarct penumbra. Because hypoxia induces HIF-1$\alpha$ in other tissues, systemic hypoxia (6% $O_2$ for 4.5 h) has also shown to increase HIF-1 protein expression in the adult rat brain. It is proposed that decreased blood flow to the penumbra decreases the supply of oxygen and that this induces HIF-1 and its target genes. These observations support the view that HIF-1$\alpha$ activation offers neuroprotection against ischemia/reperfusion injury. Collective evidence suggests that endogenous mechanisms (preconditioning) may play an important role for the treatment of hypoxic/ischemic injury (Blanco et al., 2006). The molecular mechanisms of neuroprotection that lead to ischemic tolerance are not fully understood. However, two distinct mechanisms are closely associated with neuroprotective process. The first mechanism involves the initiation of cellular defense function against ischemic injury through mechanisms inherent to neurons, such as post-translational modification of proteins or expression of new proteins via a signal transduction system to the nucleus. This phase either strengthens the influence of survival factors or inhibits apoptosis. The second mechanism includes the induction and activation of a stress response that is accompanied by the synthesis of stress proteins (heat shock proteins) or chaperones, which mediate the unfolding of misfolded cellular proteins and help cells to dispose of unneeded denatured proteins (Blanco et al., 2006). HIF-1$\alpha$ also plays a role in necrotic cell death through its interactions with calcium and calpain system (Fan et al., 2009). HIF-1$\alpha$ also exacerbates brain edema via increasing the permeability of the blood–brain barrier (BBB). Given these properties, unraveling of the complex functions of HIF-1$\alpha$ may be important when designing neuroprotective therapies for hypoxic-ischemic brain injury. Full understanding of molecular mechanisms and genes involved in hypoxic/ischemic tolerance may provide new therapeutic targets to treat ischemic injury and enhance recovery (Sharp et al., 2004; Shi, 2009).

### *3.2.10 Vaccine and Stroke Therapy*

Development of stroke vaccine is a novel neuroprotective strategy. Stroke vaccine is in the initial stages of development (During et al., 2000; Takeda et al., 2002).

In one study an adeno-associated virus (AAV) vaccine generating autoantibodies has been used to target a specific brain protein, the NR1 subunit of the *N*-methyl-D-aspartate (NMDA) receptor. After peroral administration of the AAV vaccine, transgene expression persists for at least 5 months and is involved in robust humoral response in the absence of a significant cell-mediated response. This single-dose vaccine prevents epileptic seizures in kainic acid neurotoxicity and shows neuroprotective activity in a middle cerebral artery occlusion stroke model in rats at 1–5 months following vaccination (During et al., 2000). Thus, a vaccination strategy targeting brain proteins is feasible and may have therapeutic potential for stroke, epilepsy, and other neurological disorders. In the second study, a nasal spray is used for delivering a protein (E-selectin) that, under normal circumstances, is associated with inflammation of the cells that line the inner walls of blood vessels of hypertensive, genetically stroke-prone rats (Takeda et al., 2002). Inflammation not only plays an important role in stroke but also makes cerebral blood vessels more vulnerable to formation of a clot. Exposing rats to E-selectin programs its lymphocytes to monitor the blood vessel lining for the inflammatory protein, when these lymphocytes detect E-selectin, they produce mediators that prevent inflammation. Thus, nasal instillation of E-selectin, which is specifically expressed on activated endothelium, potently prevents the development of ischemic and hemorrhagic strokes in spontaneously hypertensive stroke-prone rats with untreated hypertension (Takeda et al., 2002). It must be mentioned that the single course of vaccine treatment does not maintain the animal's resistance to stroke, but repeated treatment with the vaccine is needed for long-term stroke prevention. Suppression of delayed-type hypersensitivity to E-selectin and increased numbers of transforming growth factor-β1-positive splenocytes indicate that intranasal exposure to E-selectin mediates immunologic tolerance. E-Selectin tolerization also reduces endothelial activation and immune responses after intravenous lipopolysaccharide, as shown by marked suppression of intercellular adhesion molecule-1 expression, anti-endothelial cell antibodies on luminal endothelium, and plasma interferon-gamma levels compared with the control condition. The vaccine has no effect on blood pressure indicating that its beneficial effects are not linked to reduction of high blood pressure. While these vaccines work in rats, no one knows whether they will produce similar effects in humans (During et al., 2000; Takeda et al., 2002). Clinical trial (phase I) to test the effects of bovine E-selectin vaccine on human with high risk of stroke has been planned. That trial will provide information not only on the beneficial effects of vaccine on stroke but also on side effects of vaccine in humans.

### *3.2.11 Pipeline Developments on Drugs for Stroke Therapy*

Stroke is a vascular condition that precipitates neurological damage and paralysis. Detailed investigations on understanding of its pathogenic mechanism have not only promoted its prevention by eliminating risk factors but also facilitated the development of new drugs that are in pipeline. These drugs include Traxoprodil, Branosyn, SUN-N4057, ONO-2506, monoester of DP-b99, Tacolimus, and BIII-890-CL (Fig. 3.10 and Table 3.2). Some of these drugs have been tried in small

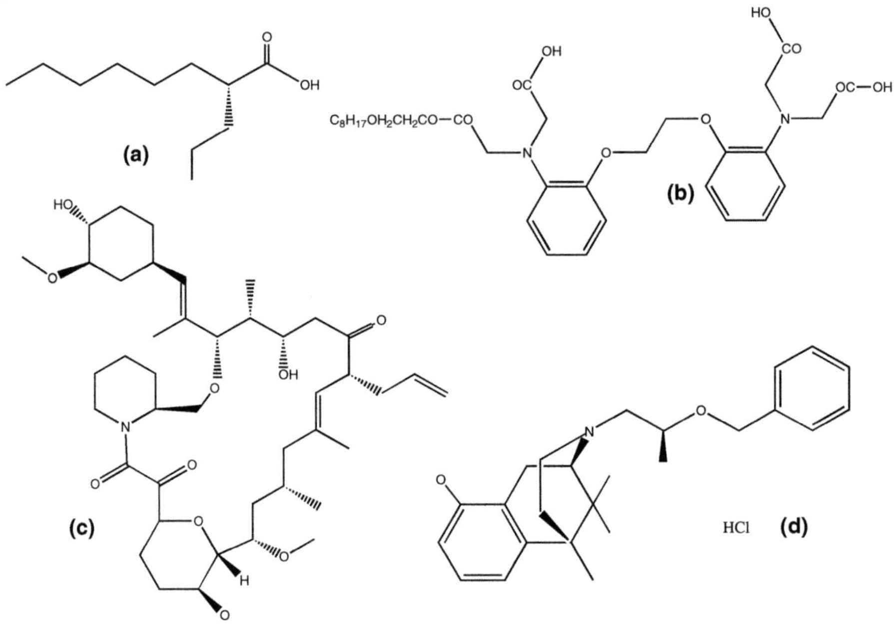


**Fig. 3.10** Chemical structures of more new compounds that are in pipeline for the treatment of stroke. ONO-2506 (**a**); monoester of DP-b99 (**b**); tacolimus (**c**); and BIII-890-CL (**d**)

**Table 3.2** Drugs that are in pipeline for the treatment of ischemic injury

| Drug | Nature/mechanism | References |
|---|---|---|
| Citicoline | A glycerophospholipid metabolism intermediate | Adibhatla et al. (2006) |
| Cerovive (NXY-059) | A free radical-trapping agent | Lees et al. (2006), Shuaib et al. (2007) |
| Tacrolimus | An immunosuppressant | Zhou et al. (2009) |
| ONO-2506 | An astroglia-modulating agent | Ohtani et al. (2007) |
| Semax | A neuropeptide | Bashkatova et al. (2001) |
| Branosyn (repinotan) (BAY x3702) | A serotonin receptor agonist | Teal et al. (2005) |
| DP-b99 | A lipophilic selective chelators for calcium and zinc | Diener et al. (2008) |
| SUN-N4057 | A serotonin (5-HT) 1A receptor agonist | Kamei et al. (2001) |
| Traxoprodil (CP-101606) | A NR2B NMDA antagonist | Yang et al. (2003) |
| BIII-890-CL | Sodium channel blocker | Carter et al. (2000) |

trials that have failed. Large clinical trials are planned on many of above drugs to protect brain tissues after the stroke-mediated brain injury. In addition, clinical trials have been planned on magnesium, 5-HT1A agonist, metal chelation, and albumin. Preliminary studies with techniques that chill the brain have shown that inducing hypothermia may reduce stroke damage.

### *3.2.12 Intracellular Cell Therapy in Stroke*

Brain damage restoration approaches including cell-based therapies have attracted considerable attention for the treatment of stroke in recent years (Hicks and Jolkkonen, 2008). A large number of experimental transplantation studies have been performed with embryonic stem, fetal neural stem, and human umbilical cord blood. Two main approaches have been utilized for the stem cell delivery (Hicks and Jolkkonen, 2008). The first approach involves a stereotaxic transplantation of cells into the brain and the second is the intravascular administration. Since stroke results in large ischemic brain damage, it remains to be seen whether stereotaxic transplantation of cells can provide efficient and wide cell engraftment for brain recovery. Another concern about stereotaxic transplantation of cells is the invasive nature of intracerebral transplantation. The second approach is intravascular administration. This approach does not necessary rely on the cellular replacement but is based on the activation of the brain's endogenous repair mechanisms through the involvement of insulin-like growth factor-1 and brain-derived growth factor. The trophic factors mediate neuroplasticity, angiogenesis and neurogenesis, and attenuation of scar formation. It is also shown that entry of intravenously injected cells into the brain tissue is not required for therapeutic effects, indicating that peripheral mechanisms may contribute to the recovery process (Borlongan et al., 2004). Human umbilical cord blood (HUCB) is now regarded as a valuable source for stem cell-based therapies. Stem cells can differentiate into neural lineages to replace lost neurons. Stem cells are pluripotential cells that not only provide delivery brain-derived neurotrophic factor (BDNF) or glial-derived neurotrophic factor (GDNF) to tissue at risk in the penumbra surrounding the infarct area and enhance vasculogenesis but also promote survival, migration, and differentiation of the endogenous precursor cells after stroke. Stem cells are highly migratory and seem to be attracted to areas of brain pathology such as ischemic regions (Chang et al., 2007; Hess and Borlongan, 2008). HUCB cells are enriched for stem cells that have the potential to initiate and maintain tissue repair following stroke. Thus, intravenous injections of HUCBCs after a middle cerebral artery occlusion produce behavioral and anatomical recovery that protects neural tissue from progressive changes (Vendrame et al., 2004, 2005). HUCBCs are recruited to the injury site and reduce inflammation in the brain and thereby enhancing neuroprotection (Vendrame et al., 2005). In addition, the transplantation of HUCB decreases CD45/CD11b- and CD45/B220-positive (+) cells. This decrease is accompanied not only by a decrease in mRNA and protein expression of pro-inflammatory cytokines but also by downregulation of nuclear factor kappaB (NF-κB) DNA binding activity in the brain of injured animals. In addition to modulating the inflammatory response, the cord blood cells increase neuronal survival through non-immune mechanisms (Vendrame et al., 2005). It is also shown that injection of HUCB cell not only improves the behavioral defects of rats but also results in extension of therapeutic window from 3 h to 24–72 h post-stroke (Newman et al., 2005). Very little is known about the molecular mechanism associated with homing of stem cells in humans and discovery of the molecular pathways that facilitate the homing of stem cells into the ischemic areas and may facilitate the development of new treatment regimens, perhaps using small molecules, designed to enhance endogenous mobilization of stem cells in the chronic stroke. For maximal functional recovery, however, regenerative therapy may need to follow combinatorial approaches, which may include cell replacement, trophic support, protection from oxidative stress, and the neutralization of the growth-inhibitory components for endogenous neuronal stem cells (Chang et al., 2007). Thus, understanding the exact molecular basis of stem cell plasticity in relation to local ischemic signals may offer new insights to permit better management of stroke and other ischemic disorders. Altogether, a number of studies support the view that potential of systemic delivery of stem cells is a novel therapeutic approach for stroke. Although stem cell transplantation is an important development for stroke therapy, only few studies have been performed using a single dose and at a single time point post-stroke (Yu et al., 2009). Due to the rapid degeneration and low survival rate of neurons at the damage or injury site and partial behavioral recovery, new strategies are needed to improve the quality and beneficial effects of stem cell transplantation in stroke. For greater behavioral benefits in stroke patients, detailed investigation is needed on types of stem cells, their optimal number for transplantation, therapeutic window, and blood–brain barrier opening agents. Furthermore, long-term studies are required to determine whether the stem cell-enhanced recovery is sustained and translate into beneficial behavioral and functional outcome (Chang et al., 2007; Yu et al., 2009). There is a possibility that stem cell transplantation may initiate tumorigenesis in brain that may be fetal for recovering stroke patients. Thus, additional preclinical studies are warranted to reveal the optimal stem transplant regimen that is safe and efficacious prior to proceeding to large-scale clinical application of these cells for stroke therapy (Yu et al., 2009).

## 3.3 Mechanism of Neuroprotection in Ischemic Injury

Ischemic injury is a multifactorial process. Although drugs targeting a single enzyme target may show some efficacy for the treatment of ischemic damage, it is becoming increasingly evident that clinical trials with a cocktail of free radical scavenger and anti-inflammatory agents may provide better efficacy for ischemic damage than a single drug. Thus, a more complex multitargeted approach may prove more successful in stroke patients than single-targeted drug. The mechanistic basis of the neuroprotective effects of various drugs may depend on their chemical nature and pathway that these drugs block. Thus, calcium blockers prevent calcium entry in neurons, NMDA antagonists block calcium entry in neurons through NMDA channels, and anti-inflammatory and antioxidant agents may not only depend on the general free radical trapping or antioxidant activity per se in neurons but also on the downregulation of NF-κB activity (Shen et al., 2003) and suppression of genes induced by proinflammatory cytokines and other mediators released by glial cells (Gilgun-Sherki et al., 2006; Wang et al., 2006).

In response to ischemia-mediated glutamate release, NF-κB translocates to the nucleus, where it binds to target sequences in the genome and facilitates the expression of a number of proteins including many enzymes ($sPLA_2$, COX-2, NADPH oxidase, and inducible nitric oxide synthase) and cytokines (TNF-α, IL-1β, and IL-6) (Table 3.3) (Farooqui et al., 2008). Ischemic injury-mediated oxidative damage is a complex therapeutic target. In addition to ROS and RNS generation at several subcellular sites, ischemic injury is also accompanied by the production of 4-hydroxynonenal and peroxynitrite. These metabolites interact with DNA and proteins and make the ischemic injury a very complex process. It is proposed that ischemic injury requires interplay among excitotoxicity, inflammation, oxidative stress, and apoptosis. The efficacy of a cocktail of anti-inflammatory and antioxidant agents for neuroprotection in stroke depends on their ability to cross the blood–brain barrier, their subcellular distribution in mitochondria, plasma membrane, and cytoplasm, their multifunctional capacity, as well as their synergistic actions (Gilgun-Sherki et al., 2006). Furthermore, spatial and temporal parameters of ischemic injury site must be to elucidate and used for the best response of anti-inflammatory and antioxidant agents cocktail for neuroprotection in stroke. Inclusion of agents that increase the production of ATP in degenerating neurons may improve the therapeutic outcome following stroke. A clearer appreciation of the potential therapeutic ability of anti-inflammatory and antioxidant cocktails will emerge only when the importance in vivo of interplay among excitotoxicity, neuroinflammation, and oxidative stress is realized and fully understood at the molecular level (Farooqui et al., 2006; Farooqui and Horrocks, 2007). By gaining

**Table 3.3** Modulation of enzymic activities, cytokines, and adhesion molecules by NF-κB

| Enzyme/cytokine/adhesion molecule | Effect | References |
|---|---|---|
| Secretory $PLA_2$ | Upregulation | Pizzi et al. (2005), Farooqui (2009a) |
| Cyclooxygenase | Upregulation | Pizzi et al. (2005), Farooqui (2009a) |
| Nitric oxide synthase | Upregulation | Sun et al. (2007), Farooqui (2009a) |
| NADPH oxidase | Upregulation | Sun et al. (2007), Farooqui (2009a) |
| Matrix metalloproteinase | Upregulation | Farooqui (2009a) |
| Tumor necrosis factor-α | Upregulation | Farooqui (2009a) |
| Interleukin-1β | Upregulation | Farooqui (2009a) |
| Interleukin-6 | Upregulation | Farooqui (2009a) |
| Vascular adhesion molecule-1 | Upregulation | Farooqui (2009a) |
| Cell adhesion molecule-1 | Upregulation | Farooqui (2009a) |

a greater understanding of interplay among excitotoxicity, neuroinflammation, and oxidative stress and timelines between injury and neuronal death, one may discover multitargeted drugs with potential for treating stroke and also to gain information about the appropriate timing, when these drugs can be administered in the degenerative cascade for better recovery.

### 3.3.1 Prevention of Stroke Through the Modulation of Risk Factors

Since neuroprotective therapy has either failed or provided limited benefits, clinicians have focused on preventive strategies to limit its first onset and recurrence. Prevention begins with awareness of risk factors by patients and clinicians. According to the American Stroke Association and American Heart Association guidelines, there are three types of risk factors for stroke: nonmodifiable, modifiable, and potentially modifiable (Glodstein et al., 2006; Rincon and Sacco, 2008; Romero et al., 2008). Nonmodifiable risk factors include age, sex, low birth weight, race/ethnicity, and genetic factors. Well-documented and modifiable risk factors include hypertension, exposure to cigarette smoke, diabetes, atrial fibrillation and certain other cardiac conditions, dyslipidemia, carotid artery stenosis, sickle cell disease, postmenopausal hormone therapy, poor diet, physical inactivity, and obesity and body fat distribution (Glodstein et al., 2006; Rincon and Sacco, 2008). Less well-documented or potentially modifiable risk factors include the metabolic syndrome, alcohol abuse, drug abuse, oral contraceptive use, sleep-disordered breathing, migraine headache, hyperhomocysteinemia, elevated lipoprotein(a), elevated lipoprotein-associated phospholipase, hypercoagulability, inflammation, and infection (Glodstein et al., 2006; Rincon and Sacco, 2008). These guidelines not only provide comprehensive and timely evidence-based recommendations on the prevention of ischemic stroke among survivors of stroke or transient ischemic attack but also guide health-care providers a potential explanation for the causes of stroke in an individual patient to select therapies that reduce the risk of recurrent events and other vascular events (Glodstein et al., 2006; Rincon and Sacco, 2008; Romero, 2007; Romero et al., 2008). Current stroke prevention strategies include high blood pressure (hypertension) control and retarding the formation of blood clots using drugs such as aspirin and warfarin.

### 3.3.2 Selection of Diet and Stroke

In order to protect the aging population from stroke, it is crucial to explore methods that may retard or slow the molecular cascade associated with oxidative stress, excitotoxicity, and neuroinflammation. Diet enriched in antioxidant and anti-inflammatory agents (curcumin, green tea, and ferulic acid) (Fig. 3.11) may lower the risk of stroke. Many studies indicate that dietary supplementation with fruit or colored vegetable extracts can decrease the age-enhanced vulnerability to oxidative

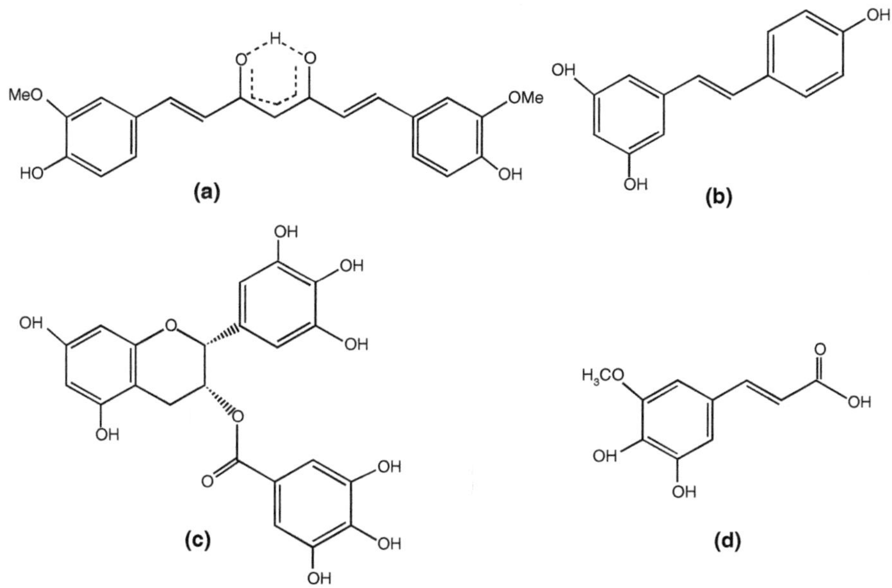


**Fig. 3.11** Chemical structure of anti-aging remedies that should be included in diet to prevent stroke. Cucurmin (**a**); resveratrol (**b**); green tea catechin (epigallocatechin gallate) (**c**); and ferulic acid (**d**)

stress and inflammation (Farooqui and Farooqui, 2009). Additional studies indicate that the polyphenolic compounds found in red wine and fruits such as blueberries may exert their beneficial effects through signal transduction and neuronal communication (Lau et al., 2007; Joseph et al., 2007). Other food-based antioxidants (such as vitamins C, E; β carotene) may also modulate processes associated with secondary injury by neutralizing free radicals.

Another important dietary factor is the ratio between arachidonic acid (ARA) and docosahexaenoic acid (DHA) (Fig. 3.12). Both polyunsaturated fatty acids are essential for human health, but cannot be synthesized de novo by mammals; ARA, DHA, or their precursors must be ingested from dietary sources and transported to the brain (Horrocks and Farooqui, 2004; Farooqui, 2009b). ARA is found in vegetable oil, whereas DHA in enriched in fatty fish and fish oil. The present-day Western diet has a ratio of ARA to DHA fatty acids of about 18:1. The Paleolithic diet on which human beings evolved and lived for most of their existence had a ratio of 1:1 (Simopoulos, 2006; Cordain et al., 2005; Farooqui, 2009b). Changes in eating habits, natural versus processed food, and agriculture development within the past 100–150 years have caused these changes in the *n*–6 to *n*–3 ratio, which has affected human health remarkably. The consumption of fish and fish oil has numerous beneficial effects on the health of the human brain (Horrocks and Farooqui, 2004; Farooqui, 2009b). The beneficial effects of docosahexaenoic acid on human brain are not only due to its effect on the physicochemical properties of neural

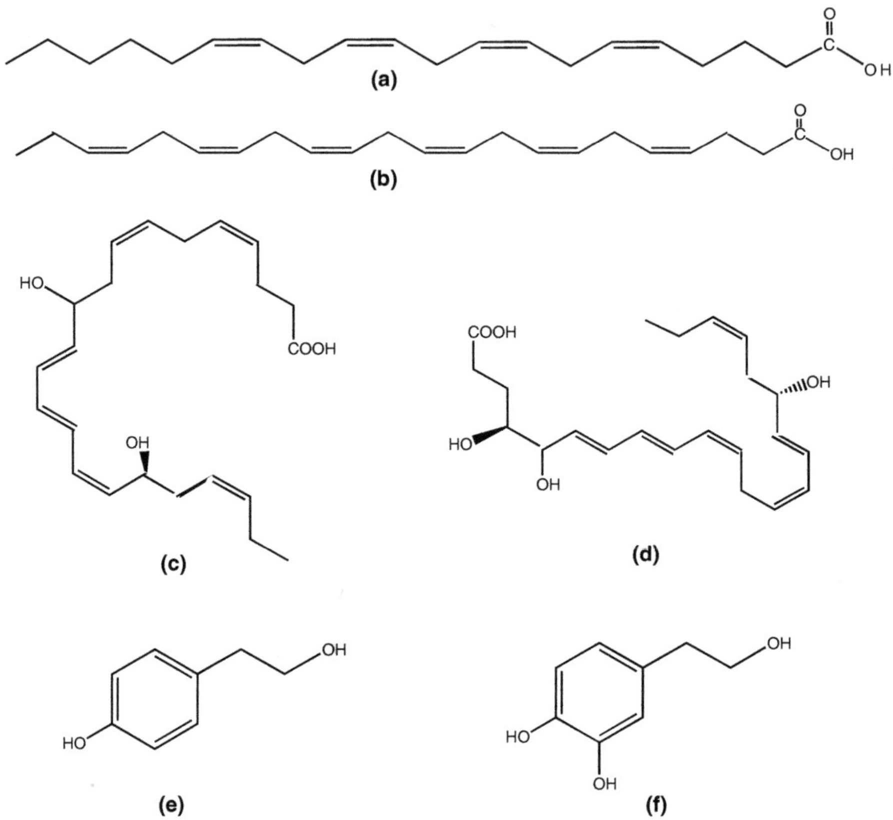


**Fig. 3.12** Chemical structures of arachidonic acid (**a**); docosahexaenoic acid (**b**); 10,17S docosatrienes (**c**); 4S,5,17S-resolvin (**d**); tyrosol [2-(4-hydroxyphenyl)ethanol] (**e**); hydroxytyrosol (**f**)

membranes but also due to modulation of neurotransmission, gene expression, activities of enzymes, ion channels, receptors, and immunity (Farooqui, 2009b). Chronic administration of DHA reduces spatial cognitive deficit following transient ischemia in rats. Neuroprotective effects of DHA in ischemic injury are controversial. Some studies show beneficial effects in CA1 region, while others indicate DHA does not protect from ischemic hippocampal damage in areas CA1, CA2, or CA4 region. It is suggested that long-term DHA or fish oil intake facilitates functional recovery after ischemic brain damage, an effect that was distinct from hippocampal damage (Okada et al., 1996; Fernandes et al., 2008).

Like pleiotropic effects of statins, DHA also produces antiexcitotoxic, antioxidant, anti-inflammatory effects through the generation of its lipid mediators (Table 3.4) (Simopoulos, 2006; Farooqui et al., 2008; Farooqui, 2009b). These lipid mediators include resolvins and neuroprotectins. These metabolites are very important from stroke therapeutic point of view. They antagonize the effects of

**Table 3.4** Comparison of properties of statins and fish oil that may be beneficial for ischemic injury

| Parameter | Statins | Fish oil | References |
|---|---|---|---|
| Antiexcitotoxic effects | Yes | Yes | Farooqui et al. (2007b) |
| Anti-inflammatory effects | Yes | Yes | Farooqui et al. (2007b) |
| Antioxidant effects | Yes | Yes | Farooqui et al. (2007b) |
| Antithrombotic effects | Yes | Yes | Farooqui et al. (2007b) |
| Proplaque stability effects | Yes | Yes | Farooqui et al. (2007b) |

eicosanoids, which are metabolites of arachidonic acid metabolism. Resolvins and neuroprotectins (Fig. 3.11) modulate leukocyte trafficking and downregulate the expression of cytokines in glial cells and modulate interactions among neurons, astrocytes, oligodendrocytes, microglia, and cells of the microvasculature (Serhan, 2005a; Bazan, 2007). The infusion of neuroprotectin $D_1$ ($NPD_1$), following ischemic reperfusion injury or during oxidative stress in cell culture, downregulates oxidative stress and apoptotic DNA damage. $NPD_1$ also upregulates the anti-apoptotic Bcl-2 proteins, Bcl-2 and bclxL and decreases the expression of the pro-apoptotic proteins, Bax and Bad (Bazan, 2007). In addition, DHA also downregulate NF-κB activity (Farooqui and Horrocks, 2007) and suppress genes induced by proinflammatory cytokines and other mediators released by glial cells. Extra-virgin olive oil (unprocessed olive oil) contains micronutrients and polyphenolic antioxidants including tyrosol [2-(4-hydroxyphenyl)ethanol], hydroxytyrosol, oleuropein, and oleocanthal (Fig. 3.11). These constituents also retard stroke-mediated brain injury (Lopez-Miranda et al., 2007).

### *3.3.3 Physical Exercise and Stroke*

In the brain, exercise produces both acute and long-term changes, such as increased levels of various neurotrophic factors or enhanced cognition. Although the signals and molecular mechanisms associated with exercise-induced changes in the brain are not yet well understood, it is becoming increasingly evident that physical exercise mediates signals through increased metabolic activity induced neuroadaptive and neuroprotective changes in brain function (Trejo et al., 2002). It is proposed that regular exercise-mediated neuroadaptations may have beneficial effects not only on depression but also on stroke and other neurological disorders (Dishman et al., 2006). Chronic voluntary physical activity not only reduces hypertension and decreases chances of heart failure but also decreases elevated sympathetic nervous system activity.

Exercise upregulates brain-derived neurotrophic factor (BDNF), a molecule that plays an important role in synaptic plasticity, neuroprotection, and learning and memory (Vaynman et al., 2004; Vaynman and Gomez-Pinilla, 2006; Ploughman et al., 2007). In addition, exercise also produces changes in the brain that are essential for optimal brain function. These changes are mediated by insulin-like growth factor I (IGF-I), a 79 amino acids containing circulating hormone that induces physical exercise-mediated potent neurotrophic activities (Carro et al., 2001). Interactions of IGF-1 with its receptor (IGF-1R) result in tyrosine phosphorylation of insulin receptor substrate-1 (IRS-1) and subsequent activation of PtdIns 3-kinase, PtdIns-dependent kinase, and protein kinase B/AKT as well as phosphorylated cAMP response-element binding protein (pCREB) (Okajima and Harada, 2008; Ploughman et al., 2007). These neurotropic activities are blocked by IGF-I antibody and IGF-I receptor antagonists and CREB phosphorylation inhibitors. Together, these results support the view that exercise prevents and protects from brain damage through increased uptake of circulating IGF-I by the brain.

Blocking the IGF-I receptor significantly reverses the exercise-mediated increase in the levels of BDNF mRNA and protein and pro-BDNF protein, suggesting that the effects of IGF-I may be partially mediated by modulation of BDNF synthesis from its precursors. Molecular analysis indicates that exercise significantly upregulates proteins downstream to BDNF activation important for synaptic function, such as synapsin I, phosphorylated calcium/calmodulin protein kinase II, and phosphorylated mitogen-activated protein kinase II (Ding et al., 2006). Blocking the IGF-I receptor retards these exercise-induced increases in BDNF. These results provide information on the molecular mechanisms by which IGF-I modulates the BDNF system to mediate exercise-induced synaptic and cognitive plasticity. BDNF not only facilitates long-term potentiation, an electrophysiological correlate of learning and memory, but also increases the activities of free radical scavenging enzymes and hence protect neurons against oxidative stress (Pelleymounter et al., 1996). Exercise also upregulates the expression of the mitochondrial uncoupling protein 2, an energy-balancing factor concerned with ATP production and free radical management (Vaynman et al., 2006), supporting the view that in brain tissue physical exercise promotes a fundamental mechanism by which key elements of energy metabolism may modulate the substrates of hippocampal synaptic plasticity (Ploughman et al., 2007). Collectively, these studies suggest that physical exercise upregulates brain-derived neurotrophic factor (BDNF), phosphorylated cAMP response-element binding protein (pCREB), insulin-like growth factor (IGF-I, and synapsin-I, each of which play some role in neuroplastic processes underlying recovery from ischemia.

### *3.3.4 Transcranial Magnetic Stimulation and Stroke Rehabilitation*

Following stroke-mediated injury, bilateral motor regions of the brain undergoes substantial reorganization, including changes in the strength of interhemispheric inhibitory interactions. This reorganization is known to contribute to behavioral gains in the rehabilitative process (Webster et al., 2006; Bolognini et al., 2009). Transcranial magnetic stimulation (TMS), transcranial direct current stimulation (tDCS), and magnetoencephalography (MEG) are noninvasive brain stimulation techniques that modulate cortical excitability in both healthy individuals and stroke patients. TMS and tDCS provide bidimensional scalp maps and MEG depicts three-dimensional spatial characteristics of virtual neural generators obtained by the use of a mathematical model of the head and brain. Repetitive transcranial magnetic stimulation (rTMS) of human primary motor cortex alters cortical excitability at the site of stimulation and at distant sites without affecting simple motor performance. Thus, rTMS and tDCS represent powerful methods for priming cortical excitability for a subsequent motor task, demand, or stimulation. Their mutual use can optimize the plastic changes mediated by motor practice, leading to more remarkable and outlasting clinical gains in rehabilitation. TMS, tDCS, and MEG have been shown to enhance the effect of training on performance of various motor tasks, including those that mimic activities of daily living (Webster et al., 2006; Bolognini et al., 2009). There has been considerable development in imaging technology enabling noninvasive exploration of brain structure and function to such an intricate degree as to enable measurements of very small spatial and short temporal cerebral operations responsible for neurological and functional recovery after stroke. Thus, combination of TMS, tDCS, and MEG with functional MRI (fMRI) and positron emission tomography (PET) will allow the excellent resolution of neural network that may facilitate the development of rehabilitation protocol, providing maximum benefits to individual stroke patient.

### *3.3.5 Occupational Therapy and Rehabilitation After Stroke*

The most common outcome of stroke is unilateral paralysis followed by alterations in coordination, balance, and movements, which are rarely recovered. Due to alterations in coordination, balance, and movements, activities among stroke patient self-care, cognition, and communication become difficult. Stroke patients require assistance and care provided by care givers. Rehabilitation after stroke is based on the concept of brain plasticity (endogenous brain repair mechanisms), which encompasses that it is possible to modulate or promote cerebral reorganization through external input or stimulus (Govender and Kalra, 2007). This reorganization may involve the recruitment of pathways that are functionally homologous to, but anatomically distinct from, the damaged ones (e.g., non-pyramidal corticospinal pathways), synaptogenesis, dendritic arborization and reinforcement of existing but functionally silent synaptic connections (particularly at the periphery of core lesion). Occupational therapy activities are specifically designed to promote this re-education process and encourage the development of lost skills while accommodating for specific physical, cognitive, or affective impairments. Principles of motor, sensory, cognitive, and affective rehabilitation are incorporated into effective task-specific activities, and environments are adapted to create the optimum conditions for successful rehabilitation (Govender and Kalra, 2007). Data on molecular aspects of rehabilitation after stroke are scarce. Thus, very little is known about molecular aspects and effectiveness of occupational therapy on motor, cognitive, and psychosocial dysfunctions (Rossini et al., 2007). However, rehabilitation after stroke is an active process beginning during hospitalization, progressing to a systematic program of rehabilitation services, and continuing after the individual returns home. Based on neuropsychology and technological advances, several promising new rehabilitation approaches have been made to complement therapy inputs and exploit the brain's capacity to recover from stroke. Neuroimaging studies in stroke patients indicate altered post-stroke activation patterns, which suggest some functional reorganization, which may be the principle process responsible for recovery after stroke (Rossini et al., 2007). It is suggested that different post-ischemic interventions like physiotherapy, occupational therapy, speech therapy, electrical stimulation facilitate functional reorganization (Aichner et al., 2002).

## 3.4 Conclusion

Stroke is a complex neurological disorder that involves multiple pathological factors, including excitotoxicity, oxidative stress, neuroinflammation, gene expression. Present-day neuroprotection strategies disrupt the cellular, biochemical, and metabolic processes that lead to brain injury, either during or after ischemia, and encompass a wide and continually expanding array of drug-mediated interventions. Most stroke trials using one drug against one specific mechanism of oxidative damage have failed. Since the pathogenesis of ischemic injury involves multiple factors and interplay among excitotoxicity, oxidative stress, and neuroinflammation, the use of a cocktail of inhibitors, free radical scavengers, and anti-inflammatory agents at the earliest stages of ischemic injury may be required to substantively and persistently alter gene expression and interplay among excitotoxicity, oxidative stress, and neuroinflammation (Morimoto et al., 2002). Thus, a combination of inhibitors, free radical scavengers, and anti-inflammatory agents may modulate the neurochemical events associated with ischemic injury and result in a successful outcome from the ischemic insult.

To date, the neuroprotectant therapy is essentially restricted to prevent or limit neuronal damage in penumbra. The use of neural stem cells may provide the possibility of two new approaches: the transplantation of stem cells and the recruitment of endogenous stem cells for generating new neurons by means of proliferation/differentiation factors. In the latter approach, key regulators of stem cell survival, proliferation, and differentiation into neurons are proteins called "neurotrophic factors." Endogenous neurotrophic factors are actually produced in the penumbra, but this process is evidently insufficient or inadequate for providing the endogenous stem cells with the proper cues to correctly proliferate, differentiate into neurons, and migrate in the correct position to restore function. Therefore, modulating the levels of neurotrophic factors in penumbra areas through stem cell transplantation represents a new approach for the stroke therapy. Since a large number of neuroprotectants have failed in clinical trials and stem cell therapy for stroke is in initial stages, therefore prevention has become an important strategy to limit the onset and recurrence of stroke. Targets for prevention include modifiable risk factors such as hypertension, diabetes mellitus, dyslipidemia, cigarette smoking, obesity, alcohol use, and physical inactivity. Brain functional imaging studies show that partial recovery from strokes during rehabilitation is associated with a marked reorganization of the activation patterns of specific brain structures. Development of neuroimaging techniques has allowed the understanding of brain physiology during the stroke recovery process to provide a solid rationale for development of rehabilitation protocols, which can provide maximum benefit for stroke patients. Through neuroimaging, it will be possible to design, optimize, and synchronize functional training of brain regions ascribed to those areas innately undergoing neuronal plasticity change responsible for stroke recovery.

## References

Adibhatla RM, Hatcher JF, Dempsey RJ (2004) Cytidine-5′-diphosphocholine affects CTP-phosphocholine cytidylyltransferase and lyso-phosphatidylcholine after transient brain ischemia. J Neurosci Res 76:390–396

Adibhatla RM, Hatcher JF, Larsen EC, Chen X, Sun D, Tsao FH (2006) CDP-choline significantly restores phosphatidylcholine levels by differentially affecting phospholipase $A_2$ and CTP: phosphocholine cytidylyltransferase after stroke. J Biol Chem 281:6718–6725

Aichner F, Adelwohrer C, Haring HP (2002) Rehabilitation approaches to stroke. J Neural Transm Suppl 63:59–73

Alberts MJ (1999) Diagnosis and treatment of ischemic stroke. Am J Med 106:211–221

Alvarez XA, Mouzo R, Pichel V, Pérez P, Laredo M, Fernández-Novoa L, Corzo L, Zas R, Alcaraz M, Secades JJ, Lozano R, Cacabelos R (1999) Double-blind placebo-controlled study with citicoline in APOE genotyped Alzheimer's disease patients. Effects on cognitive performance, brain bioelectrical activity and cerebral perfusion. Methods Find Exp Clin Pharmacol 21: 633–644

Arita M, Oh SF, Chonan T, Hong S, Elangovan S, Sun YP, Uddin J, Petasis NA, Serhan CN (2006) Metabolic inactivation of resolvin $E_1$ and stabilization of its anti-inflammatory actions. J Biol Chem 281:22847–22854

Arita M, Ohira T, Sun YP, Elangovan S, Chiang N, Serhan CN (2007) Resolvin E1 selectively interacts with leukotriene $B_4$ receptor $BLT_1$ and ChemR23 to regulate inflammation. J Immunol 178:3912–3917

Asahi M, Huang Z, Thomas S, Yoshimura S, Sumii T, Mori T, Qiu J, Amin-Hanjani S, Huang PL, Liao JK, Lo E, Moskowitz MA (2005) Protective effects of statins involving both eNOS and tPA in focal cerebral ischemia. J Cereb Blood Flow Metab 25:722–729

Avrova NF, Zakharova IO, Tyurin VA, Tyurina YY, Gamaley IA, Schepetkin IA (2002) Different metabolic effects of ganglioside GM1 in brain synaptosomes and phagocytic cells. Neurochem Res 27:751–759

Barone FC (2009) Ischemic stroke intervention requires mixed cellular protection of the penumbra. Curr Opin Invest Drugs 10:220–223

Barrachina M, Secades J, Lozano R, Gómez-Santos C, Ambrosio S, Ferrer I (2002) Citicoline increases glutathione redox ratio and reduces caspase-3 activation and cell death in staurosporine-treated SH-SY5Y human neuroblastoma cells. Brain Res 957:84–90

Barrachina M, Domínguez I, Ambrosio S, Secades J, Lozano R, Ferrer I (2003) Neuroprotective effect of citicoline in 6-hydroxydopamine-lesioned rats and in 6-hydroxydopamine-treated SH-SY5Y human neuroblastoma cells. J Neurol Sci 215:105–110

Bas O, Songur A, Sahin O, Mollaoglu H, Ozen OA, Yaman M, Eser O, Fidan H, Yagmurca M (2007) The protective effect of fish n-3 fatty acids on cerebral ischemia in rat hippocampus. Neurochem Int 50:548–554

Bashkatova VG, Koshelev VB, Fadyukova OE, Alexeev AA, Vanin AF, Rayevsky KS, Ashmarin IP, Armstrong DM (2001) Novel synthetic analogue of ACTH 4-10 (Semax) but not glycine prevents the enhanced nitric oxide generation in cerebral cortex of rats with incomplete global ischemia. Brain Res 94:145–149

Bazan NG (2005) Neuroprotectin $D_1$ ($NPD_1$): a DHA-derived mediator that protects brain and retina against cell injury-induced oxidative stress. Brain Pathol 15:159–166

Bazan NG (2007) Omega-3 fatty acids, pro-inflammatory signaling and neuroprotection. Curr Opin Clin Nutr Metab Care 10:136–141

Bazan NG (2009) Neuroprotectin $D_1$-mediated anti-inflammatory and survival signaling in stroke, retinal degenerations, and Alzheimer's disease. J Lipid Res 50(Suppl):S400–S405

Bergeron M, Yu AY, Soleay K, Semenza GL, Sharp FR (1999) Induction of hypoxia-inducible factor-1 (HIF-1) and its target genes following focal ischaemia in rat brain. Eur J Neurosci 11:4159–4170

Blanco M, Lizasoain I, Sobrino T, Vivancos J, Castillo J (2006) Ischemic preconditioning: a novel target for neuroprotective therapy. Cerebrovasc Dis 21(Suppl 2):38–47

Bolognini N, Pascual-Leone A, Fregni F (2009) Using non-invasive brain stimulation to augment motor training-induced plasticity. J Neuroeng Rehabil 6:8

Borlongan CV, Hadman M, Sanberg CD, Sanberg PR (2004) Central nervous system entry of peripherally injected umbilical cord blood cells is not required for neuroprotection in stroke. Stroke 35:2385–2389

Candelise L, Ciccone A (2001) Gangliosides for acute ischaemic stroke. Cochrane Database Syst Rev 4:CD000094

Cao D, Yang B, Hou L, Xu J, Xue R, Sun L, Zhou C, Liu Z (2007) Chronic daily administration of ethyl docosahexaenoate protects against gerbil brain ischemic damage through reduction of arachidonic acid liberation and accumulation. J Nutr Biochem 18:297–304

Carter AJ, Grauert M, Pschorn U, Bechtel WD, Bartmann-Lindholm C, Qu Y, Scheuer T, Catterall WA, Weiser T (2000) Potent blockade of sodium channels and protection of brain tissue from ischemia by BIII 890 CL. Proc Natl Acad Sci USA 97:4944–4949

Carro E, Trejo JL, Busiguina S, Torres-Aleman I (2001) Circulating insulin-like growth factor I mediates the protective effects of physical exercise. J Neurosci 21:5678–5684

Chacon MR, Jensen MB, Sattin JA, Zivin JA (2008) Neuroprotection in cerebral ischemia: emphasis on the SAINT trial. Curr Cardiol Rep 10:37–42

Chang YC, Shyu WC, Lin SZ, Li H (2007) Regenerative therapy for stroke. Cell Transplant 16:171–181

Chiang N, Serhan CN, Dahlén SE, Drazen JM, Hay DW, Rovati GE, Shimizu T, Yokomizo T, Brink C (2006) The lipoxin receptor ALX: potent ligand-specific and stereoselective actions in vivo. Pharmacol Rev 58:463–487

Chudzik W, Kaczorowska B, Chiemlewski H, Przybyla M, Galka M (2005) Statins and stroke. Pol Merkur Lekarski 19:591–595

Clark WM (2009) Efficacy of citicoline as an acute stroke treatment. Exp Opin Pharmacother 10:839–846

Cordain L, Eaton SB, Sebastian A, Mann N, Lindeberg S, Watkins BA, O'Keefe J, Brand-Miller JH (2005) Origins and evolution of the Western diet: health implications for the 21st century. Am J Clin Nutr 81:341–354

Culman J, Zhao Y, Gohlke P, Herdegen T (2007) PPAR-gamma: therapeutic target for ischemic stroke. Trends Pharmacol Sci 28:244–249

Diener HC, Schneider D, Lampl Y, Bornstein NM, Kozak A, Rosenberg G (2008) DP-b99, a membrane-activated metal ion chelator, as neuroprotective therapy in ischemic stroke. Stroke 39:1774–1778

Ding O, Vaynman S, Akhavan M, Ying Z, Gomez-Pinilla F (2006) Insulin-like growth factor I interfaces with brain-derived neurotrophic factor-mediated synaptic plasticity to modulate aspects of exercise-induced cognitive function. Neurosci 140:823–833

Dishman RK, Berthoud HR, Booth FW, Cotman CW, Edgerton VR, Fleshner MR, Gandevia SC, Gomez-Pinilla F, Greenwood BN, Hillman CH, Kramer AF, Levin BE, Moran TH,

Russo-Neustadt AA, Salamone JD, Van Hoomissen JD, Wade CE, York DA, Zigmond MJ (2006) Neurobiology of exercise. Obesity (Silver Spring) 14:345–356
During MJ, Symes CW, Lawlor PA, Lin J, Dunning J, Fitzsimons HL, Poulsen D, Leone P, Xu R, Dicker BL, Lipski J, Young D (2000) An oral vaccine against NMDAR1 with efficacy in experimental stroke and epilepsy. Science 287:1453–1460
Endres M (2005) Statins and stroke. J Cereb Blood Flow Metab 25:1093–1110
Fagan SC, Bowes MP, Berri SA, Zivin JA (1999) Combination treatment for acute ischemic stroke: a ray of Hope? J Stroke Cerebrovasc Dis 8:359–367
Fan X, Heijnen CJ, van der Kooij MA, Groendaal F, van Bel F (2009) The role and regulation of hypoxia-inducible factor-1alpha expression in brain development and neonatal hypoxic-ischemic brain injury. Brain Res Rev 62:99–108
Farooqui AA, Haun SE, Horrocks LA (1994) Ischemia and hypoxia. In: Siegel GJ, Agranoff BW, Albers RW, Molinoff PB (eds) Basic neurochemistry: molecular, cellular, and medical aspects. Raven Press, New York, NY, pp 867–883
Farooqui AA, Ong WY, Horrocks LA (2006) Inhibitors of brain phospholipase $A_2$ activity: their neuropharmacological effects and therapeutic importance for the treatment of neurologic disorders. Pharmacol Rev 58:591–620
Farooqui AA, Horrocks LA (2007) Glycerophospholipids in Brain. Springer, New York, NY
Farooqui AA, Horrocks LA, Farooqui T (2007a) Modulation of inflammation in brain: a matter of fat. J Neurochem 101:577–599
Farooqui AA, Ong WY, Horrocks LA, Chen P, Farooqui T (2007b) Comparison of biochemical effects of statins and fish oil in brain: the battle of the titans. Brain Res Rev 56: 443–471
Farooqui AA, Ong WY, Horrocks LA (2008) Neurochemical aspects of excitotoxicity. Springer, New York, NY
Farooqui AA (2009a) Hot topics in neural membrane lipidology. Springer, New York, NY
Farooqui AA (2009b) Beneficial effects of fish oil on human brain. Springer, New York, NY
Farooqui T, Farooqui AA (2009) Aging: an important factor for the pathogenesis of neurodegenerative diseases. Mech Ageing Dev 130:203–215
Fernandes JS, Mori MA, Ekuni R, Oliveira RM, Milani H (2008) Long-term treatment with fish oil prevents memory impairments but not hippocampal damage in rats subjected to transient, global cerebral ischemia. Nutr Res 28:798–808
Feuerstein GZ, Ruffolo RR (2007) Discontinued drugs in 2006: cardiovascular drugs translational medicine perspective. Expert Opin Invest Drugs 16:1315–1326
Feuerstein GZ, Zaleska MM, Krams M, Wang X, Day M, Rutkowski JL, Finklestein SP, Pangalos MN, Poole M, Stiles GL, Ruffolo RR, Walsh FL (2008) Missing steps in the STAIR case: a translational medicine perspective on the development of NXY-059 for treatment of acute ischemic stroke. J Cereb Blood Flow Metab 28:217–219
Fioravanti M, Yanagi M (2005) Cytidinediphosphocholine (CDP-choline) for cognitive and behavioural disturbances associated with chronic cerebral disorders in the elderly. Cochrane Database Syst Rev 2:CD000269
Ford GA (2008) Clinical pharmacological issues in the development of acute stroke therapies. Br J Pharmacol 153(Suppl 1):S112–S119
Gilgun-Sherki Y, Melamed E, Offen D (2006) Anti-inflammatory drugs in the treatment of neurodegenerative diseases: current state. Curr Pharmaceut Des 12:3509–3519
Gilroy DW, Lawrence T, Perretti M, Rossi AG (2004) Inflammatory resolution: new opportunities for drug discovery. Nat Rev Drug Discov 3:401–416
Giménez R, Aguilar J (2003) Effects of cytidine 5′-diphosphocholine on plasma homocysteine levels in rat. Comp Biochem Physiol B Biochem Mol Biol 134:271–276
Glodstein LB, Adams R, Alberts MJ, Degraba TJ, Gorelick PB, Guyton JR, Hart RG, Howard G, Kelly-Hayes M, Nixon JV, Sacco RL (2006) Primary prevention of ischemic stroke: a guideline from the American Heart Association/American Stroke Association Stroke Council: cosponsored by the Atherosclerotic Peripheral Vascular Disease Interdisciplinary Working Group; Cardiovascular Nursing Council; Clinical Cardiology Council; Nutrition, Physical Activity,

and Metabolism Council; and the Quality of Care and Outcomes Research Interdisciplinary Working Group. Stroke 37:1583–1633

Goettl VM, Zhang H, Burrows AC, Wemlinger TA, Neff NH, Hadjiconstantinou M (2003) GM1 enhances dopaminergic markers in the brain of aged rats. Exp Neurol 183: 665–672

Goldberg WJ, Dorman RV, Dabrowiecki Z, Horrocks LA (1985) The effects of ischemia and CDPamines on $Na^{+}$, $K^{+}$-ATPase and acetylcholinesterase activities in rat brain. Neurochem Pathol 3:237–248

Govender P, Kalra L (2007) Benefits of occupational therapy in stroke rehabilitation. Expert Rev Neurother 7:1013–1019

Green AR, Ashwood T (2005) Free radical trapping as a therapeutic approach to neuroprotection in stroke: experimental and clinical studies with NXY-059 and free radical scavengers. Curr Drug Targets CNS Neurol Disord 4:109–118

Green AR (2008) Pharmacological approaches to acute ischaemic stroke: reperfusion certainly, neuroprotection possibly. Br J Pharmacol 153(Suppl 1):S325–S338

Greisenegger S, Müllner M, Tentschert S, Lang W, Lalouschek W (2004) Effect of pretreatment with statins on the severity of acute ischemic cerebrovascular events. J Neurol Sci 221:5–10

Grotta JC, Picone CM, Ostrow PT, Strong RA, Earls RM, Yao LP, Rhoades HM, Dedman. JR (1990) CGS-19755, a competitive NMDA receptor antagonist, reduces calcium-calmodulin binding and improves outcome after global cerebral ischemia. Ann Neurol 27: 612–619

Hess DC, Borlongan CV (2008) Stem cells and neurological diseases. Cell Prolif 41(Suppl 1): 94–114

Hicks A, Jolkkonen J (2008) Challenges and possibilities of intravascular cell therapy in stroke. Acta Neurobiol Exp 68:1–10

Higashi Y (2009) Edaravone for the treatment of acute cerebral infarction: role of endothelium-derived nitric oxide and oxidative stress. Expert Opin Pharmacother 10:323–331

Ho PW, Reutens DC, Phan TG, Wright PM, Markus R, Indra I, Young D, Donnan GA (2005) Is white matter involved in patients entered into typical trials of neuroprotection? Stroke 36: 2742–2744

Horrocks LA, Farooqui AA (2004) Docosahexaenoic acid in the diet: its importance in maintenance and restoration of neural membrane function. Prostaglandins Leukot Essent Fatty Acids 70:361–372

Hossain MS, Hashimoto M, Masumura S (1998) Influence of docosahexaenoic acid on cerebral lipid peroxide level in aged rats with and without hypercholesterolemia. Neurosci Lett 244:157–160

Hossain MS, Hashimoto M, Gamoh S, Masumura S (1999) Antioxidative effects of docosahexaenoic acid in the cerebrum versus cerebellum and brainstem of aged hypercholesterolemic rats. J Neurochem 72:1133–1138

Jang YG, Ilodigwe D, Macdonald RL (2009) Metaanalysis of tirilazad mesylate in patients with aneurysmal subarachnoid hemorrhage. Neurocrit Care 10:141–147

Joseph JA, Shukitt-Hale B, Lau FC (2007) Fruit polyphenols and their effects on neuronal signaling and behavior in senescence. Ann N Y Acad Sci 1100:470–485

Kamei K, Maeda N, Ogino R, Koyama M, Nakajima M, Tatsuoka T, Ohno T, Inoue T (2001) New 5-HT1A receptor agonists possessing 1,4-benzoxazepine scaffold exhibit highly potent anti-ischemic effects. Bioorg Med Chem Lett 11:595–598

Kantarci A, Van Dyke TE (2003) Lipoxins in chronic inflammation. Crit Rev Oral Biol Med 14: 4–12

Kapadia R, Yi JH, Vemuganti R (2008) Mechanisms of anti-inflammatory and neuroprotective actions of PPAR-gamma agonists. Front Biosci 13:1813–1826

Kawasaki-Yatssugi S, Ichiki C, Yatsugi S, Takahashi M, Shimizu-Sasamata M, Yamaguchi T, Minematsu K (2000) Neuroprotective effects of an AMPA receptor antagonist YM872 in a rat transient middle cerebral artery occlusion model. Neuropharmacol 39:211–217

Kim D, Jahan R, Starkman S, Abolian A, Kidwell CS, Vinuela F, Duckwiler GR, Ovbiagele B, Vespa B, Seico S, Rajajee V, Saver JL (2006) Endovascular mechanical clot retrieval in a broad ischemic stroke cohort. AJNR Am J Neuroradiol 27:2048–2052

Krupinski J, Ferrer I, Barrachina M, Secades JJ, Mercadal J, Lozano R (2002) CDP-choline reduces pro-caspase and cleaved caspase-3 expression, nuclear DNA fragmentation, and specific PARP-cleaved products of caspase activation following middle cerebral artery occlusion in the rat. Neuropharmacology 42:846–854

Labiche LA, Grotta JC (2004) Clinical trials for cytoprotection in stroke. NeuroRx 1:46–70

Lapchak PA, Zivin JA (2009) The lipophilic multifunctional antioxidant edaravone (radicut) improves behavior following embolic strokes in rabbits: a combination therapy study with tissue plasminogen activator. Exp Neurol 215:95–100

Lau FC, Shukitt-Hale B, Joseph JA (2007) Nutritional intervention in brain aging: reducing the effects of inflammation and oxidative stress. Subcell Biochem 42:299–318

Ledeen RW, Wu G (2002) Ganglioside function in calcium homeostasis and signaling. Neurochem Res 27:637–647

Lees KR, Zivin JA, Ashwood T, Davalos A, Davis SM, Diener HC, Grotta J, Lyden P, Shuaib A, Hårdemark HG, Wasiewski WW (2006) Stroke-Acute Ischemic NXY Treatment (SAINT I) Trial Investigators. NXY-059 for acute ischemic stroke. N Engl J Med 354:588–600

Lo EH, Dalkara T, Moskowitz MA (2003) Mechanisms, challenges and opportunities in stroke. Nat Rev Neurosci 4:399–415

Loor G, Schumacker PT (2008) Role of hypoxia-inducible factor in cell survival during myocardial ischemia-reperfusion. Cell Death Differ 15:686–690

Lopez-Miranda J, Delgado-Lista J, Perez-Martinez P, Jimenez-Gómez Y, Fuentes F, Ruano J, Marin C (2007) Olive oil and the haemostatic system. Mol Nutr Food Res 51:1249–1259

McCulloch J, Dewar D (2001) A radical approach to stroke therapy. Proc Natl Acad Sci USA 98:10989–10991

Mir C, Clotet J, Aledo R, Durany N, Argemí J, Lozano R, Cervós-Navarro J, Casals N (2003) CDP-choline prevents glutamate-mediated cell death in cerebellar granule neurons. J Mol Neurosci 20:53–60

Morimoto K, Murasugi T, Oda T (2002) Acute neuroinflammation exacerbates excitotoxicity in rat hippocampus in vivo. Exp Neurol 177:95–104

Newman MB, Willing AE, Manresa JJ, Davis-Sanberg C, Sanberg PR (2005) Stroke-induced migration of human umbilical cord blood cells: time course and cytokines. Stem Cell Dev 14:576–586

No author listed (2000) Tirilazad mesylate in acute ischemic stroke: a systematic review. Tirilazad Int Steering Comm Stroke 31:2257–2265

Norel X, Brink C (2004) The quest for new cysteinyl-leukotriene and lipoxin receptors: recent clues. Pharmacol Ther 103:81–94

Okada M, Amamoto T, Tomonaga M, Kawachi A, Yazawa K, Mine K, Fujiwara M (1996) The chronic administration of docosahexaenoic acid reduces the spatial cognitive deficit following transient forebrain ischemia in rats. Neuroscience 71:17–25

Okajima K, Harada N (2008) Promotion of insulin-like growth factor-I production by sensory neuron stimulation; molecular mechanism(s) and therapeutic implications. Curr Med Chem 15:3095–3112

Ohtani R, Tomimoto H, Wakita H, Kitaguchi H, Nakaji K, Takahashi R (2007) Expression of S100 protein and protective effect of arundic acid on the rat brain in chronic cerebral hypoperfusion. Brain Res 1135:195–200

Papademetriou V, Doumas M (2009) Treatment strategies to prevent stroke: focus on optimal lipid and blood pressure control. Expert Opin Pharmacother 10:955–966

Parnetti L, Caso V, Lanari A, Saggese E, Sebastianelli M, Tayebati SK, Amenta F (2006) Stroke prevention and statin treatment. Clin Exp Hyperten 28:335–344

Patzer A, Zhao Y, Stock I, Gohlke P, Ferdegen T, Culman J (2008) Peroxisome proliferator-activated receptors gamma (PPARgamma) differently modulate the interleukin-6 expression in the peri-infarct cortical tissue in the acute and delayed phases of cerebral ischaemia. Eur J Neurosci 28:1786–1794

Pelleymounter MA, Cullen MJ, Baker MB, Gollub M, Wellman C (1996) The effects of intrahippocampal BDNF and NGF on spatial learning in aged Long Evans rats. Mol Chem Neuropathol 29:211–226

Pizzi M, Sarinico I, Boroni F, Benetti A, Benarese M, Spano PF (2005) Inhibition of IkappaBalpha phosphorylation prevents glutamate-induced NF-kappaB activation and neuronal cell death. Acta Neurochir Suppl 93:59–63

Plataras C, Angelogianni P, Tsakiris S (2003) Effect of CDP-choline on hippocampal acetylcholinesterase and $Na^+$,$K^+$-ATPase in adult and aged rats. Z Naturforsch 58:277–281

Ploughman M, Granter-Button S, Chernenko G, Attwood Z, Tucker BA, Mearow KM, Corbett D (2007) Exercise intensity influences the temporal profile of growth factors involved in neuronal plasticity following focal ischemia. Brain Res 1150:207–216

Ran R, Xu H, Lu A, Bernaudin M, Sharp FR (2005) Hypoxia preconditioning in the brain. Dev Neurosci 27:87–92

Rincon F, Sacco RL (2008) Secondary stroke prevention. J Cardiovasc Nurs 23:34–41

Romero JR (2007) Prevention of ischemic stroke: overview of traditional risk factors. Curr Drug Targets 8:794–801

Romero JR, Morris J, Pikula A (2008) Stroke prevention: modifying risk factors. Ther Adv Cardiovasc Dis 2:287–303

Rossini PM, Altamura C, Ferreri F, Melgari JM, Tecchio M, Pasgualetti P, Vernieri F (2007) Neuroimaging experimental studies on brain plasticity in recovery from stroke. Eura Midicophys 43:241–254

Sare GM, Geeganage C, Bath PM (2009) High blood pressure in acute ischaemic stroke-broadening therapeutic horizons. Cerebrovasc Dis 27(Suppl 1):156–161

Saver JL (2008) Citicoline: update on a promising and widely available agent for neuroprotection and neurorepair. Rev Neurol Dis 5:167–177

Schäbitz WR, Li F, Fisher M (2000) The *N*-methyl-D-aspartate antagonist CNS 1102 protects cerebral gray and white matter from ischemic injury following temporary focal ischemia in rats. Stroke 31:1709–1714

Secades JJ, Lorenzo JL (2006) Citicoline pharmacological and clinical review, update. Methods Find Exp. Clin Pharmacol 28(Suppl B):1–56

Serhan CN, Arita M, Hong S, Gotlinger K (2004) Resolvins, docosatrienes, and neuroprotectins, novel omega-3-derived mediators, and their endogenous aspirin-triggered epimers. Lipids 39:1125–1132

Serhan CN (2005a) Novel ω-3-derived local mediators in anti-inflammation and resolution. Pharmacol Ther 105:7–21

Serhan CN (2005b) Novel eicosanoid and docosanoid mediators: resolvins, docosatrienes, and neuroprotectins. Curr Opin Clin Nutr Metab Care 8:115–121

Serhan CN, Brain SD, Buckley CD, Gilroy DW, Haslett C, O'Neill LAJ, Perretti M, Rossi AG, Wallace JL (2007) Resolution of inflammation: state of the art, definitions and terms. FASEB J 21:325–332

Sharp FR, Ran R, Lu A, Tang Y, Strauss KI, Glass T, Ardizzone T, Bernaudin M (2004) Hypoxic preconditioning protects against ischemic brain injury. NeuroRx 1:26–35

Shen WH, Zhang CY, Zhang GY (2003) Antioxidants attenuate reperfusion injury after global brain ischemia through inhibiting nuclear factor-kappa B activity in rats. Acta Pharmacol Sin 24:1125–1130

Shi H (2009) Hypoxia inducible factor-1 as a therapeutic target in ischemic stroke. Curr Med Chem 16:4593–4600

Shimizu-Sasamata M, Kawasaki-Yatsugi S, Okada M, Sakamoto S, Yatsugi S, Togami J, Hatanaka K, Ohmori J, Koshiya K, Usuda S, Murase K (1996) YM90K: pharmacological characterization as a selective and potent alpha-amino-3-hydroxy-5-methylisoxazole-4-propionate/kainate receptor antagonist. J Pharmacol Exp Ther 276:84–92

Shuaib A, Lees KR, Lyden P, Grotta J, Davalos A, Davis SM, Diener HC, Ashwood T, Wasiewski WW, Emeribe U SAINT II Trial Investigators (2007) NXY-059 for the treatment of acute ischemic stroke. N Engl J Med 357:562–571

Shuaib A, Hussain MS (2008) The past and future of neuroprotection in cerebral ischaemic stroke. Eur Neurol 59:4–14

Smith WS (2006) Safety of mechanical thrombectomy and intravenous tissue plasminogen activator in acute ischemic stroke. Results of the multi Mechanical Embolus Removal in Cerebral Ischemia (MERCI) trial, part I. AJNR Am J. Neuroradiol 27:1177–1182

Simopoulos AP (2006) Evolutionary aspects of diet, the omega-6/omega-3 ratio and genetic variation: nutritional implications for chronic diseases. Biomed Pharmacother 60:502–507

Singhal AB (2006) Oxygen therapy in stroke: past, present, and future. Int J Stroke 1:191–200

Singhal AB (2007) A review of oxygen therapy in ischemic stroke. Neurol Res 29:173–183

Sobrado M, López MG, Carceller F, García AG, Roda JM (2003) Combined nimodipine and citicoline reduce infarct size, attenuate apoptosis and increase bcl-2 expression after focal cerebral ischemia. Neuroscience 118:107–113

Strokin M, Chechneva O, Reymann KG, Reiser G (2006) Neuroprotection of rat hippocampal slices exposed to oxygen-glucose deprivation by enrichment with docosahexaenoic acid and by inhibition of hydrolysis of docosahexaenoic acid-containing phospholipids by calcium independent phospholipase $A_2$. Neuroscience 140:547–553

Su EJ, Fredriksson L, Geyer M, Folestad E, Cale J, Andrae J, Gao Y, Pietras K, Mann K, Yepes M, Strickland DK, Betsholtz C, Eriksson U, Lawrence DA (2008) Activation of PDGF-CC by tissue plasminogen activator impairs blood-brain barrier integrity during ischemic stroke. Nat Med 14:731–737

Sun GY, Horrocks LA, Farooqui AA (2007) The roles of NADPH oxidase and phospholipases $A_2$ in oxidative and inflammatory responses in neurodegenerative diseases. J Neurochem 103:1–16

Takeda H, Spatz M, Ruetzler C, McCarron R, Becker K, Hallenbeck J (2002) Induction of mucosal tolerance to E-selectin prevents ischemic and hemorrhagic stroke in spontaneously hypertensive genetically stroke-prone rats. Stroke 33:2156–2163

Tan DX, Manchester LC, Sainz R, Mayo JC, Alvares FL, Reiter RJ (2003) Antioxidant strategies in protection against neurodegenerative disorders. Expert Opin Ther Patents 13:1513–1543

Teal P, Silver FL, Simard D (2005) The BRAINS study: safety, tolerability, and dose-finding of repinotan in acute stroke. Can J Neurol Sci 32:61–67

Terano T, Fujishiro S, Ban T, Yamamoto K, Tanaka T, Noguchi Y, Tamura Y, Yazawa K, Hirayama T (1999) Docosahexaenoic acid supplementation improves the moderately severe dementia from thrombotic cerebrovascular diseases. Lipids 34 Suppl:S345–S346

Trejo JL, Carro E, Nunez A, Torres-Aleman I (2002) Sedentary life impairs self-reparative processes in the brain: the role of serum insulin-like growth factor-I. Rev Neurosci 13:365–374

Vaynman S, Ying Z, Gomez-Pinilla F (2004) Hippocampal BDNF mediates the efficacy of exercise on synaptic plasticity and cognition. Eur J Neurosci 20:2580–2590

Vaynman S, Gomez-Pinilla F (2006) Revenge of the "sit": how lifestyle impacts neuronal and cognitive health through molecular systems that interface energy metabolism with neuronal plasticity. J Neurosci Res 84:699–715

Vaynman S, Ying Z, Wu A, Gomez-Pinilla F (2006) Coupling energy metabolism with a mechanism to support brain-derived neurotrophic factor-mediated synaptic plasticity. Neuroscience 139:1221–1234

Vendrame M, Cassady J, Newcomb J, Butler T, Pennypacker KR, Zigova T, Sanberg CD, Sanberg PR, Willing AE (2004) Infusion of human umbilical cord blood cells in a rat model of stroke dose-dependently rescues behavioral deficits and reduces infarct volume. Stroke 35:2390–2395

Vendrame M, Gemma C, de Mesquita D, Collier L, Bickford PC, Sanberg CD, Sanberg PR, Pennypacker KR, Willing AE (2005) Anti-inflammatory effects of human cord blood cells in a rat model of stroke. Stem Cell Dev 14:595–604

Wang JY, Wen LL, Huang YN, Chen YT, Ku MC (2006) Dual effects of antioxidants in neurodegeneration: direct neuroprotection against oxidative stress and indirect protection via suppression of glia-mediated inflammation. Curr Pharmaceut Design 12:3521–3533

Wardlaw JM, Zoppo G, Yamaguchi T, Berge E (2003) Thrombolysis for acute ischaemic stroke. Cochrane Database Syst Rev 3:CD000213

Webster BR, Celnik PA, Cohen LG (2006) Noninvasive brain stimulation in stroke rehabilitation. NeuroRx 3:474–481

Wen W, Sachdev PS (2004) Extent and distribution of white matter hyperintensities in stroke patients: the Sydney Stroke Study. Stroke 35:2813–2819

Xing B, Xin T, Hunter RL, Bing G (2008) Pioglitazone inhibition of lipopolysaccharide-induced nitric oxide synthase is associated with altered activity of p38 MAP kinase and PI3K/Akt. J Neuroinflammation 5:4

Yang Y, Li Q, Yang T, Hussain M, Shuaib A (2003) Reduced brain infarct volume and improved neurological outcome by inhibition of the NR2B subunit of NMDA receptors by using CP101,606-27 alone and in combination with rt-PA in a thromboembolic stroke model in rats. J Neurosurg 98:397–403

Yoshida H, Yanai H, Namiki Y, Fukatsu-Sasaki K, Furutani N, Tada N (2006) Neuroprotective effects of edaravone: a novel free radical scavenger in cerebrovascular injury. CNS Drug Rev 12:9–20

Yu G, Borlongan CV, Stahl CE, Hess DC, Ou Y, Kaneko Y, Yu SJ, Yang T, Fang L, Xie X (2009) Systemic delivery of umbilical cord blood cells for stroke therapy: a review. Restor Neurol Neurosci 27:41–54

Zhao Y, Patzer A, Herdegen T, Gohlke P, Culman J (2006) Activation of cerebral peroxisome proliferator-activated receptors gamma promotes neuroprotection by attenuation of neuronal cyclooxygenase-2 overexpression after focal cerebral ischemia in rats. FASEB J 20:1162–1175

Zhou Y, Xiong Y, Yuan SY (2009) Effect of tacrolimus on growth-associated protein-43 expression in the hippocampus of neonatal rats with hypoxic-ischemic brain damage. Zhongguo Dang Dai Er Ke Za Zhi 11:65–68

Zivin JA (1998) Factors determining the therapeutic window for stroke. Neurology 50:599–603

# Chapter 4
# Neurochemical Aspects of Spinal Cord Injury

## 4.1 Introduction

Spinal cord injury (SCI) is a catastrophic event resulting in the loss of motor and sensory functions of the body innervated by the spinal cord below the injury site. Trauma to the spinal cord induces autodestructive changes that lead to varying degrees of tissue necrosis and paralysis, depending on the severity of the injury, which consists of two broadly defined events: a primary event, attributable to the mechanical insult itself, and a secondary event, attributable to the series of systemic and local neurochemical and pathophysiological changes that occur in spinal cord after the initial traumatic insult (Klussmann and Martin-Villalba, 2005). Unlike the primary event, which is instantaneous and beyond therapeutic management, the secondary event develops over the hours and days after SCI, causing behavioral and functional impairments. At the core of primary injury site, SCI causes a rapid deformation of spinal cord tissue due to compression, contusion, and laceration due to penetrating injury along with acute stretching of the spinal cord as a result of iatrogenic vertebral distraction, rupturing of neural cell membranes resulting in the release of neuronal intracellular contents (Sekhon and Fehlings, 2001). In contrast, secondary event that occurs into rostral/caudal spinal levels include many neurochemical alterations (ischemia, edema, increase in excitatory amino acids, and reactive oxygen species). These neurochemical alterations not only effect neuronal activities and glial cell reaction associated with astrocytic activation, and demyelination involving oligodendrocytes, but also modulate leukocyte infiltration, and activation of macrophages and vascular endothelial cells (Bramlett and Dietrich, 2004). Among non-neural cells following SCI, macrophages are present at the injury site in large numbers and for the longer duration.

Interactions between neural and non-neural cells are essential for endogenous restructuring and repairing injured spinal cord tissue (Popovich et al., 1999). In fact, maintenance and repair of injured neurons at the injury site surrounding area depends on the active assistance from immune cells. SCI also triggers a systemic, neurogenic immune depression syndrome characterized by a rapid and drastic decrease of CD14+ monocytes, CD3+ T-lymphocytes, and CD19+ B-lymphocytes and MHC class II (HLA-DR)+ cells within 24 h reaching minimum levels within the

A.A. Farooqui, *Neurochemical Aspects of Neurotraumatic and Neurodegenerative Diseases*, DOI 10.1007/978-1-4419-6652-0_4,

first week (Reigger et al., 2009). This suggests that SCI is associated with an early onset of immune suppression and secondary immune deficiency syndrome (SCI-IDS). In addition, SCI also induces the synthesis of autoantibodies that bind nuclear antigens including DNA and RNA (Ankeny and Popovich, 2009). This observation is similar to the elevation of anti-DNA antibodies in systemic lupus erythematosus. It is likely that SCI-induced antibodies may show a similar pathologic potential as that of autoantibodies in systemic lupus erythematosus (Ankeny et al., 2006). During restructuring and repairing process, released glutamate, reactive oxygen and nitrogen species (ROS and RNS), cytokine, and proteases initiate damage to surrounding healthy neurons in the vicinity of injury site. Thus, accumulating evidence suggests that secondary event associated with SCI involves interactions among excitotoxicity (a process by which high levels of glutamate induce neurodegeneration), oxidative stress (a process involving cytotoxic consequences initiated and caused by oxygen-free radicals), and neuroinflammation (a neuroprotective mechanism whose prolonged presence is injurious to neurons) (Farooqui and Horrocks, 1994; Leker and Shohami, 2002; Block and Hong, 2005; Farooqui et al., 2007; Farooqui and Horrocks, 2007; Farooqui et al., 2008; Chan, 2008; Farooqui and Horrocks, 2009) (Fig. 4.1). In SCI, commencement of excitotoxicity, oxidative stress, and neuroinflammation is supported by alterations in ion homeostasis, changes in cellular

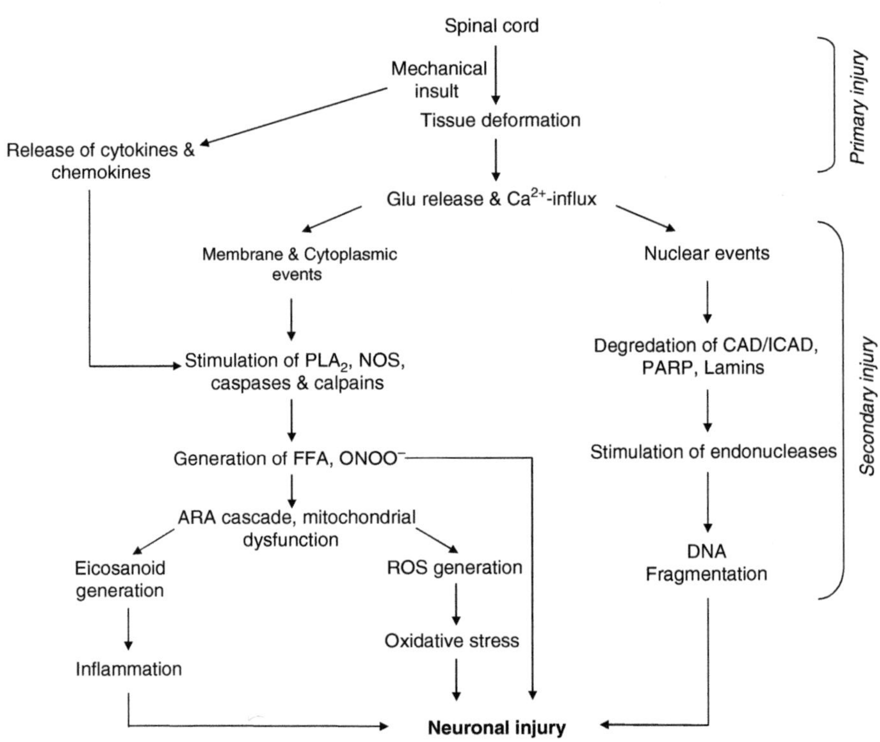


**Fig. 4.1** Diagram showing neurochemical changes associated with primary and secondary events in spinal cord injury

redox, mitochondrial dysfunction, induction of neurodestructive and neuroprotective genes, alterations in enzymic activities, and changes in neurotrophic factor expression. In addition, SCI also triggers an early and prolonged inflammatory response, with increased TNF-α and interleukin-1β levels. Transient changes are observed in subunit populations of the transcription factor nuclear factor-kappaB (NF-κB), which plays a key role in regulating inflammation in brain and spinal cord pathologies (Fig. 4.2) (Farooqui and Horrocks, 2009).

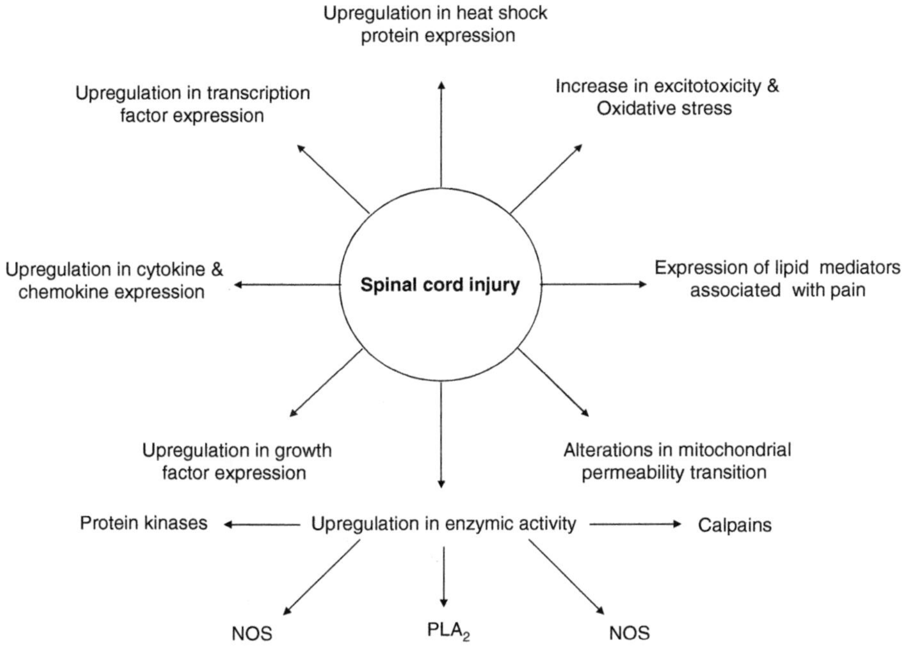


**Fig. 4.2** Neurochemical changes associated with spinal cord injury

## 4.2 Regeneration and Neuritogenesis in SCI

Functional recovery in SCI victims is very limited because injured axons within the spinal cord do not regenerate spontaneously and do not respond to therapeutic strategies. This is because of induction of myelin-associated glycoproteins, MAG and Nogo, at the injury site. These proteins obstruct axonal regeneration of injured neurons (Skaper et al., 2001; McKerracher and Winton, 2002; Watkins and Barres, 2002; Filbin, 2003; Eftekharpour et al., 2008). Although the exact molecular mechanism associated with the obstruction of axonal regeneration is not fully understood, interactions among MAG and its receptor (MAG receptor), Nogo and its receptor (NgR1), and the neurotrophin receptor $p75^{NTR}$ modulate axonal growth and growth cone mortality through the involvement of RhoA activity. In addition, LINGO-1 (a nervous system-specific transmembrane protein) also binds to NgR1–$p75^{NTR}$

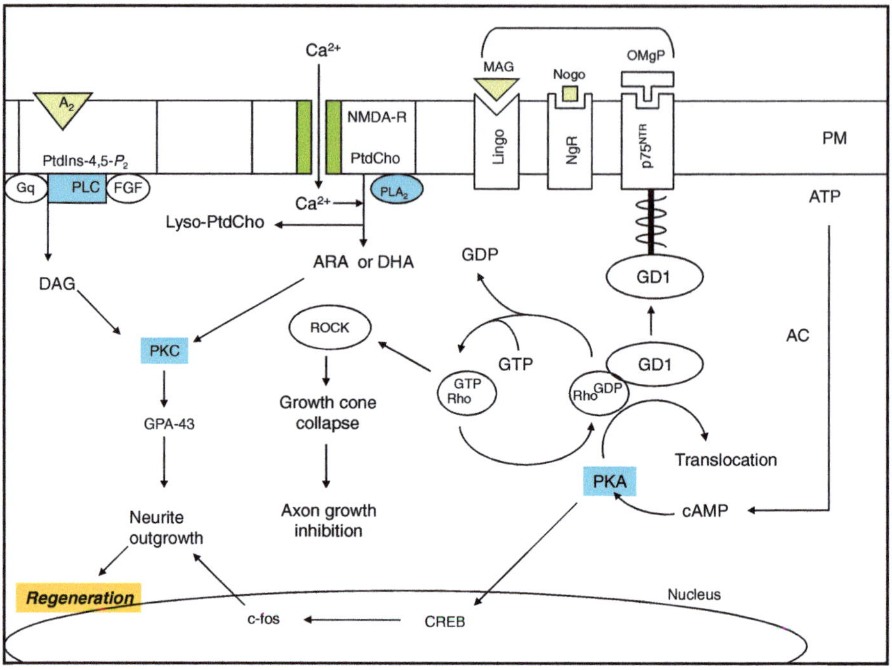


**Fig. 4.3** Extracellular signals, factors, and their receptors that modulate axonal regeneration and neurite outgrowth formation. Phosphatidylinositol 4,5-bisphosphate (PtdIns-4,5-$P_2$); phospholipase C (PLC); phospholipase $A_2$ ($PLA_2$); *N*-methyl-D-aspartate receptor (NMDA-R); phosphatidylcholine (PtdCho); lyso-phosphatidylcholine (lyso-PtdCho); fibroblast growth factor (FGF); specific transmembrane protein that binds NgR1 and $p75^{NTR}$ (Lingo); Nogo receptor (NgR); low-affinity neurotrophin receptor p75 ($p75^{NTR}$); arachidonic acid (ARA); docosahexaenoic acid (DHA); diacylglycerol (DAG); Rho-GDP dissociation inhibitor (GDI); serine/threonine kinases (ROCK); cyclic AMP (cAMP); cAMP-activated protein kinase (PKA); protein kinase C (PKC); growth-associated protein-43 (GAP-43); guanosine 5′-triphosphate (GTP); guanosine 5′-diphosphate (GDP); adenosine triphosphate (ATP); and adenylyl cyclase (AC)

complex, and impedes the axonal regeneration (Fig. 4.3). Collective evidence suggests that MAG, Nogo, and $p75^{NTR}$ receptors interact with each other and modulate downstream signal transduction net work. Nogo interacts with NgR1, and Rho-GDP dissociation inhibitor (Rho-GDI) is associated with $p75^{NTR}$. The dissociation of Rho-GDI with $p75^{NTR}$ allows the exchange of GTP with GDP resulting in activation of the Rho protein. Rho-GTP, a Rho GTPase, then activates ROCK, which phosphorylates other proteins involved in blocking neurite outgrowth formation and depolymerization of F-actin (Skaper et al., 2001; Ruff et al., 2008). In the absence of Nogo and Nogo receptor interactions, $p75^{NTR}$ is not activated and Rho-GDI remains bound to Rho-GDP. The Rho protein remains bound with GDP and remains inactive. Therefore, ROCK is not activated and cannot change transcription patterns to inhibit neuronal outgrowth. In contrast, induction of neurite outgrowth is facilitated by the activation of isoforms of $PLA_2$ activity. $PLA_2$ isoforms release arachidonic acid or docosahexaenoic acid. These fatty acids and their metabolites induce the expression of genes related to neurite outgrowth formation and differentiation (Ikemoto et al., 1997; Calderon and Kim, 2004; Cao et al., 2005; Farooqui, 2009).

Astrocytes contribute to the inhibition of regeneration by synthesizing multiple inhibitory proteoglycans, such as chondroitin sulfate proteoglycans (CSPGs) and facilitating the formation of a glial scar, a major obstacle to axonal growth after injury to the adult CNS and PNS. The inhibition of regeneration results in significant functional deficits and, depending on the severity of injury, may contribute to permanent paralysis or loss of senses distal to the site of injury. In addition, 40% of SCI patients develop persistent neuropathic pain, which has a detrimental impact on the patient's quality of life and is a major specific health-care problem in its own right. Nearly 250,000 Americans are currently living with SCI and approximately 12,000 new injuries occur every year. While SCI account for 2–3% death in USA per year, the victims that survive suffer from physical, cognitive, and emotional stress and the rehabilitation cost of these patients goes as high as $10–20 billion per year. The most frequent victims (40%) of spinal cord trauma are young men (18–30 years) injured in automobile and motor cycle accidents, sporting accidents, falls, and gunshot wounds.

## 4.3 Necrosis and Apoptosis in SCI

Apoptosis and necrosis are two basic mechanisms of cell death that occur in SCI (Liu et al., 1997; Farooqui et al., 2004; Farooqui, 2009). Mechanical trauma to spinal cord ruptures neuronal membranes and releases of intracellular components that induce inflammatory reaction. This type of cell death is a passive process characterized by the intense mitochondrial damage, rapid loss of ATP, sudden loss of ion homeostasis, and high levels of ROS. In contrast, apoptosis is an active process, where neurochemical changes occur in an orderly fashion. Apoptosis is characterized by the stimulation of enzymic activities (proteases, phospholipases, cyclooxygenases, nitric oxide synthases, and kinases), cell shrinkage, dynamic membrane blebbing, chromatin condensation, DNA laddering, loss of plasma membrane phospholipid asymmetry, maintenance of ATP, mitochondrial oxy-radical generation, and mild calcium overload (Sastry and Rao, 2000; Farooqui et al., 2004; Farooqui, 2009). Major components of the apoptotic pathway include, Bcl-2, an oncogene that antagonizes the apoptotic response, and caspases, a group of apoptotic-specific cysteine proteases that cleave their substrates with specificity after aspartic acid residues. Caspase activation promotes chromatin condensation, DNA fragmentation into nucleosomal fragments, and generation of apoptotic bodies. At least two pathways are involved in caspase activation. First pathway involves death receptors, such as Fas and a TNF receptor, and the second pathway for caspase activation is triggered by the release of cytochrome *c* from the mitochondria (Keane et al., 2006; Davis et al., 2007). Apoptotic proteins such as p53, Bax, and

Par-4 induce mitochondrial membrane permeability changes resulting in the release of cytochrome *c*. This release not only initiates caspase activation through Apaf-1 activation, but it also breaks the electron transfer chain causing reduction in ATP production. The dead cells are removed from the tissue through apoptotic body formation and phagocytosis (Sastry and Rao, 2000; Farooqui, 2009; Farooqui and Horrocks, 2009). It is becoming increasingly evident that at the end point, apoptosis and necrosis are two extremes of a wide spectrum of cell death processes with different mechanistic and morphological features (Farooqui et al., 2004; Farooqui, 2009). However, they also share some common lipid mediators and signal transduction processes that are often inseparable. Apoptotic cell death at the injury site occurs as early as 6 h and as long as 3 weeks after moderately severe SCI (Beattie et al., 2000). Majority of cells dying through apoptotic cell death are found in the white matter and increased in number both rostral and caudal to the injury site. Many of the degenerating oligodendrocytes and microglia are observed in the dying tracts. Since SCI blocks transmission of information through the lesion site, degeneration of oligodendrocytes may significantly contribute to neurological deficit after SCI. Necrosis normally occurs at the core of injury site in minutes following SCI, whereas neural cells undergo apoptosis several hours or days after injury in the surrounding area (Farooqui et al., 2004; Farooqui, 2009).

## 4.4 Contribution of Excitotoxicity in Spinal Cord Injury

Morphologically, excitotoxicity is characterized by neuronal swelling, vacuolization, and eventual neural cell death (Farooqui and Horrocks, 1994). In SCI, released glutamate interacts with glutamate receptors (excitatory amino acid receptors) (Panter et al., 1990; Farooqui and Horrocks, 1991; Auger and Attwell, 2000). As stated in Chapter 2, glutamate receptors are classified into *N*-methyl-D-aspartate (NMDA), α-amino-3-hydroxy-5-methyl-4-isoxazole propionate (AMPA), kainate (KA), and metabotropic glutamate receptors (Farooqui et al., 2008). Hyperstimulation of glutamate receptors allows calcium influx that initiates a cascade of events involving mitochondrial dysfunction, activation of enzymes associated with the release and oxidation of arachidonic acid, and generation of free radicals ($O_2^{\bullet-}$) and peroxynitrite, a toxic reaction product of NO and superoxide (Fig. 4.4) (Farooqui and Horrocks, 1991, 1994; Park et al., 2004; Farooqui et al., 2008). Thus, an uncontrolled and sustained increase in cytosolic calcium levels triggers the activation of phospholipase $A_2$ ($PLA_2$), cyclooxygenase-2 (COX-2), lipoxygenases (LOX), calpains, nitric oxide synthase (NOS), calcineurin, MAP kinase, matrix metalloproteinase (MMP), caspases, and poly(ADP-ribose) polymerase (PARP) (Table 4.1) (Bao et al., 2009; Pavel et al., 2001; Ray et al., 2003; Ellis et al., 2004; Knoblach et al., 2005; McEwen and Springer, 2005; Genovese et al., 2005; Lee et al., 2009; Gris et al., 2008). Increased degradation of neural membrane glycerophospholipid and accumulation of oxygenated arachidonic acid metabolites along with abnormal ion homeostasis, changes in redox status,

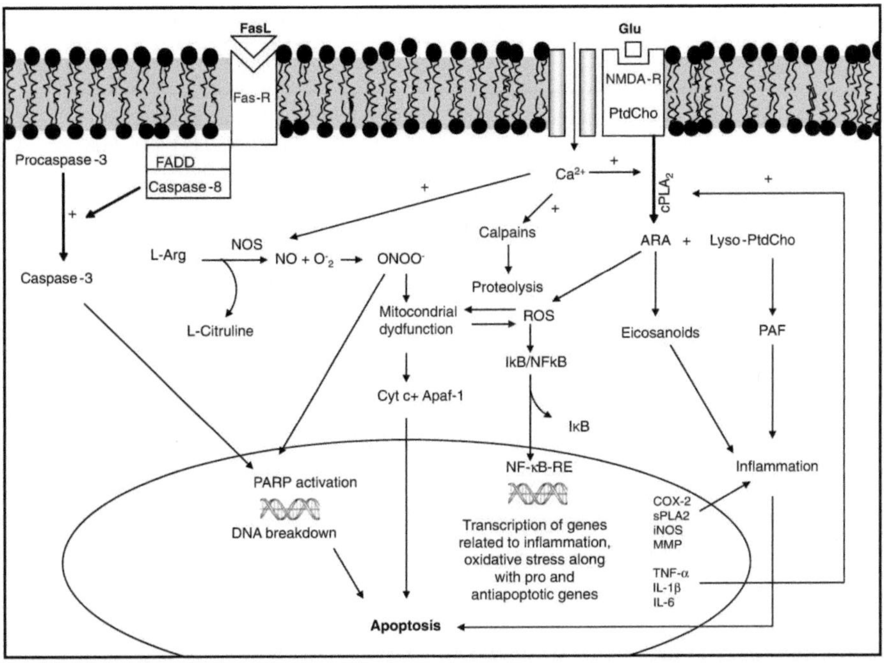


**Fig. 4.4** Involvement of Fas and NMDA receptors in apoptotic and necrotic cell death. Fas ligand (FasL); Fas receptor (Fas-R); *N*-methyl-D-aspartate receptor (NMDA-R); phosphatidylcholine (PtdCho); cytosolic phospholipase $A_2$ ($cPLA_2$); arachidonic acid (ARA); arginine (Arg); nitric oxide synthase (NOS); nitric oxide (NO); superoxide ($O_2^-$); peroxynitrite ($ONOO^-$); arachidonic acid (ARA); lyso-phosphatidylcholine (lyso-PtdCho); platelet-activating factor (PAF); cytochrome *c* (Cytc); apoptosome complex with apoptosis-activating factor-1 (Apaf-1); and poly(ADP)ribose polymerase (PARP); secretory phospholipase ($sPLA_2$); inducible nitric oxide synthase (iNOS); cyclooxygenase-2 (COX-2); matrix metalloproteinase (MMP); nuclear factor-kappa B (NF-κB); inhibitory form of nuclear factor kappa B (I-κB/NF-κB); nuclear factor κB-response element (NF-κB-RE); inhibitory subunit of NF-κB (I-κB); tumor necrosis factor-α (TNF-α); interleukin-1β (IL-1β); and interleukin-6 (IL-6)

cytoskeletal degradation, induction of extrinsic and intrinsic apoptotic pathways, and lack of energy generation are closely associated with neural cell injury in spinal cord trauma (Demediuk et al., 1985; Horrocks et al., 1985; Park et al., 2004; Farooqui et al., 2008).

Glutamate also mediates injury to glial cells through mechanism that does not involve glutamate receptor activation, but rather glutamate uptake (Matute et al., 2006). High concentration of glutamate interferes with cystine/glutamate antiporter, which normally transports cystine into the cell. Inhibition of cystine uptake leads to a decrease in the level of intracellular cystine and its reduction product cysteine, with consequent decrease in glutathione synthesis, accumulation of cellular oxidants, and eventual cell death (Matute et al., 2006; Farooqui et al., 2008). Collectively, these studies indicate that glutamate-mediated brain damage not only involves

**Table 4.1** Effect of $Ca^{2+}$ influx on enzymic activities in injured spinal cord

| Enzyme | Effect | References |
|---|---|---|
| $PLA_2$ | Increased | Liu et al. (2006), Huang et al. (2009) |
| COX-2 | Increased | Adachi et al. (2005), Gris et al. (2008) |
| Calpain | Increased | Ray et al. (2003) |
| Calcineurin | Increased | Springer et al. (2000) |
| MAP kinase | Increased | Esposito et al. (2009) |
| NOS | Increased | Chatzipanteli et al. (2002) |
| MMP | Increased | Buss et al. (2007), Esposito et al. (2008) |
| PARP | Increased | Genovese et al. (2005) |

Phospholipase $A_2$ ($PLA_2$); phospholipase C (PLC); cyclooxygenase-2 (COX-2); nitric oxide synthase (NOS); matrix metalloproteinase (MMP); and poly(ADP-ribose) polymerase (PARP).

interactions among excitotoxicity, oxidative glutamate toxicity but also mitochondrial dysfunction, decrease in ATP levels, and changes in neural cell redox (Farooqui and Horrocks, 1991, 1994).

## 4.5 Enzymic Activities in Spinal Cord Injury

SCI induces changes in activities of number of enzymes, including isoforms of $PLA_2$, calpains, nitric oxide synthases, cyclooxygenases, lipoxygenases, calcineurin, caspases, and matrix metalloproteinases (Fig. 4.4). Activation of isoforms of $PLA_2$, cyclooxygenase-2, and lipoxygenase contribute to the production of eicosanoids and other non-enzymic arachidonic acid-derived products, such as 4-hydroxynonenal and isoprostane (Farooqui and Horrocks, 2009). Activation of calpains not only causes a breakdown of the cytoskeleton proteins (spectrin and MAP2) but also mediates the conversion of xanthine dehydrogenase to xanthine oxidase. This helps in production of more free radicals (Farooqui and Horrocks, 1994). Activation of nitric oxide synthase (NOS) promotes the generation of peroxynitrite and finally activation of caspases in SCI is associated with apoptotic cell death (Farooqui et al., 2008) (Fig. 4.4).

### *4.5.1 Activation of $PLA_2$ in Spinal Cord Injury*

High levels of calcium-dependent cytosolic phospholipase $A_2$ ($cPLA_2$) activity have been reported to occur in rat and monkey spinal cords. At the cellular level, dense immunoreactivity is present in motor neurons from cervical, thoracic, lumbar, and sacral regions (Ong et al., 1999). Traumatic injury to spinal cord stimulates activities of lipases and phospholipases (Taylor et al., 1988). SCI significantly stimulates $cPLA_2$ activity and its expression in injured spinal cord. This increase in $cPLA_2$ activity can be blocked by the $PLA_2$ inhibitor, mepacrine (Liu et al., 2006).

Treatment with $PLA_2$ or melittin, an activator of endogenous $PLA_2$, increases spinal neuronal death in vitro, and this process is also substantially reversed by mepacrine. Results on microinjections of $PLA_2$ or melittin into the normal spinal cord cause harmful effects. $PLA_2$ mediates demyelination, whereas melittin diffuses and promotes tissue necrosis. Thus, melittin induces inflammation and oxidative stress-mediated tissue damage. Furthermore, $PLA_2$-mediated demyelination in SCI can be significantly reversed by mepacrine (Liu et al., 2006).

At the injury site, $PLA_2$ catalyzed reaction product, arachidonate (ARA), is metabolized to neuroactive pro-inflammatory compounds such as prostaglandin $E_2$ ($PGE_2$), which facilitates macrophage and microglial recruitment at the injury site (see below). In addition, $PGE_2$ also increases local blood flow and leukocyte infiltration and enhances vascular permeability and proinflammatory cytokine production (Amar and Levy, 1999).

It is shown that intravenous injections of $cPLA_2$ inhibitor, arachidonyl trifluoromethyl ketone ($AACOCF_3$) not only has neuroprotective effect but also improves Basso, Beattie, and Bresnahan (BBB) score indicating recovery locomotion parameter (Huang et al., 2009). Studies on chronic constriction nerve injury (CCI) in rats indicate that CCI decreases spinal glutamate uptake activity and increases ARA and extracellular and glutamate levels in spinal microdialysates on postoperative day 8 (Sung et al., 2007). Treatment with $AACOCF_3$ (intrathecally twice a day) for postoperative day 1–7 not only reverses CCI-induced spinal ARA generation but also prevents the reduction in spinal glutamate uptake activity and elevates extracellular glutamate concentration. Conversely, alteration of spinal ARA metabolism by diclofenac (a cyclooxygenase 1/2 inhibitor) further decreases spinal glutamate uptake activity and increases extracellular glutamate concentration in CCI rats (Sung et al., 2007). Thus, $AACOCF_3$ reduces the development of both thermal hyperalgesia and mechanical allodynia, whereas diclofenac exacerbates thermal hyperalgesia, in CCI rats, suggesting that in spinal cord CCI-mediated alterations in regional glutamate uptake activity, glutamate homeostasis, and neuropathic pain behaviors may be modulated by ARA turnover, and regulation of spinal ARA turnover may be a useful approach for improving the clinical management of neuropathic pain in SCI (Sung et al., 2007). Collective evidence suggests that $PLA_2$ isoforms may act as convergence molecules that mediate multiple key mechanisms associated with the secondary injury. $PLA_2$ isoforms are modulated by multiple factors, including inflammatory cytokines, free radicals, and excitatory amino acids. Blocking $PLA_2$ isoforms may represent a novel and efficient strategy to block multiple injury pathways associated with the brain and spinal cord secondary injuries (Farooqui et al., 2006; Titsworth et al., 2008).

Annexin A1 (ANXA1), a family of structurally related calcium- and phospholipid-binding and $PLA_2$ inhibitory protein, is known to facilitate anti-inflammatory action of glucocorticoids. SCI upregulates the expression of annexins I, II, and V in the injured spinal cord. Thus, annexin I expression increases at 3 days after SCI, peaks at 7 days, start to decline at 14 days, and return to the baseline level at and beyond 28 days post-injury (Liu et al., 2004). Similarly, the expression of annexin II begins to increase at 3 days, reaches its maximal level at 14 days, remain at a high level up to 28 days, and then decline to the basal level by 56 days after injury. Annexin V expression starts to elevate at day 3 and reaches its maximal level at day 7 and remains at this level until 56 days after injury. Real-time polymerase chain reaction (RT-PCR) studies confirm the expression of all three annexins at the mRNA level after SCI. Injections of ANXA1 (Ac 2–26) into the acutely injured spinal cord prevent SCI-induced increase in $PLA_2$ and myeloperoxidase activities (Liu et al., 2007a). In addition, ANXA1 administration reduces the expression of interleukin-1β and activated caspase-3 at 24 h, and glial fibrillary acidic protein at 4 weeks post-injury. In addition, ANXA1 administration not only reverses $PLA_2$-mediated spinal cord neuronal death in vitro but also reduces tissue damage and increases white matter sparing in vivo compared to the vehicle-treated controls (Liu et al., 2007a). Fluoro-Gold retrograde tracing patterns demonstrate that ANXA1 administration protects axons of long descending pathways at 6 weeks post-SCI. ANXA1 administration also increases the number of animals that responded to transcranial magnetic motor-evoked potentials. However, no measurable behavioral improvement is observed following ANXA1 treatments. These results along with electrophysiologic measures support the view that ANXA1 has a neuroprotective effect in SCI (Liu et al., 2004, 2007a).

### *4.5.2 Activation of COX-2 in Spinal Cord Injury*

Cyclooxygenase, or prostaglandin G/H synthase, the rate-limiting enzyme for the production of prostaglandins is known to occur in spinal cord. SCI stimulates the activity and expression of cyclooxygenase-2 (COX-2) and synthesis of prostaglandins 2 h following injury (Resnick et al., 1998). COX-2 levels peak at 48 h following traumatic SCI. Selective inhibition of COX-2 activity with SC58125 results not only in neuroprotection from SCI but also in improvement in mean BBB scores in injured animals. Reverse transcriptase-polymerase chain reaction (RT-PCR) studies reveal that COX-2 transcription in the spinal cord starts to increase within 30 min, peaks at 3 h after SCI (Adachi et al., 2005). Western blotting analysis demonstrates that the deglycosylated COX-2 protein is significantly increased 6 h after injury. Similarly, in severe clip compression model of SCI in the rat, the expression of COX-2 and formation of 8-OHdG and protein carbonyl groups are markedly increased after SCI while APE/Ref-1 expression is decreased (Bao et al., 2004). Anti-CD11d mAb treatment clearly attenuates COX-2 expression and 8-OHdG and protein carbonyl formation and rescues APE/Ref-1 expression after SCI, demonstrating that anti-CD11d mAb treatment significantly reduces intraspinal free radical formation after SCI, thereby reducing protein and DNA oxidative damage (Bao et al., 2004). Accumulating evidence suggests that COX-2 mRNA and protein expression are induced by spinal cord injury, and that selective inhibition of COX-2 or CD11d mAB improves functional outcome following experimental SCI.

### 4.5.3 Activation of NOS in Spinal Cord Injury

Nitric oxide (NO) is a diffusible molecule closely associated within the pathogenesis of spinal cord trauma (Chatzipanteli et al., 2002; Marsala et al., 2007). It is generated through the conversion of arginine to citrulline. This reaction is catalyzed by nitric oxide synthase (NOS). Several forms of NOS occur in brain and spinal cord tissues. SCI differentially affects activities of isoforms of NOS. Thus, constitutive nitric oxide synthase (cNOS) activity is significantly decreased in the traumatized rat spinal cord. cNOS activity returns to control levels within 6–24 h after injury. In contrast, inducible NOS (iNOS) enzymic activity is elevated and peaks at 24 h. Detailed investigation has revealed that a significant cellular source of iNOS protein is from invading polymorphonuclear leukocytes (PMNLs), macrophages, and endothelial cells. Generation of NO and prostaglandins induces changes in the vascular tone and/or permeability, which result in increased infiltration of cells of the immune system. These immune system cells intensify spinal cord tissue damage. Following SCI, NO concentration rises into the micromolar range and swamps the available iron ions, and begins to interact with sulfhydryl groups in glutathione and cysteine and reactive hydroxyl moieties such as tyrosine residues. NO is an important modulator of axon outgrowth and guidance, synaptic plasticity, neural precursor proliferation and neuronal survival. Excessive NO synthesis as that evoked by inflammatory signals has been reported to be as one of the major causative step for the pathogenesis of traumatic injuries to brain and spinal cord. Treatment of rats with aminoguanidine results in significant improvement in hind limb function up to 7 weeks after SCI. Histopathological analysis of contusion volume indicates that aminoguanidine treatment decreases lesion volume by 37% (Chatzipanteli et al., 2002; Marsala et al., 2007). As stated above, the reaction between NO and superoxide radical generates peroxynitrite, a metabolite that induces lipid peroxidation (Fig. 4.5). Peroxynitrite mediates nitrosative stress and is a potent inducer of cell death through its reaction with lipids, proteins, and DNA. Particularly DNA damage caused by both oxidative and nitrosative stresses results in activation of poly(ADP-ribose) polymerase (PARP), a nuclear enzyme implicated in DNA repair. In response to excessive DNA damage, massive PARP activation results in energetic depletion and apoptotic cell death. Peroxynitrite may also be involved in myelin damage. This metabolite is a crucial player in post-traumatic oxidative damage after SCI (Xiong et al., 2007). It diffuses through neural cell membranes without specific release or uptake mechanisms producing changes in signal-related functions by several means. In particular, the activation of guanylyl cyclases, the synthesis of cGMP, the action of cGMP-dependent protein kinases, phosphodiesterases, and ion channels has been reported to be the major signal transduction pathways of NO in the spinal cord, where the activation of membrane-bound guanylyl cyclase has only been demonstrated for natriuretic peptides, which stimulate cGMP accumulation in GABA-ergic structures in laminae I–III of the rat cervical spinal cord. These neurons are involved in controlling the action of the locomotor circuit (Marsala et al., 2007). Generation of peroxynitrite induces reversible conduction deficits within axons of the spinal cord by inducing alterations in $Na^+$ channel conductance in

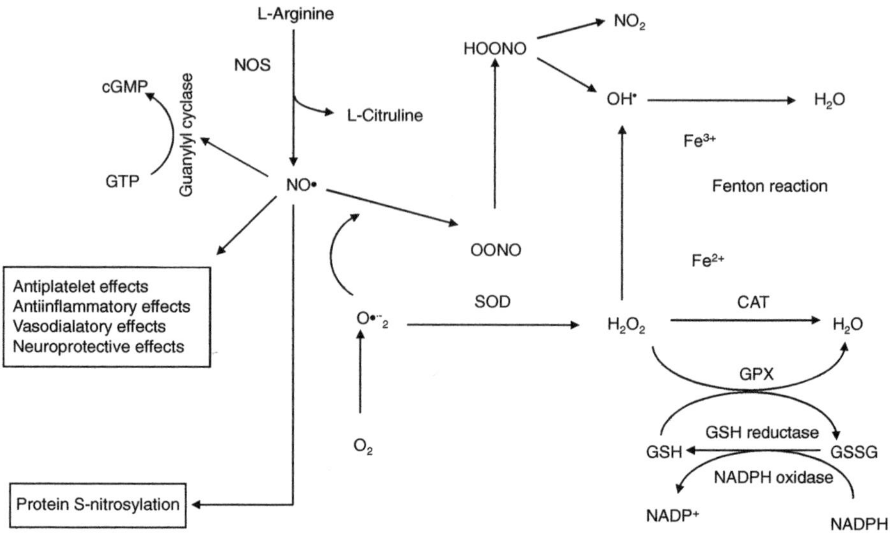


**Fig. 4.5** Reactions showing the generation of nitrite and peroxynitrite. Nitric oxide synthase (NOS); nitric oxide (NO); peroxynitrite ($OONO^-$); superoxide dismutase (SOD); catalase (CAT); and glutathione peroxydase (GPX)

the axolemma. These results support the view that peroxynitrite contributes to both reversible and non-reversible neurologic deficits following SCI (Ashki et al., 2008).

3-Nitrotyrosine (3-NT), a specific marker for peroxynitrite-mediated damage rapidly accumulates at all time points and is significantly increased in injured rats compared with sham rats after SCI. Accumulation of 3-NT is accompanied by significant increase in the levels of protein oxidation-related protein carbonyl and lipid peroxidation product, 4-hydroxynonenal (4-HNE). Highest increases in 3-NT and 4-HNE are seen at 24 h post-injury. Immunohistochemical studies indicate that 3-NT and 4-HNE are co-localized in degenerating neurons and peroxynitrite is closely associated with peroxidative as well as protein nitrosative damage after SCI. The consequences of oxidative damage to spinal cord include overloading of intracellular calcium, which may activate the cysteine protease, calpain leading to the degradation of cytoskeletal protein (α-spectrin). Western blot analysis of α-spectrin breakdown products show that the 145 kDa fragments of α-spectrin, which are specifically generated by calpain, are significantly increased within 1 h following injury and peak after 72 h post-injury (Xiong et al., 2007). Based on these results, it is proposed that activation of calpain is most likely linked to peroxynitrite-mediated secondary oxidative damage (Xiong et al., 2007). Involvement of peroxynitrite in SCI is also supported by the effect of ww-85, a metalloporphyrinic peroxynitrite decomposition catalyst. In a vascular clips model of SCI in mice, treatment with ww-85 significantly reduces (a) the degree of spinal cord inflammation and tissue injury, (b) neutrophil infiltration (myeloperoxidase activity), (c) nitrotyrosine formation and PARP activation, (d) pro-inflammatory cytokines expression,

(e) NF-κB activation, and (f) apoptosis. Furthermore, ww-85 significantly improves the recovery of limb function (evaluated by motor recovery score) in a dose-dependent manner. These results demonstrate that ww-85 treatment reduces the development of inflammation and tissue injury associated with spinal cord trauma (Genovese et al., 2009a).

### *4.5.4 Activation of Calcineurin in Spinal Cord Injury*

Calcineurin, a serine/threonine phosphatase is modulated by cellular $Ca^{2+}$ and calmodulin, a $Ca^{2+}$-binding protein. It is linked to dopamine and NMDA receptors and has been implicated in a wide variety of biological responses including lymphocyte activation, neuroinflammation, neurite outgrowth development, and apoptosis. It occurs as a complex with Bcl-2 in various regions of rat and mouse brain and spinal cord. Activation of the caspase-3 apoptotic cascade in SCI is regulated, in part, by calcineurin-induced BAD dephosphorylation (Springer et al., 2000). BAD, a proapoptotic member of the bcl-2 gene family, is rapidly dephosphorylated after SCI, dissociates from 14-3-3 in the cytosol, and translocates to the mitochondria of neurons where it binds to Bcl-x(L) and triggers cytochrome *c* release in the cytosol. Cytochrome *c* binds to Apaf-1 and dATP and recruits and cleaves pro-caspase-9 in the apoptosome. Both caspase-8 and caspase-9 activate caspase-3, among other caspases, which in turn cleave several crucial substrates, including the DNA-repairing enzyme PARP, into fragments of 89 and 28 kDa. Pretreatment of animals with FK506, an immunosuppressant and potent inhibitor of calcineurin activity inhibit BAD dephosphorylation and prevents activation of the caspase-3 apoptotic cascade (Springer et al., 2000). Calcineurin also dephosphorylates the nuclear factor of activated T cells (NF-AT). This dephosphorylation allows it to enter the nucleus and interact with NF-AT through distinct binding motifs: the PxIxIT and LxVP sites. Alterations in NF-AT binding motif interfere with calcineurin-immunosuppressant binding, and an LxVP-based peptide competes with immunosuppressant–immunophilin complexes for binding tocalcineurin (Rodriguez et al., 2009). Thus, NF-AT and calcineurin interactions not only promote immune activation and development of the vascular and nervous systems but also play an essential role in lymphocyte activation. Cyclosporin A and FK-506 block the above process. Calcineurin also activates neuronal NOS through dephosphorylation resulting in increase in generation of NO, which as described above may lead to neurotoxicity.

### *4.5.5 Activation of Matrix Metalloproteinases in Spinal Cord Injury*

Matrix metalloproteinases (MMPs) are a family of extracellular zinc-dependent endopeptidases that hydrolyze the extracellular matrix and other extracellular proteins (Malemud, 2006). These enzymes are involved in both injury and repair mechanisms in brain and spinal cord. Three members of the MMP family, MMP-2,

MMP-9, and MMP-12, are transiently upregulated in the spinal cord wound following SCI (Hsu et al., 2006, 2008; Wells et al., 2002; Yu et al., 2008). The expression of MMP-12 is increased 189-fold over normal levels (Wells et al., 2002). SCI studies in wild-type (WT) and MMP-12 null mice indicate that these mice show significant improvement in functional recovery compared with WT controls. Twenty-eight days after injury, the BBB score in the MMP-12 group is 7, representing extensive movement of all three hind limb joints, compared with 4 in the WT group, representing only slight movement of these joints. Furthermore, MMP-12 null mice exhibits recovery of hind limb strength more rapidly than control mice, with significantly higher inclined plane scores on days 14 and 21 after SCI. Mechanistically, there is a decrease in permeability of the blood–spinal barrier and reduction in microglial and macrophage density in MMP-12 null mice compared to WT controls (Wells et al., 2002). Similarly studies on MMP-9 expression after SCI in copper/zinc-superoxide dismutase (SOD1) transgenic (Tg) rats indicate that MMP-9 activity is significantly increased after SCI in both SOD1 Tg rats and their wild-type (Wt) littermates, although the increase is less in the SOD1 Tg rats (Yu et al., 2008). In situ zymography demonstrate that gelatinolytic activity is increased after SCI in the Wt rats, while the increase is less in the Tg rats. Intrathecal injection of SB-3CT (a selective MMP-2/MMP-9 inhibitor) results in significant decrease in apoptotic cell death after SCI, suggesting that increased oxidative stress after SCI may cause MMP-9 upregulation, BBB disruption, and apoptosis; the overexpression of SOD1 in Tg rats decreases oxidative stress that further attenuates MMP-9-mediated BBB disruption (Yu et al., 2008).

Unlesioned human spinal cord shows very low MMP immunoreactivity. The involvement of MMP-1, -2, -9, and -12 has been reported in the post-traumatic events after human SCI (Buss et al., 2007). With an expression pattern MMPs is similar to experimental studies in animals. MMPs are mainly expressed during the first weeks after SCI and are most likely associated with the destructive inflammatory events of protein breakdown and phagocytosis carried out by infiltrating neutrophils and macrophages, as well as being involved in enhanced permeability of the blood spinal cord barrier (Buss et al., 2007). Collective evidence suggests that MMPs play a key role in abnormal vascular permeability and inflammation within the first 3 days after SCI, and that blockade of MMPs during this critical period attenuates these vascular events and leads to improved locomotor recovery. MMPs also modulate neuropathic pain following SCI. Involvement of MMPs in the development of mechanical allodynia through myelin protein degradation in L5 spinal nerve crush (L5 SNC) model of nerve injury in rat and MMP-9–/– mouse indicates that MMPs promote selective degradation of myelin basic protein (MBP), with MMP-9 regulating initial Schwann cell-induced MBP processing after L5 SNC. Acute and long-term treatment with GM6001 (broad-spectrum MMP inhibitor) not only protects nerve from injury-mediated MBP degradation of caspase-induced apoptosis and macrophage infiltration in the spinal nerve but also blocks astrocyte activation in the spinal cord (Kobayashi et al., 2008). In SCI, upregulation of MMPs also contribute to apoptotic cell death, which can be reduced with MMP2/MMP9 inhibition (Dang et al., 2008).

SCI is accompanied by glial scar formation, which becomes a major obstacle to axonal growth and regeneration. Formation of the glial scar involves the migration of astrocytes toward the injury site. MMP-9 and MMP-2 are two proteases that govern cell migration through their ability to degrade constituents of the extracellular matrix. Thus, these enzymes may be promising therapeutic target to reduce glial scar formation during wound healing after SCI (Hsu et al., 2008). Collective evidence suggests that MMPs contribute to the pathogenesis of SCI and their inhibitors may not only decrease pain and apoptotic cell death but also reduce glial scar formation leading to better recovery process.

### *4.5.6 Activation of Poly (ADP-Ribose) Polymerase in Spinal Cord Injury*

PARP is a DNA-binding protein that is primarily activated by nicks in the DNA molecule. It regulates the activity of various enzymes, including itself, that are associated with the control of DNA metabolism. Upon binding to DNA breaks, activated PARP degrades $NAD^+$ into nicotinamide and ADP-ribose and promotes the polymerization of the ADP-ribose on nuclear acceptor proteins including histones, transcription factors, and PARP itself. Poly(ADP-ribosylation) facilitates DNA repair and the maintenance of genomic stability. This process results in at least three important consequences in the brain, depending on the cell type and the extent of DNA damage: (1) Poly(ADP-ribose) formation on histones and on enzymes involved in DNA repair retard sister chromatid exchange and facilitate base-excision repair, (2) poly(ADP-ribose) formation modulates transcription factors, notably NF-κB, and thereby facilitating neuroinflammation, and (3) marked PARP-1 activation induces neuronal death through mechanisms involving $NAD^+$ depletion and release of apoptosis-inducing factor from the mitochondria (Kauppinen and Swanson, 2007).

Stimulation of PARP activity in SCI increases poly(ADP-ribose) (PAR) immunoreactivity at the injury site in the injured spinal cord (Genovese et al., 2005). Although the molecular mechanism of PARP-mediated spinal cord damage is not fully defined, the generation of peroxynitrite is known to mediate overactivation of PARP resulting in the depletion of $NAD^+$ and ATP and the release of apoptosis-inducing factor (AIF) from the mitochondria leading to cell death in traumatic situations. PARP also upregulates numerous proinflammatory genes and adhesion molecules through the activation of NF-κB and AP-1 (Komjati et al., 2005), supporting the view that PARP is closely associated with the pathogenesis of SCI (Genovese et al., 2005; Genovese and Cuzzocrea, 2008). PAR and PARP are also involved in the transcriptional regulation through their ability to modify chromatin-associated proteins. In traumatic situations, PARP-mediated poly(ADP-ribosyl)ation also correlates directly with induction of 4-hydroxy-2-nonenal-induced apoptosis. Treatment of the mice with the PARP inhibitors 3-aminobenzamide (3-AB) or 5-aminoisoquinolinone (5-AIQ) not only reduces the intensity of inflammation, PAR immunoreactivity, and neutrophil infiltration but also reduces apoptotic cell death in the injured spinal cord. In addition, PARP inhibitors also significantly ameliorate the recovery of limb function (Genovese et al., 2005).

### 4.5.7 Activation of RhoA and RhoB in Spinal Cord Injury

Inhibition of the small GTPase Rho or of its downstream target, Rho-associated kinase, ROCK, not only facilitates axon regeneration, neurite growth, gene expression, cell proliferation, but also promotes functional recovery following SCI in adult rats. Myelin-derived inhibitory proteins inhibit Rho-mediated signaling associated with the regeneration. Control spinal cords has no $RhoA^+$ cells, but contains few $RhoB^+$, microglial cells and some dissociated neurons. SCI results in $RhoA^+$ and $RhoB^+$ cells accumulation not only in perilesional areas but also in area with developing necrotic core 1 day after the injury (Fig. 4.3). The number of $RhoA^+$ and $RhoB^+$ cells reaches maximum levels at day 3 and day 1, respectively. $RhoA^+$ and $RhoB^+$ cell numbers remain significantly elevated until day 28 (Conrad et al., 2005). In other studies, upregulation of RhoA mRNA and expression of Rho proteins are observed in the injured spinal cord 1 week after surgery. Treatment with C3 exozyme (RhoA inhibitor), Y-27632 (selective Rho kinase inhibitor), and fasudil (non-selective protein kinase inhibitor) not only promotes long-distance regeneration of anterogradely labeled corticospinal axons and increase in levels of GAP-43 mRNA in the motor cortex but also improves BBB scores, and promotes locomotion recovery as well as progressive recuperation of fore limb–hind limb coordination (Dergham et al., 2002; Sung et al., 2003; Conrad et al., 2005). Collectively, these studies suggest that Rho-ROCK pathway is involved in many aspects of neuronal functions, including neurite outgrowth, retraction, and axon regeneration. This pathway has become a potentially important target for the development of drugs for treating SCI in recent years.

### 4.5.8 Activation of Caspases in Spinal Cord Injury

Caspases, a family of aspartate-specific cysteine proteases, are essential in the initiation and execution of apoptosis (Creagh et al., 2003; Cohen, 1997). They are expressed as inactive proenzymes (zymogens) that become active during apoptosis. Out of 14 caspase enzymes, caspase-3 appears to be the major effector of neuronal apoptosis induced by a variety of stimuli as well as traumatic injuries (Fig. 4.4). A role for caspase-3 in injury-induced neuronal cell death has been established using semispecific peptide caspase inhibitors. Caspases not only cleave other downstream caspases but also a variety of enzymes, cytokines, cytoskeletal, nuclear, and cell cycle regulatory proteins (Cohen, 1997). Their activities in brain and spinal cord tissues are regulated by the occurrence in zymogens form, by members of Bcl-2 family, and certain cellular inhibitor of apoptosis proteins (cIAPs).

Caspases are closely associated with apoptotic cell death in experimental SCI (Yakovlev et al., 2005). Thus, SCI is accompanied by a rapid upregulation of caspase-3 gene expression along with localization of active caspase-3 in neurons and activated microglia (Citron et al., 2008). Determination of enzymic activity in injured spinal cord tissue indicates that caspase-3, caspase-8, and caspase-9 are activated from 1 to 72 h after SCI. Intrathecal injection of the pan-caspase inhibitor, Boc-Asp (OMe)-fluoromethylketone (Boc-d-fmk), and treatment with *N*-benzyloxycarbonyl-Asp-Glu-Val-Asp-fluoromethyl ketone (z-DEVD-fmk), a selective caspase-3 inhibitor, improves locomotion function after SCI (Yakovlev et al., 2005; Barut et al., 2005; Citron et al., 2008). Molecular mechanisms associated with the activation of caspases in SCI are not fully understood. However, release of glutamate and stimulation of NMDA receptors, release of cytochrome *c* and Apaf-1 from mitochondria, activation of Fas receptors, and peroxynitrite may play an important role (Farooqui, 2009). The activation of the caspase-3-mediated apoptotic cascade in SCI is modulated, in part, by calcineurin-induced BAD dephosphorylation. BAD, a pro-apoptotic member of the bcl-2 gene family, is rapidly dephosphorylated after injury, dissociates from 14-3-3 protein in the cytosol, and translocates to the mitochondria of neurons where it binds to Bcl-$x_L$ (Springer et al., 2000). Pretreatment of animals with FK506, a potent inhibitor of calcineurin activity, or an NMDA glutamate receptor antagonist (MK-801) inhibits BAD dephosphorylation and abolishes activation of the caspase-3 and apoptotic cascade (Springer et al., 2000).

Recent studies indicate involvement of molecular platforms or intracellular sensors NALP1 (NAcht leucine-rich repeat protein 1) or inflammasomes in SCI. These platforms consist of caspase-1, caspase-11, ASC (apoptosis-associated speck-like protein containing a caspase-activating recruitment domain), and NALP1. In SCI, inflammasomes promote the processing of IL-1β, IL-18, activation of caspase-1, cleavage of X-linked inhibitor of apoptosis protein (XIAP) and facilitate the assembly of multiprotein complex (de Rivero et al., 2008). Administration of anti-ASC neutralizing antibodies not only reduces caspase-1 activation and X-linked inhibitor of apoptosis protein cleavage but also increases the processing of interleukin-1β in head injury (de Rivero et al., 2009). This treatment also results in a significant decrease in contusion volume. These studies show that the NLRP1 inflammasome constitutes an important component of the innate central nervous system inflammatory response after traumatic brain and spinal cord injuries and may be a novel therapeutic target for reducing the damaging effects of post-traumatic inflammation (de Rivero et al., 2008).

### *4.5.9 Activation of Calpains and Other Proteases in Spinal Cord Injury*

Calpains are a family of calcium-dependent cysteine proteases that are widely expressed in brain and spinal cord tissues. These enzymes have been implicated in cell death in spinal cord injury (Ray et al., 2003; Buki et al., 2003). Calpain activity is modulated by an endogenous protein inhibitor called calpastatin. Overactivation of calpains degrades calpastatin, limiting its regulatory efficiency. Although the precise physiological function of calpains remains elusive, association of calpains with spinal cord injury suggests that calpains participate in neurodegenerative process via increase in intracellular free $Ca^{2+}$, which promotes the degradation of key cytoskeletal, membrane, and myelin proteins. Cleavage of these key proteins by calpain is an irreversible process that perturbs the integrity and stability of neural cells, leading to neuronal cell death. It is proposed that calpain in conjunction with caspases promotes neuronal apoptosis in brain tissue (Wang, 2000).

Kallikrein 6 (K6) is a member of the kallikrein gene family that comprises 15 structurally and functionally related serine proteases. This trypsin-like enzyme is preferentially expressed in neurons and oligodendroglia of the adult central nervous system (CNS). It is upregulated not only at the site of injury due to expression by infiltrating immune and resident CNS cells but also in spinal cord segment above and below the injury site (Scarisbrick et al., 2006). At the cellular level, elevation in K6 activity is particularly prominent in macrophages, microglia, and reactive astrocytes. It is proposed that K6 enzymic cascades mediate events secondary to spinal cord trauma, including dynamic modification of the capacity for axon outgrowth (Scarisbrick et al., 2006).

## 4.6 Activation of Cytokines and Chemokines in Spinal Cord Injury

Cytokines are heterogeneous group of proteins and polypeptides associated with the regulation of cell–cell interactions. They include interleukins (IL-1, IL-2, IL-6, and IL-12), interferons (IFN-γ), tumor necrosis factor (TNF-α), tumor growth factors (TGF-α and β), and colony-stimulating factors (Sun et al., 2004; Kim et al., 2001). In brain and spinal cord, cytokines mediate cellular intercommunication through autocrine, paracrine, or endocrine mechanisms (Wilson et al., 2002). Their actions are mediated through a complex network, which is linked to feedback loops and cascades. Expression of cytokines is very low in normal spinal cord and brain. SCI induces significant increases in the synthesis of multiple cytokines, including tumor necrosis factor-α (TNF-α), interleukin-6 (IL-6), and interleukin-1β (IL-1β) (Yang et al., 2004; Farooqui and Horrocks, 2009). Macrophages and neutrophils, astrocytes, and microglial cells are the major sources of these cytokines. Increased immunoreactivities of IL-1β, IL-6, and TNF-α are detected in neurons 30 min after SCI, and in neurons and microglia 5 h after injury, but the expression of these pro-inflammatory cytokines is short lived and declines sharply to baseline by 2 days after injury. As early as 30 min after SCI, activated microglial cells are detected along with axonal swellings at the injury site. In addition, axons are surrounded by microglial processes. Numerous neutrophils appear in the injured cord 1 day after injury, and then their number declines dramatically, whereas macrophages progressively increase after day 1 (Yang et al., 2004). Thus, SCI is characterized by edema, neutrophil infiltration, and cytokine production. These processes are followed by recruitment of other inflammatory cells, which generate a range of inflammatory mediators that cause apoptosis. Increased cytokine levels in SCI play an important role not only in neurodegeneration but also in metabolic responses during compensatory regenerative processes (Vitkovic et al., 2000). Cytokines produce their effects by interacting with specific membrane-associated cytokine receptors (Rothwell and Relton, 1993). The magnitude and persistence of the elevations in cytokine levels may be related to the severity of trauma (Rothwell and Relton, 1993). TNF-α and IL-1β trigger biologically indistinguishable effects by binding to their specific receptors using the same set of transcription factors. The role of cytokine in SCI is quite complex. An early administration (at 1 day) of TNF-α has detrimental effects on the spinal cord, whereas delayed administration (at 4 days) reduces the extent of the lesions (Klusman and Schwab, 1997). TNF-α is a master pro-inflammatory cytokine that modulates the induction of a large number of other inflammatory cytokines, chemokines, and adhesion molecules under traumatic conditions. Cytokines produce their effect by modulating a number of signaling pathways, including phosphatases, kinases, phospholipases, sphingomyelinases, oxygen radicals, and transcription factors (Jupp et al., 2003; Gomes-Leal et al., 2004). SCI induces the TNF-α-mediated activation of NF-κB (Xu et al., 1998; Bethea et al., 1998), which in turn modulates the transcription of other proinflammatory cytokines, chemokines, and proinflammatory enzymes, such as isoforms $PLA_2$, COX-2, iNOS, SMase, and MMP. These factors further intensify SCI. Cytokine antagonists and cell cycle inhibitor, olomoucine, significantly suppress microglial proliferation and produce a remarkable reduction of tissue edema formation. In the olomoucine-treated group, a significant reduction of activated and/or proliferated microglial-induced IL-1β expression is observed 24 h after SCI. Moreover, cytokine antagonists and olomoucine attenuate the number of apoptotic neurons after SCI (Tian et al., 2007).

In addition, SCI also induces the expression of chemokines, which are small structurally similar proteins released locally at the site of inflammation. They recruit immune cells and include MCP-1 and MIP-1α. mRNA levels of IP-10 peak around 6 h post-injury and are upregulated up to 7 days post-injury. MCP-1 mRNA can be detected at 1 h post-injury and its levels returned to baseline by 14 days post-injury. An increase in MCP-1 staining is observed from 1 to 7 days post-injury (Lee et al., 2000). Infusion of the broad-spectrum chemokine receptor antagonist (vMIPII) in the contused spinal cord initially attenuates leukocyte infiltration, suppresses gliotic reaction, and reduces neuronal damage after injury (Ghirnikar et al., 2000, 2001). These changes are accompanied by the upregulation in expression of bcl-2, the endogenous apoptosis inhibitor, and reduced neuronal apoptosis. Two and four weeks after vMIPII infusion, the injured spinal cord shows a reduction in myelin breakdown in the dorsal and ventral funiculi. Immunohistochemical studies indicate an increase in calcitonin gene-related peptide, choline acetyl transferase, and tyrosine hydroxylase positive fibers as well as increase in GAP43 staining in treated cords. These results suggest that sustained reduction in post-traumatic cellular infiltration is beneficial for tissue survival. In contrast, infusion of MCP-1 (9–76), a N-terminal analog of the MCP-1 chemokine, produces only a modest reduction in cellular infiltration at 14 and 21 days post-injury without significant tissue survival after spinal cord contusion injury. Comparison of results on tissue survival provided by vMIPII and MCP-1 (9–76) validates the importance of the use of broad-spectrum antagonists in the treatment of SCI.

## 4.7 Fas/CD95 Receptor–Ligand System in Spinal Cord Injury

The Fas/CD95 receptor–ligand system plays an important role in apoptotic cell death after SCI (Casha et al., 2005; Davis et al., 2007; Yu et al., 2009) (Fig. 4.4). Studies on the involvement of Fas/CD95 receptor–ligand system in moderately injured animals and sham operation controls indicate that in sham-operated animals, a portion of FasL but not Fas is present in membrane rafts (Davis et al., 2007). SCI induces the translocation of FasL and Fas into membrane raft microdomains where Fas interacts with the adaptor proteins Fas-associated death domain (FADD), caspase-8, cellular FLIP long form (cFLIPL), and caspase-3, forming a death-inducing signaling complex (DISC). Moreover, SCI also results in the expression of Fas in clusters around the nucleus in both neurons and astrocytes (Davis et al., 2007). The formation of the DISC signaling platform causes a rapid activation of initiator caspase-8 and effector caspase-3 and the modification of signaling intermediates such as FADD and cFLIP(L). This observation supports the view that FasL-/Fas-mediated signaling after SCI is similar to Fas-induced cellular apoptosis (Davis et al., 2007). Although it is generally believed that Fas activation mediates apoptotic cell death predominantly through the extrinsic pathway, involvement of intrinsic mitochondrial signaling in Fas-induced apoptosis after SCI has also been reported (Yu et al., 2009). In the Fejota clip compression model of SCI in C57BL/6 Fas-deficient (lpr) and wild-type mice, it is shown that lpr mice show a downregulation in several parameters involved in apoptosis, including a decrease in numbers of terminal deoxynucleotidyl transferase dUTP nick end labeling (TUNEL)-positive cells at the injury site, reduction in expression of truncation of Bid (tBid), apoptosis-inducing factor, activated caspase-9, and activated caspase-3, and upregulation in expression of the anti-apoptotic proteins Bcl-2 and Bcl-xL. This suggests that the induction of intrinsic mitochondrial signaling pathways also contribute to Fas-mediated apoptosis after SCI (Yu et al., 2009). FAS-deficient mice not only show decrease in apoptotic cell death in neurons but also exhibit improved locomotor recovery, axonal sparing, and preservation of oligodendrocytes and myelin. FAS-deficient mice do not show a significant increase in surviving neurons in the spinal cord at 6 weeks after injury, supporting the involvement of other cell death mechanisms for neurodegeneration in SCI (Casha et al., 2005).

## 4.8 Activation of Transcription Factors in Spinal Cord Injury

Transcription factors are proteins involved in the regulation of gene expression that bind to the promoter elements upstream of genes and either promote or block transcription. Through this process they modulate gene expression. Transcription factors consist of two essential functional domains: a DNA-binding domain and an activator domain. The DNA-binding domain consists of amino acids that recognize specific DNA bases near the start of transcription. Transcription factors not only interact with RNA polymerase but also bind to other transcription factors and *cis*-acting DNA sequences. SCI causes alterations in several transcription factors, including NF-κB, AP-1, and members of STAT family (Xu et al., 1998; Bethea et al., 1998; Rafati et al., 2008).

### *4.8.1 NF-κB in Spinal Cord Injury*

The transcription factor NF-κB is a critical regulator of cell survival, immune function, inflammatory responses, and secondary injury processes. It is also required for the transcriptional activation of numerous genes regulating excitotoxicity, apoptosis, and numerous other injury responses. NF-κB is composed of homo- and heterodimeric complexes of proteins containing a Rel-homology domain: RelA (p65), RelB (p38), c-Rel ($p75^{NTR}$); NF-κB1 (p50/p105); and NK-κB2 (p52/p100). In its inhibited form (complexed with I-κB) NF-κB is retained in the cytoplasm. Phosphorylation of I-κB at two conserved serines ($Ser^{32/36}$) by the I-κB kinase complex and subsequent I-κB degradation by proteasomes results in translocation of active NF-κB to the nucleus (Schultz et al., 2006), where it induces gene transcription by binding cognate 5′ GGGRNNYYCC 3′ κB sequence motif in target gene promoters (Yamamoto and Gaynor, 2004). In the nucleus NF-κB mediates the transcription of more than 150 genes that not only influence the survival of neural cells but also maintain their normal functional integrity. NF-κB also induces many genes implicated in inflammation, oxidative stress, and immune responses (Fig. 4.4). These genes include COX-2, iNOS, $sPLA_2$, MMPs, TNF-α, IL-1β, IL-6, intracellular adhesion molecule-1 (ICAM-1), and vascular adhesion molecule-1 (VCAM-1). NF-κB is also stimulated by reactive oxygen species, which are derived from mitochondrial dysfunction and NADPH oxidase (Sun et al., 2007). These studies suggest that the diverse interactions of NF-κB with co-activators, co-repressors, and other signaling networks that influence NF-κB-mediated gene expression.

SCI induces transient changes in subunit populations of NF-κB. Detailed investigation indicates that there are decreases in neuronal c-Rel levels and inverse increases in p65 and p50 levels. No changes are observed in neuronal p52 or RelB subunits after SCI (Rafati et al., 2008). NF-κB inactivation in astrocytes results in improved functional recovery following SCI. This not only correlates with reduction in expression of pro-inflammatory mediators and chondroitin sulfate proteoglycans but also with increased white matter preservation (Brambilla et al., 2005). It is proposed that inactivation of astrocytic NF-κB creates a more permissive environment for axonal sprouting and regeneration. Studies on contusive and complete transection SCI in GFAP-inhibitor of κB-dominant negative (GFAP-I-κBα-dn) and wild-type (WT) mice indicate that inhibition of astroglial NF-κB leads to a growth-supporting terrain promoting sparing and sprouting, rather than regeneration, of supraspinal and propriospinal circuitries essential for locomotion.

This process contributes to the improved functional recovery observed after SCI in GFAP-I-κBα-dn mice (Brambilla et al., 2009).

The use of synthetic double-stranded "decoy" deoxyoligonucleotides containing selective NF-κB protein dimer binding consensus sequences indicates that decoy targets the p65/p50 binding site on the COX-2 promoter and decreases SCI-mediated neural cell damage and losses. NF-κB p65/p50 decoy not only improves early locomotor recovery after moderate SCI but also ameliorate SCI-mediated hypersensitization (Rafati et al., 2008). Activation of NF-κB also leads to the local generation of more cytokines and chemokines, which in turn promulgate glutamate-mediated signals and potentiates the activation of NF-κB activity (Block and Hong, 2005).

### *4.8.2 Peroxisome Proliferator-Activated Receptor in Spinal Cord Injury*

PPARs are members of the nuclear hormone receptor family. Several forms, PPAR-α, PPAR-γ, and PPAR-δ, are known to occur in brain and spinal cord (Drew et al., 2005). PPAR-α plays a role in controlling inflammatory processes associated with SCI (Genovese et al., 2009a, b). The role of PPAR-α in glucocorticoid-mediated anti-inflammatory activity is studied by testing the efficacy of dexamethasone, a synthetic glucocorticoid specific for glucocorticoid receptor in an experimental model of spinal cord trauma induced in PPAR-αKO (mice lacking PPAR-α) and wild type (WT) mice. Results indicate that compared to WT controls, dexamethasone-mediated anti-inflammatory activity is weakened in PPAR-αKO mice. In particular, dexamethasone is less effective in PPAR-αKO compared to WT mice as evaluated by inhibition of the degree of spinal cord inflammation and tissue injury, neutrophil infiltration, nitrotyrosine formation, pro-inflammatory cytokine expression, NF-κB activation, iNOS expression, and apoptosis. These observations suggest that PPAR-α contribute to the anti-inflammatory activity of glucocorticoids in SCI.

### *4.8.3 STAT in Spinal Cord Injury*

Signal transducers and activators of transcription (STAT), a group of novel transcription factors, that orchestrate the downstream events propagated by cytokine/growth factor interactions with their cognate receptors (Rane and Reddy, 2002). Injury to neural tissue induces STAT activation, and STATs are increasingly recognized for their role in neuronal survival (Dziennis and Alkayed, 2008). These factors are activated by the Janus kinase. The dysregulation of this pathway is associated with angiogenesis and immunosuppression. Unphosphorylated STAT proteins are monomers, which are translocated from cytoplasm to the nucleus, where in response to specific stimuli they are phosphorylated and bind to the promoter region of target genes and are thereby involved in regulating the transcription of target genes. As stated above, spinal cord response to injury includes expression of genes encoding cytokines and chemokines. These genes regulate entry of immune cells to the injured tissue. The synthesis of many cytokines and chemokines not only involves NF-κB, but also STAT. Studies on the expression of STATs and the chemokine CCL2 and their relationship to astroglial NF-κB signaling in the CNS following axonal transaction indicate that STAT1 is upregulated and phosphorylated in neurons and astrocytes and upregulation and phosphorylation of STAT2 in astrocytes depends on NF-κB (Khorooshi et al., 2008). Lack of NF-κB signaling significantly reduces chemokine CCL2-mediated injury as well as leukocyte infiltration. This suggests that NF-κB signaling in astrocytes controls expression of both STAT2 and CCL2 and thus regulates infiltration of leukocytes into lesion-reactive hippocampus after axonal injury (Khorooshi et al., 2008). Uninjured adult STAT3 knock-out mice (STAT3-CKO) have morphologically similar astrocytes to those in $STAT3^{+/+}$ mice except for a partially decrease in expression of GFAP (Herrmann et al., 2008). In $STAT3^{+/+}$ mice, phosphorylated STAT3 (pSTAT3) is not detectable in astrocytes in uninjured spinal cord. SCI markedly increases pSTAT3 in astrocytes and other cell types near the injury. In addition astrocytes show hypertrophy and pronounced disruption of astroglial scar formation. These changes may be involved in increase in spread of inflammation, increase in lesion volume, and partial attenuation of motor recovery over the first 28 day after SCI. Accumulating evidence suggests that STAT3 signaling is a critical regulator of certain aspects of reactive astrogliosis (Herrmann et al., 2008). It is also proposed that increase levels of activator of STAT after SCI may represent an early attempt of spinal cord repair and regeneration. Methylprednisolone (MP), a synthetic glucocorticoid, interacts with glucocorticoid receptor (GR) and produces beneficial effects in SCI. It is shown that GR forms a complex with STAT5. This complex is present on the STAT5-binding site of the bcl-x promoter region in oligodendrocytes. The overexpression of an activated form of STAT5 prevents α-amino-3-hydroxy-5-methyl-4-isoxazolepropionic acid (AMPA)-mediated oligodendrocyte cell death, which can be blocked when the STAT5 gene is knocked down (Xu et al., 2009). Collective evidence suggests that interactions of glucocorticoid signaling pathway with STAT5 and upregulation of bcl-X(L) may protect oligodendrocytes in SCI.

### *4.8.4 AP-1 in Spinal Cord Injury*

Activator protein 1 (AP-1) is a group of dimeric complexes consisting of several proteins belonging to the c-fos, c-jun, ATF, and JDP families. It binds to a specific site in the promotor region of a wide variety of genes involved in cell proliferation, differentiation, and survival. Phosphorylation of the existing Jun and Fos proteins by MARK kinases at specific serine and threonine residues regulates AP-1 activity. AP-1 modulates gene expression in response to cytokines, growth, oxidative stress (ROS), and infections. It controls differentiation, proliferation, and apoptosis (Hsu et al., 2000). SCI involves significant increase in activation of redox-sensitive transcription factor, NF-κB, and AP-1, as well as overexpression of MCP-1 and TNF-α in both the thoracic and lumbar regions (Xu et al., 2001; Ravikumar et al., 2004). Increase in AP-1 binding is observed in 1 h after SCI. Increase in AP-1 peaks at 8 h after SCI and declines to basal value 7 days after SCI. Methylprednisolone reduces post-traumatic AP-1 activation and RU486, a glucocorticoid receptor antagonist, and reverses methylprednisolone-mediated inhibition of AP-1 activation. Significant increases also occur in the expression of the Fos-B and c-jun components of AP-1 in the injured cord. A c-fos antisense oligodeoxynucleotide (ODN) blocks SCI-mediated increase in AP-1 but not NF-κB. Collective evidence suggests that inhibition of AP-1 activity attenuates processes propagating pathogenesis of SCI.

## 4.9 Gene Transcription in Spinal Cord Injury

SCI leads to induction and/or suppression of many genes, the interplay of which governs the neuronal death and subsequent loss of motor function (Song et al., 2001; Bareyre and Schwab, 2003). Early stages after SCI result in the potent upregulation of genes associated with transcription and inflammation and a general downregulation of genes modulating expression of structural proteins and proteins involved in neurotransmission. Later stages of SCI are characterized by the upregulation of genes responsible for the modulation of growth factors, axonal guidance factors, extracellular matrix molecules, and angiogenic factors and downregulation of cytoskeletal proteins. These genes have been implicated in repair, recovery, and survival processes after SCI (Bareyre and Schwab, 2003). In addition, upregulation of immediate early genes, genes regulation of heat shock proteins (Hsp-70), and pro-inflammatory genes (interleukin-6) have been reported to occur after SCI. In SCI, induction of Hsp has beneficial effects. These proteins are expressed by acutely stressed microglial, endothelial, and ependymal cells. They assist in the protection of motor neurons and to prevent chronic inflammation after SCI (Reddy et al., 2008). Hsps promote cell survival by preventing mitochondrial outer membrane permeabilization or apoptosome formation as well as via regulation of Akt and JNK activities (Beere, 2005). This up- and downregulation of gene transcription persists for many hours (more than 24 h) after SCI (Song et al., 2001). In SCI, changes in gene expression have been confirmed using Genechip and real-time quantitative PCR studies. In addition, changes in gene involved cell cycle (gadd45a, c-myc, cyclin D1 and cdk4, pcna, cyclin G, Rb, and E2F5) also occur after SCI (Di Giovanni et al., 2003). Collective evidence suggests that transcription of various genes after SCI not only modulates oxidative stress and inflammation but also controls neurotransmitter dysfunction, ionic imbalance, and redox status in the injured spinal cord.

## 4.10 Mitochondrial Permeability Transition in Spinal Cord Injury

Mitochondria are the powerhouse of neural cells. They play a critical role in initiating both apoptotic and necrotic cell death (Fig. 4.6). By maintaining ratios of ATP:ADP that thermodynamically favor the hydrolysis of ATP to ADP + Pi, they

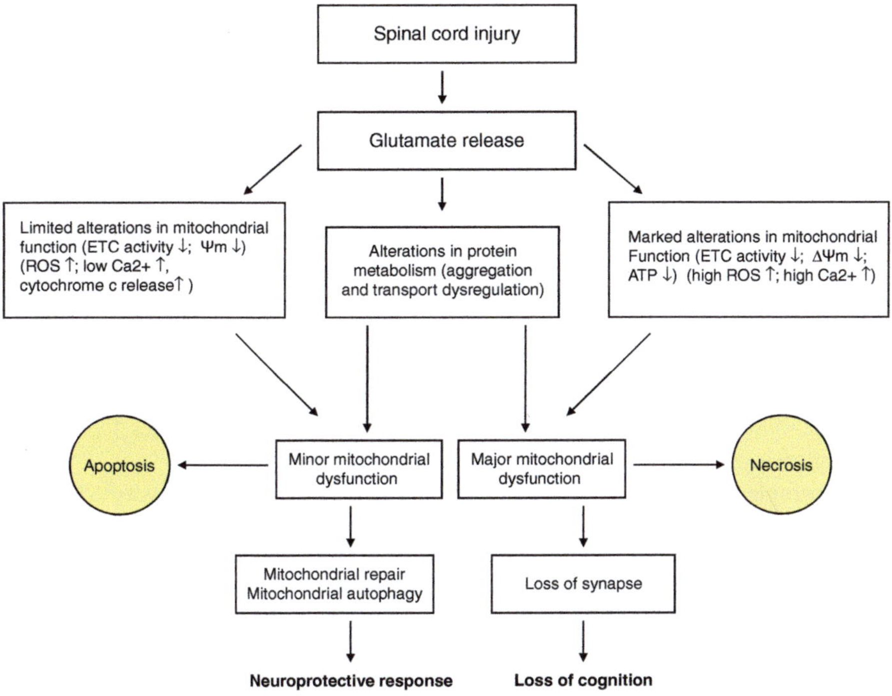


**Fig. 4.6** Involvement of mitochondrial dysfunction in spinal cord injury

also generate ROS. Proton pumping by components of the electron transport system (ETS) generates a membrane potential (DeltaPsi) that can then be utilized to phosphorylate ADP or sequester $Ca^{2+}$ out of the cytosol into the mitochondrial matrix. This allows mitochondria to act as cellular $Ca^{2+}$ sinks and to be in phase with alterations in cytosolic $Ca^{2+}$ levels. Under extreme $Ca^{2+}$ load, elevated phosphate concentrations and adenine nucleotide depletion may cause the opening of the mitochondrial permeability transition pore (mPTP) which produces the extrusion of mitochondrial $Ca^{2+}$ and other high- and low molecular weight components. This catastrophic event discharges DeltaPsi and uncouples the ETS from ATP synthesis and results in neuronal cell death (Tsujimoto and Shimizu, 2003; Sullivan et al., 2005). Thus, the mitochondrial permeability transition (mPT) involves the opening of a non-specific pore in the inner mitochondrial membrane, converting them from organelles, which produce and sustain ATP, to instruments of cell death. The anti-apoptotic proteins Bcl-2 and Bcl-$x_L$ block the mPT and can therefore block mPT-dependent cell death. Collective evidence suggests that the inhibition of the mPT has a therapeutic potential for treating SCI and other neurodegenerative conditions.

Cyclosporin A (CsA), a potent immunosuppressive drug, blocks mitochondrial permeability transition (mPT) through its interactions with matrix cyclophilin D. Binding of cyclophilin D is increased in response to oxidative stress and some thiol reagents that sensitize the mPT to $Ca^{2+}$. Peripherally administered CsA attenuates mitochondrial dysfunction and neuronal damage in an experimental rodent model of traumatic brain injury (TBI), in a dose-dependent manner (Sullivan et al., 2005). The underlying mechanism of neuroprotection mediated by CsA may involve interactions with the mPTP because FK506, which blocks mPT, has some neuroprotective effects. Another mechanism associated with CsA effect may involve the inhibition of calcineurin-mediated dephosphorylation of BAD through an interaction with CYP A (Waldmeier et al., 2003). Similarly, NIM811 is a non-immunosuppressive CsA derivative that inhibits mPT at nanomolar concentrations and with significantly less cytotoxicity than CsA has been used to study the involvement of mPT in SCI. Pretreatment with NIM811 not only improves the mitochondrial respiratory control ratios but also maintains maximal electron transport capacity of complex I and II, as well as their ATP-producing capacity. Consistent with the improvements in mitochondrial function, NIM811 pretreatment significantly reduces free radical generation in isolated mitochondria (McEwen et al., 2007).

In complete spinal cord transaction model of SCI, neurons, astrocytes, and microglia undergo two phases of apoptotic cell death (Wu et al., 2007). The early phase is characterized by high molecular weight DNA fragmentation with nuclear translocation of apoptosis-inducing factor, reduction in mitochondrial respiratory chain enzyme activity, and decrease in cellular levels of ATP. The delayed phase is associated with low molecular weight DNA fragmentation, release of cytochrome *c* from mitochondria into the cytoplasm, activation of caspase-9 and caspase-3, and resumption of mitochondrial respiratory functions and restoration of ATP contents (Wu et al., 2007). Microinfusion of coenzyme $Q_{10}$ into the epicenter of the transected spinal cord not only attenuates both phases of induced apoptosis but also reverses the alterations in mitochondrial dysfunction, bioenergetic failure, and activation of apoptosis-inducing factor, cytochrome *c*, or caspase-9 and caspase-3. It is suggested that mitochondrial dysfunction after spinal cord transection represents the initiating cellular events that trigger the sequential activation of apoptosis-inducing factor-dependent and caspase-dependent signaling cascades, leading to apoptotic cell death in the injured spinal cord (Wu et al., 2007).

## 4.11 Heat Shock Proteins in Spinal Cord Injury

Heat shock proteins (HSPs) are normal intracellular proteins that are expressed in greater amounts when cells are under stress or subjected to traumatic injury (Reddy et al., 2008). These proteins are expressed and released by acutely stressed microglial, endothelial, and ependymal cells and play a key role in the downregulation following SCI. Several HSPs (HSP90, HSP70, and HSP60) are present in brain and spinal cord tissues. These proteins act as molecular chaperones and are called protein guardians because they act to repair partially damaged proteins. They are released in the systemic circulation to act as important anti-inflammatory and anti-apoptotic mediators and provide protection from signaling pathways leading to cell death. After their release into the extracellular fluid, HSP interacts with the surfaces of adjacent cells and initiates signal transduction cascades as well as the transport of cargo molecules, such as antigenic peptides (Chen et al., 2007). By entering bloodstream, HSP60 and HSP70 possess the ability to act at distant sites in the body. Many of the effects of HSPs are mediated through cell surface receptors, including Toll-like receptors (TLRs) 2 and 4, CD40, CD91, CCR5, and members of the scavenger receptor family, such as LOX-1 and SREC-1. The occurrence of a wide range of receptors for the HSP allows their interactions with a diverse range of cells associated with complex multicellular functions particularly in immune cells and neural cell (Chen et al., 2007). At the molecular level, HSP90 interacts with RIP and Akt and promotes NF-κB-mediated downregulation of apoptosis. HSP70 is mostly anti-apoptotic and acts at several levels like prevention of translocation of Bax into mitochondria, release of cytochrome *c* from mitochondria, formation of apoptosome, and inhibition of activation of initiator caspases. HSP70 also modulates JNK, NF-κB, and Akt signaling pathways in the apoptotic cascade (Arya et al., 2007). In contrast, HSP60 has both anti- and pro-apoptotic roles. Cytosolic HSP60 prevents translocation of the pro-apoptotic protein Bax into mitochondria and thus promotes not only cell survival but also promotes maturation of procaspase-3, essential for caspase-mediated cell death (Arya et al., 2007). Collective evidence suggests that spinal cord HSPs assist in the protection of motor neurons and to prevent chronic inflammation and apoptosis following SCI.

## 4.12 Growth Factors in Spinal Cord Injury

Neurotrophins are critical for the survival of neurons not only during development but also after acute neural trauma. Astroglial cells respond to SCI and become reactive, forming scar, a physical and chemical barrier to axonal regeneration. Astrocytic response involves well-described morphological alterations and less characterized functional changes. The functional consequences of astrocyte reactivity seem to depend on the molecular pathway involved and may result in the enhancement of several neuroprotective and neurotrophic functions. Epidermal growth factor (EGF) receptor is upregulated in astrocytes after SCI and facilitates resting astrocyte transformation into reactive astrocytes (Table 4.2). EGF receptor inhibitors enhance axon regeneration promote recovery after SCI. The signaling pathways associated with above processes involves mTOR pathway, a key regulator of astrocyte physiology (Codeluppi et al., 2009). mTOR pathway integrates signals from multiple upstream pathways, including insulin, insulin-like growth factor-1 (IGF-1) and IGF-2, and mitogens. mTOR also functions as a sensor of nutrients, energy status, and cellular redox. These processes occur through Akt-mediated phosphorylation of the GTPase-activating protein tuberin, which blocks tuberin's ability to inhibit the small GTPase Rheb. Indeed, Rheb is necessary for EGF-dependent mTOR activation in spinal cord astrocytes. The astrocytic growth and EGF-dependent chemoattraction are blocked by the mTOR-selective drug rapamycin (Codeluppi

**Table 4.2** Effect of the spinal cord injury on growth factors

| Growth factor | Effect | References |
|---|---|---|
| EGF | Increased | Codeluppi et al. (2009) |
| FGF | Increased | Tassi et al. (2007) |
| TGFβ | Increased | Wang et al. (2009) |
| VEGF | Decreased | Herrera et al. (2009) |
| BDNF | Decreased | Hajebrahimi et al. (2008) |
| NGF | Decreased | Hajebrahimi et al. (2008) |
| NT-3 | Decreased | Hajebrahimi et al. (2008) |
| $P75^{NTR}$ | Increased | Chu et al. (2007) |

et al., 2009). In ischemic model of spinal cord injury, elevation in levels of activated EGF receptor and mTOR signaling occurs in reactive astrocytes in vivo. Furthermore, increased Rheb expression likely contributes to mTOR activation in the injured spinal cord. Treatment of injured rats with rapamycin shows reduced signs of reactive gliosis, suggesting that rapamycin can be used to promote more permissive environment for axon regeneration (Codeluppi et al., 2009). Like the expression of EGF in SCI, unilateral hemisection and contusion injury to adult rat spinal cord cause increased expression of fibroblast growth factor (FGF) and fibroblast growth factor-binding protein (FGF-BP) (Tassi et al., 2007) (Table 4.2). Increase in expression of FGF-BP occurs at all post-injury time points peaking at day 4, a time when injury-mediated increase in levels of FGF2 levels has been reported to be maximal. Although the molecular mechanism associated with the involvement of FGF-BP/FGF2 is not fully understood, FGF-BP is known to enhance FGF2-induced protein tyrosine phosphorylation and AKT/PKB activation. Altogether, these results indicate that FGF-BP is an early response gene after SCI and that its upregulation in regenerating spinal cord tissue may be associated with enhancing the initial FGF2-mediated neurotrophic effects after SCI. Similarly, SCI also increases the expression of thrombospondin-1 (TSP-1) and transforming growth factor-β (TGF-β) in the injured segment of rat spinal cord. After 12 h, levels of TSP-1 increase more rapidly and dramatically than TGF-β levels in the injured segment. Elevations in TSP-1 and TGF-β concentrations persist for 24 h after injury (Wang et al., 2009). Vascular endothelial growth factor (VEGF), a potent mitogen for endothelial cells, plays an important role in vessel outgrowth, arterial and venous differentiation, and vascular remodeling and patternings involved in angiogenesis. Three major isoforms of VEGF (VEGF120, VEGF188, and VEGF164) are known to occur in vascular system. They differ from each other in their solubility (VEGF120 is freely soluble and VEGF188 is completely matrix-bound, while VEGF164 has intermediate properties) and receptor-binding properties. SCI decreases the levels of VEGF165 and other VEGF isoforms at the lesion epicenter 1 day after injury, which was maintained up to 1 month after injury, indicating that VEGF may be associated with the pathophysiology of SCI (Herrera et al., 2009) (Table 4.2).

In addition, SCI markedly effects the expression of several members of neurotrophin family including nerve-growth factor (NGF), brain-derived neurotrophic factor (BDNF), and neurotrophin-3 (NT-3). All these neurotrophins are significantly reduced in the injured spinal cord, as early as 6 h after the induction of the contusion (Hajebrahimi et al., 2008). The expression of other neurotrophin receptors (high-affinity Trk receptors) is severely reduced after the contusion. The expression of TrkA and TrkC is completely blocked after injury along with decrease in expression of TrkB receptor. In contrast to expression of Trk receptors, the expression of $p75^{NTR}$ receptor is significantly upregulated after SCI. $p75^{NTR}$ cooperates with trkA to promote survival. Detailed investigations on the role of the $p75^{NTR}$ in a clip compression model of SCI in $p75^{NTR}$ null mice with an exon III mutation indicate that compared to the functionally deficient $p75^{NTR}$ mice, $p75^{NTR}$ mice functional show an increase in caspase-9 activation at 3 days after SCI. No differences in the activation of the effector caspases (caspase-3 and caspase-6) are observed in the spinal cord lesion at 7 days following SCI (Chu et al., 2007). SCI produces an increase in terminal deoxynucleotidyl transferase-mediated dUTP nick-end (TUNEL) positive cell death in p75NTR-deficient mice at the injury site at 7 days after SCI. Double labeling with TUNEL and cell specific markers indicates that the deficiency of $p75^{NTR}$ increases the extent of neuronal but not oligodendroglial cell death at the injury site. This selective loss of neuronal cells after SCI is accompanied by a decrease in levels of microtubule-associated protein 2 in the $p75^{NTR}$ null mice. Furthermore, the wild-type mice show a dramatic improvement in survival and enhancement in locomotor recovery at 8 weeks after SCI when compared with the $p75^{NTR}$ null mice (Chu et al., 2007). Also at 8 weeks, more neurons present at the injury site of wild-type mice when compared with $p75^{NTR}$ null mice, supporting the view that $p75^{NTR}$ receptor is an integral part of neuronal cell survival in compressive/contusive SCI (Chu et al., 2007).

## 4.13 Other Neurochemical Changes in Spinal Cord Injury

Spinal cord injury and/or regeneration related protein 1 (SCIRR1 protein) is a transcribed product of scirr1 gene. SCIRR 1 protein is upregulated and expressed very highly in spinal cord neurons farther from the epicenter of injury (Liu et al., 2007a). Although the molecular mechanism and precise function of scirr1 gene and SCIRR1 protein are unknown, its upregulation after spinal cord injury suggests that SCIRR1 protein may be closely associated with repair processes in the injured spinal cord. In addition, the typical F-box and leucine-rich repeat (LRR) architecture of rat SCIRR1 protein indicate that it may be involved in substrate recruiting role in the pleiotropic ubiquitin/proteasome pathway (Liu et al., 2007b).

Erythropoietin (EPO), a hematopoietic cytokine, is a glycoprotein that mediates cytoprotection in brain and spinal cord through activation of multiple signaling pathways (Grasso et al., 2006). In addition, EPO also has a crucial hormonal role in red cell production. In the brain and spinal cord, EPO and its receptor (EPO-R) are modulated by metabolic stressors, and provide anti-inflammatory functions. RT-PCR and immunocytochemical studies indicate that rat microglial cells and the murine microglia cell line BV-2 express the EPO-R. However, EPO has no effect on the release of the pro-inflammatory mediators' nitric oxide and TNF-α. Moreover, EPO does not reduce the LPS (lipopolysaccharide)-mediated translocation of the pro-inflammatory transcription factor NF-κB into the nucleus of murine microglia, but induce $^3$H-thymidine incorporation into DNA of microglial cells (Wilms et al., 2009). These results show that microglial cells are target cells for EPO, which possesses mitogenic, but not anti-inflammatory effects on microglia (Wilms et al., 2009).

SCI markedly increases the expression of EPO and EPO-R in neurons, vascular endothelium, and glial cells 8 h after injury. Expression peaks at 8 days, after which it gradually decreases (Grasso et al., 2006). Two weeks after injury, EPO immunoreactivity is scarcely detected in neurons, whereas in glial cells and vascular endothelium, EPO-R immunoreactivity is strongly expressed suggesting that the local EPO and EPO-R system is markedly engaged in the early stages after SCI (Grasso et al., 2006; Matis and Birbilis, 2009). In addition to the above neurochemical changes, SCI involves alterations in mitogen-activated protein kinase pathways, including ASK1, JNK, and p38, which are activated in destructive spinal cord under chronic compression (Takenouchi et al., 2008). Activation of these kinases facilitates both secondary degeneration around the site of injury and chronic demyelination.

SCI is accompanied by alterations in ceramide metabolism (Cuzzocrea et al., 2009). Inhibitors of ceramide synthase (fumonisin B1), acid sphingomyelinase (tyclodecan-9-xanthogenate, D609), and the secretory form of acid sphingomyelinase (3-carbazol-9-yl-propyl)-[2-(3,4-dimethoxy-phenyl)-ethyl]-methylamine (NB6) not only reduce the degree of ceramide synthesis, and tissue injury, but also block neutrophil infiltration, inhibit nitrotyrosine generation, TNF-α and IL-β release, and apoptosis (TUNEL staining and Bax and Bcl-2 expression). Significant improvement of motor function occurs in mice treated with fumonisin B1 and D609, NB6. Collective evidence suggests that ceramide participates in pathogenesis of spinal cord injury (Cuzzocrea et al., 2009)

## 4.14 Neuropathic Pain in SCI

Neuropathic pain is a spontaneous persistent pain characterized by a range of abnormally evoked responses, e.g., allodynia (pain evoked by normally non-noxious stimuli) and hyperalgesia (an increased response to noxious stimuli). Neuropathic pain following SCI is usually present at or below the level of injury (Yiu and He, 2006). The molecular mechanism associated with neuropathic pain is not fully understood. However, recent studies indicate the involvement of interactions between non-enzymic oxidation product of arachidonate and transient receptor potential cation channel A1 (TRPA1, an excitatory ion channel) may contribute to neuropathic pain following SCI (Trevisani et al., 2007). TRPA1 channels function as a mechanical stress sensor and play an important role in inflammatory pain. This suggestion is based on TRPA1 knock-out mice, which show near complete attenuation of formalin-induced pain behaviors (McNamara et al., 2007). TRPA1 channels are expressed in a subpopulation of primary afferent somatosensory neurons that contain substance P and calcitonin gene-related peptide. 4-HNE and acrolein have been reported to activate TRPA1 provoke the release of substance P and calcitonin gene-related peptide from central (spinal cord) and peripheral (esophagus) nerve endings. These processes are closely associated with acute pain and neurogenic plasma protein extravasation in peripheral tissues along with inflammation (Trevisani et al., 2007). Moreover, injection of 4-HNE into the rodent hind paw elicits pain-related behaviors that are blocked by TRPA1 antagonists and absent in animals lacking functional TRPA1 channels. Collective evidence indicates that arachidonate-derived 4-HNE may activate TRPA1 on nociceptive neurons to promote acute pain, neuropeptide release, and neurogenic inflammation in SCI (Trevisani et al., 2007).

Oral pain killer used for the treatment of neuropathic pain act in several ways: (a) by depressing neuronal activity, (b) by blocking sodium channels or inhibiting calcium channels, (c) by increasing inhibition via GABA agonists, (d) by serotonergic and noradrenergic reuptake inhibition, and (e) by decreasing activation via glutamate receptor inhibition, especially by blocking the NMDA receptor (Yezierski, 2005). At present, only ten randomized, double-blind, controlled trials have been performed on oral drug treatment of pain after SCI, but results have been negative.

## 4.15 Contribution of Oxidative Stress in Spinal Cord Injury

Oxidation of arachidonic acid after SCI generates high levels of reactive oxygen species (ROS), which include oxygen free radicals (superoxide, hydroxyl, and alkoxyl radicals), and peroxides (hydrogen peroxide and lipid hydroperoxides) (Fig. 4.7). ROS are also produced by mitochondrial dysfunction. At high levels ROS after SCI contribute to neural membrane damage when the balance between reducing and oxidizing (redox) forces shifts toward oxidative stress. The biological targets of ROS include membrane proteins, unsaturated lipids, and DNA (Farooqui and Horrocks, 2009). The increase in carbonyl groups in proteins is an important index of protein oxidation and ROS-mediated damage. The reaction between ROS and proteins or unsaturated lipids in the plasma membrane leads to chemical cross-linking of membrane proteins and lipids and a reduction in membrane unsaturation. The depletion of unsaturation in membrane lipids is associated with

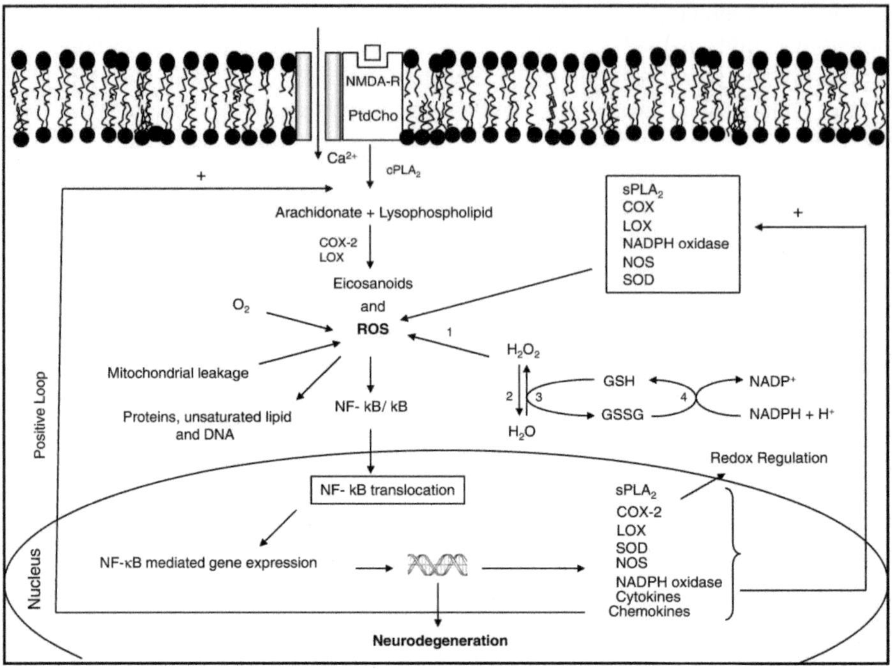


**Fig. 4.7** Involvement of ROS-induced activation of NF-κB, redox status, and gene expression in spinal cord injury. (1) superoxide dismutase; (2) catalase; (3) glutathione peroxidase; (4) glutathione reductase; $cPLA_2$, cytosolic phospholipase $A_2$; $sPLA_2$, secretory phospholipase $A_2$; COX-2, cyclooxygenase-2; LOX, lipoxygenase; SOD, superoxide dismutase; NOS, nitric oxide; cytokines, TNF-α, and IL-1β; and $O_2^-$, superoxide radical. These interactions facilitate the transcription of $sPLA_2$ and COX-2 in the nucleus. The expression of cytokines upregulates activities of $cPLA_2$ and $sPLA_2$ through a positive loop type of mechanism in cytoplasm and neural membranes

alteration in membrane fluidity and decrease in the activity of membrane-bound enzymes, ion channels, and receptors (Farooqui and Horrocks, 2009). ROS also attack DNA bases causing damage through hydroxylation, ring opening, and fragmentation. This attack generates 8-hydroxy-2′-deoxyguanosine (8-OHdG) and 2, 6-diamino-4-hydroxy-5-formamidopyrimidine (FapyGua) (Farooqui and Horrocks, 2007, 2009). Among the neural cells, astrocytes are most resistant to ROS attack. Astrocytes protect neurons from oxidative stress because they have higher glutathione content than other neural cells. During ROS scavenging, the reduced form of glutathione is oxidized. Collective evidence suggests that the oxidation of glycerophospholipids, chemical cross-linking of neural membrane proteins, and oxidation of neural cell DNA are significant chemical events associated with oxidative stress and disruption of ion homeostasis during injury-mediated spinal cord damage.

## 4.16 Inflammation in Spinal Cord Injury

Inflammation is a protective mechanism associated with neutralization of an insult and restoration of normal structure and function of brain and spinal cord tissues (Farooqui et al., 2007). The main mediators of inflammation are microglial cells and macrophages, which release a variety of factors that initiate and support inflammation (Fig. 4.8). Inflammatory process also recruits polymorphonuclear leukocytes (PMN) from the bloodstream into brain and spinal cord tissues. This PMN migration is a coordinated multistep process involving chemotaxis, adhesion of PMN to endothelial cells in the area of inflammation (Farooqui et al., 2007). PMN eliminate dead cells by phagocytosis and release free radicals and lytic enzymes into phagolysosomes. Thus, both neural and non-neural cells are activated after SCI and initiate many process facilitating cell migration, proliferation, and release of cytokines/chemokines and trophic and/or toxic effects. Cytokines/chemokines stimulate $PLA_2$, COX-2, NOS, and SMases (Tian et al., 2007) (Figs. 4.4 and 4.7). This results in breakdown of glycerophospholipids and sphingolipids with release of arachidonic acid and ceramide. Oxidation of arachidonic acid generates pro-inflammatory prostaglandins, leukotrienes, and thromboxanes and metabolites of ceramide metabolism are associated with apoptosis (Table 4.3). Glycerophospholipid- and sphingolipid-derived pro-inflammatory mediators intensify inflammation and apoptotic cell death. Studies on the comparison of inflammation in the brain and spinal cord following mechanical injury

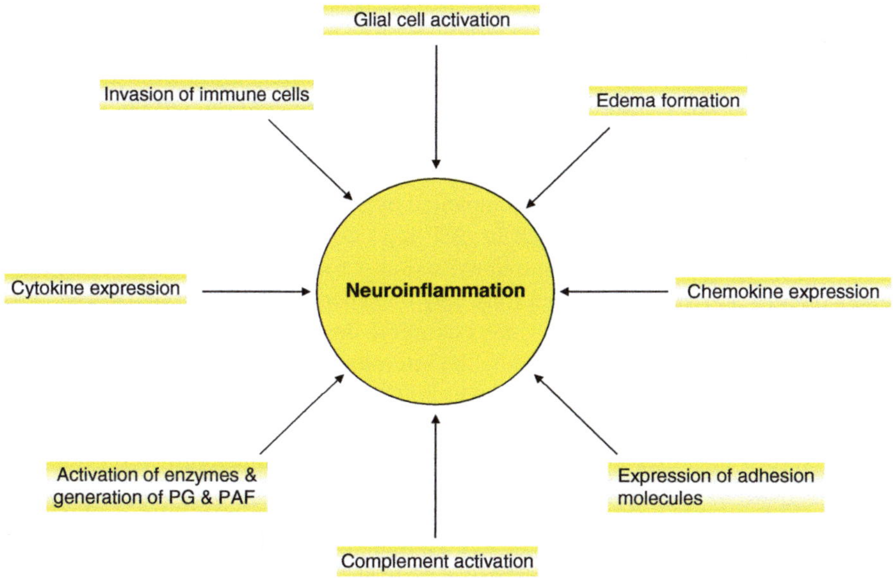


**Fig. 4.8** Factors promoting inflammation in injured spinal cord

**Table 4.3** Status of lipid mediators in SCI

| Neurochemical parameter | Head injury | References |
|---|---|---|
| Glycerophospholipid metabolism | Enhanced | Farooqui et al. (2004) |
| Free fatty acid levels | Increased | Phillis et al. (2006) |
| Eicosanoids levels | Increased | Phillis et al. (2006) |
| Lipid peroxidation rate | Increased | Phillis et al. (2006) |
| 4-Hydroxynonenal levels | Increased | Phillis et al. (2006) |
| Isoprostanes | Increased | Oner-Iyidogan et al. (2004) |
| Excitotoxicity intensity | Involved | Farooqui and Horrocks (2009) |
| Oxidative stress intensity | Increased | Farooqui and Horrocks (2009) |
| Neuroinflammation intensity | Increased | Farooqui and Horrocks (2009) |
| Neurodegeneration rate | Increased | Farooqui and Horrocks (2009) |
| Apoptosis | Increased | Beattie et al. (2000) |

indicate that 1 week after injury, the microglial and macrophage response is significantly greater in the spinal cord compared to the brain (Batchelor et al., 2008). Moreover, a greater inflammatory response occurs in white matter compared to gray matter within the brain and spinal cord injuries. Because activated microglia and macrophages are the effectors of secondary damage, a greater degree of inflammation in the spinal cord is likely to result in more extensive secondary damage mediated by eicosanoids, cytokines, and chemokines (Batchelor et al., 2008). It is suggested that inflammation facilitates the development of scar formation following SCI. Collective evidence suggests that inflammation has beneficial as well as detrimental effects after spinal cord injury (Chan, 2008).

## 4.17 Interactions Among Excitotoxicity, Oxidative Stress, and Inflammation in Spinal Cord Injury

It is well established that SCI is accompanied by excitotoxicity, oxidative stress, and inflammation. Neuronal cell death in SCI is a coordinated multistep process that involves interplay among excitotoxicity, oxidative stress, and neuroinflammation. The effect of this interplay on neurons may be synergistic or cumulative (Fig. 4.9). Terminally differentiated neurons may commit to death in response to abnormal signal transduction processes initiated by the interplay among excitotoxicity, oxidative stress, and inflammation (Farooqui et al., 2007). Initially, the coordinated interplay among excitotoxicity, oxidative stress, and neuroinflammation in SCI may cause abnormalities in motor and cognitive performance. An enhanced rate (upregulation) of interplay among the above processes may be associated with the increased vulnerability of neurons in SCI. This interplay may be a common mechanism of brain damage in acute neural trauma, which include SCI, TBI, and stroke (Farooqui and Horrocks, 2007; Farooqui, 2009; Farooqui and Horrocks, 2009). Environmental factors such as diet (enrichment of ω-6 fatty acids) and lifestyle (lack of exercise)

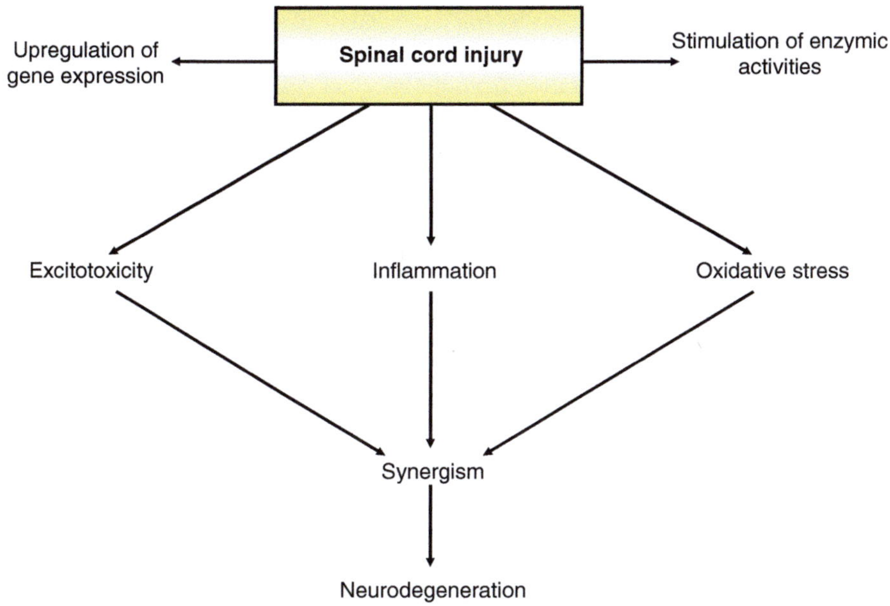


**Fig. 4.9** Excitotoxicity, inflammation, and oxidative stress-mediated neurodegeneration in spinal cord injury

may also play a prominent role in modulating the interplay among excitotoxicity, oxidative stress, and neuroinflammation. Thus, long- and short-term locomotor activity of moderate intensity induce stimuli sufficient to recruit a majority of spinal cells to increased BDNF synthesis, suggesting that continuous tuning of pro-BDNF and BDNF levels permits spinal networks to undergo trophic modulation without requiring changes in TrkB mRNA supply. In SCI, neurons die rapidly, a matter of hours to days, because of the sudden lack of oxygen, decrease in ATP level, sudden collapse of ion gradients, and the rapid upregulation of interplay among excitotoxicity, oxidative stress, and neuroinflammation. In contrast, in neurodegenerative diseases, oxygen, nutrients and ATP continue to be available to the nerve cells, and ionic homeostasis is maintained to a limited extent. The interplay among excitotoxicity, oxidative stress, and neuroinflammation occurs at a slow rate, resulting in a neurodegenerative process that takes several years to develop (Farooqui, 2009).

## 4.18 Conclusion

SCI is an irreversible condition that causes damage to myelinated fiber tracts that carry sensation and motor signals to and from the brain. It involves primary and secondary mechanisms. Primary mechanism of SCI refers to the initial mechanical damage due to local deformation of the spine. Direct compression and trauma to neural elements and blood vessels by fractured and displaced bone fragments or disc material occur after mechanical insult. The secondary mechanism is initiated by the primary injury. Neurodegeneration from the mechanical injury is predominated by necrosis. Secondary injury triggers neurodegeneration through necrotic and apoptotic cell death. The secondary mechanism includes a cascade of biochemical and cellular processes, such as release of glutamate; overstimulation of glutamate receptors and calcium influx; stimulation of $PLA_2$, COX-2, NOS, calpains, caspases, and MMP; formation of free radicals, oxidative stress, vascular ischemia, edema; activation of transcription factors; induction of cytokines and chemokines, post-traumatic inflammatory reaction, activation of the complement system; and apoptotic cell death. Apoptotic cascade also involves the mitochondrial dysfunction and release of cytochrome *c*, activation of caspases, and ultimately induction of nuclear DNA condensation and fragmentation. Anti-apoptotic signaling pathways involve the activation of neurotrophic factors and certain cytokines. Neuroprotective pathways following SCI involve the activation of the transcription factors (NF-κB) that induce expression of stress proteins, antioxidant enzymes, and calcium-regulating proteins; phosphorylation-mediated modulation of ion channels and membrane transporters; cytoskeletal alterations that modulate calcium homeostasis; and modulation of proteins that stabilize mitochondrial function (e.g., Bcl-2). Blunt trauma to spinal cord results not only in primary membrane damage to neuronal cell bodies but also to white matter structures. Severe traumatic insult to spinal cord produces mitochondrial dysfunction, alteration in ion homeostasis, and changes in redox status of spinal cord tissue ultimately resulting in neuronal death. Thus, accumulating evidence suggests that SCI is characterized by the upregulation of genes involved in transcription, inflammation, excitotoxicity, oxidative stress, and a general downregulation of neural function-related genes. These changes result in edema, apoptosis, and recruitment of peripherally derived immature cells. Following SCI, apoptotic cell death continues, and scarring and demyelination accompany Wallerian degeneration. These processes are reflected in a general failure of normal neural functions and a stage of signal shock that lasts for several days in experimental SCI. Strong expression of transcription factor, STAT, may represent an early attempt of spinal cord repair and regeneration.

## References

Adachi K, Yimin Y, Satake K, Matsuyama Y, Ishiguro N, Sawada M, Hirata Y, Kiuchi K (2005) Localization of cyclooxygenase-2 induced following traumatic spinal cord injury. Neurosci Res 51:73–80

Amar AP, Levy ML (1999) Pathogenesis and pharmacological strategies for mitigating secondary damage in acute spinal cord injury. Neurosurgery 44:1027–1040

Ankeny DP, Lucin KM, Sander VM, McGaughy VA, Popovich PG (2006) Spinal cord injury triggers systemic autoimmunity: evidence for chronic B lymphocyte activation and lupus-like autoantibody synthesis. J Neurochem 99:1073–1087

Ankeny DP, Popovich PG (2009) Mechanisms and implications of adaptive immune responses after traumatic spinal cord injury. Neuroscience 158:1112–1121

Arya R, Mallik M, Lakhotia SC (2007) Heat shock genes – Integrating cell survival and death. J BioSci 32:595–610

Ashki N, Hayes KC, Bao F (2008) The peroxynitrite donor 3-morpholinosydnonimine induces reversible changes in electrophysiological properties of neurons of the guinea-pig spinal cord. Neurosci 156:107–117

Auger C, Attwell D (2000) Fast removal of synaptic glutamate by postsynaptic transporters. Neuron 28:547–558

Bao F, Bailey CS, Gurr KR, Bailey SI, Rosas-Arellano MP, Dekaban GA, Weaver LC (2009) Increased oxidative activity in human blood neutrophils and monocytes after spinal cord injury. Exp Neurol 215:308–316

Bareyre FM, Schwab ME (2003) Inflammation, degeneration and regeneration in the injured spinal cord: insights from DNA microarrays. Trends Neurosci 26:555–563

Barut S, Unlu YA, Karaoglan A, Tuncdemir M, Dagistanli FK, Ozturk M, Colak A (2005) The neuroprotective effects of z-DEVD.fmk, a caspase-3 inhibitor, on traumatic spinal cord injury in rats. Nurg Neurol 64:213–220

Batchelor PE, Tan S, Wills TE, Porritt MJ, Howells DW (2008) Comparison of inflammation in the brain and spinal cord following mechanical injury. J Neurotrauma 25:1217–1225

Beattie MS, Farooqui AA, Bresnahan JC (2000) Review of current evidence for apoptosis after spinal cord injury. J Neurotrauma 17:915–925

Beere HM (2005) Death versus survival: functional interaction between the apoptotic and stress-inducible heat shock protein pathways. J Clin Invest 115:2633–2639

Bethea JR, Castro M, Keane RW, Lee TT, Dietrich WD, Yezierski RP (1998) Traumatic spinal cord injury induces nuclear factor-ćB activation. J Neurosci 18:3251–3260

Brambilla R, Bracchi-Ricard V, Hu WH, Frydel B, Bramwell A, Karmally S, Green EJ, Bethea JR (2005) Inhibition of astroglial nuclear factor kappaB reduces inflammation and improves functional recovery after spinal cord injury. J Exp Med 202:145–156

Brambilla R, Hurtado A, Persaud T, Esham K, Pearse DD, Oudega M, Bethea JR (2009) Transgenic inhibition of astroglial NF-kappaB leads to increased axonal sparing and sprouting following spinal cord injury. J Neurochem 110:765–778

Bramlett HM, Dietrich WD (2004) Pathophysiology of cerebral ischemia and brain trauma: similarities and differences. J Cereb Blood Flow Metab 24:133–150

Block ML, Hong JS (2005) Microglia and inflammation-mediated neurodegeneration: multiple triggers with a common mechanism. Prog Neurobiol 76:77–98

Buki A, Farkas O, Doczi T, Povlishock JT (2003) Preinjury administration of the calpain inhibitor MDL-28170 attenuates traumatically induced axonal injury. J Neurotrauma 20:261–268

Buss A, Pech K, Kakulas BA, Martin D, Schoenen J, Noth J, Brook GA (2007) Matrix metalloproteinases and their inhibitors in human traumatic spinal cord injury. BMC Neurol 26:7–17

Calderon F, Kim HY (2004) Docosahexaenoic acid promotes neurite growth in hippocampal neurons. J Neurochem 90:979–988

Cao DH, Xue RH, Xu J, Liu ZL (2005) Effects of docosahexaenoic acid on the survival and neurite outgrowth of rat cortical neurons in primary cultures. J Nutr Biochem 16:538–546

Casha S, Yu WR, Fehlings MG (2005) FAS deficiency reduces apoptosis, spares axons and improves function after spinal cord injury. Exp Neurol 196:390–400

Chan CC (2008) Inflammation: beneficial or detrimental after spinal cord injury? Recent Pat CNS Drug Discov 3:189–199

Chatzipanteli K, Garcia R, Marcillo AE, Loor KE, Kraydieh S, Dietrich WD (2002) Temporal and segmental distribution of constitutive and inducible nitric oxide synthases after traumatic spinal cord injury: effect of aminoguanidine treatment. J Neurotrauma 19:639–651

Chen Y, Voegeli TS, Liu PP, Nobel EG, Currie RW (2007) Heat shock paradox and a new role of heat shock proteins and their receptors as anti-inflammation targets. Inflamm Allergy Drug Targets 6:91–100

Chu GK, Yu W, Fehlings MG (2007) The p75 neurotrophin receptor is essential for neuronal cell survival and improvement of functional recovery after spinal cord injury. Neuroscience 148:668–682

Citron BA, Arnold PM, Haynes NG, Ameenuddin S, Farooque M, Santacruz K, Festoff BW (2008) Neuroprotective effects of caspase-3 inhibition on functional recovery and tissue sparing after acute spinal cord injury. Spine 33:2269–2277

Codeluppi S, Svensson CI, Hefferan MP, Valencia F, Silldorff MD, Oshiro M, Marsala M, Pasquale EB (2009) The Rheb-mTOR pathway is upregulated in reactive astrocytes of the injured spinal cord. J Neurosci 29:1093–1104

Cohen GM (1997) Caspases: the executioners of apoptosis. Biochem J 326:1–16

Conrad S, Schluesener HJ, Trautmann K, Joannin N, Meyermann R, Schwab JM (2005) Prolonged lesional expression of RhoA and RhoB following spinal cord injury. J Comp Neurol 487: 166–175

Creagh EM, Conroy H, Martin SJ (2003) Caspase-activation pathways in apoptosis and immunity. Immunol Rev 193:10–21

Cuzzocrea S, Deigner HP, Genovese T, Mazzon E, Esposito E, Crisafulli C, Di Paola R, Bramanti P, Matuschak G, Salvemini D (2009) Inhibition of ceramide biosynthesis ameliorates pathological consequences of spinal cord injury. Shock 31:634–644

Dang AB, Tay BK, Kim HT, Nauth A, Alfonso-Jaume MA, Lovett DH (2008) Inhibition of MMP2/MMP9 after spinal cord trauma reduces apoptosis. Spine 33:E576–E579

Davis AR, Lotocki G, Marcillo AE, Dietrich WD, Keane RW (2007) FasL, Fas, and death-inducing signaling complex (DISC) proteins are recruited to membrane rafts after spinal cord injury. J Neurotrauma 24:823–834

Demediuk P, Saunders RD, Anderson DK, Means ED, Horrocks LA (1985) Membrane lipid changes in laminectomized and traumatized cat spinal cord. Proc Natl Acad Sci USA 82:7071–7075

Dergham P, Ellezam B, Essagian C, Avedissian H, Lubell WD, McKerracher L (2002) Rho signaling pathway targeted to promote spinal cord repair. J Neurosci 22:6570–6577

de Rivero Vaccari JP, Lotocki G, Marcillo AE, Dietrich WD, Keane RW (2008) A molecular platform in neurons regulates inflammation after spinal cord injury. J Neurosci 28:3404–3414

de Rivero Vaccari JP, Lotocki G, Alonsoo F, Bramlett HM, Dietrich WD, Keane RW (2009) Therapeutic neutralization of the NLRP1 inflammasome reduces the innate immune response and improves histopathology after traumatic brain injury. J Cerebr Blood Flow Metab 29 Apr [Epub ahead of print]

Di Giovanni S, Knoblach SM, Brandoli C, Aden SA, Hoffman EP, Faden AI (2003) Gene profiling in spinal cord injury shows role of cell cycle in neuronal death. Ann Neurol 53:454–468

Drew PD, Storer PD, Xu J, Chavis JA (2005) Hormone regulation of microglial cell activation: relevance to multiple sclerosis. Brain Res Brain Res Rev 48:322–327

Dziennis S, Alkayed NJ (2008) Role of signal transducer and activator of transcription 3 in neuronal survival and regeneration. Rev Neurosci 19:341–361

Eftekharpour E, Karimi-Abdolrezaee S, Fehlings MG (2008) Current status of experimental cell replacement approaches to spinal cord injury. Neurosurg Focus 24:E19

Ellis RC, Earnhardt JN, Hayes RL, Wang KKW, Anderson DK (2004) Cathepsin B mRNA and protein expression following contusion spinal cord injury in rats. J Neurochem 88:689–697

Esposito E, Genovese T, Caminiti R, Bramanti P, Meli R, Cuzzocrea S (2008) Melatonin regulates matrix metalloproteinases after traumatic experimental spinal cord injury. J Pineal Res 45: 149–156

Farooqui AA, Horrocks LA (1991) Excitatory amino acid receptors, neural membrane phospholipid metabolism and neurological disorders. Brain Res Rev 16:171–191

Farooqui AA, Horrocks LA (1994) Excitotoxicity and neurological disorders: involvement of membrane phospholipids. Int Rev Neurobiol 36:267–323

Farooqui AA, Ong WY, Horrocks LA (2004) Biochemical aspects of neurodegeneration in human brain: involvement of neural membrane phospholipids and phospholipases $A_2$. Neurochem Res 29:1961–1977

Farooqui AA, Ong WY, Horrocks LA (2006) Inhibitors of brain phospholipase $A_2$ activity: their neuropharmacological effects and therapeutic importance for the treatment of neurologic disorders. Pharmacol Rev 58:591–620
Farooqui AA, Horrocks LA (2007) Glycerophospholipids in brain. Springer, New York, NY
Farooqui AA, Horrocks LA, Farooqui T (2007) Modulation of inflammation in brain: a matter of fat. J Neurochem 101:577–599
Farooqui AA, Ong WY, Horrocks LA (2008) Neurochemical aspects of excitotoxicity. Springer, New York, NY
Farooqui AA (2009) Hot topics in neural membrane lipidology. Springer, New York, NY
Farooqui AA, Horrocks LA (2009) Glutamate and cytokine-mediated alterations of phospholipids in head injury and spinal cord trauma. In: Banik NK Ray SK (eds) Handbook of neurochemistry and molecular neurobiology, vol 24. Springer, New York, NY, pp 71–89
Filbin MT (2003) Myelin-associated inhibitors of axonal regeneration in the adult mammalian CNS. Nat Rev Neurosci 4:703–713
Genovese T, Mazzon E, Esposito E, Di Paola R, Bramanti P, Cuzzocrea S (2005) Inhibitors of poly(ADP-ribose) polymerase modulate signal transduction pathways and secondary damage in experimental spinal cord trauma. J Pharmacol Exp Ther 312:449–457
Genovese T, Cuzzocrea S (2008) Role of free radicals and poly(ADP-ribose)polymerase-1 in the development of spinal cord injury: new potential therapeutic targets. Curr Med Chem 43: 763–780
Genovese T, Mazzon E, Esposito E, Di Paola R, Murthy K, Neville L, Bramanti P, Cuzzocrea S (2009a) Effects of a metalloporphyrinic peroxynitrite decomposition catalyst, ww-85, in a mouse model of spinal cord injury. Free Radic Res 5:1–15
Genovese T, Esposito E, Mazzon E, Crisafulli C, Paterniti I, Di Paola R, Galuppo M, Bramanti P, Cuzzocrea S (2009b) PPAR-? modulate the anti-inflammatory effect of glucocorticoids in the secondary damage in experimental spinal cord trauma. Pharmacol Res 59:338–350
Ghirnikar RS, Lee YL, Eng LF (2000) Chemokine antagonist infusion attenuates cellular infiltration following spinal cord contusion injury in rat. J Neurosci Res 59:63–73
Ghirnikar RS, Lee YL, Eng LF (2001) Chemokine antagonist infusion promotes axonal sparing after spinal cord contusion injury in rat. J Neurosci Res 64:582–589
Gomes-Leal W, Corkill DJ, Freire MA, Picanço-Diniz CW, Perry VH (2004) Astrocytosis, microglia activation, oligodendrocyte degeneration, and pyknosis following acute spinal cord injury. Exp Neurol 190:456–467
Grasso DG, Sfacteria A, Passalacqua M, Morabito A, Buemi M, Marcri B, Brines M, Tomasello F (2006) Erythropoietin and erythropoietin receptor expression after experimental spinal cord injury encourages therapy by exogenous erythropoietin. Neurosurg 56:821–827
Gris D, Hamilton EF, Weaver LC (2008) The systemic inflammatory response after spinal cord injury damages lungs and kidneys. Exp Neurol 211:259–270
Hajebrahimi Z, Mowla SJ, Movahedin M, Tavallaei M (2008) Gene expression alterations of neurotrophins, their receptors and prohormone convertases in a rat model of spinal cord contusion. Neurosci Lett 441:261–266
Herrera JJ, Nesic-Taylor DO, Narayana PA (2009) Reduced Vascular Endothelial Growth Factor Expression In Contusive Spinal Cord Injury. J Neurotrauma 26:995–1003
Herrmann JE, Imura T, Song B, Oi J, Ao Y, Nguyen TK, Korsak RA, Takeda K, Akira S, Sifroniew MV (2008) STAT3 is a critical regulator of astrogliosis and scar formation after spinal cord injury. J Neurosci 28:7231–7243
Horrocks LA, Demediuk P, Saunders RD, Dugan L, Clendenon NR, Means ED, Anderson DK (1985) The degradation of phospholipids, formation of metabolites of arachidonic acid, and demyelination following experimental spinal cord injury. Cent Nerv Syst Trauma 2:115–120
Hsu TC, Young MR, Cmarik J, Colburn NH (2000) Activator protein 1 (AP-1)- and nuclear factor kappaB (NF-kappaB)-dependent transcriptional events in carcinogenesis. Free Radic Biol Med 28:1338–1348

Hsu JY, Mckeon R, Goussev S, Werb Z, Lee JU, Trivedi A, Noble-Haeusslein LJ (2006) Matrix metalloproteinase-2 facilitates wound healing events that promote functional recovery after spinal cord injury. J Neurosci 26:9841–9850

Hsu JY, Bourguignon LY, Adams CM, Peyrollier K, Zhang H, Fandel T, Cun CL, Werb Z, Noble-Haeusslein LJ (2008) Matrix metalloproteinase-9 facilitates glial scar formation in the injured spinal cord. J Neurosci 28:13467–13477

Huang W, Bhavsar A, Ward RE, Hall JC, Priestley JV, Michael-Titus AT (2009) Arachidonyl trifluoromethyl ketone is neuroprotective after spinal cord injury. J Neurotrauma 16 Apr 2009 [Epub ahead of print]

Ikemoto A, Kobayashi T, Watanabe S, Okuyama H (1997) Membrane fatty acid modifications of PC12 cells by arachidonate or docosahexaenoate affect neurite outgrowth but not norepinephrine release. Neurochem Res 22:671–678

Jupp OJ, Vandenabeele P, MacEwan DJ (2003) Distinct regulation of cytosolic phospholipase $A_2$ phosphorylation, translocation, proteolysis and activation by tumour necrosis factor-receptor subtypes. Biochem J 374:453–461

Kauppinen TM, Swanson RA (2007) The role of poly(ADP-ribose) polymerase-1 in CNS disease. Neuroscience 145:1267–1272

Keane RW, Davis AR, Dietrich WD (2006) Inflammatory and apoptotic signaling after spinal cord injury. J Neurotrauma 23:335–344

Khorooshi R, Babcock AA, Owens T (2008) NF-kappaB-driven STAT2 and CCL2 expression in astrocytes in response to brain injury. J Immunol 181:7284–7291

Klusman I, Schwab ME (1997) Effects of pro-inflammatory cytokines in experimental spinal cord injury. Brain Res 762:173–184

Klussmann S, Martin-Villalba A (2005) Molecular targets in spinal cord injury. J Mol Med 83: 657–671

Kim GM, Xu J, Xu JM, Song SK, Yan P, Ku G, Xu XM, Hsu CY (2001) Tumor necrosis factor receptor deletion reduces nuclear factor-kappa B activation, cellular inhibitor of apoptosis protein 2 expression, and functional recovery after traumatic spinal cord injury. J Neurosci 21:6617–6625

Knoblach SM, Huang X, VanGelderen J, Calva-Cerqueira D, Faden AE (2005) Selective caspase activation may contribute to neurological dysfunction after experimental spinal cord trauma. J Neurosci Res 80:369–380

Kobayashi H, Chattopadhyay S, Kato K, Dolkas J, Kikuchi S, Myers RR, Shubayev VI (2008) MMPs initiate Schwann cell-mediated MBP degradation and mechanical nociception after nerve damage. Mol Cell Neurosci 39:619–627

Komjati K, Besson VC, Szabo C (2005) Poly (adp-ribose) polymerase inhibitors as potential therapeutic agents in stroke and neurotrauma. Curr Drug Targets CNS Neurol Disord 4: 179–194

Lee YL, Bao P, Ghirnikar RS, Eng LF (2000) Cytokine chemokine expression in contused rat spinal cord. Neurochem Int 36:417–425

Lee MY, Chen L, Toborek M (2009) Nicotine attenuates iNOS expression and contributes to neuroprotection in a compressive model of spinal cord injury. J Neurosci Res 87:937–947

Leker RR, Shohami E (2002) Cerebral ischemia and trauma-different etiology yet similar mechanisms: neuroprotective opportunities. Brain Res Rev 39:55–73

Liu XZ, Xu XM, Hu R, Du C, Zhang SX, McDonald JW, Dong HX, Wu YJ, Fan GS, Jacquin MF, Hsu CY, Choi DW (1997) Neuronal and glial apoptosis after traumatic spinal cord injury. J Neurosci 17:5395–5406

Liu N, Han S, Lu PH, Xu XM (2004) Upregulation of annexins I, II, and V after traumatic spinal cord injury in adult rats. J Neurosci Res 77:391–401

Liu NK, Zhang YP, Titsworth WL, Jiang X, Han S, Lu PH, Shields CB, Xu XM (2006) A novel role of phospholipase $A_2$ in mediating spinal cord secondary injury. Ann Neurol 59:606–619

Liu NK, Zhang YP, Han S, Pei J, Xu LY, Lu PH, Shields CB, Xu XM (2007a) Annexin A1 reduces inflammatory reaction and tissue damage through inhibition of phospholipase A2 activation in adult rats following spinal cord injury. J Neuropathol Exp Neurol 66:932–943

Liu T, Ma Z, Que H, Li X, Ni Y, Jing S, Liu S (2007b) Identification and characterization of scirr1, a novel gene up-regulated after spinal cord injury. Exp Mol Med 39:255–266

Malemud CJ (2006) Matrix metalloproteinases (MMPs) in health and disease: an overview. Front BioSci 11:1696–1701

Marsala J, Orendacova J, Lukacova N, Vanicky I (2007) Traumatic injury of the spinal cord and nitric oxide. Prog Brain Res 161:171–183

Matis GK, Birbilis TA (2009) Erythropoietin in spinal cord injury. Eur Spine J 18:314–323

Matute C, Domercq M, Sánchez-Gómez MV (2006) Glutamate-mediated glial injury: mechanisms and clinical importance. Glia 53:212–224

McEwen ML, Springer JE (2005) A mapping study of caspase-3 activation following acute spinal cord contusion in rats. J Histochem Cytochem 53:809–819

McEwen ML, Sullivan PG, Springer JE (2007) Pretreatment with the cyclosporin derivative, NIM811, improves the function of synaptic mitochondria following spinal cord contusion in rats. J Neurotrauma 24:613–624

McKerracher L, Winton MJ (2002) Nogo on the go. Neuron 36:345–348

McNamara CR, Mandel-Brehm J, Bautista DM, Siemens J, Deranian KL, Zhao M, Hayward NJ, Chong JA, Julius D, Moran MM, Fanger CM (2007) TRPA1 mediates formalin-induced pain. Proc Natl Acad Sci USA 104:13525–13530

Oner-Iyidoğan Y, Koçak H, Gürdöl F, Koçak T, Erol B (2004) Urine 8-isoprostane F2alpha concentrations in patients with neurogenic bladder due to spinal cord injury. Clin Chim Acta 339:43–47

Ong WY, Horrocks LA, Farooqui AA (1999) Immunocytochemical localization of $cPLA_2$ in rat and monkey spinal cords. J Mol Neurosci 12:123–130

Panter SS, Yum SW, Faden AI (1990) Alteration in extracellular amino acids after traumatic spinal cord injury. Ann Neurol 27:96–99

Park E, Velumian AA, Fehling MS (2004) The role of excitotoxicity in secondary mechanisms of spinal cord injury: a review with an emphasis on the implications for white matter degeneration. J Neurotrauma 21:754–774

Pavel J, Lukácová N, Maršala J, Maršala M (2001) The regional changes of the catalytic NOS activity in the spinal cord of the rabbit after repeated sublethal ischemia. Neurochem Res 26:833–839

Phillis J, Horrocks LA, Farooqui AA (2006) Cyclooxygenases, lipoxygenases, and epoxygenases in CNS: their role and involvement in neurological disorders. Brain Res Rev 52:201–243

Popovich PG, Guan Z, Wei P, Huitinga I, van Rooijen N, Stokes BT (1999) Depletion of hematogenous macrophages promotes partial hindlimb recovery and neuroanatomical repair after experimental spinal cord injury. Exp Neurol 158:351–365

Rafati DS, Geissler K, Johnson K, Unabia G, Hulsebosch C, Nesic-Taylor O, Perez-Polo JR (2008) Nuclear factor-kappaB decoy amelioration of spinal cord injury-induced inflammation and behavior outcomes. J Neurosci Res 86:566–568

Rane SG, Reddy EP (2002) JAKs, STATs and Src kinases in hematopoiesis. Oncogenes 21: 3334–3358

Ravikumar R, Flora G, Geddes JW, Hennig B, Toborek M (2004) Nicotine attenuates oxidative stress, activation of redox-regulated transcription factors and induction of proinflammatory genes in compressive spinal cord trauma. Brain Res Mol Brain Res 124:188–198

Ray SK, Hogan EL, Banik NL (2003) Calpain in the pathophysiology of spinal cord injury: neuroprotection with calpain inhibitors. Brain Res Rev 42:169–185

Reddy SJ, La Marca F, Park P (2008) The role of heat shock proteins in spinal cord injury. Neurosurg Focus 25:E4

Resnick DK, Graham SH, Dixon CE, Marion DW (1998) Role of cyclooxygenase 2 in acute spinal cord injury. J Neurotrauma 15:1005–1013

Riegger T, Conrad S, Schluesener HJ, Kaps HP, Badke A, Baron C, Gerstein J, Dietz K, Abdizahdeh M, Schwab JM (2009) Immune depression syndrome following human spinal cord injury (SCI): a pilot study. Neuroscience 158:1194–1199

Rodríguez A, Roy J, Martínez-Martínez S, López-Maderuelo MD, Niño-Moreno P, Ortí L, Pantoja-Uceda D, Pineda-Lucena A, Cyert MS, Redondo JM (2009) A conserved docking surface on calcineurin mediates interaction with substrates and immunosuppressants. Mol Cell 33: 616–626

Rothwell NJ, Relton JK (1993) Involvement of cytokines in acute neurodegeneration in the CNS. Neurosci Biobehav Rev 17:217–227

Ruff RL, McKerracher L, Selzer ME (2008) Repair and neurorehabilitation strategies for spinal cord injury. Ann N Y Acad Sci 1142:1–20

Sastry PS, Rao KS (2000) Apoptosis and the nervous system. J Neurochem 74:1–20

Scarisbrick IA, Sabharwal P, Cruz H, Larsen N, Vandell AG, Blaber SI, Ameenuddin S, Papke LM, Fehlings MG, Reeves RK, Blaber M, Windebank AJ, Rodriguez M (2006) Dynamic role of kallikrein 6 in traumatic spinal cord injury. Eur J Neurosci 24:1457–1469

Schultz C, König HG, Del Turco D, Politi C, Eckert GP, Ghebremedhin E, Prehn JH, Kögel D, Deller T (2006) Coincident enrichment of phosphorylated IkappaBalpha, activated IKK, and phosphorylated p65 in the axon initial segment of neurons. Mol Cell Neurosci 33:68–80

Sekhon LH, Fehlings MG (2001) Epidemiology, demographics, and pathophysiology of acute spinal cord injury. Spine 26(24 Suppl):S2–S12

Song G, Cechvala C, Resnick DK, Dempsey RJ, Rao VL (2001) GeneChip analysis after acute spinal cord injury in rat. J Neurochem 79:804–815

Skaper SD, Moore SE, Walsh FS (2001) Cell signalling cascades regulating neuronal growth-promoting and inhibitory cues. Prog Neurobiol 65:593–608

Springer JE, Azbill RD, Nottingham SA, Kennedy SE (2000) Calcineurin-mediated BAD dephosphorylation activates the caspase-3 apoptotic cascade in traumatic spinal cord injury. J Neurosci 20:7246–7251

Sullivan PG, Rabchevsky AG, Waldmeier PC, Spiringer JE (2005) J. Mitochondrial permeability transition in CNS trauma: cause or effect of neuronal cell death? Neurosci Res 79:231–239

Sun D, Newman TA, Perry VH, Weller RO (2004) Cytokine-induced enhancement of autoimmune inflammation in the brain and spinal cord: implications for multiple sclerosis. Neuropathol Appl Neurobiol 30:374–384

Sun GY, Horrocks LA, Farooqui AA (2007) The roles of NADPH oxidase and phospholipases $A_2$ in oxidative and inflammatory responses in neurodegenerative diseases. J Neurochem 103:1–16

Sung JK, Miao L, Calvert JW, Huang L, Louis Harkey H, Zhang JH (2003) A possible role of RhoA/Rho-kinase in experimental spinal cord injury in rat. Brain Res 959:29–38

Sung B, Wang S, Zhou B, Lim G, Yang L, Zeng O, Lim JA, Wang JD, Kang JX, Mao J (2007) Altered spinal arachidonic acid turnover after peripheral nerve injury regulates regional glutamate concentration and neuropathic pain behaviors in rats. Pain 131:121–131

Takenouchi T, Setoguchi T, Yone K, Komiya S (2008) Expression of apoptosis signal-regulating kinase 1 in mouse spinal cord under chronic mechanical compression: possible involvement of the stress-activated mitogen-activated protein kinase pathways in spinal cord cell apoptosis. Spine 33:1943–1950

Tassi E, Walter S, Aigner A, Cabal-Manzano RH, Ray R, Reier PJ, Wellstein A (2007) Effects on neurite outgrowth and cell survival of a secreted fibroblast growth factor binding protein upregulated during spinal cord injury. Am J Physiol Regul Integr Comp Physiol 293: R775–R783

Taylor WA (1988) Effects of impact injury of rat spinal cord on activities of some enzymes of lipid hydrolysis. Dissertation, The Ohio State University, Columbus, Ohio

Tian DS, Xie MJ, Yu ZY, Zhang O, Wang YH, Chen B, Chen C, Wang Y (2007) Cell cycle inhibition attenuates microglia induced inflammatory response and alleviates neuronal cell death after spinal cord injury in rats. Brain Res 1135:177–185

Titsworth WL, Liu NK, Xu XM (2008) Role of secretory phospholipase $A_2$ in CNS inflammation: implications in traumatic spinal cord injury. CNS Neurol Disord Drug Targets 7:254–269

Trevisani M, Siemens J, Materazzi S, Bautista DM, Nassini R, Campi B, Imamachi N, Andrè E, Patacchini R, Cottrell GS, Gatti R, Basbaum AI, Bunnett NW, Julius D, Geppetti P (2007)

4-Hydroxynonenal, an endogenous aldehyde, causes pain and neurogenic inflammation through activation of the irritant receptor TRPA1. Proc Natl Acad Sci USA 104:13519–13524

Tsujimoto Y, Shimizu S (2003) Role of the mitochondrial membrane permeability transition in cell death. Apoptosis 12:835–840

Vitkovic L, Bockaert J, Jacque C (2000) "Inflammatory" cytokines: neuromodulators in normal brain? J Neurochem 74:457–471

Waldmeier PC, Zimmermann K, Oian T, Tintelnot-Blomley M, Lemasters JJ (2003) Cyclophilin D as a drug target. Curr Med Chem 10:1485–1506

Wang KKW (2000) Calpain and caspases. Trends Neurosciences 23:20–26

Wang X, Chen W, Liu W, Wu J, Shao Y, Zhang X (2009) The role of thrombospondin-1 and transforming growth factor-beta after spinal cord injury in the rat. J Clin Neurosci 16:818–821

Watkins TA, Barres BA (2002) Nerve regeneration: regrowth stumped by shared receptor. Curr Biol 12:R654–R656

Wells JE, Rice TK, Nuttall RK, Edwards DR, Zekki H, Rivest S, Yong VW (2002) An adverse role for matrix metalloproteinase 12 after spinal cord injury in mice. J Neurosci 23: 10107–10115

Wilms H, Schwabedissen B, Sievers J, Lucius R (2009) Erythropoietin does not attenuate cytokine production and inflammation in microglia – Implications for the neuroprotective effect of erythropoietin in neurological diseases. J Neuroimmunol [Epub ahead of print]

Wilson CJ, Finch CE, Cohen HJ (2002) Cytokines and cognition – The case for a head-to-toe inflammatory paradigm. J Am Geriatr Soc 50:2041–2056

Wu KL, Hsu C, Chan JY (2007) Impairment of the mitochondrial respiratory enzyme activity triggers sequential activation of apoptosis-inducing factor-dependent and caspase-dependent signaling pathways to induce apoptosis after spinal cord injury. J Neurochem 101: 1552–1566

Xiong Y, Rabchevsky AG, Hall ED (2007) Role of peroxynitrite in secondary oxidative damage after spinal cord injury. J Neurochem 100:639–649

Xu J, Fan GS, Chen SW, Wu YJ, Xu XM, Hsu CY (1998) Methylprednisolone inhibition of TNF-α expression and NF-ćB activation after spinal cord injury in rats. Mol Brain Res 59: 135–142

Xu J, Kim GM, Ahmed SH, Xu J, Yan P, Xu XM, Hsu CY (2001) Glucocorticoid receptor-mediated suppression of activator protein-1 activation and matrix metalloproteinase expression after spinal cord injury. J Neurosci 21:92–97

Xu J, Chen S, Chen H, Xiao Q, Hsu CY, Michael D, Bao J (2009) STAT5 mediates antiapoptotic effects of methylprednisolone on oligodendrocytes. J Neurosci 29:2022–2026

Yakovlev AG, Huang X, VanGelderen J, Calva-Cerqueira D, Faden AI (2005) Selective caspase activation may contribute to neurological dysfunction after experimental spinal cord trauma. J Neurosci Res 80:369–380

Yamamoto Y, Gaynor RB (2004) IkappaB kinases: key regulators of the NF-kappaB pathway. Trends Biochem Sci 29:72–79

Yang L, Blumbergs PC, Jones NR, Manavis J, Sarvestani GT, Ghabriel MN (2004) Early expression and cellular localization of proinflammatory cytokines interleukin-1beta, interleukin-6, and tumor necrosis factor-alpha in human traumatic spinal cord injury. Spine 29:966–971

Yezierski RP (2005) Spinal cord injury: a model of central neuropathic pain. Neurosignals 14: 182–193

Yiu G, He Z (2006) Glial inhibition of CNS axon regeneration. Nat Rev Neurosci 7:617–627

Yu F, Kamada H, Niizuma K, Endo H, Chan PH (2008) Induction of mmp-9 expression and endothelial injury by oxidative stress after spinal cord injury. J Neurotrauma 25:184–195

Yu WR, Liu T, Fehlings TK, Fehlings MG (2009) Involvement of mitochondrial signaling pathways in the mechanism of Fas-mediated apoptosis after spinal cord injury. Eur J Neurosci 29:114–131

# Chapter 5
# Potential Neuroprotective Strategies for Experimental Spinal Cord Injury

## 5.1 Introduction

Spinal cord injury (SCI) is a complex and devastating clinical condition that produces loss of motor and sensory functions below the injury site, often affecting young and healthy individuals throughout the world (Beattie et al., 2000). Functional recovery is very limited because injured axons within the brain and spinal cord are unable to regenerate spontaneously and therapeutic strategies to reestablish lost neuronal connections in spinal cord injury patients are currently unavailable (Schwab et al., 2006; Fouad and Pearson, 2004; Fouad and Tse, 2008). Several factors, including myelin-associated neurite growth inhibitors3, myelin-associated glycoprotein (MAG), myelin-associated glycoprotein (Nogo), and oligodendrocyte-myelin glycoprotein (OMgp), block the regeneration of injured neurons (McKerracher and Winton, 2002; Watkins and Barres, 2002; Filbin, 2003; Watkins and Barres, 2002). Canonical axon guidance molecules belonging to the semaphorin, ephrin, slits, and netrin families and bone morphogenetic proteins (BMPs) and Wnts also contribute to the growth-hostile environment of injured spinal cord tissue (Yiu and He, 2006). Astrocytes also play a crucial role in the failure to regenerate by synthesizing multiple inhibitory proteoglycans, such as chondroitin sulfate proteoglycans (CSPGs), which are upregulated around the injury site (Pizzi and Crowe, 2007; Kwok et al., 2008). In addition, SCI also results in increased immunolabeling of neurocan, brevican, and versican within days in injured spinal cord parenchyma surrounding the lesion site. The neurocan and verican immunolabeling peaks at 2 weeks and remains elevated from weeks to months. These molecules also contribute in limiting axonal regeneration (Jones et al., 2003). After SCI, astrocytes become hypertrophic and proliferative and form a dense network of astroglial processes at the site of lesion constituting a physical and biochemical barrier called glial scar. The hydrolysis of CSPG chains by the addition of exogenous chondroitinase ABC promotes axon regeneration and reactivates plasticity (Kwok et al., 2008).

A.A. Farooqui, *Neurochemical Aspects of Neurotraumatic and Neurodegenerative Diseases*, DOI 10.1007/978-1-4419-6652-0_5,

## 5.2 Metalloproteinases and Glial Scar Formation

During glial scar formation, astrocytes migrate toward the lesion and this process involves matrix metalloproteinases (MMPs) (Noble et al., 2002; Wells et al., 2002; Pizzi and Crowe, 2007). MMPs not only facilitate the migration of astrocytes by degrading the core protein of some CSPGs as well as other growth-inhibitory molecules, such as Nogo and tenascin-C, but also play an important role in blood–spinal cord barrier dysfunction, inflammation, and locomotor recovery (Noble et al., 2002; Hsu et al., 2008).

MMP-9 null mice exhibit significantly less disruption of the blood–spinal cord barrier, attenuation of neutrophil infiltration, and significant locomotor recovery compared with wild-type mice, suggesting that MMP-9 plays a key role in abnormal vascular permeability and inflammation within the first 3 day after SCI (Noble et al., 2002). Detailed investigation on the SCI in wild-type mice expressing MMPs and MMP-9 null mice indicates that wild-type mice expressing MMPs develop a more severe glial scar and enhanced expression of chondroitin sulfate proteoglycans, indicating the existence of a more inhibitory environment for axonal regeneration/plasticity, than MMP-9 null mice (Hsu et al., 2008). Treatment of MMP-9 null astrocytes and wild-type astrocytes with MMP-9 inhibitor results in impairment of astrocytes migration compared to untreated wild-type controls. MMP-9 null astrocytes show abnormalities in the actin cytoskeletal organization and function but no detectable untoward effects on proliferation, cellular viability, or adhesion (Hsu et al., 2008). Interestingly, MMP-2 null astrocytes show increased migration, which can be attenuated in the presence of an MMP-9 inhibitor. Collective evidence suggests that MMP-9 contribute to inhibitory glial scar formation and cytoskeleton-mediated astrocyte migration. Downregulations of astroglial proliferation and inhibitory CSPG production may facilitate axonal regeneration. MMP-9 may thus be a promising therapeutic target for reducing glial scarring during wound healing after SCI (Hsu et al., 2008; Wells et al., 2002). Studies on the expression of astrocytic gliosis 10 days after SCI by using gliosis-specific microdissection, genome-wide microarray, and MetaCore (trade mark) pathway analysis indicate that SCI-induces proliferation of reactive astrocytes in the lesion in accordance with the increase in the expression and phosphorylation of MEK-ERK. Administration of liposomes containing the interferon-β (IFN-β) reduces reactive gliosis after SCI. At 14 days after this treatment, GFAP-positive intensity and MEK-ERK phosphorylation at the lesion are decreased, indicating that liposome-mediated IFN-β gene delivery inhibits glial scar formation after SCI and promotes functional recovery (Ito et al., 2009).

## 5.3 Other Inhibitory Molecules Contributing to Axonal Growth Inhibition

The activation of small GTPase RhoA and its effector Rho-kinase (a serine/threonine kinase) has been shown to be a key element for neurite growth inhibition and growth cone collapse elicited by receptor complex comprising of the Nogo receptor, the p75$^{NTR}$ receptor, and LINGO-1. This suggestion is supported by the observation that inhibition of RhoA or Rho-kinase promotes axon growth and functional recovery after SCI (Yamashita, 2007). The presence of above factors and lack of neuronal regeneration and repair often lead to the failure of the injured and disrupted spinal cord axons to regenerate and form functional synapses (Domeniconi and Filbin, 2005). Recently, the repulsive guidance molecule (RGM) has been included in the list of potent myelin-derived neurite outgrowth inhibitors in vitro and in vivo (Kubo et al., 2008). The discovery of the receptors and downstream signals of these inhibitors may enable further understanding of the mechanism underlying the failure of axonal regeneration. The activation of RhoA and its effector Rho kinases (ROCK) after the ligation of these inhibitors to the corresponding receptors has been reported to contribute axonal growth inhibition. Blockade of the Rho-ROCK pathway reverses the inhibitory effects of these inhibitors in vitro and promotes axonal regeneration in vivo (Kubo et al., 2008). Three ROCK inhibitors (Y-27632), fasudil (HA-1077), and dimethylfasudil (H-1152) partially restore neurite outgrowth of Ntera-2 neurons on the inhibitory chondroitin sulfate proteoglycan substrate. In the rat optic nerve crush model, Y-27632 dose dependently increases regeneration of retinal ganglion cell axons in vivo. Application of dimethylfasudil results in a trend toward increased axonal regeneration in an intermediate concentration (Lingor et al., 2007). Collective evidence suggests that Rho-ROCK inhibitors have a therapeutic potential against head and spinal cord injuries (Kubo; et al., 2008).

Wnts are a large family of axon guidance diffusible molecules (all 19 Wnts) that can attract ascending axons and repel descending axons along the length of the developing spinal cord (Liu et al., 2008b). Their expression is not detectable in

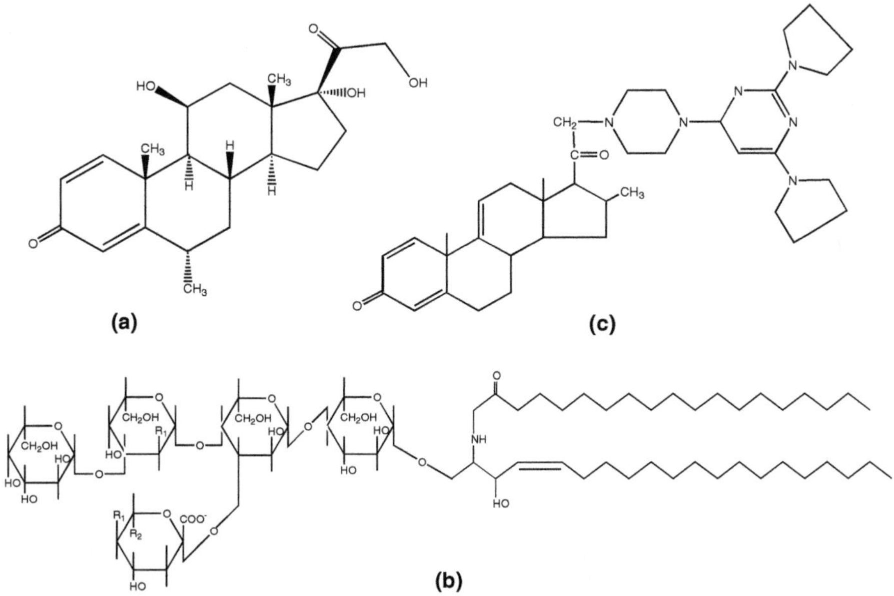


**Fig. 5.1** Chemical structures of methylprednisolone (**a**); GM1 ganglioside (**b**); and tirilazad (**c**)

normal adult spinal cord by in situ hybridization. However, three of them are upregulated following SCI. Wnt1 and Wnt5a, encoding potent repellents of the descending corticospinal tract (CST) axons, are robustly and acutely upregulated broadly in the spinal cord gray matter after unilateral hemisection (Liu et al., 2008b). Wnts interact with receptor related to tyrosine kinase (Ryk) and guide corticospinal axons down the spinal cord during development (Miyashita et al., 2009). Ryk-Wnt signaling mediates the inhibition of corticospinal axon growth in the adult spinal cord. In reactive astrocytes following SCI, the expression of Wnt-5a is increased significantly around the injury site. In vitro, Wnt-5a retards the neurite growth of postnatal

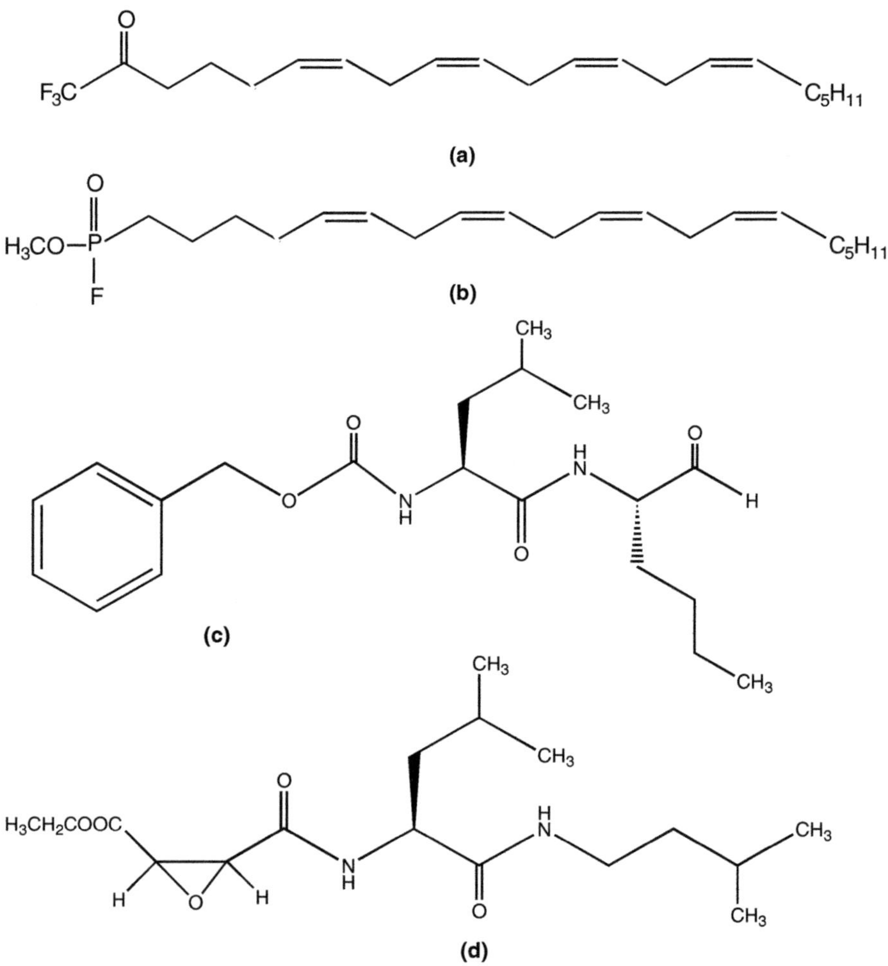


**Fig. 5.2** Chemical structures of arachidonyltrifluoromethyl ketone (**a**); methyl arachidonyl fluorophosphonate (**b**); calpeptin (**c**); and E-64-d ((2S,3S)-trans-epoxysuccinyl-L-3-methylbutaneethyl ester) (**d**)

cerebellar neurons by activating RhoA/Rho-kinase. In rats with thoracic spinal cord contusion, intrathecal administration of a neutralizing antibody to Ryk results in significant axonal growth of the corticospinal tract and enhanced functional recovery (Miyashita et al., 2009). Collectively, these studies suggest that re-expression of the embryonic repulsive cues in adult tissues contributes to the failure of axon regeneration in the central nervous system. Despite these obstacles to axonal growth, some recovery of motor and sensory function has been observed in both patients with incomplete SCI and its animal models, supporting the view that the potential for "repair" exists at some level in brain and spinal cord.

As stated in Chapter 4, numerous neurochemical changes in spinal cord tissue occur following SCI. There is no treatment available that restores the injury-induced loss of function to a degree that an independent life can be guaranteed. Three fundamental strategies have been developed in animal models of SCI. They include neuroprotection (pharmacological prevention of some of the damaging intracellular cascades that lead to secondary tissue loss) to reduce the progressive secondary injury processes that occur during the first few weeks after the initial trauma. The second strategy, which is initiated not long after the trauma,

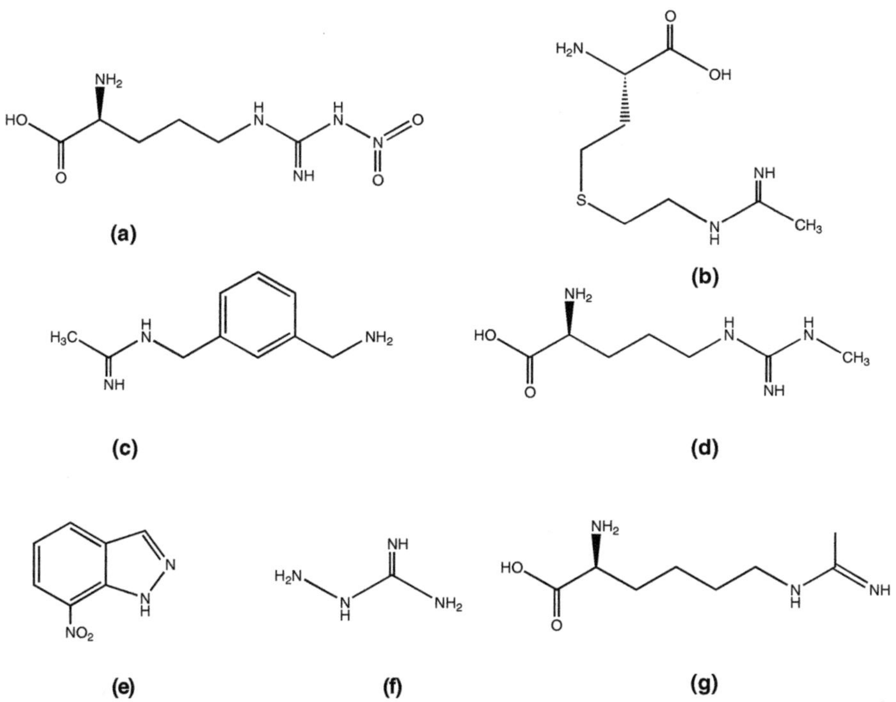


**Fig. 5.3** Chemical structures of nitric oxide synthase inhibitors. *N*G-nitro-L-arginine (L-NNA) (**a**); *S*-[2-[(1-iminoethyl)amino]ethyl]-L-homocysteine (GW274150) (**b**); *N*-[3-(aminomethyl)benzyl] acetamidine (1,400 W) (**c**); *N*G-monomethyl-L-arginine (L-NMMA) (**d**); 7-nitroindazole (7-NI) (**e**); aminoguanidine (**f**); *N*6-iminoethyl-L-lysine (L-NIL) (**g**)

aims at promoting axonal regeneration by acting on the main barrier to regeneration of lesioned axons: the glial scar (cell transplantation, genetic engineering to increase growth factors, neutralization of inhibitory factors, reduction in scar formation). The third strategy includes the management of the sublesional spinal cord by sensorimotor stimulation and/or supply of missing key afferents as a part of rehabilitation (Bunge, 2008). The main objective of investigators in SCI field is to discover the effective combination strategies to improve outcome after SCI to the adult rat thoracic spinal cord. Combination interventions not only include implantation of Schwann cells (SCs) plus neuroprotective drugs (methylprednisolone sodium succinate (MP), monosialoganglioside $GM_1$, tirilazad, calpain inhibitors, nitric oxide inhibitors, $PLA_2$ inhibitors, antioxidants, ω-3 or n-3 fatty acids (Figs. 5.1, 5.2, and 5.3), but administration of growth factors (BDNF, bFGF, EGF, GDNF, IGF-1), treatment with chondroitinase, elevation of cyclic AMP, and injections of stem/progenitor cells (Bunge, 2008). All these are known to promote behavioral and functional recovery in animal models of SCI.

## 5.4 Neuroprotective Strategies

In past years, major advances in understanding molecular mechanism of primary and secondary injury have led to the preclinical trials of many promising pharmacological therapies, all with the goal of improving behavioral and neurologic outcome. These trials involve the treatment with neuroprotective agent, surgery, treatment with agents inducing regeneration, and facilitation of rehabilitation care (Hawryluk et al., 2008). In neuroprotection trials, methylprednisolone, thyrotropin-releasing hormone, gangliosides, and tirilazad have been used as major therapeutic agents. Many randomized controlled trials on the use of methylprednisolone sodium succinate, tirilazad mesylate, monosialoganglioside, thyrotropin-releasing hormone, gacyclidine, naloxone, and nimodipine have been completed. The primary outcome in these trials has been negative. However, administration of methylprednisolone sodium succinate within 8 h after SCI shows some beneficial effects. A drawback of these SCI trials has been the use of drugs that block single pathway. Because neurodegeneration in SCI is multifactorial (oxidative stress, mitochondrial breakdown, and inflammation) process, effective therapies for SCI must include drugs that modulate multiple pathophysiological pathways.

Regeneration involves stem cell transplantation and similar rehabilitative restorative approaches designed to optimize spontaneous regeneration by mobilizing endogenous stem cells and facilitating other cellular mechanisms of regeneration, such as axonal growth and myelination. It includes the use of pluripotent human stem cells, embryonic stem cells, and a number of adult-derived stem and progenitor cells, such as mesenchymal stem cells, Schwann cells, olfactory ensheathing cells, and adult-derived neural precursor cells. Although current strategies to repair the subacutely injured cord appear promising, many obstacles continue to render the treatment of acute and chronic SCI challenging, therefore, more research is required on the treatment of SCI (Eftekharpour et al., 2008; Hawryluk et al., 2008).

Strategies for rehabilitation include passive exercise, active exercise with some voluntary control, and use of neuroprostheses. These activities enhance sensorimotor recovery after SCI by promoting adaptive structural and functional plasticity while mitigating maladaptive changes at multiple levels of the neuraxis. Following SCI, the degree and extent of neuroplasticity and recovery depend not only on the level and extent of injury but also on post-injury medical and surgical care and rehabilitative interventions. Rehabilitation strategies are focused less on repairing lost connections and more on modulating neuroplasticity, which may promote regaining of neural cell function (Lynskey et al., 2008). The mechanism of plasticity and neural adaptation is not fully understood. However, basic mechanisms of plasticity include neurogenesis, programmed cell death, and activity-dependent synaptic plasticity. Repetitive stimulation of synapses may result in long-term potentiation or long-term depression of neurotransmission. These changes are associated with physical changes in dendritic spines and neuronal circuits. There are four major types of plasticity: adaptive plasticity, impaired plasticity, excessive plasticity, and the "Achilles heel" in the developing brain. Plasticity is modulated by genetic factors, such as mutations in brain-derived neuronal growth factor. Induction of neural plasticity may facilitate endogenous recovery. The reorganization of injured tissue is rapidly induced by acute injury and is likely based on unmasking of latent synapses resulting from modulation of neurotransmitters, while the long-term changes after chronic injury involve changes of synaptic efficacy modulated by long-term potentiation and axonal regeneration and sprouting (Ding et al., 2005). The functional significance of neural plasticity after SCI remains unclear. It indicates that in some situations plasticity changes can result in functional improvement, while in other situations they may have harmful consequences. Thus, more studies and better understanding of the molecular mechanisms of plasticity may lead to better ways of promoting useful reorganization and preventing undesirable consequences (Ding et al., 2005).

### 5.4.1 Methylprednisolone and SCI

Methylprednisolone (MP), a glucocorticoid (Fig. 5.1), is the only drug used for treating and improving the neurologic and behavioral functions after human SCI (Bracken et al., 1990; Anderson and Hall, 1994). The standard MP treatment protocol requires the spine immobilization, management of neurogenic shock for perfusion and oxygenation, intravenous injection of 250 mg methylprednisolone sodium succinate on admission and 125 mg every 6 h for 72 h, surgical interventions to stabilize and decompress the spinal cord, prompt anatomic alignment of the spine bony elements along with continuous intravenous injection of dopamine hydrochloride to reverse the neurogenic shock, and maintenance of normal to high blood pressure (Geisler, 1998). However, improvements as a result of MP treatments are modest and are associated with myopathy and immunosuppression resulting in an increased risk of infectious and metabolic complications. In addition, MP administration also causes gastrointestinal hemorrhage and respiratory complication. Magnetic resonance imaging (MRI) studies indicate that MP therapy in the acute phase of cervical spinal cord injury patients decreases the extent of intramedullary spinal cord hemorrhage. MP should be given within 6–8 h after SCI significantly improves neurological function. MP acts through glucocorticoid receptor (GR). Immunohistochemistry and western blot analysis in a weight-drop SCI model in adult rats show upregulation in GR protein expression as early as 15 min after injury. GR expression is markedly increased at 4 h (22-fold), peaked at 8 h (56-fold), rapidly declined at 1 day, and returned to the baseline level at and after 3 days (Yan et al., 1999). During its peak expression, GR is localized in neural somata and dendrites but not in axons and their terminals. GR immunoreactivity is also found in oligodendrocytes and astrocytes, but no immunoreactivity is observed in endothelial cells. An increase in the binding activity of nuclear proteins to the glucocorticoid-responsive element is also seen after SCI, indicating a functional element of GR activation. Furthermore, colocalization of GR and TNF-α occurs in neurons and glial cells. This observation is consistent with MP-mediated regulation of TNF-α in weight-drop model of SCI. The use of high-dose MP for the treatment of acute SCI is controversial because of significant dose-related side effects and relatively modest improvements in neurological function. This has made treating SCI with MP controversial. Recently attempts have been made to develop novel, minimally invasive, and localized drug delivery systems for delivering MP to the injury site in adult rat spinal cord. This may minimize potential side effects and deleterious consequences of systemic corticosteroid therapy. MP has been encapsulated in biodegradable PLGA-based nanoparticles, and these nanoparticles have been embedded in an agarose hydrogel for localization to the site of contusion injury. Studies on the delivery of MP through hydrogel-nanoparticle system indicate that MP enters the injured spinal cord and diffuses up to 1.5 mm deep and up to 3 mm laterally into the injured spinal cord within 2 days (Chvatal et al., 2008; Kim et al., 2009). Topical delivery of MP significantly reduces early inflammation inside the contusion injured spinal cord as evidenced by a significant decrease in the number of ED-1(+) macrophages/activated microglia. This decrease in early inflammation is accompanied by downregulation in the expression of pro-inflammatory proteins, such as calpain and iNOS. Hydrogel-nanoparticle system-mediated delivery of MP significantly reduces lesion volume 7 days after contusion injury. It is suggested that this delivery has the potential to enhance the effectiveness of high doses of MP therapy in SCI with minimal side effects (Chvatal et al., 2008; Kim et al., 2009).

Studies on the effect of MP on hippocampal progenitor cells indicate that MP treatment reduces the number of cells proliferating acutely after SCI in the hippocampus. Besides reducing activation and proliferation of microglia/macrophages in the spinal cord, MP also decreases the number of oligodendrocyte progenitor cells (Schröter et al., 2009). Treatment of neuronal and oligodendroglial cell cultures with α-amino-3-hydroxy-5-methyl-4-isoxazolepropionic acid (AMPA) or staurosporine results in neuronal and oligodendroglial cell death in 24 h. MP protects oligodendrocyte from death in a dose-dependent manner, but neurons are not protected by the same doses of MP (Lee et al., 2008). This neuroprotective effect of MP can be reversed by the glucocorticoid receptor antagonist (11, 17)-11-[4-(dimethylamino) phenyl]-17-hydroxy-17-(1-propynyl)estra-4,9-dien-3-one (RU486) and small interfering RNA directed against glucocorticoid receptor, suggesting the involvement of a receptor-mediated mechanism. Detailed investigations have shown that MP reverses AMPA-mediated decrease in anti-apoptotic Bcl-$x_L$ expression, caspase-3 activation, and DNA laddering. All these processes are closely linked to anti-apoptotic activity of MP in oligodendrocytes (Xu et al., 2001; Lee et al., 2008). The treatment of methylprednisolone also increases the Bcl-2/Bax ratio and prevents neuronal death for 1–7 days after spinal cord injury. These findings suggest that rats with spinal cord injury show ascending brain injury that can be restricted through methylprednisolone management (Chang et al., 2009). Treatment of traumatized rats with MP indicates that this drug significantly increases number of oligodendrocytes, but neuronal number remains unchanged. RU486 abolishes the protective effect of MP. MP also blocks SCI-mediated decreases in Bcl-$x_L$ and caspase-3 activation (Lee et al., 2008). This process involves STAT5, which mediates anti-apoptotic effects of MP on oligodendrocytes by interacting glucocorticoid receptor and upregulating bcl-$X_L$ (Xu et al., 1998, 2009). Collective evidence suggests that MP selectively inhibits oligodendrocyte but not neuronal cell death via a receptor-mediated action and may be a mechanism for its limited protective effect after SCI (Xu et al., 1998; Lee et al., 2008; Xu et al., 2009). Treatment of astrocytes with AMPA and cyclothiazide, a diuretic, produces an increase in expression of glial fibrillary acidic protein (GFAP) and CSPG (neurocan and phosphacan). Similar neurochemical changes occur in SCI. Treatment with MP downregulates expression of GFAP and CSPG expression in adult rats following SCI. Additionally, both the glucocorticoid receptor (GR) antagonist RU486 and GR siRNA reverse the inhibitory effects of MP on GFAP and neurocan expression. These results indicate that MP may improve neuronal repair and promote neurite outgrowth after excitotoxic insult via GR-mediated downregulation of astrocyte reactivation and inhibition of CSPG expression (Liu et al., 2008a). Collectively, these studies indicate that molecular mechanism of MP action can be attributed to anti-inflammatory, antioxidant, and antiexcitotoxic properties. MP not only prevents neurofilament degradation but also reduces edema and modulates blood flow. All these effects may contribute to neuroprotective properties of MP.

Infections are major cause of death in SCI patients. They are associated with hampered wound healing, prolonged hospitalization, and impaired neurological recovery. SCI injury studies in rat model indicate that SCI induces early onset of an immune suppression that may result in SCI-immune depression syndrome (Riegger et al., 2007). Iatrogenic application of methylprednisolone in patients suffering from SCI worsens the immune suppression (Riegger et al., 2009). A thorough understanding of the molecular mechanisms of SCI-immune depression syndrome is essential for decreasing mortality, costs (time of hospitalization), and protecting the intrinsic neurological recovery potential following SCI.

### 5.4.2 $GM_1$ Ganglioside and SCI

Gangliosides are carbohydrate-rich complex lipids (Fig. 5.1). They are a major component of neuronal cells and are essential for brain function. They contain sphingosine, fatty acid, and an oligosaccharide chain that vary in size from one to four or more monosaccharides. They are present in the external leaflet of the neural membrane (Shioiri et al., 2009). The hydrophobic moiety of ganglioside consists of sphingosine and fatty acid (stearic acid, 95%). This moiety is inserted into the neural membrane, while the hydrophilic moiety, consisting of sialic acid (NANA) and other carbohydrates, protrudes toward the extracellular fluid.

Microglial cells are brain-resident macrophages. They are present in brain tissue in resting state. Following brain injury or in neurodegenerative diseases, microglial cells are activated and undergo the process of ramification (Pyo et al., 1999). $GM_1$ ganglioside not only induces ramification of cultured rat primary microglia but also mediates the expression of neurotrophin-3 (NT-3), which has no effect on the morphology of cultured rat primary microglial cells. It is suggested that ganglioside effects on microbial cells is mediated through the activation of mitogen-activated protein kinase and NF-κB.

SB203580 (an inhibitor of p38) and paclitaxel and nocodazole (microtubule-disrupting drugs) block $GM_1$-mediated microglial ramification, but Jaki (an inhibitor of JAK), PD98059 (an inhibitor of Erk1/2), SP600125 (an inhibitor of JNK), and cytochalasin B and latrunculin B (actin polymerization inhibitors) have no effect indicating that $GM_1$ induces ramification of microglia in p38- and microtubule-dependent manner (Park et al., 2008). Other gangliosides, $GD_1a$ and $GT_1b$, have no effects on microbial cell ramification in cell culture. Gangliosides not only regulate $Ca^{2+}$ influx channels and $Ca^{2+}$ exchange proteins (Ledeen and Wu, 2002) but also modulate activities of enzymes involved in signal transduction. These enzymes include adenylate cyclases, protein kinases, phospholipases $A_2$, PLC, and $Na^+$, $K^+$ATPases (Leon et al., 1981; Partington et al., 1979; Yang et al., 1994a, b; Yates et al., 1989). Gangliosides also induce the generation of nitric oxide (NO), production and release of TNF-α, and expression of cyclooxygenase-2 (COX-2) (Pyo et al., 1999).

$GM_1$ ganglioside (Fig. 5.1) has neuroprotective effects in glutamate-mediated neurotoxicity (Favaron et al., 1988; Phillis and O'Regan, 1995), acute neural trauma such as ischemia and spinal cord injury, and neurodegenerative diseases such as Alzheimer disease and Parkinson disease (Ala et al., 1990; Geisler et al., 1991; Svennerholm, 1994). Collective evidence suggests that $GM_1$ ganglioside acts as a membrane stabilizer, an antiexcitotoxic agent, and antioxidant in brain tissue. In contrast, addition of $GM_3$ ganglioside or overexpression of the $GM_3$ synthase gene produces glutamate-mediated cell death in hippocampal cell line HT22 (Sohn et al., 2006). $GM_3$ ganglioside strongly inhibits the plasmalogen-selective $PLA_2$ in brain. This prevents the release of DHA, the precursor of protective compounds (Yang et al., 1994; Latorre et al., 2003). Accumulating evidences suggest that $GM_1$ and $GM_3$ gangliosides produce differential effects in the brain tissue.

The biological effects of exogenously administered gangliosides have been extensively investigated in vitro and in experimental animal models where they have neurotrophic and neuritogenic properties. In SCI neurological deficit varies widely with the severity of injury. Light SCI causes transient abnormal reflexes where as severe SCI produces complete absence of motor and sensory function. Administration of $GM_1$ ganglioside produces an effective locomotive function recovery in rats (Carvalho et al., 2008). Combined administration of ganglioside and MP produces better outcomes than administration of MP alone. In humans, $GM_1$ enhances the recovery of neurological function of 1 year after major spinal cord injury (Geisler et al., 1991, 1993; Geisler, 1998; Walker and Harris, 1993). In addition to $GM_1$ ganglioside, SCI patients also receive aggressive medical and surgical treatment, as well as MP. Results indicate the enhancement in motor recovery compared with placebo in the lower extremities, but not in the upper extremities, over time. This corresponds to improved function of axons passing through the site of injury. In contrast, other studies indicate that ganglioside neither reduce the death rate in SCI patients nor have any effect on the recovery or quality of life in survivors (Chinnock and Roberts, 2005).

### *5.4.3 Tirilazad Mesylate and SCI*

Tirilazad (U-74006F, Freedox; Pharmacia & Upjohn, Kalamazoo, Michigan) is a non-glucocorticoid, 21-aminosteroid (Fig. 5.1) that inhibits lipid peroxidation. It inhibits lipid peroxidation by scavenging lipid peroxyl and hydroxyl groups of free radicals. In addition, it decreases membrane phospholipid fluidity and maintains endogenous antioxidant levels (especially vitamins E and C) (Villa and Gorini, 1997). In addition, it improves neuronal survival reducing cerebral edema in animal models of focal cerebral ischemia. The pharmacokinetic studies indicate that tirilazad mesylate is well tolerated between doses of 0.5–4.0 mg/kg. No clinically significant effects of tirilazad mesylate on the cardiovascular function or on clinical laboratory determinations have been observed. Thus, single doses of tirilazad mesylate appear to be devoid of glucocorticoid and mineralocorticoid activity in healthy male volunteers, and no safety concerns for single-dose tirilazad mesylate are identified (Villa and Gorini, 1997).

Antioxidant activities of tirilazad have been compared with pyrrolopyrimidine lazaroid PNU-101033E and glucocorticoid methylprednisolone on mitogen-induced respiration rate and ATP consumption in activated human peripheral blood mononuclear cells (PBMC) (Schmid et al., 2001). It is shown that tirilazad inhibits concanavalin A-stimulated respiration rate and sodium cycling across the plasma membrane. MP produces similar effect indicating the involvement of same cellular mechanisms. However, unlike MP, tirilazad has no significant effect on calcium cycling across the plasma membrane. The other lazaroid, PNU-101033E, has cytotoxic effects on PBMC. These results indicate that although tirilazads mimick the immunosuppressive effect of MP, but produce its therapeutic effect through their antioxidant effects and inducing reduction in membrane fluidity (Schmid et al., 2001). In addition, tirilazads have potent membrane stabilizing effects. The compounds have high affinity for the lipid bilayer because of their lipophilic nature and are incorporated into the lipid bilayer, where they occupy strictly defined positions and orientations (Villa and Gorini, 1997). Collective evidence suggests that tirilazads are very lipophilic compounds that are localized in membranes and protect cell membranes from peroxidative damage. They also exert positive effects on endothelial cell membranes. Although some studies indicate that tirilazad can penetrate the blood–brain barrier (BBB), other studies indicate that these compounds have limited penetration into brain parenchyma. This may be the reason why tirilazads have generally failed to protect from delayed neuronal damage to the selectively vulnerable hippocampal CA1 and striatal regions (Hall et al., 1996). Studies on SCI in cats and rats indicate that injections of tirilazad produce better locomotion scores and behavioral recovery than vehicle-treated animals (Hall, 1988; Anderson et al., 1991, 1988). These compounds facilitate the restoration of spinal cord blood flow after SCI. This property may be related to antioxidant activity of tirilazads. As stated above, pathogenesis of SCI involves abnormality in many signal transduction pathways and therefore, one drug may not be able to provide optimal neuroprotection. A combination of MP, ganglioside, and tirilazad should be first tried to treat SCI in animal models.

### *5.4.4 Inhibitors of Calpains, Nitric Oxide Synthase, and PLA$_2$ and SCI*

As stated in Chapter 4, SCI is accompanied by an increase in intracellular free $Ca^{2+}$ level. This increase in $Ca^{2+}$ results in the stimulation of $Ca^{2+}$-dependent enzymes including calpains, nitric oxide synthases, phospholipases $A_2$, and protein kinase C (Farooqui et al., 2004). Calpains are markedly increased at the injury site and contribute to neuronal death (Ray et al., 2003; Buki et al., 2003). As stated earlier that calpain activity is modulated by calpastatin, overactivation of calpains promotes the degradation of key cytoskeletal, membrane, and myelin proteins. Cleavage of these key proteins by calpain is an irreversible process that perturbs the integrity and stability of neural cells, leading to neuronal cell death. It is proposed that calpains in conjunction with caspases and kallikrein 6 promote neuronal apoptosis in the brain tissue. Many cell permeable calpain inhibitors, such as calpeptin, MD1-28170, peptide epoxide, aldehyde, and ketoamid inhibitors (Fig. 5.2), target the active site of calpains and have been effective against the enzymes and are under evaluation in animal models of SCI. Some calpain inhibitors (MDL-28170), *N*-acetyl-Leu-Leu-Met-CHO (ALLM), calpain inhibitor III (CI III) (MDL28170, and CEP-4143) have shown to be significantly neuroprotective in animal models of spinal cord trauma and head injury suggesting their therapeutic potential (Ray et al., 2003; Buki et al., 2003; Schumacher et al., 2000; Moriwaki et al., 2005).

At least three nitric oxide synthase (NOS) isoforms: a neuronal NOS or type 1 NOS (nNOS), an immunologic NOS or type 2 NOS (iNOS), and an endothelial

NOS or type 3 NOS (eNOS) occur in brain and spinal cord (Marsala et al., 2007). The activities of eNOS or nNOS are modulated by phosphorylation triggered by $Ca^{2+}$ entering cells and binding to calmodulin. In contrast, the regulation of iNOS depends on de novo synthesis of the enzyme in response to a variety of cytokines, such as TNF-$\alpha$ and interferon-$\gamma$. SCI produces upregulation of nNOS activity in neurons, eNOS in glial cells and vascular endothelium, and later an increase in iNOS activity has been observed in a range of cells, including infiltrating neutrophils and macrophages, activated microglia and astrocytes. Studies on expression of inducible iNOS and/or neuronal NOS (nNOS) in injured spinal cords indicate that SCI dramatically increases iNOS (but not nNOS) mRNA and protein levels in microglial cells in the thoracic and lumbar regions of spinal cords. iNOS overexpression causes an increased nitrotyrosine formation, decreased number of NeuN (neuronal nuclei)-immunoreactive cells, and upregulation of inflammatory genes (Lee et al., 2009). The effects of NO on the spinal cord depend not only on concentration of produced NO and activity of different synthase isoforms but also on cellular source of NO generation and time of release. Low NO concentrations may play a role in physiologic processes, whereas large amounts of NO may be detrimental by increasing oxidative stress. Thus, roles of nitric oxide are very complex, as NO can be cytotoxic or cytoprotective (Marsala et al., 2007). As stated in Chapter 4, excessive amounts of NO in neural cells give arise to highly toxic oxidant (peroxynitrite, nitric dioxide, nitron ion) that is associated with apoptotic and necrotic cell death in SCI. The inducible nitric oxide synthase (iNOS) isoform is a mediator in inflammatory reactions that involve the synthesis of nitric oxide in the injured spinal cord. iNOS inhibitors (L-*N*-iminoethyl-lysine, *N*(G)-nitro-l-arginine methyl ester, *N*(omega)-propyl-l-arginine, 2-amino-5,6-dihydro-6-methyl-4H-1,3-thiazine hydrochloride, L-NNA; L-NMMA, and pimagedine) reduce apoptotic cell death and provide protection against SCI (Sharma et al., 2005; Lukáčová et al., 2005; Lukacova et al., 2008) (Fig. 5.3). It is recently shown that chronic nicotine administration improves the recovery of the locomotor functions following SCI. Indeed, nicotine-treated animals scored consistently higher on the BBB scale indicating that the treatment altered animal behavior. Based on this observation it is proposed that agonists of neuronal nicotinic receptors can be attractive candidates for SCI therapy (Ravikumar et al., 2005).

Dynorphins (Dyn), endogenous opioid neuropeptides derived from the prodynorphin gene, not only protect neurons and oligodendroglia via their opioid receptor-mediated effects but are also involved in antinociception and neuroendocrine signaling. Dyn-induced signaling is closely associated with cross talk between NMDA type of glutamate and opioid receptors and involves the participation of isoforms of NOS (Fig. 5.4). Antiserum to dynorphin A (1–17) induces marked neuroprotection in SCI, indicating an interaction between dynorphin and NOS regulation (Sharma et al., 2006). Overexpression or overactivation of nNOS in the ventral spinal cord is closely associated with Dyn spinal neurotoxicity, whereas as the reduction of nNOS activities in the dorsal spinal cord may be involved in Dyn-mediated pain modulation. These observations support the view that the opioid-active peptide dynorphin A may be involved in the mechanisms underlying

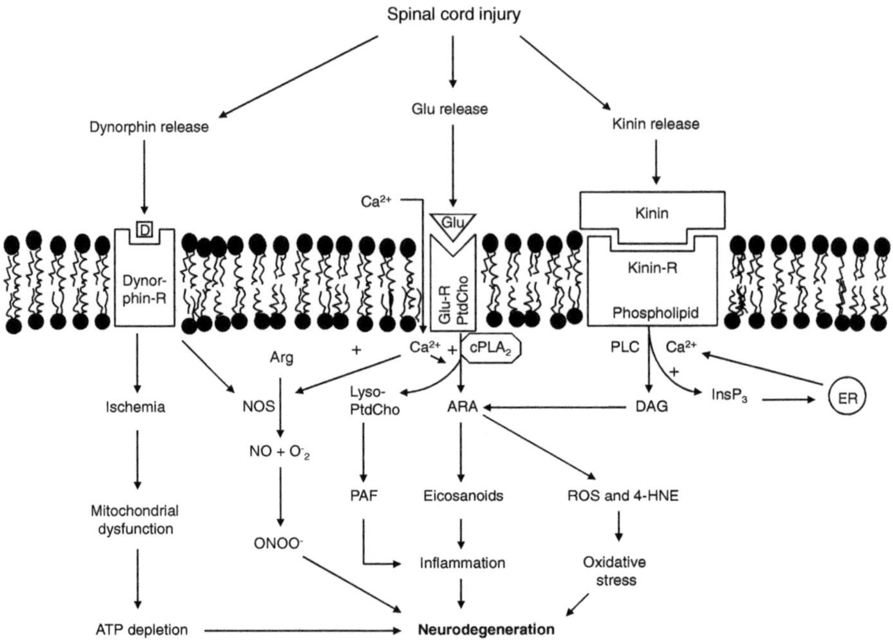


**Fig. 5.4** Interactions among dynorphin, glutamate, and kinin receptors in spinal cord injury. Dinorphin (D); glutamate (Glu); arginine (Arg); nitric oxide synthase (NOS); nitric oxide (NO); superoxide ($O_2^-$); peroxynitrite ($ONOO^-$); phosphatidylcholine (PtdCho); cytosolic phospholipase $A_2$ ($cPLA_2$); arachidonic acid (ARA); phospholipase C (PLC); diacylglycerol (DAG); inositol 1,4,5-trisphosphate ($InP_3$); endoplasmic reticulum (ER); platelet-activating factor (PAF); reactive oxygen species (ROS); and 4-hydroxynonenal (4-HNE)

the NOS regulation in the spinal cord after injury and confirms the hypothesis that upregulation of neuronal NOS is injurious to the cord (Sharma et al., 2005, 2006; Hu et al., 2000).

$PLA_2$ activity is increased significantly after SCI suggesting that this enzyme may play a key role in mediating neuronal death and oligodendrocyte demyelination following SCI and inhibition of $PLA_2$ action may represent a novel repair strategy to reduce tissue damage and increase function after SCI. Injections of $cPLA_2$ inhibitor arachidonyl trifluoromethyl ketone ($AACOCF_3$) (Fig. 5.2) not only results in increased number of surviving neurons and oligodendrocytes but also better BBB scores supporting the view that $cPLA_2$ is critically involved in acute spinal injury (Huang et al., 2009; Liu et al., 2006). In fact $PLA_2$ inhibitors have emerged as major drugs for preventing inflammation and oxidative stress (Farooqui et al., 1997, 1999, 2006; Farooqui and Horrocks, 2007; Olivas and Noble-Haeusslein, 2006). They modulate the expression of cytokines, growth factors, nuclear factor-κB, and adhesion molecules and thus can be used for the treatment of endogenous oxidative stress and neuroinflammation in SCI animal models.

### 5.4.5 *Minocycline and SCI*

Minocycline is a lipophilic second-generation tetracycline analog (Fig. 5.5) that crosses BBB and produces neuroprotection in animal models of acute neural trauma and neurodegenerative diseases. Although precise molecular mechanism of its action and primary target still remain elusive, recent studies indicate that minocycline-mediated neuroprotection may involve signaling pathway associated with inhibition of mitochondrial permeability transition-mediated cytochrome *c* release from mitochondria, the inhibition of caspase-1 and caspase-3 expressions, upregulation of iNOS, and the suppression of microglial activation (Kim et al., 2004) (Fig. 5.6). In addition, minocycline inhibits expression and activities of phospholipase $A_2$ ($PLA_2$), cyclooxygenase-2 (COX-2), 5-lipoxygenase (LOX), MMP-2, MMP-9, p38 mitogen-activated protein kinase (MARK) and decreases the expression of c-fos in brain (Pruzanski et al., 1992; Song et al., 2004; Hua et al., 2005; Machado et al., 2006; Marchand et al., 2009). Above-mentioned enzymes and events are closely associated with nociception, neuroinflammation, and apoptotic cell death. In SCI, minocycline treatment modulates expression of cytokines (IL-1β and TNF-α), attenuates cell death and the size of lesions, and improves functional recovery in the injured rat (Lee et al., 2003; Teng et al., 2004; Stirling et al., 2004, 2005; Festoff et al., 2006). Minocycline also inhibits microglial cell activation, reduces microglial OX-42 expression, attenuates reductions in O1- and O4-positive oligodendrocyte progenitor cells, and long-term pain phenomenon following SCI

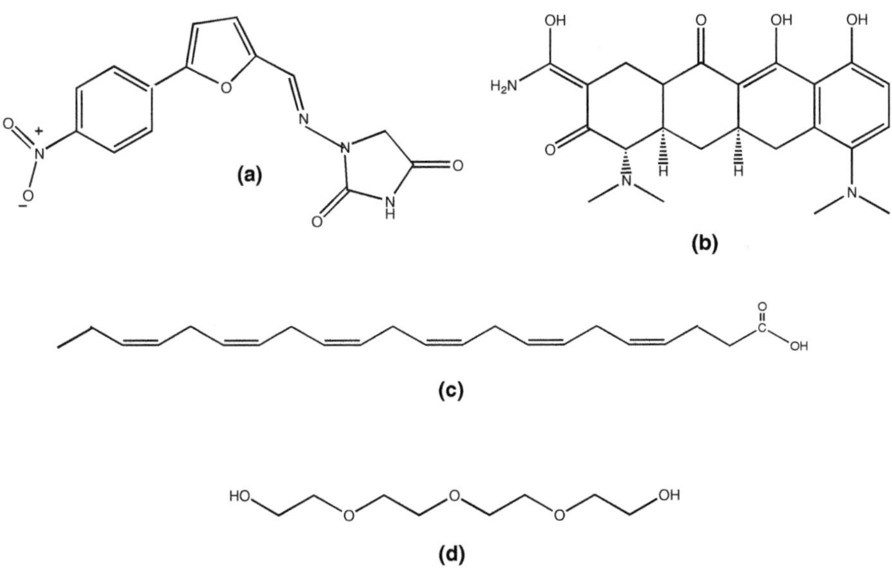

**Fig. 5.5** Chemical structures of dantrolene, a muscle relaxant (**a**); monocycline, a broad-spectrum tetracyclin antibiotic (**b**); docosahexaenoic acid, a arachidonic acid oxidation inhibitor (**c**); and polyethylene glycol (**d**)

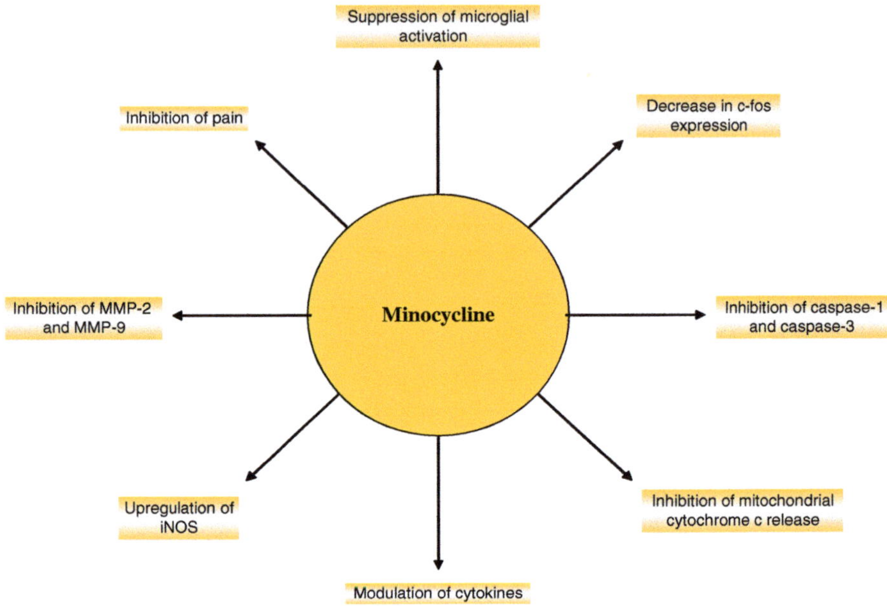


**Fig. 5.6** Effect of monocyline on neurochemical activities in brain and spinal cord

supporting the view that modulation of microglial signaling may provide a new therapeutic strategy for patients suffering from post-SCI pain (Tan et al., 2009). At concentrations higher than those shown to block inflammation and inflammation-induced neuronal death, minocycline prevents NMDA-mediated cytosolic and mitochondrial increases in $Ca^{2+}$ concentrations in a reversible manner (Melero-Fernández de Mera et al., 2008). Minocycline also blocks $Ca^{2+}$-mediated increase in ROS in isolated brain mitochondria. Although the molecular mechanisms associated with these processes are not fully understood, there is some evidence that minocycline inhibits NADH-cytochrome *c* reductase and cytochrome c oxidase activities without affecting the activity of succinate-cytochrome *c* reductase. This suggests that mitochondria are a critical factor in minocycline-mediated neuroprotection (Yrjanheikki et al., 1999; Garcia-Martinez et al., 2010). Collectively, these studies suggest that minocycline produces neuroprotective and nociceptive effects in SCI not only through its anti-inflammatory and anti-apoptotic effects but also by inhibiting MMP-2 and MMP-9, caspase-1, caspase-3, and p38 MARK. Minocycline spares white matter and increases ventral horn motor neuron survival in spinal cord adjacent to the injury site, where neurodegeneration occurs following SCI (Teng et al., 2004). Minocycline reduces the number of reactive astrocytes and augment survival of oligodendrocytes in the spared white matter. Thus, minocycline is a multifaceted therapeutic agent that has proven clinical safety and efficacy during a clinically relevant therapeutic window. It can be effective in treating acute SCI. Because of the high tolerance and the excellent penetration through blood–brain barrier, minocycline has been used for the treatment of many neurological disorders, including stroke, multiple sclerosis, SCI, amyotropic lateral sclerosis, Huntington disease, and Parkinson disease (Kim and Suh, 2009).

### *5.4.6 Thyrotropin-Releasing Hormone and SCI*

Thyrotropin-releasing hormone (TRH), a hypothalamic orally active neuropeptide, regulates the pituitary–thyroid axis by simulating the release of thyrotropin. TRH elicits its biological response through two G protein coupled receptors, namely TRH-R1 and TRH-R2 (Monga et al., 2008). Autocrine/paracrine cellular signaling motifs of TRH and TRH receptors are expressed through the body and organs of the immune system.

Considerable evidence supports a pivotal role for TRH in the pathophysiology of the inflammatory process with specific relevance to the "cytokine-induced sickness behavior" paradigm (Monga et al., 2008). Studies on the treatment of rat SCI with TRH or naloxone indicate that subcutaneous injections of TRH (2.5, 10, and 40 mg/kg/day) once daily for 7 consecutive days starting 24 h or 7 days after injury improve the neurologic function in the rats with SCI in a dose-related manner, with a minimum effective dose of less than 2.5 mg/kg/day in both cases. However, subcutaneous treatment with naloxone (40 mg/kg/day) once daily for 7 consecutive days starting 24 h after injury does not produce any beneficial effects on neurologic function (Hashimoto and Fukuda, 1991). These results indicate that TRH, but not naloxone treatment after SCI, is effective in rats with the severest neurologic impairment. It is suggested that the duration of the effectiveness of late treatment with TRH on the neurologic impairment in rats with spinal cord injury is more than 1 week, while the duration with naloxone is less than 24 h (Hashimoto and Fukuda, 1991). The effects of nimodipine and thyrotropin-releasing hormone (TRH) have been compared in a clip-compression model of experimental spinal cord injuries (SCI) in rats. TRH treatment improves somatosensory-evoked potential (SEPs) and mean arterial blood pressures, whereas nimodipine treatment has no effect on these variables, supporting the beneficial effects of TRH in SCI (Ceylan et al., 1992). This indicates that TRH not only promotes electrophysiological recovery and neurobehavioral outcome but also preserves spinal cord tissue by improving blood flow and modulating levels of cytokines.

### *5.4.7 Dantrolene and SCI*

Dantrolene (DNT), a long-acting muscle relaxant (Fig. 5.4), acts through ryanodine receptor and abolishes excitation–contraction coupling in muscle cells. Ryanodine-sensitive receptors (RyRs) are involved in the release of intracellular $Ca^{2+}$, and this release can be blocked by DNT. Based on electrophysiological studies, it is suggested that injurious effects of $Ca^{2+}$ in white matter injury may be mediated both by RyRs and through Ins*P*3Rs calcium-induced calcium release receptors (Thorell et al., 2002). DNT also produces neuroprotective effects through its antioxidant and anti-apoptotic effects. Treatment of SCI in rat model results in significant improvement in DNT-treated rats, 24 h after SCI, with respect to control. SCI-mediated increase in the lipid peroxidation, decrease in enzymic or non-enzymic endogenous antioxidative defense systems, and increase in apoptotic cell numbers can be prevented by DNT. DNT treatment blocks lipid peroxidation and augments endogenous enzymic or non-enzymic antioxidative defense systems, and significantly decreases the apoptotic cell death following SCI (Aslan et al., 2009). In addition DNT treatment also prevents hemorrhage, edema, and decrease in GSH levels.

### *5.4.8 ω-3 Fatty Acids and SCI*

Injections of α-linolenic acid (ALA) and DHA 30 min after SCI not only result in significant improvement in locomotor performance and neuroprotection but also reduce lesion size, inhibit apoptosis, and increase neuronal and oligodendrocyte survival (Lang-Lazdunski et al., 2003; King et al., 2006; Michael-Titus, 2007). The molecular mechanism associated with neuroprotective effects of DHA (Fig. 5.5) and other ω-3 fatty acids in SCI remains unknown. However, based on reduction of oxidation of proteins and RNA/DNA, and inhibition of lipid peroxidation, it is proposed that the neuroprotective effect of ω-3 fatty acids may involve their antioxidant activity, and generation of resolvins and neuroprotectins, which protect neuronal cells from apoptotic cell death (King et al., 2006; Michael-Titus, 2007; Huang et al., 2007). In addition, neuroprotective effect of ω-3 fatty acid may be associated with interactions among TWIK-related $K^+$ channel (TREK), TWEK-related $K^+$ channel (TRAAK), and ω-3 fatty acids (Lauritzen et al., 2000). These fatty acids may also downregulate NF-κB and pro-apoptotic protein, Bax immunoreactivity, and block apoptotic and necrotic neuronal death (Lang-Lazdunski et al., 2003). Furthermore, ALA is metabolized to EPA, which is known to generate anti-inflammatory series-3 prostaglandins. These metabolites prevent the generation of ARA-derived pro-inflammatory eicosanoids. In contrast, injections of ARA injections in rats produce a significantly worse outcome of injured animals following SCI than controls. These studies indicate that there is a striking difference in efficacy of ω-3 and ω-6 fatty acids on the outcome of spinal cord with ω-3 fatty acids being neuroprotective and *n*–6 fatty acids having damaging effects (King et al., 2006; Michael-Titus, 2007; Huang et al., 2007). Thus, treatment with ω-3 fatty acids may serve as promising therapeutic agents for the management of spinal cord injury.

### *5.4.9 Polyethylene Glycol and SCI*

Polyethylene glycol (PEG) is a fusogen (Fig. 5.5). It has been shown to mechanically repair damaged cellular membranes and reduce secondary axotomy after traumatic brain injury (TBI) and SCI. This repair is achieved following spontaneous reassembly of cell membranes made possible by the action of targeted hydrophilic polymers, which first seal the compromised portion of the plasmalemma, and secondarily allow the lipidic core of the compromised membranes to resolve into each other (Koob et al., 2008). Although the molecular mechanism of PEG-mediated neuroprotection after SCI remains unknown, it is proposed that this fusogen reduces apoptotic cell death following SCI (Baptiste et al., 2009). In clip compression model of SCI at C8, intravenous injections of PEG indicate that this fusogen also reduces 200 kd neurofilament degradation. It also promotes spinal cord tissue sparing. This proposal is based on retrograde axonal Fluoro-Gold tracing and morphometric histological assessment. Polyethylene glycol also induces significant and modest, neurobehavioral recovery after SCI. In another study, intravenous injections of PEG + $MgSO_4$ improve locomotor recovery and reduce pain but do not provide additional benefit compared with either treatment alone. Neither treatment nor their combination attenuate mean arterial pressure (MAP) increases during autonomic dysreflexia (Ditor et al., 2007). PEG + $MgSO_4$ treatment causes significant increases in dorsal myelin sparing, and the latter results in significant reductions in lesion volume, compared with saline-treated controls. Furthermore, mean lesion volumes correlate negatively with the corresponding mean locomotion BBB scores and positively with the corresponding mean pain scores (Ditor et al., 2007; Kwon et al., 2009). Collective evidence suggests that PEG protects key axonal cytoskeletal proteins after SCI, and that the protection is associated with axonal preservation. The modest extent of locomotor recovery after treatment with PEG suggests that this compound may not confer sufficient neuroprotection to be used clinically as a single treatment (Baptiste et al., 2009; Kwon et al., 2009). Derivatization of protein with PEG (pegylation) not only improves pharmacokinetic and pharmacodynamic properties of the proteins but also improves efficacy and minimize the dose. Attachment of PEG with brain-derived neurotrophic factor (BDNF) and its intrathecal administration results in enhanced delivery of PEG-bound BDNF to the spinal cord. The biological activity of BDNF-PEG conjugate mixture has assessed with the goal of identifying a relationship between the number of PEG molecules attached to BDNF and biological activity. These preparations have been used to study their effects on SCI (Soderquist et al., 2008).

### *5.4.10 Opioid Receptor Antagonists, Glutamate Receptor Antagonists, and Calcium Channel Blockers in SCI*

The cerebrovascular and metabolic changes associated with SCI are closely associated with pathologic alterations in endogenous neurochemical systems, including those involved with normal neurotransmission. These processes may include alterations in neurotransmitter synthesis, release, and reuptake mechanisms or changes in pre- or postsynaptic receptor activity. Although the timing of the precise cascade of neurochemical events following SCI is poorly understood, identification of alterations in glutamate and $Ca^{2+}$ levels following SCI provides an opportunity for the development and employment of therapeutic agents, such as glutamate receptor antagonists, calcium blockers, and opioid receptor antagonists designed to modulate glutamate receptors, $Ca^{2+}$ channels, and opioid receptors respectively. This process may not only result in attenuation of local secondary tissue damage to spinal cord tissue but also in improvement of outcome and promotion of functional recovery. Many studies on the treatment of SCI in animal models have been performed. Although encouraging results have been obtained in animal models of SCI, many human trials of glutamate receptor antagonists, calcium blockers, and opioid receptor antagonists have been stopped due to side effects.

### *5.4.11 Growth Factors and SCI*

It is well known that SCI induces upregulation in the expression of BDNF mRNA (Ikeda et al., 2001), which reaches maximum levels of 24 h after the spinal cord trauma. Expression of BDNF mRNA comes back to the levels of sham-operated control animals within 3 days of the injury. In situ hybridization studies indicate that BDNF is expressed in motor and sensory neurons, glia cells (astrocytes and oligodendrocyte), and putative macrophages and/or microglia, but not until day 7 following the SCI. It is suggested that BDNF is synthesized in both neurons and astrocytes during the acute response to SCI to perform a neuroprotective role in earlier phases. This is followed by a later phase of expression in which the expression of BDNF occurs in macrophages and/or microglia, apparently for neural cell restoration and survival (Ikeda et al., 2001). Glial cell line-derived neurotrophic factor (GDNF) is another member of growth factor family, which is expressed on a variety of neurons that project from brain into the spinal cord, including supraspinal neurons, dorsal root ganglia, and local neurons. It acts through GDNF-receptor α-1 using an ex vivo gene delivery approach that provides both trophic support and a cellular substrate for axonal growth. It is shown that implants of primary fibroblasts can be genetically modified to secrete GDNF into complete and partial mid-thoracic spinal cord transection sites. Compared to recipients of control grafts expressing a reporter gene, GDNF-expressing grafts promote significant regeneration of several spinal systems, including dorsal column sensory, regionally projecting propriospinal, and local motor axons. Local GDNF expression also induces Schwann cell migration to the lesion site, leading to remyelination of regenerating axons. Thus, GDNF exerts tropic effects on adult spinal axons and Schwann cells that contribute to axon growth after injury (Blesch and Tuszynski, 2003). Vascular endothelial growth factor (VEGF) also produces multifaceted therapeutic effects in a rat spinal cord injury (SCI) model. It acts by stimulating proliferation of endogenous glial progenitor cells. VEGF increases the density of blood vessels in the injured spinal cord and enhances tissue sparing. These anatomical results are accompanied by improved BBB locomotor scores. It is proposed that multifaceted effects of VEGF on endogenous gliogenesis, angiogenesis, and tissue sparing can be utilized to improve functional outcomes following SCI (Kim et al., 2009).

## 5.5 Regeneration and SCI

The majority of current spinal cord repair and regeneration therapies are at the experimental stage in vitro or animal models. As stated earlier, very slow regeneration and reconstruction of spinal cord after SCI is caused by a number of factors including inflammation, cavitation, secondary axonal demyelination, and glial scar formation. Consequently, functional deficits persist after SCI. Recovery from SCI is a big challenge. It not only requires survival of transplanted cells and axonal regeneration but also physiological targeting by growing axons and establishment of correct and functional synaptic appositions. As stated earlier, after acute SCI, there is a therapeutic window of opportunity within which the devastating consequences of the secondary injury can be ameliorated. Cellular replacement (neural transplantation) and axon guidance are both necessary for the repair of injured spinal cord in animal model of SCI (Bartolomei and Greer, 2000). Two types of stem cells, namely stem/progenitor cells and human umbilical cord blood stem cells (hUCB), have been used for SCI treatment in animal models.

### *5.5.1 Stem/Progenitor Cell Transplants*

Stem/progenitor cells provide a valuable cellular source for promoting repair following SCI. Stem/progenitor cells are multipotent and dynamic cells that have the capacity to expand in vitro. They not only can self-renew and differentiate into CNS cell lineages but are capable of long-term survival following transplantation (Webber et al., 2007). Thus, they can be directed to differentiate into neurons or glia in vitro, which can be used for the replacement of neural cells lost, after SCI. Transplantation of stem/progenitor cells has been shown to promote neuroprotective and axon regeneration-promoting effects in the spinal cord. Promising results can be obtained in experimental models of SCI (Kim et al., 2007).

Four types of embryonic cells and other neural cells have been used for neural cell therapy in animal models of SCI: stem/progenitor cells, bone marrow mesenchymal stem cells, Schwann cells, and olfactory ensheathing glia (Lu and Ashwell, 2002; Kim et al., 2007). These cells are preferred because they have clear capacity to become neurons or glial cells after transplantation into the injured spinal cord. Directed differentiation of stem/progenitor cells to oligodendrocyte lineage prior to transplantation may promote oligodendroglial differentiation. It is stated that this may be an effective strategy to increase the extent of remyelination. Transplanted stem/progenitor cells can also contribute to axonal regeneration by functioning as cellular scaffolds for growing axons. The combinatorial approaches using polymer scaffolds to fill the lesion cavity or introducing regeneration-promoting genes can greatly increase the efficacy of cellular transplantation strategies for SCI (Kim et al., 2007; Webber et al., 2007). The use of olfactory ensheathing cells (OECs) is another procedure for neural transplantation. Olfactory ensheathing glial cells (OEG) are a specialized type of glial cells that guide primary olfactory axons from the neuroepithelium in the nasal cavity to the brain (Franssen et al., 2007). The ability of olfactory neurons to grow axons in the mature brain milieu has been attributed to the presence of OECs. It has been shown that transplanted OECs are capable of migrating into and through astrocytic scars and thereby facilitating axonal regrowth through an injury barrier. It is suggested that cotransplantation of stem/progenitor cells and OECs into an injured spinal cord may have a synergistic effect, promoting neural regeneration and functional reconstruction. The lost neurocytes can be replaced by stem/progenitor cells, while the OECs can promote the formation of "bridges" crossing the glial scaring that conduct axon elongation and promote myelinazation, simultaneously (Bartolomei and Greer, 2000). It is suggested that two types of cells may first be seeded into a bioactive scaffold and then the cell seeded construct can be implanted into the injury site. This may facilitate treatment that may lead to improved neural regeneration and functional reconstruction after SCI (Ao et al., 2007). Therapeutic approaches using stem/progenitor cells transplants in animal models of SCI have provided mixed results. Some studies have provided positive results on behavioral recovery, whereas other investigators have reported that stem/progenitor cell transplants fail to promote significant functional recovery, with a small improvement observed in only one of the four tasks employed, primarily related to improvements in sensory function. Tracing of the corticospinal tract and ascending dorsal column pathway reveals no regeneration of the axons beyond the lesion site (Webber et al., 2007). In spite of this challenge, stem/progenitor cell therapy is likely to remain within the experimental arena for the foreseeable future.

### *5.5.2 Human Umbilical Cord Blood Stem Cells Transplants*

hUCB cells provide great promise for therapeutic repair after SCI. Ultrastructural analysis of axons has revealed that hUCB can be transformed into morphologically normal appearing myelin sheaths around axons in the injured areas of spinal cord. Stereotactic transplantation of hUCB into the injury epicenter of spinal cord 7 days after weight-drop injury results in the survival of transplanted cells for at least 2 weeks. These cells differentiate into oligodendrocytes and neurons and cause improvement in hind limb locomotor function as judged by better Basso–Beattie–Bresnahan (BBB) scores (Dasari et al., 2007). RT-PCR microarray studies indicate the upregulation of genes involved in inflammation and apoptosis in injured spinal cords of rats, whereas genes associated with neuroprotection are upregulated in the hUCB-treated rats (Dasari et al., 2009). These studies emphasize the therapeutic potential of hUCB in inhibiting the neuronal apoptosis during the repair of injured spinal cord.

Although stem/progenitor cell and hUCB biotechnologists have realized commercial potentials of human stem cell research, clinical applications of human cell for the treatment of neurological disorders have been the subject of intense ethical and legislative considerations. Very little is known about neurochemical aspects of stem/progenitor cells and hUCB transplant-mediated regeneration, which may involve neurotrophic factors, modulation of neuroinflammatory processes, and participation of internal protrusive forces generated by microtubules either through their own elongation or by transporting other cytoskeletal elements, such as neurofilaments into the axon tip (Ruff et al., 2008; Song et al., 2008; Furukawa and Furukawa, 2007).

Another strategy to induce regeneration is facilitation of axonal regeneration and growth after spinal cord injury is the implantation of autologous Schwann cells into sites of spinal cord injury to support and guide axonal growth (Jones et al., 2001). Furthermore, recent experiments have shown that neurotrophic factors can also promote axonal growth, and when combined with Schwann cell grafts they can further amplify axonal extension after injury. Collective evidence suggests that due to the complexity of the regenerative processes, it is likely that above approaches may not be enough to achieve functional restoration of neuronal circuits and recovery from SCI. This is tempting to suggest that many refinements, practical considerations, and risk factors that must be addressed before the use of above method for human SCI treatment.

## 5.6 Rehabilitation and SCI

Patients with SCI exhibit deficits in volitional motor control and sensation that limit not only the performance of daily tasks but also the overall activity level of these individuals. These patients have extremely sedentary lifestyle with an increased incidence of secondary complications including diabetes mellitus, hypertension, and atherogenic lipid profiles (Jacob and Nash, 2004). As the daily lifestyle of SCI patients is without physical exercise, structured exercise activities must be added to the regular schedule if the individual is to reduce the likelihood of secondary complications and/or to enhance their physical capacity. Thus, physical rehabilitation following SCI traditionally focuses on teaching compensatory techniques, thus enabling the patient to achieve day-to-day function despite significant neurological and behavioral deficits (Sadowsky and McDonald, 2009). Rehabilitation requires a comprehensive, highly integrated, and intensive program that involves a combined neural and mechanical measurement approach that assists in the determination of arm movement as well as conditioning of lower limbs after SCI. SCI patients perform three rhythmic arm movement tasks, such as (a) hand and foot cycling, (b) swinging while standing, and (c) swinging while treadmill walking. Any difference in neural control between tasks (i.e., pattern of muscle activity) may reflect changes in the mechanical constraints unique to each task (Sadowsky and McDonald, 2009). Rehabilitation plans include mobility, activities of daily living, equipment needs, such as braces, assistive devices, and multi-sensor activity monitoring wheelchairs to achieve upright and seated mobility, and adjustment issues after SCI (Behrman et al., 2006). Dealing the health-care needs of SCI patient is an immense challenge and responsibility. It requires multidisciplinary team of highly trained orthopedic surgeon, neurosurgeon, therapists, nurses, and psychosocial support for patient and his or her family (Mitcho and Kanko, 1999; Murphy, 1999). Physical therapists help SCI patients with lower extremity function and locomotion difficulties. Occupational therapists deal with upper extremity dysfunction and problems with activities of daily living. Rehabilitation nurses educate and help with the issues of bowel and bladder dysfunction and the management of pressure ulcers. Psychologists focus on the emotional and behavioral concerns of the newly injured patient and with any potential cognitive dysfunction. Speech language pathologists address with issues of communication and swallowing (Mitcho and Kanko, 1999; Murphy, 1999). The rehabilitation team operates under the direction of rehabilitation specialist physician who specializes in physical medicine and rehabilitation. Rehabilitation after SCI is complicated by autonomic dysreflexia, heterotropic ossification, neurogenic bowel, and orthostasis.

## 5.7 Conclusion

SCI is a most survivable and yet disabling condition that happens to animals and patients. Significant advances have been made in understanding the pathophysiology of SCI and a number of therapeutic agents have been discovered and tried in animal models. Furthermore, several randomized controlled trials examining therapeutic agents including methylprednisolone sodium succinate, tirilazad mesylate, monosialotetrahexosyl-ganglioside, thyrotropin-releasing hormone, gacyclidine, naloxone, and nimodipine have been performed in animals and humans. The primary outcome of trials with above therapeutic agents has been largely negative. However, administration of methylprednisolone sodium succinate within 8 h after SCI has emerged as a drug with some clinical benefits in SCI. New clinical trials on neuroprotective effects of riluzole and minocycline, the inactivation of myelin inhibition by blocking Nogo and Rho, and the transplantation of various cellular substrates into the injured cord have been planned. A number of strategies have also been developed to facilitate regeneration (axonal growth) across the lesion with a variety of cellular substrates. These include fetal tissue transplants, stem/progenitor cells, olfactory ensheathing cells, and human umbilical cord blood stem cells. Promising results have been obtained in experimental models of SCI with stem cells, which can differentiate into neurons or glia and used for the replacement of neural cells lost after SCI. Neuroprotective and axon regeneration-promoting effects have also been credited to transplanted stem cells.

## References

Ala T, Romero S, Knight F, Feldt K, Frey WH 2nd (1990) GM-1 treatment of Alzheimer's disease. A pilot study of safety and efficacy. Arch Neurol 47:1126–1130

Anderson DK, Braughler JM, Hall ED, Waters TR, McCall JM, Means ED (1988) Effects of treatment with U-74006F on neurological outcome following experimental spinal cord injury. J Neurosurg 69:562–567

Anderson DK, Hall ED, Braughler JM, McCall JM, Means ED (1991) Effect of delayed administration of U74006F (tirilazad mesylate) on recovery of locomotor function after experimental spinal cord injury. J Neurotrauma 8:187–192

Anderson DK, Hall ED (1994) Lipid hydrolysis and free radical formation in central nervous system trauma. In: Salzman SK, Faden AI (eds) The neurobiology of central nervous system trauma. Oxford University Press, Oxford, pp 131–138

Ao Q, Wang AJ, Chen GO, Wang ST, Zuo HC, Zhang XF (2007) Combined transplantation of neural stem cells and olfactory ensheathing cells for the repair of spinal cord injuries. Med Hypotheses 69:1234–1237

Aslan A, Cemek M, Buyukokuroglu ME, Altunbas K, Bas O, Yurumez Y, Cosar M (2009) Dantrolene can reduce secondary damage after spinal cord injury. Eur Spine J 26 May 2009 [Epub ahead of print]

Baptiste DC, Austin JW, Zhao W, Nahirny A, Sugita S, Fehlings MG (2009) Systemic polyethylene glycol promotes neurological recovery and tissue sparing in rats after cervical spinal cord injury. J Neuropathol Exp Neurol 18 May 2009 [Epub ahead of print]

Bartolomei JC, Greer CA (2000) Olfactory ensheathing cells: bridging the gap in spinal cord injury. Neurosurgery 47:1057–1069

Beattie MS, Farooqui AA, Bresnahan JC (2000) Review of current evidence for apoptosis after spinal cord injury. J Neurotrauma 17:915–925

Behrman AL, Bowden MG, Nair PM (2006) Neuroplasticity after spinal cord injury and training: an emerging paradigm shift in rehabilitation and walking recovery. Phys Ther 86: 1406–1425

Blesch and Tuszynski (2003) Cellular GDNF delivery promotes growth of motor and dorsal column sensory axons after partial and complete spinal cord transections and induces remyelination. J Comp Neurol 467:403–417

Bracken MB, Shepard MJ, Collins WF, Holford TR, Young W, Baskin DS, Eisenberg HM, Flamm E, Leo-Summers L, Maroon J (1990) A randomized, controlled trial of methylprednisolone or naloxone in the treatment of acute spinal-cord injury. Results of the second national acute spinal cord injury study. N Engl J Med 322:1405–1411

Buki A, Farkas O, Doczi T, Povlishock JT (2003) Preinjury administration of the calpain inhibitor MDL-28170 attenuates traumatically induced axonal injury. J Neurotrauma 20:261–268

Bunge MB (2008) Novel combination strategies to repair the injured mammalian spinal cord. J Spinal Cord Med 31:262–269

Carvalho MO, Barros Filho TE, Tebet MA (2008) Effects of methylprednisolone and ganglioside GM-1 on a spinal lesion: a functional analysis. Clinics (Sao Paulo) 63:375–380

Ceylan S, Ilbay K, Baykal S, Ceylan S, Sener U, Ozmenoğlu M, Kalelioğlu M, Aktürk F, Komsuoğlu SS, Ozoran A (1992) Treatment of acute spinal cord injuries: comparison of thyrotropin-releasing hormone and nimodipine. Res Exp Med (Berl) 192:23–33

Chang CM, Lee MH, Wang TC, Weng HH, Chung CY, Yang JT (2009) Brain protection by methylprednisolone in rats with spinal cord injury. Neuroreport 20:968–972

Chinnock P, Roberts I (2005) Gangliosides for acute spinal cord injury. Cochrane Database Syst Rev 2:CD004444

Chvatal SA, Kim YT, Bratt-Leal AM, Lee H, Bellamkonda RV (2008) Spatial distribution and acute anti-inflammatory effects of methylprednisolone after sustained local delivery to the contused spinal cord. Biomaterials 29:1967–1975

Dasari VR, Spomar DG, Gondi CS, Sloffer CA, Saving KL, Gujrati M, Rao JS, Dinh DH (2007) Axonal remyelination by cord blood stem cells after spinal cord injury. J Neurotrauma 24: 391–410

Dasari VR, Veeravalli KK, Tsung AJ, Gondi CS, Gujrati M, Dinh D, Rao JS (2009) Neuronal apoptosis inhibited by cord blood stem cells after spinal cord injury. J Neurotrauma 26 May 2009 [Epub ahead of print]

Ding Y, Kastin AJ, Pan W (2005) Neural plasticity after spinal cord injury. Curr Pharm Des 11:1441–1450

Ditor DS, John SM, Roy J, Marx JC, Kittmer C, Weaver LC (2007) Effects of polyethylene glycol and magnesium sulfate administration on clinically relevant neurological outcomes after spinal cord injury in the rat. J Neurosci Res 85:1458–1467

Domeniconi M, Filbin MT (2005) Overcoming inhibitors in myelin to promote axonal regeneration. J Neurol Sci 233:43–47

Eftekharpour E, Karimi-Abdolrezaee S, Fehlings MG (2008) Current status of experimental cell replacement approaches to spinal cord injury. Neurofocus 24:E19

Farooqui AA, Yang HC, Horrocks LA (1997) Involvement of phospholipase $A_2$ in neurodegeneration. Neurochem Int 30:517–522

Farooqui AA, Litsky ML, Farooqui T, Horrocks LA (1999) Inhibitors of intracellular phospholipase $A_2$ activity: their neurochemical effects and therapeutical importance for neurological disorders. Brain Res Bull 49:139–153

Farooqui AA, Ong WY, Horrocks LA (2004) Biochemical aspects of neurodegeneration in human brain: involvement of neural membrane phospholipids and phospholipases A2. Neurochem Res 29:1961–1977

Farooqui AA, Ong WY, Horrocks LA (2006) Inhibitors of brain phospholipase $A_2$ activity: their neuropharmacological effects and therapeutic importance for the treatment of neurologic disorders. Pharmacol Rev 58:591–620

Farooqui AA, Horrocks LA (2007) Glycerophospholipids in brain. Springer, New York, NY

Favaron M, Manev H, Alho H, Bertolino M, Ferret B, Guidotti A, Costa E (1988) Gangliosides prevent glutamate and kainate neurotoxicity in primary neuronal cultures of neonatal rat cerebellum and cortex. Proc Natl Acad Sci USA 85:7351–7355

Festoff BW, Ameenuddin S, Arnold PM, Wong A, Santacruz KS, Citron BA (2006) Minocycline neuroprotects, reduces microgliosis, and inhibits caspase protease expression early after spinal cord injury. J Neurochem 97:1314–1326

Filbin MT (2003) Myelin-associated inhibitors of axonal regeneration in the adult mammalian CNS. Nat Rev Neurosci 4:703–713

Fouad K, Pearson K (2004) Restoring walking after spinal cord injury. Prog Neurobiol 73:107–126

Fouad K, Tse A (2008) Adaptive changes in the injured spinal cord and their role in promoting functional recovery. Neurol Res 30:17–27

Franssen EH, de Bree FM, Verhaagen J (2007) Olfactory ensheathing glia: their contribution to primary olfactory nervous system regeneration and their regenerative potential following transplantation into the injured spinal cord. Brain Res Rev 56:236–258

Furukawa S, Furukawa Y (2007) FGF-2-treatment improves locomotor function via axonal regeneration in the transected rat spinal cord. Brain Nerve 59:1333–1339

Garcia-Martinez EM, Sanz-Blasco S, Karachitos A, Bandez MJ, Fernandez-Gomez FJ, Perez-Alvarez S, de Mera RM, Jordan MJ, Aguirre N, Galindo MF, Villalobos C, Navarro A, Kmita H, Jordán J (2010) Mitochondria and calcium flux as targets of neuroprotection caused by minocycline in cerebellar granule cells. Biochem Pharmacol 79:239–250

Geisler FH, Dorsey FC, Coleman WP (1991) Recovery of motor function after spinal-cord injury–a randomized, placebo-controlled trial with GM-1 ganglioside. N Engl J Med 324:1829–1838

Geisler FH, Dorsey FC, Coleman WP (1993) Past and current clinical studies with GM-1 ganglioside in acute spinal cord injury. Ann Emerg Med 22:1041–1047

Geisler FH (1998) Clinical trials of pharmacotherapy for spinal cord injury. Ann NY Acad Sci 845:374–381

Hall ED (1988) Effects of the 21-aminosteroid U74006F on posttraumatic spinal cord ischemia in cats. J Neurosurg 68:462–465

Hall ED, Andrus PK, Smith SL, Oostveen JA, Scherch HM, Lutzke BS, Raub TJ, Sawada GA, Palmer JR, Banitt LS, Tustin JS, Belonga KL, Ayer DE, Bundy GL (1996) Neuroprotective efficacy of microvascularly-localized versus brain-penetrating antioxidants. Acta Neurochir Suppl 66:107–113

Hashimoto T, Fukuda N (1991) Effect of thyrotropin-releasing hormone on the neurologic impairment in rats with spinal cord injury: treatment starting 24 h and 7 days after injury. Eur J Pharmacol 203:25–32

Hawryluk GW, Rowland J, Kwon BK, Fehlings MG (2008) Protection and repair of the injured spinal cord: a review of completed, ongoing, and planned clinical trials for acute spinal cord injury. Nurosurg Focus 25:E14

Hsu JY, Bourguignon LY, Adams CM, Peyrollier K, Zhang H, Fandel T, Cun CL, Werb Z, Noble-Haeusslein LJ (2008) Matrix metalloproteinase-9 facilitates glial scar formation in the injured spinal cord. J Neurosci 28:13467–13477

Hu WH, Qiang WA, Li F, Liu N, Wang GO, Wang HY, Wan XS, Liao WH, Liu JS, Jen MF (2000) Constitutive and inducible nitric oxide synthases after dynorphin-induced spinal cord injury. J Chem Neuroanat 17:183–197

Hua XY, Svensson CI, Matsui T, Fitzsimmons B, Yaksh TL, Webb M (2005) Intrathecal minocycline attenuates peripheral inflammation-induced hyperalgesia by inhibiting p38 MAPK in spinal microglia. Eur J Neurosci 22:2431–2440

Huang WL, King VR, Dyall SC, Ward RE, Lal N, Priestley JV, Michael-Titus AT (2007) A combination of intravenous and dietary docosahexaenoic acid significantly improves outcome after spinal cord injury. Brain 130:3004–3019

Huang W, Bhavsar A, Ward RE, Hall JC, Priestley JV, Michael-Titus AT (2009) Arachidonyl trifluoromethyl ketone is neuroprotective after spinal cord injury. J Neurotrauma 16 Apr 2009 [Epub ahead of print]

Ikeda O, Murakami M, Ino H, Yamazaki M, Nemoto T, Koda M, Nakayama C, Moriya H (2001) Acute up-regulation of brain-derived neurotrophic factor expression resulting from experimentally induced injury in the rat spinal cord. Acta Neuropathol 102:239–245

Ito M, Natsume A, Takeuchi H, Shimato S, Ohno M, Wakabayashi T, Yoshida J (2009) Type I interferon inhibits astrocytic gliosis and promotes functional recovery after spinal cord injury by deactivation of the MEK/ERK Pathway. J Neurotrauma 26:41–53

Jacob PL, Nash MS (2004) Exercise recommendations for individuals with spinal cord injury. Sport Med 34:727–751

Jones LL, Oudega M, Bunge MB, Tuszynski MH (2001) Neurotrophic factors, cellular bridges and gene therapy for spinal cord injury. J Physiol 533:83–89

Jones LL, Margolis RU, Tuszynski MH (2003) The chondroitin sulfate proteoglycans neurocan, brevican, phosphacan, and versican are differentially regulated following spinal cord injury. Exp Neurol 182:399–411

Kim SS, Kong PJ, Kim BS, Sheen DH, Nam SY, Chun W (2004) Inhibitory action of minocycline on lipopolysaccharide-induced release of nitric oxide and prostaglandin E2 in BV2 microglial cells. Arch Pharm Res 27:314–318

Kim HM, Hwang DH, Lee JE, Kim SU, Kim BG (2007) Stem cell-based cell therapy for spinal cord injury. Cell Transplant 16:355–364

Kim HM, Hwang DH, Lee JE, Kim SU, Kim BG (2009) Ex vivo VEGF delivery by neural stem cells enhances proliferation of glial progenitors, angiogenesis, and tissue sparing after spinal cord injury. PLoS One 4:e4987

Kim YT, Caldwell JM, Bellamkonda RV (2009) Nanoparticle-mediated local delivery of Methylprednisolone after spinal cord injury. Biomaterials 30:2582–2590

Kim HS, Suh YH (2009) Minocycline and neurodegenerative diseases. Behav Brain Res 196: 168–179

King VR, Huang WL, Dyall SC, Curran OE, Priestley JV, Michael-Titus AT (2006) Omega-3 fatty acids improve recovery, whereas omega-6 fatty acids worsen outcome, after spinal cord injury in the adult rat. J Neurosci 26:4672–4680

Koob AO, Colby JM, Borgens RB (2008) Behavioral recovery from traumatic brain injury after membrane reconstruction using polyethylene glycol. J Biol Eng 2:9

Kubo T, Hata K, Yamaguchi A, Yamashita T (2008) Rho-ROCK inhibitors as emerging strategies to promote nerve regeneration. Curr Pharm Des 13:2493–2499

Kwok JC, Afshari F, Garcia-Alias G, Fawcett JW (2008) Proteoglycans in the central nervous system: plasticity, regeneration and their stimulation with chondroitinase ABC. Restor Neurol Neurosci 26:131–145

Kwon BK, Roy J, Lee JH, Okon EB, Zhang H, Marx JC, Kindy MS (2009) Magnesium chloride in a polyethylene glycol formulation as a neuroprotective therapy for acute spinal cord injury: preclinical refinement and optimization. J Neurotrauma 24 Mar 2009 [Epub ahead of print]

Lang-Lazdunski L, Biondeau N, Jarretou G, Heurteaux C (2003) Linolenic acid prevents neuronal cell death and paraplegia after transient spinal cord ischemia in rats. J Vasc Surg 38:564–575

Latorre E, Collado MP, Fernández I, Aragonés MD, Catalán RE (2003) Signaling events mediating activation of brain ethanolamine plasmalogen hydrolysis by ceramide. Eur J Biochem 270: 36–46

Lauritzen I, Blondeau N, Heurteaux C, Widmann C, Romey G, Lazunski M (2000) Polyunsaturated fatty acids are potent neuroprotectors. EMBO J 19:1784–1793

Ledeen RW, Wu G (2002) Ganglioside function in calcium homeostasis and signaling. Neurochem Res 27:637–647

Lee SM, Yune TY, Kim SJ, Park DW, Lee YK, Kim YC, Oh YJ, Markelonis GJ, Oh TH (2003) Minocycline reduces cell death and improves functional recovery after traumatic spinal cord injury in the rat. J Neurotrauma 20:1017–1027

Lee JM, Yan P, Xiao Q, Chen S, Lee KY, Hsu CH, Xu J (2008) Methylprednisolone protects oligodendrocytes but not neurons after spinal cord injury. J Neurosci 28:3141–3149

Lee MY, Chen L, Toborek M (2009) Nicotine attenuates iNOS expression and contributes to neuroprotection in a compressive model of spinal cord injury. J Neurosci Res 87:937–947

Leon A, Facci L, Toffano G, Sonnino S, Tettamanti G (1981) Activation of $Na^+$, $K^+$-ATPase by nanomolar concentrations of $GM_1$ ganglioside. J Neurochem 37:350–357

Lingor P, Teusch N, Schwarz K, Mueller R, Mack H, Bahr M, Mueller BK (2007) Inhibition of Rho kinase (ROCK) increases neurite outgrowth on chondroitin sulphate proteoglycan in vitro and axonal regeneration in the adult optic nerve in vivo. J Neurochem 103:181–191

Liu NK, Zhang YP, Titsworth WL, Jiang X, Han S, Lu PH, Shields CB, Xu XM (2006) A novel role of phospholipase $A_2$ in mediating spinal cord secondary injury. Ann Neurol 59:606–619

Liu Y, Wang X, Lu CC, Kerman R, Steward O, Xu XM, Zou Y (2008a) Repulsive Wnt signaling inhibits axon regeneration after CNS injury. J Neurosci 28:8376–8382

Liu WL, Lee YH, Tsai SY, Hsu CY, Sun YY, Yang LY, Tsai SH, Yang WC (2008b) Methylprednisolone inhibits the expression of glial fibrillary acidic protein and chondroitin sulfate proteoglycans in reactivated astrocytes. Glia 56:1390–1400

Lu J, Ashwell K (2002) Olfactory ensheathing cells: their potential use for repairing the injured spinal cord. Spine (Phila Pa 1978) 27:887–892

Lukáčová N, Kolesárová M, Kuchárová K, Pavel J, Kolesár D, Radoňák J, Maršala M, Chalimoniuk M, Langfort J, Maršala J (2005) The effect of a spinal cord hemisection on changes in nitric oxide synthase pools in the site of injury and in regions located far away from the injured site. Cell Mol Neurobiol 26:1365–1383

Lukacova N, Davidova A, Kolesar D, Kolesarova M, Schreiberova A, Lackova M, Krizanova O, Marsala M, Marsala J (2008) The effect of N-nitro-L-arginine and aminoguanidine treatment on changes in constitutive and inducible nitric oxide synthases in the spinal cord after sciatic nerve transection. Int J Mol Med 21:413–421

Lynskey JV, Belanger A, Jung R (2008) Activity-dependent plasticity in spinal cord injury. J Rehabil Res Dev 45:229–240

Machado LS, Kozak A, Erqul A, Hess DC, Borlougan CV, Fagan SC (2006) Delayed minocycline inhibits ischemia-activated matrix metalloproteinases 2 and 9 after experimental stroke. BMC 7:56

Marchand F, Tsantoulas C, Singh D, Grist J, Clark AK, Bradbury EJ, McMahon SB (2009) Effects of Etanercept and Minocycline in a rat model of spinal cord injury. Eur J Pain 13:673–681

Marsala J, Orendacova J, Lukacova N, Vanicky I (2007) Traumatic injury of the spinal cord and nitric oxide. Prog Brain Res 161:171–183

McKerracher L, Winton MJ (2002) Nogo on the go. Neuron 36:345–348

Melero-Fernández de Mera RM, García-Martínez E, Fernández-Gómez FJ, Hernández-Guijo JM, Aguirre N, Galindo MF, Jordán J (2008) Rev. Neurol 47:31–38

Michael-Titus AT (2007) Omega-3 fatty acids and neurological injury. Prost Leukot Essent Fatty Acids 77:295–300

Mitcho K, Kanko JR (1999) Acute care management of spinal cord injuries. Crit Care Nurse 22: 60–79

Miyashita T, Koda M, Kitajo K, Yamazaki M, Takahashi K, Kikuchi A, Yamashita T (2009) Wnt-Ryk signaling mediates axon growth inhibition and limits functional recovery after spinal cord injury. J Neurotrauma 27 May 2009 [Epub ahead of print]

Monga V, Meena CL, Kaur N, Jain R (2008) Chemistry and biology of thyrotropin-releasing hormone (TRH) and its analogs. Curr Med Chem 15:18–33

Moriwaki A, Nishida K, Matsushita M, Ozaki T, Kunisada T, Yoshida A, Inoue H, Matsui H (2005) Calpain inhibitors prevent neuronal cell death and ameliorate motor disturbances after compression-induced spinal cord injury in rats. J Neurotrauma 22:398–406

Murphy M (1999) Traumatic spinal cord injury: an acute care rehabilitation perspective. Crit Care Nurse Q 22:51–59

Noble LJ, Donovan F, Igarashi T, Gousseo S, Werb Z (2002) Matrix metalloproteinases limit functional recovery after spinal cord injury by modulation of early vascular events. J Neurosci 22:7526–7535

Olivas AD, Noble-Haeusslein LJ (2006) Phospholipase $A_2$ and spinal cord injury: a novel target for therapeutic intervention. Ann Neurol 59:577–579

Park JY, Kim HY, Jou I, Park SM (2008) GM1 induces p38 and microtubule dependent ramification of rat primary microglia in vitro. Brain Res 1244:13–23

Partington CR, Daly JW (1979) Effect of gangliosides on adenylate cyclase activity in rat cerebral cortical membranes. Mol Pharmacol 15:484–491

Phillis JW, O'Regan MH (1995) $GM_1$ ganglioside inhibits ischemic release of amino acid neurotransmitters from rat cortex. Neuroreport 6:2010–2012

Pizzi MA, Crowe MJ (2007) Matrix metalloproteinases and proteoglycans in axonal regeneration. Exp Neurol 204:496–511

Pruzanski W, Greenwald RA, Street IP, Laliberte F, Stefanski E, Vadas P (1992) Inhibition of enzymatic activity of phospholipases $A_2$ by minocycline and doxycycline. Biochem Pharmacol 44:1165–1170

Pyo H, Joe E, Tung S, Lee SH, Jou I (1999) Gangliosides activate cultured rat brain microglia. J Biol Chem 274:34584–34589

Ravikumar R, Fugaccia I, Scheff SW, Geddes JW, Srinivasan C, Toborek M (2005) Nicotine attenuates morphological deficits in a contusion model of spinal cord injury. J Neurotrauma 22:240–251

Ray SK, Hogan EL, Banik NL (2003) Calpain in the pathophysiology of spinal cord injury: neuroprotection with calpain inhibitors. Brain Res Brain Res Rev 42:169–185

Riegger T, Conrad S, Liu K, Schluesener HJ, Adibzahdeh M, Schwab JM (2007) Spinal cord injury-induced immune depression syndrome (SCI-IDS). Eur J Neurosci 25:1743–1747

Riegger T, Conrad S, Schluesener HJ, Kaps HP, Badke A, Baron C, Gerstein J, Dietz K, Abdizahdeh M, Schwab JM (2009) Immune depression syndrome following human spinal cord injury (SCI): a pilot study. Neuroscience 158:1194–1199

Ruff RL, McKerracher L, Selzer ME (2008) Repair and neurorehabilitation strategies for spinal cord injury. Ann NY Acad Sci 1142:1–20

Sadowsky CL, McDonald JW (2009) Activity-based restorative therapies: concepts and applications in spinal cord injury-related neurorehabilitation. Dev Disabil Res Rev 15:112–126

Schmid D, Burmester GR, Tripmacher R, Fici G, von Voigtlander P, Buttgereit F (2001) Short-term effects of the 21-aminosteroid lazaroid tirilazad mesylate (PNU-74006F) and the pyrrolopyrimidine lazaroid PNU-101033E on energy metabolism of human peripheral blood mononuclear cells. Biosci Rep 21:101–110

Schröter A, Lustenberger RM, Obermair FJ, Thallmair M (2009) High-dose corticosteroids after spinal cord injury reduce neural progenitor cell proliferation. Neuroscience 161: 753–763

Schwab JM, Brechtel K, Mueller CA, Failli V, Kaps HP, Tuli SK, Schluesener HJ (2006) Experimental strategies to promote spinal cord regeneration—an integrative perspective. Prog Neurobiol 78:91–116

Schumacher PA, Siman RG, Fehlings MG (2000) Pretreatment with calpain inhibitor CEP-4143 inhibits calpain I activation and cytoskeletal degradation, improves neurological function, and enhances axonal survival after traumatic spinal cord injury. J Neurochem 74:1646–1655

Sharma HS, Badgaiyan RD, Alm P, Mohanty S, Wiklund L (2005) Neuroprotective effects of nitric oxide synthase inhibitors in spinal cord injury-induced pathophysiology and motor functions: an experimental study in the rat. Ann NY Acad Sci 1053:422–434

Sharma HS, Nyberg F, Gordh T, Alm P (2006) Topical application of dynorphin A (1–17) antibodies attenuates neuronal nitric oxide synthase up-regulation, edema formation, and cell injury following focal trauma to the rat spinal cord. Acta Neurochir Suppl 96:309–315

Shioiri Y, Kurimoto A, Ako T, Daikoku S, Ohtake A, Ishida H, Kiso M, Suzuki K, Kanie O (2009) Energy-resolved structural details obtained from gangliosides. Anal Chem 81:139–145

Soderquist RG, Milligan ED, Sloane EM, Harrison JA, Douvas KK, Potter JM, Huges TS, Chavez RA, Johnson K, Watkins LR, Mahoney MJ (2008) PEGylation of brain-derived neurotrophic factor for preserved biological activity and enhanced spinal cord distribution. J Biomed Mater Res [Epub ahead of print]

Sohn H, Kim YS, Kim HT, Kim CH, Cho EW, Kang HY, Kim NS, Kim CH, Ryu SE, Lee JH, Ko JH (2006) Ganglioside GM3 is involved in neuronal cell death. FASEB J 20:1248–1250

Song Y, Wei EQ, Zhang WP, Zhang L, Liu JR, Chen Z (2004) Minocycline protects PC12 cells from ischemic-like injury and inhibits 5-lipoxygenase activation. Neuroreport 15:2181–2184

Song XY, Li F, Zhang FH, Zhong JH, Zhou XF (2008 Mar 5) 2008. Peripherally-derived BDNF promotes regeneration of ascending sensory neurons after spinal cord injury. PLoS One 3(3):e1707

Stirling DP, Khodarahmi K, Liu J, McPhail LT, McBride CB, Steeves JD, Ramer MS, Tetzlaff W (2004) Minocycline treatment reduces delayed oligodendrocyte death, attenuates axonal dieback, and improves functional outcome after spinal cord injury. J Neurosci 24:2182–2190

Stirling DP, Koochesfahani KM, Steeves JD, Tetzlaff W (2005) Minocycline as a neuroprotective agent. Neuroscientist 11:308–322

Svennerholm L (1994) Gangliosides—a new therapeutic agent against stroke and Alzheimer's disease. Life Sci 55:2125–2134

Tan AM, Zhao P, Waxman SG, Hains BC (2009) Early microglial inhibition preemptively mitigates chronic pain development after experimental spinal cord injury. J Rehabi Res Dev 46:123–133

Teng YD, Choi H, Onario RC, Zhu S, Desilets FC, Lan S, Woodard EJ, Snyder EY, Eichler ME, Friedlander RM (2004) Minocycline inhibits contusion-triggered mitochondrial cytochrome c release and mitigates functional deficits after spinal cord injury. Proc Natl Acad Sci USA 101:3071–3076

Thorell WE, Leibrock LG, Agrawal SK (2002) Role of RyRs and $IP_3$ receptors after traumatic injury to spinal cord white matter. J Neurotrauma 19:335–342

Villa RF, Gorini A (1997) Pharmacology of lazaroids and brain energy metabolism: a review. Pharmacol Rev 49:99–136

Walker JB, Harris M (1993) GM-1 ganglioside administration combined with physical therapy restores ambulation in humans with chronic spinal cord injury. Neurosci Lett 161:174–178

Watkins TA, Barres BA (2002) Nerve regeneration: regrowth stumped by shared receptor. Curr Biol 12:R654–R656

Webber DJ, Bradbury EJ, McMohan SB, Minger SL (2007) Transplanted neural progenitor cells survive and differentiate but achieve limited functional recovery in the lesioned adult rat spinal cord. Regen Med 2:929–945

Wells JE, Rice TK, Nuttall RK, Edwards DR, Zekki H, Rivest S, Yong VW (2002) An adverse role for matrix metalloproteinase 12 after spinal cord injury in mice. J Neurosci 23:10107–10115

Xu J, Fan G, Chen S, Wu Y, Xu XM, Hsu CY (1998) Methylprednisolone inhibition of TNF-alpha expression and NF-kB activation after spinal cord injury in rats. Brain Res Mol Brain Res 59:135–142

Xu J, Kim GM, Ahmed SH, Xu J, Yan P, Xu XM, Hsu CY (2001) Glucocorticoid receptor-mediated suppression of activator protein-1 activation and matrix metalloproteinase expression after spinal cord injury. J Neurosci 21:92–97

Xu J, Chen S, Chen H, Xiao Q, Hsu CY, Michael D, Bao J (2009) STAT5 mediates antiapoptotic effects of methylprednisolone on oligodendrocytes. J Neurosci 29:2022–2026

Yamashita T (2007) Molecular mechanism and regulation of axon growth inhibition. Brain Nerve 59:1347–1353

Yan P, Xu J, Li Q, Chen S, Kim GM, Hsu CY, Xu XM (1999) Glucocorticoid receptor expression in the spinal cord after traumatic injury in adult rats. J Neurosci 19:9355–9363

Yang H-C, Farooqui AA, Horrocks LA (1994a) Effects of glycosaminoglycans and glycosphingolipids on cytosolic phospholipases A2 from bovine brain. Biochem J 299:91–95

Yang H-C, Farooqui AA, Horrocks LA (1994b) Effects of sialic acid and sialoglycoconjugates on cytosolic phospholipases $A_2$ from bovine brain. Biochem Biophys Res Commun 199: 1158–1166

Yates AJ, Walters JD, Wood CL, Johnson JD (1989) Ganglioside modulation of cyclic AMP-dependent protein kinase and cyclic nucleotide phosphodiesterase in vitro. J Neurochem 53:162–167

Yiu G, He Z (2006) Glial inhibition of CNS axon regeneration. Nat Rev Neurosci 7:617–627

Yrjanheikki J, Tikka T, Keinanen R, Goldsteins G, Chan PH, Koistinaho J (1999) A tetracycline derivative, minocycline, reduces inflammation and protects against focal cerebral ischemia with a wide therapeutic window. Proc Natl Acad Sci USA 96:13496–13500

# Chapter 6
# Neurochemical Aspects of Traumatic Brain Injury

## 6.1 Introduction

Traumatic brain injury is a silent epidemic and major source of death and disability worldwide in modern society. The Centers for Disease Control and Prevention estimates that approximately 1.4 million US individuals sustain traumatic brain injuries (TBIs) per year of which, approx 50,000 people die from TBI each year and 85,000 people suffer long-term disabilities. In the USA, more than 5.3 million people live with long-term disability with dramatic impacts on their own and their families' lives. The socioeconomic cost of treating and rehabilitating TBI patients exceeds $56 billion. This economic cost and rate of mortality has generated considerable interest in elucidating the complex molecular mechanism underlying cell death and dysfunction after TBI. Most common causes of TBI are car accidents, bicycle accidents (more than 50%), falls and sport injuries (20–25%), and violence and domestic abuse (including shaken baby syndrome) (20–25%). TBI produces physical, cognitive, emotional, and behavioral effects in the traumatized subject. The outcome of TBI ranges from complete recovery to permanent disability or death.

Like spinal cord injury (SCI), TBIs consist of two broadly defined components: a primary component, attributable to the mechanical insult itself, and a secondary component, attributable to the series of systemic and local neurochemical and pathophysiological changes that occur in the brain after the initial insult (Raghupathi, 2004). The primary injury rapidly causes rapid deformation of brain tissue and rupture of neural cell membranes leading in the release of intracellular contents, disruption of blood flow, breakdown of the blood–brain barrier, and intracranial hemorrhage. In contrast, secondary injury to the brain induces neurochemical alterations, activation of microglial cells and astrocytes, and demyelination involving oligodendroglia (Raghupathi, 2004). Clinical symptoms of secondary injury appear slowly (days/week/months) after TBI (Table 6.1). Cerebral ischemia is the most important mechanism underlying secondary injury. It is caused by a decrease in cerebral blood flow within the first hours after TBI (van Santbrink et al., 2002). As mentioned in Chapter 2, decrease in cerebral blood flow not only results mitochondrial damage but also induces alterations in ion homeostasis, edema, and greater reduction in cerebral blood flow. An increase of mitochondrial membrane

A.A. Farooqui, *Neurochemical Aspects of Neurotraumatic and Neurodegenerative Diseases*, DOI 10.1007/978-1-4419-6652-0_6,

**Table 6.1** Time-dependence, neurochemical events and mode of cell death following TBI

| Time post–TBI | Pathological events | Mode of cell death |
|---|---|---|
| 1–3 h | Disruption of the BBB, deformation of brain tissue, swelling, and ischemia | Alterations in ion homeostasis |
| 6 h | Rupture of neural cells, cell/axon stretching | Start of necrotic cell death |
| 12 h | Rupture of neural cells, cell/axon stretching | Maximum necrosis |
| 1 week | Edema, vasospasm, inflammation, and oxidative stress | Start of apoptotic cell death |
| 2 weeks | Edema, vasospasm, inflammation, and oxidative stress | Maximum apoptosis |
| 1 month | Edema, inflammation, and oxidative stress | Some apoptotic cell death |
| 3 months | Start of neurogenesis | Development of neurites |
| 6–12 months | – | Neuropsychiatric symptoms |

TBI rapidly initiates a series of secondary events that induce long-term neurological consequences, such as cognitive dysfunction due to neural injury (Agoston et al., 2009).

permeability is an important process in neural cell death. The mitochondrial membrane permeability transition (mPT) is a $Ca^{2+}$-dependent increase of mitochondrial membrane permeability that leads to loss of mitochondrial membrane potential (Delta Psi), mitochondrial swelling, and rupture of the outer mitochondrial membrane. In experimental TBI, extensive cell death (necrosis) occurs at the primary injury site and is driven in part by significant mitochondrial dysfunction.

Adult brain responds to TBI not only by activating a program of cell proliferation during which many oligodendrocyte precursors, microglia, and some astrocytes proliferate but also by inducing reactive gliosis, a process by which dormant astrocytes undergo morphological changes and alter their transcriptional profiles. Very little is known about the relationship between TBI-mediated reactive gliosis and proliferation of surrounding neural cells. However, two mechanisms have been proposed. One involves mitogen sonic hedgehog (SHH) factor, which is produced in reactive astrocytes after injury to the cerebral cortex. It participates in regulating the proliferation of Olig2-expressing (Olig2$^{+}$) cells after brain injury (Amankulor et al., 2009; Tatsumi et al., 2008) and the other mechanism, supporting the participation of basic helix-loop-helix transcription factor for reactive astrocyte proliferation after cortical injury (Chen et al., 2008a).

Inflammatory reactions, oxidative stress (increase in production of reactive oxygen species, ROS), and nitrosative stress (increased generation of reactive nitrogen species, RNS) are major components of secondary injury. All these processes play a major role in regulating the pathogenesis of acute and chronic TBI (Fig. 6.1).

Neuroinflammation is a neuroprotective mechanism that involves a complex cellular and molecular response of brain tissue against neural injury. It is associated with the activation of glia, release of inflammatory mediators within the brain, and recruitment of peripheral immune cells. It not only constitutes attempts of

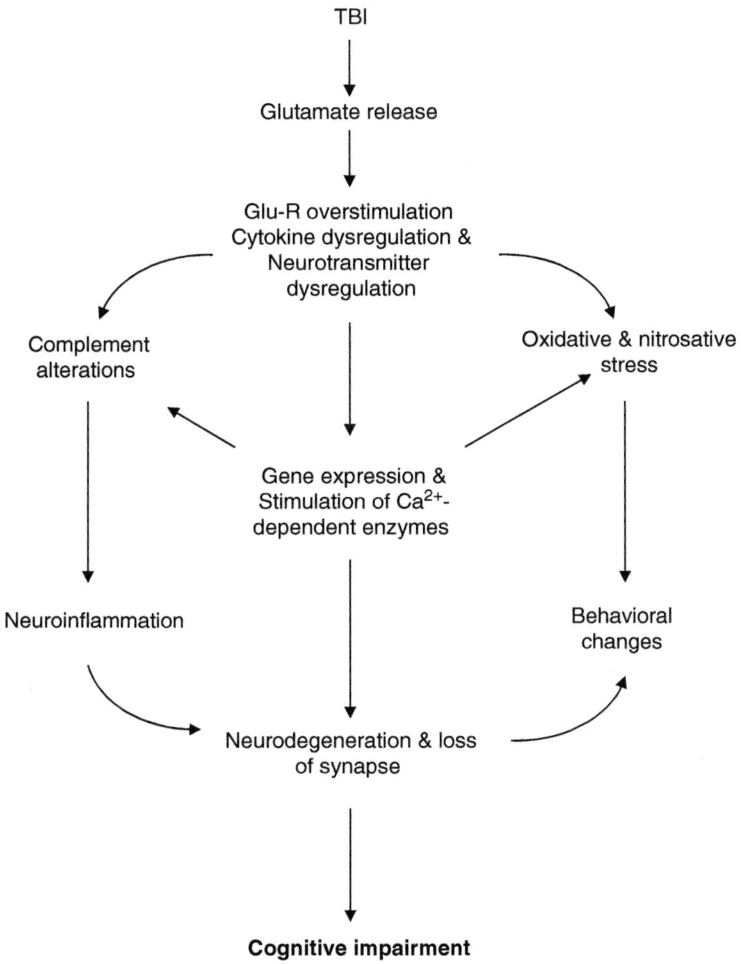


**Fig. 6.1** Hypothetical mechanism of neurodegeneration, synaptic loss, and cognitive impairment following TBI

brain tissue to defend against insults, clear dead and damaged neurons, but also facilitates the return of brain to a normal state (Farooqui and Horrocks, 2009). Inflammation in the CNS is driven by the activation of resident microglia, astrocytes, and infiltrating peripheral macrophages, which release a plethora of anti- and proinflammatory cytokines, chemokines, neurotransmitters, and ROS. The overexpression of cytokines (Hayes et al., 2002; Ahn et al., 2004), elevation in levels of S100B, glial fibrillary acidic protein (GFAP), and heat shock proteins (Hsp) (Pelinka et al., 2004; Wiesmann et al., 2010) have been reported to occur in TBI. Increase in the expression of GFAP is a characteristic feature of astrogliosis. It occurs in the brain during neurodegeneration and coincides with impairment of the ubiquitin-proteasome system. Increased expression of cytokines, S100B and GFAP protein along with a rapid decrease in ATP level, changes in ion homeostasis, oxidative damage of mitochondrial proteins, alterations in cellular redox, and induction of edema and intracranial hypertension are closely associated with increased mortality and morbidity after head injury.

## 6.2 TBI-Mediated Alterations in Glutamate and Calcium Levels

TBI is caused by a blow or jolt to the head or a penetrating head injury that disrupts the normal function of the brain. TBI releases glutamate from intracellular stores (Demediuk et al., 1988; Panter et al., 1990; Sundsrom and Mo, 2002) (Figs. 6.1 and 6.2). Glutamate causes neural cell death through several mechanisms. It hyperstimulates both NMDA (*N*-methyl-D-aspartate) and AMPA (α-amino-3-hydroxy-5-methyl-4-isoxazole propionate) types of glutamate receptors resulting in the influx of $Na^+$, efflux of $K^+$, and a large $Ca^{2+}$ influx into neurons (Farooqui et al., 2008). This process is called as excitotoxicity. It results in an uncontrolled and sustained increase in cytosolic calcium, which produces not only the uncoupling of mitochondrial electron transport but also the stimulation of many calcium-dependent enzymes, including lipases, phospholipases, calpains, nitric oxide synthase, protein phosphatases, and various protein kinases (Fig. 6.3) (Pavel et al., 2001; Ray et al., 2003; Ellis et al., 2004; Arundine and Tymianski, 2004; Atkins et al., 2007a, 2009a). Recently, excitotoxicity has been linked to autophagy (Bigford et al., 2009). It is shown that NR2B (NMDA receptor subunit) interacts with autophagic protein, Beclin-1 in membrane rafts of the normal rat cerebral cortex. Moderate TBI induces rapid recruitment and association of NR2B and $Ca^{2+}$/calmodulin-dependent protein kinase II (CaMKII) to membrane rafts and translocation of Beclin-1 out of membrane microdomains. Furthermore, TBI produces significant increases in the expression of key autophagic proteins.

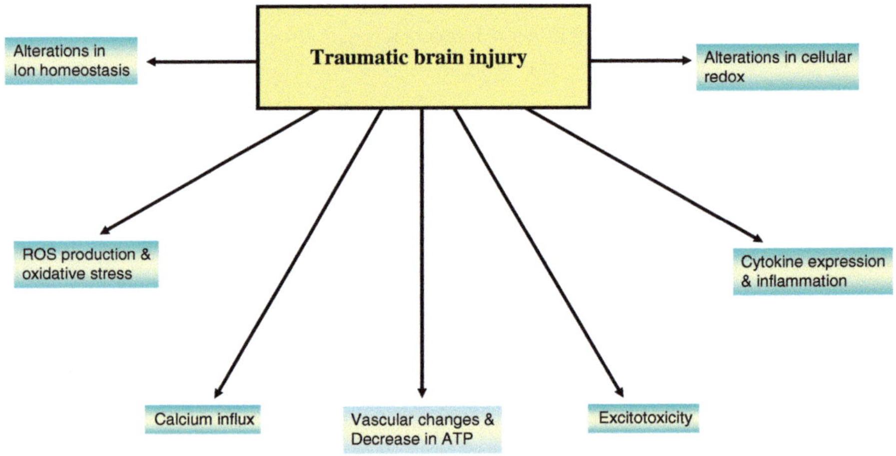


**Fig. 6.2** TBI-induced alterations in neurochemical processes

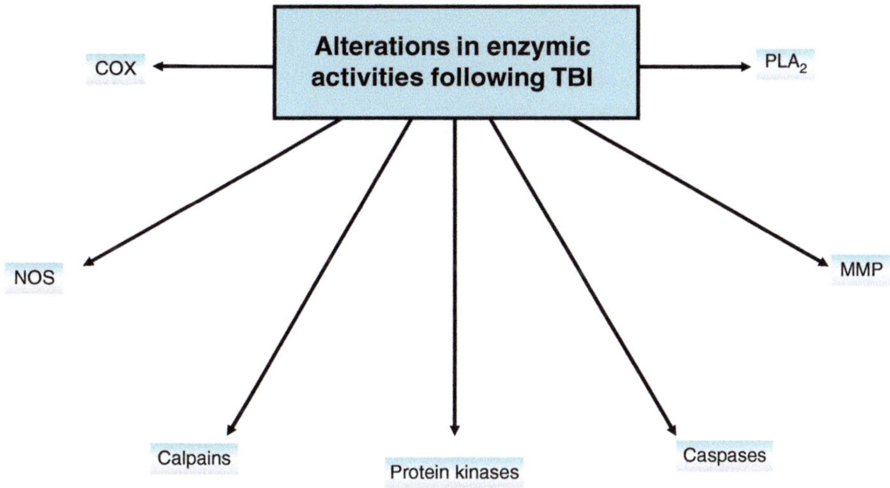


**Fig. 6.3** Enzymes that are stimulated by TBI. Cyclooxygenase (COX); phospholipase $A_2$ ($PLA_2$); nitric oxide synthase (NOS); matrix metalloproteinase (MMP)

Morphological hallmarks of autophagy that are significantly attenuated by the treatment with the NR2B antagonist Ro 25-6981, suggesting that stimulation of autophagy by NR2B signaling may be regulated by redistribution of Beclin-1 in membrane rafts after TBI (Bigford et al., 2009).

Glutamate-mediated glial cell damage does not involve glutamate receptor activation, but rather glutamate uptake (Oka et al., 1993; Matute et al., 2006). It is well known that glutamate uptake from the extracellular space by specific glutamate transporters is essential for the maintenance of excitatory post-synaptic currents (Auger and Attwell, 2000) and for blocking excitotoxic death due to overstimulation of glutamate receptors (Farooqui et al., 2008). Excitatory amino acid transporter E1 (EAAT1) and excitatory amino acid transporter E2 (EAAT2) are expressed in astrocytes, oligodendrocytes, and microglial cells (Matute et al., 2006). Glutamate produces glial cell demise by inhibiting cystine uptake, which causes a decrease in glutathione and makes glial cells vulnerable to oxidative stress (Oka et al., 1993; Matute et al., 2006; Murphy et al., 1989; Pereira and Resende de Oliveira, 2000). It is recently shown that glutamate-mediated delayed post-traumatic white matter degeneration also involves the reversal of $Na^+$-dependent glutamate transport with subsequent activation of AMPA receptors and oligodendrocyte death (Li and Stys, 2000; Park et al., 2004).

## 6.3 TBI-Mediated Alterations in Cytokines

Cytokines represent a broad, heterogeneous group of proteins and polypeptides associated with the regulation of cell–cell interactions. Levels of cytokines are markedly increased in traumatized brain (Ghirnikar et al., 1998; Sandhir et al.,

2004; Kadhim, 2008). Cytokines include interleukins (IL-1β, IL-2, IL-6, and IL-12), interferons (IFN-γ), tumor necrosis factors-α (TNF-α), tumor growth factors (TGF-α and β), and colony stimulating factors (Sun et al., 2004; Kim et al., 2001). Expression of these cytokines is very low in normal brain, where they mediate cellular intercommunication through autocrine, paracrine, or endocrine mechanisms (Wilson et al., 2002). Their actions involve a complex network linked to feedback loops and cascades. Their overall response depends on the synergistic or antagonistic actions of various components. Thus, cytokines play an important role not only in neuronal development, synaptic plasticity, survival, learning and memory, and regeneration but also in neurodegeneration. For example, TNF-α and IL-1β modulate neuronal and astroglial cell synaptic plasticity and survival at low concentrations, but at high concentrations, these cytokines also contribute to neurodegeneration. These cytokines act as important mediators for the initiation and the support of post-traumatic inflammation. In contrast, TGF-β is a potent anti-inflammatory agent, which may also have some deleterious long-term effects in the injured brain (Lenzlinger et al., 2009). Although TNF-α and IL-1β trigger biologically indistinguishable effects by activating the same set of transcription factors, these cytokines are structurally unrelated polypeptides that exert their effect through distinct and structurally unrelated cell surface receptors. The mechanism of TNF-α, IL-1β, and TGF-β actions is quite complex because they activate a number of signaling pathways, including phosphatases, kinases, phospholipases, oxygen radicals, and transcription factors (Jupp et al., 2003; Gomes-Leal et al., 2004). All these targets may participate in neurotoxicity induced by TNF-α and IL-1β in traumatized brain. Once the inflammatory cascade is initiated, these cytokines amplify their own production via autocrine induction or interact with complement proteins C1s and C1r leading to an upregulation of neuroinflammation. Collective evidence suggests that cytokines can either promote this neurotoxicity, by encouraging excitotoxicity and propagating the inflammatory response, or attenuate the damage through neuroprotective and neurotrophic mechanisms, including the induction of cell growth factors (Morganti-Kossmann et al., 2007).

## 6.4 TBI-Mediated Alterations in Chemokines

Chemokines are a group of proteins (8–12 kDa) involved in the trafficking of leukocytes in physiological immune surveillance and inflammatory cell recruitment in host defense. Chemokines include CCL2/MCP-1, CXCL12/SDF-1α, CX3CL1/fractalkine, CXCL10/IP 10, CCL3/MIP-1α, and CCL5/RANTES. They are classified into four classes based on the positions of key cysteine residues: C, CC, CXC, and CX3C. They exert their effect through both specific and shared G protein-coupled receptors expressed on microglial cells, astrocytes, and neurons. Chemokine receptors are found in brain areas such as the hypothalamus, nucleus accumbens, limbic system, hippocampus, thalamus, cortex, and cerebellum. In addition to their role in the immune system, chemokines also play a role in the brain, where their expression is increased after induction with inflammatory mediators (Bajetto et al., 2001). Chemokines are also associated with brain development neural cell migration, differentiation, and proliferation. Accumulating evidence suggests that chemokines are plurifunctional family of proteins that modulate the communication between the neuroendocrine and the immune system (Bajetto et al., 2001; Callewaere et al., 2007). Expression and levels of chemokines are markedly increased in brain following TBI (Israelsson et al., 2008).

TBI is accompanied by neutrophil infiltration of the choroid plexus (CP), a site of the blood–cerebrospinal fluid (CSF) barrier (BCSFB), and accumulation of neutrophils in the CSF space near the injury, from where they may migrate to brain parenchyma (Szmydynger-Chodobska et al., 2009). It is hypothesized that the CP functions as an entry point for neutrophils to invade the traumatized brain. The expression of CXC chemokines, such as cytokine-induced neutrophil chemoattractant (CINC)-1 or CXCL1, CINC-2α or CXCL3, and CINC-3 or CXCL2, is markedly increased in CP. It is stated that secretion of these chemokines is the prerequisite for neutrophil migration across epithelial barriers (Szmydynger-Chodobska et al., 2009). Although relative contribution of various neural cells to increased chemokine expression is not known, it is shown that neurons as well as glial cells contribute to the expression of macrophage inflammatory protein-2 (MIP-2/CXCL2) and the monocyte chemokine monocyte chemotactic protein-1 (MCP-1/CCL2). Increased levels of chemokines have been detected in the CSF of TBI patients. The increased expression of proinflammatory cytokines and chemokines may be responsible for the acute pathologic alterations (cerebral edema and intracranial hypertension) and cognitive impairment following TBI (Rhodes et al., 2009). Collective evidence suggests that chemokine activation occurs early after moderate or severe TBI and is maintained for several days after TBI. This event may contribute to neuroinflammatory exacerbation of post-traumatic brain damage.

## 6.5 TBI-Mediated Alterations in Enzymic Activities

Glutamate-mediated calcium influx results in stimulation arachidonic acid release from neural membrane glycerophospholipids. This release is catalyzed by $cPLA_2$ and PLC/DAG-lipase pathway (McIntosh et al., 1998; Schuhmann et al., 2003; Shohami et al., 1987, 1989; Wei et al., 1982; Dhillon et al., 1996; Homayoun et al., 1997, 2000). Arachidonic acid release occurs in traumatic as well as fluid percussion models of brain injury (FPI). Enzymic oxidation of arachidonic acid generates prostaglandins, leukotrienes, and thromoboxanes whereas non-enzymic oxidation produces isoprostanes and ROS which include superoxide and hydroxyl radicals (Farooqui and Horrocks, 2007).

Astroglial response to TBI is characterized by hyperplasia and upregulation in glial fibrillary acidic protein (Table 6.2). The reactive astrocytes also express neurotrophic factors, and cytokines, which modulate the generation of recognition molecules allowing the support of post-lesional axonal regrowth. Major

**Table 6.2** Biomarkers for traumatic brain injury

| Biomarker | Change in CSF | References |
|---|---|---|
| Creatine kinase | Increased | Vázquez et al. (1995) |
| Glial fibrillary protein | Increased | Wiesmann et al. (2010) |
| Lactate dehydrogenase | Increased | Osuna et al. (1992) |
| Myelin basic protein | Increased | Ottens et al. (2008) |
| Neuronal enolase | Increased | Vázquez et al. (1995) |
| S-100 proteins | Increased | Pelinka et al. (2004) |
| c-Tau | Increased | Svetlov et al. (2009) |
| NMDA-R fragments | Increased | Svetlov et al. (2009) |
| Spectrin fragments | Increased | Svetlov et al. (2009) |
| Cytokines and chemokines | Increased | Svetlov et al. (2009) |

consequences of TBI are astrocyte-mediated brain edema and increase in intracranial pressure. Exposure of cultured rat astrocytes to 5 atm of pressure induces significant cell swelling at 1–24 h following FPI with maximal swelling at 3 h. Several factors contribute to astrocytic swelling. They include oxidative stress, mitochondrial permeability transition (mPT), and mitogen-activated protein kinases (extracellular signal-regulated kinase 1/2, c-jun-N-terminal kinase, and p38-MAPK). ROS activate NF-κB, a transcription factor, which is involved in the expression of many genes, including inducible nitric oxide synthase (iNOS), secretory phospholipase $A_2$ ($sPLA_2$), and cyclooxygenase (COX-2) (Fig. 6.3).

## 6.5.1 *$PLA_2$ and DAG/PLC Pathway in TBI*

As stated above, stimulation of $cPLA_2$ and the PLC/DAG-lipase pathway results in degradation of neural membrane phospholipids and generation of arachidonic acid and diacylglycerol (Shohami et al., 1989; Wei et al., 1982; Dhillon et al., 1994; Homayoun et al., 1997, 2000; Schuhmann et al., 2003) (Table 6.3). Although exact molecular mechanisms of TBI-mediated stimulation of $cPLA_2$ and DAG/PLC pathway are fully understood, several mechanisms have been proposed. One mechanism of $cPLA_2$ stimulation involves translocation of $cPLA_2$ plasma and nuclear membranes. Another mechanism is associated with cytokines (TNF-α and IL-1β)-mediated phosphorylation of $cPLA_2$ by mitogen-activated protein kinase in the presence of $Ca^{2+}$. A third mechanism involves TNF-α-mediated activation of caspase-3 and the proteolytic cleavage of $cPLA_2$ by caspase-3 (Wissing et al., 1997; Beer et al., 2000). Acetyl-Asp-Glu-Val-Asp-aldehyde, a specific tetrapeptide inhibitor of caspase-3 blocks the proteolytic cleavage and activation of $cPLA_2$, suggesting that caspase-3-mediated $cPLA_2$ proteolysis retards cell injury and death. Activation of $cPLA_2$ increases levels of arachidonic acid and other free fatty acids in cerebrospinal fluids from patients with traumatic brain injuries are significantly elevated (Pilitsis et al., 2003). TBI patients with favorable outcome scores have lower arachidonic acid concentrations at 48 h than patients with worse Glasgow

**Table 6.3** Status of lipid mediators in TBI

| Neurochemical parameter | Head injury | References |
|---|---|---|
| Glycerophospholipid metabolism | Enhanced | Farooqui et al. (2004) |
| Free fatty acid levels | Increased | Phillis et al. (2006) |
| Eicosanoids levels | Increased | Phillis et al. (2006) |
| Lipid peroxidation rate | Increased | Phillis et al. (2006) |
| 4-Hydroxynonenal levels | Increased | Phillis et al. (2006) |
| Isoprostanes levels | Increased | Varma et al. (2003) |
| Diacylglycerols | Increased | Farooqui and Horrocks (2007) |
| Excitotoxicity intensity | Involved | Farooqui and Horrocks (2009) |
| Oxidative stress intensity | Increased | Farooqui and Horrocks (2009) |
| Neuroinflammation intensity | Increased | Farooqui and Horrocks (2009) |
| Neurodegeneration rate | Increased | Farooqui and Horrocks (2009) |
| Apoptosis | Increased | McIntosh et al. (1998) |

scores at the time of hospital discharge. Released arachidonic acid is oxidized by cyclooxygenases and lipoxygenases (Gopez et al., 2005; Hickey et al., 2007). This results in production of eicosanoids, which not only regulate many brain functions but also modulate cerebral blood flow. In addition, TBI is also accompanied by increased rate of lipid peroxidation and increased in levels of isoprostanes, a family of prostaglandin-like compounds that are generated in vivo by free radical attack of esterified arachidonic acid and are released in free form in biological fluids such as CSF. This is accompanied by concomitant reduction in tissue concentrations of ascorbate, GSH, and protein sulfhydryls (Varma et al., 2003; Praticò et al., 2002).

### 6.5.2 Cyclooxygenases (COX) and Lipoxygenases (LOX) in TBI

Changes in cyclooxygenase (COX) and lipoxygenase (LOX) activities and in levels of eicosanoids have been observed not only in brain but also in plasma and CSF following TBI (Phillis et al., 2006). Thus, levels of $PGE_1$, 6-keto-$PGF_{1\alpha}$, and $PGF_{2\alpha}$ are increased in the cerebral cortex, plasma, CSF following concussive brain injury. These increases are sustained at up to 30–60 min post-injury. Elevated leukotriene levels have also been observed in brain after concussive brain injury. Azelastine, an agent that inhibits the release of leukotrienes and $PGD_2$, protects CA1 neurons in hippocampal slices from injury elicited by fluid percussion (Girard et al., 1996).

COX-2 is an important mediator of neuroinflammation (Phillis et al., 2006). $PGE_2$ has been implicated in neuroinflammation and the apoptosis of cortical cells through the activation of the EP2 receptor, which in turn activates caspase-3, a pro-apoptotic agent (Takadera et al., 2002). Concussive injury of the rat cerebral cortex causes a bilateral induction of COX-2 mRNA in the cortex and dentate gyrus. COX-2 activity is detectable in these areas and persisted in the ipsilateral cortex for at least 72 h (Kunz et al., 2002). Furthermore, a persistent accumulation of microglial cells and macrophages expressing COX-1 is also observed in human and rat TBI (Schwab et al., 2001; Schwab, 2002). Elevation in tissue levels of leukoteienes ($LTC_4$, $LTD_4$, and $LTB_4$) occurs in the cerebral cortex, hippocampus, and CSF following percussion injury in rats, which lasts for 1–2 h (Dhillon et al., 1996; Schuhmann et al., 2003), suggesting that $LTC_4$ may also play a role in the experimental brain injury. It is proposed that changes in leukotriene levels may be related to tissue edema, leukocyte infiltration, presence of macrophages, and microglial activation.

### *6.5.3 Calpain Activity in TBI*

Overactivation of calpain, a family of ubiquitous calcium-sensitive cysteine protease, has been linked to a variety of degenerative conditions in the brain (Ray et al., 2003; Buki et al., 2003; Ray, 2006; Carragher, 2006). Proteolytic substrates for calpain include receptor and cytoskeletal proteins, signal transduction enzymes, and transcription factors. TBI-mediated activation of calpain results in the cleavage of a number of neuronal substrates that negatively affect neuronal structure and function, leading to inhibition of essential neuronal survival mechanisms. Calpastatin, an endogenous protein inhibitor, modulates calpain activity. Overactivation of calpains degrades calpastatin, limiting its regulatory efficiency. Although the precise physiological function of calpains in TBI remains elusive, their association with SCI and TBI suggests that calpains participate in the neurodegenerative process via increase in intracellular free $Ca^{2+}$, which promotes the degradation of key cytoskeletal and membrane proteins. Cleavage of these key proteins by calpain is an irreversible process that perturbs the integrity and stability of neural cells, leading to neuronal cell death. Thus, studies on the determination of neurofilament M protein degradation and $\alpha$-spectrin breakdown products (SBDP 150 and 145) pattern in male and female rats following TBI, indicating that both calpain and caspase-3 are involved in pathogenesis of TBI. In general, males incur peak protein degradation and neurodegeneration within 3 days after injury, while in females this does not occur until 14 days (Pike et al., 1998; Kupina et al., 2003). It is suggested that TBI-mediated differences in male and female may be related to the hormonal status of these animals. Many cell permeable calpain inhibitors have shown to produce neuroprotective effects in animal models of spinal cord trauma and head injury indicating their therapeutic potential (Ray and Banik, 2003; Buki et al., 2003).

### *6.5.4 Caspases in TBI*

Caspases are a family of at least 14 aspartate-specific cysteine proteases that are essential in the initiation and execution of apoptosis (Creagh et al., 2003; Cohen, 1997). Caspases are normally expressed as inactive proenzymes (zymogens) that become activated during apoptosis (Zhivotovsky et al., 1999). All members of the caspase family share a number of amino acid residues crucial for substrate binding and catalysis. Amino acid residues Cys-285 and His-237 participate in catalysis and Arg-179, Gln-283, Arg-341, and Ser-347 are associated with carboxylate-binding pocket of all caspases except caspase-8. TBI-mediated activation of caspases results in not only proteolytic cleavage of procaspases, cytokines, but also degradation of cytoskeletal, nuclear, and cell cycle regulatory proteins (Pineda et al., 2007). Cytoskeletal protein α-II-spectrin is degraded by calpain and caspase-3 to SBDPs, which are released in CSF following severe TBI. Studies on the analysis of ventricular CSF taken at different time point indicate that levels of SBDP are significantly increased in TBI patients at several time points after injury, compared to control subjects. The time course of calpain-mediated SBDP150 and SBDP145 differs from that of caspase-3-mediated SBDP120 during the post-injury period. Taken together, these results support the view that calpain- and caspase-mediated α-II-spectrin breakdown products are potentially useful biomarker of severe TBI in humans (Pineda et al., 2007; Brophy et al., 2009).

It is also shown that some caspases are associated with inflammasome, which are large multiprotein complex whose assembly leads to the activation of caspase-1. Inflammasome consist of NLRP1 (nucleotide-binding, leucine-rich repeat pyrin domain containing protein 1), caspase-1, caspase-11, apoptosis-associated speck-like protein containing a caspase recruitment domain (ASC), the X-linked inhibitor of apoptosis protein, and pannexin 1 (de Rivero et al., 2009). Moderate parasagittal fluid percussion injury (FPI) not only activates the degradation of caspase-1, X-linked inhibitor of apoptosis protein, but also promotes assembly of the NLRP1 inflammasome complex. Administration of anti-ASC neutralizing antibodies immediately after FPI to injured rats blocks caspase-1 activation and X-linked inhibitor of apoptosis protein cleavage resulting in a significant decrease in contusion volume. These studies show that the NLRP1 inflammasome is an important component of the innate central nervous system inflammatory response after traumatic brain injury (de Rivero et al., 2009).

### *6.5.5 Nitric Oxide Synthase in TBI*

Nitric oxide synthases (NOS) are enzymes that liberate nitric oxide (NO) from arginine. In brain, NO is involved in a variety of broad physiological processes, including control of cerebral blood flow, interneuronal communications, synaptic plasticity, memory formation, receptor functions, intracellular signal transmission, and release of neurotransmitters. At least three NOS isoforms have been reported to occur in the brain. They include neuronal NOS (nNOS), inducible NOS (iNOS), and endothelial NOS (eNOS). Acute neural trauma is accompanied by an upregulation in nNOS activity in neurons, eNOS activity in glial cells and vascular endothelium, and iNOS activity in a range of cells including infiltrating neutrophils and macrophages and activated microglia and astrocytes. The role of nitric oxide is very complex, as it can be cytotoxic or cytoprotective in relation to sources, time of synthesis, and medium redox state (Cherian et al., 2004). There are two periods of time after injury when NO accumulates in the brain, immediately after injury and then again several hours to days later. The initial immediate peak in NO after injury is probably due to the activity of endothelial NOS and neuronal NOS, whereas peak is due to the induction of iNOS, which is a mediator in inflammatory reactions. Under physiological conditions, low levels of NO contribute to vasodilation, neurotransmission, and synaptic plasticity. Following TBI, increased expression of iNOS generates excessive NO. NO reacts with $O_2^{\bullet-}$ and produces $ONOO^-$, which is highly toxic to neuronal proteins, lipids, and nucleic acid (Xiong et al., 2007). It not only nitrates and hydroxylates aromatic rings on amino acid residues in proteins but also oxidizes lipids and damages DNA causing activation of the nuclear DNA repair enzyme, poly(ADP-ribose) synthase (PARS) (Fig. 6.4). Prolonged activation of this enzyme depletes ATP. In addition, $ONOO^-$ also inhibits mitochondrial respiratory chain enzymes (Arundine and Tymianski, 2004). Inhibition of iNOS synthesis improves histopathological and clinical outcomes of TBI in animal models (Wada et al., 1998).

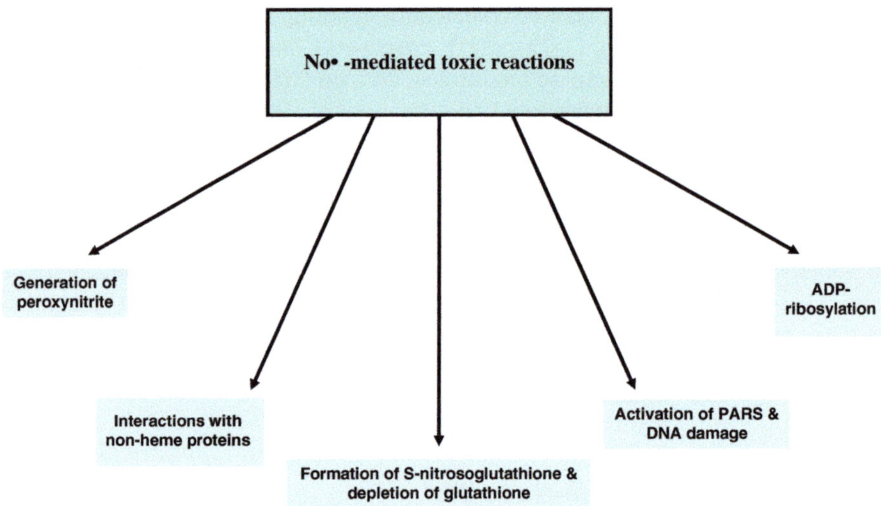


**Fig. 6.4** Effect of nitric oxide toxicity on proteins, lipids, and DNA

## 6.5.6 Kinases in TBI

TBI activates several protein kinase signaling pathways in the hippocampus that are critical for hippocampal-dependent memory formation. In particular, extracellular signal-regulated kinase (ERK), a protein kinase activated during and necessary for hippocampal-dependent learning, is transiently activated after TBI. However, TBI patients experience hippocampal-dependent cognitive deficits that occur for several months to years after the initial injury. Although basal activation levels of ERK return to sham levels within hours after TBI, it is hypothesized that activation of ERK-CREB (cAMP response element-binding protein) pathway may be impaired after TBI (Atkins et al., 2009b). Administration of ERK inhibitor, U0126 significantly reduces both CA3 neuronal damage and contusional lesion volume after TBI. In addition, U0126 treatment also ameliorates motor function recovery on days 3, 4, and 5 after injury, suggesting that ERK is closely associated with metabolic alterations in TBI (Otani et al., 2007). TBI-mediated increase in intracellular calcium alters activity and function of calcium–calmodulin-dependent protein kinase II (CaMKII), which is autophosphorylated on $Thr^{286}$($pCaMKII^{286}$) in the presence of calcium and calmodulin (Atkins et al., 2009b). Time-dependent studies indicate that activation of CaMKI and CaMKIV occurs in a more delayed manner. The increase in activated α-CaMKII in membrane fractions is accompanied by a decrease in cytosolic total α-CaMKII, suggesting redistribution to the membrane. Confocal microscopic studies indicate that activation of α-CaMKII occurs within hippocampal neurons of the dentate gyrus, CA3, and CA1 regions. One hour after TBI, CaMKII-mediated phosphorylation of two downstream substrates of α-CaMKII (AMPA-type glutamate receptor GluR1 and cytoplasmic polyadenylation element-binding protein are significantly increased in phosphorylation in the hippocampus and cortex. Collective evidence suggests that several of the biochemical cascades that subserve memory formation are activated unselectively in neurons after TBI. It is becoming increasingly evident that memory formation occurs in hippocampus and requires CaMKII-mediated signaling pathways at specific neuronal synapses. Unselective activation of CaMKII signaling in all synapses after TBI may disrupt the machinery for memory formation causing memory loss. In contrast, TBI downregulates cAMP-PKA signaling cascade and that treatment with a PDE IV inhibitor improves histopathological outcome and decreases inflammation after TBI (Atkins et al., 2007a, b; Sharma et al., 2009).

Mild TBI increases the phosphorylation of inhibitory site $serine^9$ of glycogen synthase kinase-3 (GSK-3β), which coincides with increased $serine^{473}$ phosphorylation of its upstream kinase (PKB) and accumulation of its downstream target β-catenin in the hippocampus (Shapira et al., 2007). Mild TBI also mediates a depressive behavior which is evident as early as 24 h post-injury. Pretreatment with GSK-3 inhibitors, lithium, or L803-mts retards mild TBI-induced depression. It is suggested that mild TBI elicits a prosurvival cascade of PKB/GSK-3β/β-catenin as part of a rehabilitation program (Shapira et al., 2007).

The clearance of cellular debris after TBI is a crucial step for restoration of the traumatized neural network. Microglial cells not only play an important role in the elimination of degenerating neurons and axons in the brain tissue, but also facilitate the restoration of favorable environment after the injury (Tanaka et al., 2009). Based on cell culture (primary microglia or the MG5 microglial cell line) studies, it is proposed that p38 mitogen-activated protein kinase (MAPK) plays an important role in debris clearance (Tanaka et al., 2009). Engulfment of axon debris can be prevented by the p38 MAPK inhibitor, SB203580, suggesting that p38 MAPK is required for phagocytic activity.

### *6.5.7 Matrix Metalloproteinases (MMPs) in TBI*

Matrix metalloproteinases (MMPs) degrade components of the extracellular matrix. These enzymes have been implicated in the pathophysiology of TBI by increasing blood–brain barrier permeability and exacerbating post-traumatic edema. In addition, they also play an essential role in the tissue repair, cell death, and morphogenesis. TBI not only triggers widespread cell death in the cortex, basal ganglia and white matter but also increases mRNA levels for MMP-2 and -9 in injured brain at 12–72 h after trauma (Sifringer et al., 2007). Protein expression of MMPs and activity of MMP-2 are increased at 12 h and peaked at 24 h after trauma. It is also shown that TBI increases MMPs activities in ventricular cerebrospinal fluid (CSF) and plasma (Grossetete et al., 2009). Intraperitoneal injection of GM6001 (Ilomastat), an MMP inhibitor, 2 h after TBI, substantially attenuates TBI in a dose-dependent manner. These observations causally link the MMPs to TBI-induced neuronal cell death in the immature rodent brain (Sifringer et al., 2007). In addition, TBI also results in a significant increase in gene and protein expressions of hypoxia-inducible factor-1alpha (HIF-1α), MMP-2 and -9, as well as enzyme activity of MMP-2 and -9 at the same time points. Inhibition of either MMPs or HIF-1α significantly reverses the TBI-induced decrease in synaptophysin (Ding et al., 2009). Inhibition of HIF-1α reduced expression of MMP-2 and -9. These results indicate an early detection of a correlation between synaptic loss and MMP expression after TBI. These data also support a role for HIF-1α in the MMP regulatory cascade in synapse loss after TBI (Ding et al., 2009).

### *6.5.8 Calcineurin in TBI*

Increase in calcineurin (CaN), a calcium/calmodulin-dependent phosphatase activity has been reported to occur in hippocampus following TBI (Kurz et al., 2005a, b). Changes in CaN activity persist for 2–3 weeks following TBI. Increases in CaN activity following TBI may be due to significant increase in intracellular $Ca^{2+}$ (Fineman et al., 1993; Bales et al., 2010) and is closely associated with increases in cellular death and dysfunction in both ischemic injury and TBI (Morioka et al., 1999; Bales et al., 2010). Investigations on the involvement of CaN in inflammatory processes indicate that astrocytes modulate neuronal resilience to inflammatory insults through the CaN. In quiescent astrocytes, inflammatory cytokine, (TNF-α recruits CaN to stimulate a canonical inflammatory pathway involving the NF-κB) and nuclear factor of activated T-cells (NF-AT). However, in reactive astrocytes, local neuroprotector and anti-inflammatory mediator, insulin-like growth factor I (IGF-1) also recruits CaN but utilizes it to retard NF-κB/NF-AT-mediated processes. During this process, IGF-I not only mediates a site-specific dephosphorylation of I-kBa (phospho-$Ser^{32}$) but also inhibits the nuclear translocation of NF-κB (p65) in astrocytes. This hypothesis is supported by experiments showing the expression of constitutively active CaN in astrocytes can markedly reduce the inflammatory injury in transgenic mice, in a calcineurin-dependent manner. Thus in astrocytes, calcineurin participates in a molecular pathway that determines the outcome of the neuroinflammatory process by directing it toward either its resolution or its progression (Fernandez et al., 2007).

### *6.5.9 Other Enzymes in TBI*

TBI is characterized by alterations in the mitochondrial metabolism, dysregulation in glucose metabolism, and accumulation of lactose (Xing et al., 2009). Activity of pyruvate dehydrogenase (PDH), a rate-limiting enzyme couples cytosolic glycolysis to mitochondrial citric acid cycle, is significantly decreased following TBI. Although the molecular mechanism associated with decrease in PDH activity is not known, downregulation in PDH expression and phosphorylation may alter brain PDH activity and glucose metabolism in TBI (Xing et al., 2009). Studies on mitochondrial metabolism following TBI have indicated that a decrease in the cytochrome oxidase complex of the electron transport chain (complex IV), and an immediate reduction in mitochondrial state 3 respiratory rate, persists for up to 14 days post-injury (Harris et al., 2001). Similarly, FPI also decreases the levels of mitochondrial creatine kinase and cytochrome *c* oxidase II in FPI rats as compared to the sham rats (Sharma et al., 2009). The curcumin-containing diet counteracts the effects of FPI and elevated the levels of AMP-activated protein kinase (AMPK), mitochondrial creatine kinase, cytochrome *c* oxidase II in curcumin/FPI rats as compared to regular diet consuming/sham rats, indicating the importance of curcumin in the regulation of energy homeostasis following TBI.

## 6.6 TBI-Mediated Alterations in Cytoskeletal Protein

Biomarkers are of enormous importance for the diagnosis, prognosis, and therapeutic evaluation of TBI-mediated acute brain damage. It is proposed that a panel of neuron-enriched proteins measurable in cerebrospinal fluid (CSF) and blood may be used as surrogate markers to improve clinical evaluation and therapeutic management of TBI. These surrogate biomarkers include 14-3-3β, 14-3-3ζ, 3 distinct phosphoforms of neurofilament H, ubiquitin hydrolase L1, neuron-specific enolase, α-spectrin, and three calpain- and caspase-derived fragments of α-spectrin (Siman et al., 2009). Both αII and βII spectrin have calpain target sites and the preferential cleavage of αII spectrin over βII spectrin is mediated through NMDA receptor-mediated calcium entry. It is postulated that calpain-induced proteolysis of spectrin can activate two physiologically distinct responses: one that enhances skeletal plasticity without destroying the spectrin-actin skeleton, characterized by preservation of βII spectrin or an alternative response closely correlated with non-apoptotic cell death and characterized by proteolysis of βII spectrin and complete dissolution of the spectrin skeleton (Glantz et al., 2007). αII-Spectrin is a structural protein abundant in neurons of the central nervous system and cleaved into signature fragments by proteases, such as calpains and caspases. Levels of spectrin and spectrin breakdown products (SBDPs) are significantly increased in CSF from rats and patients with severe TBI (Cardali and Mangeri, 2006). SBDPs are highly stable. Detailed investigations on severe TBI patients indicate that concentrations of 150-, 145-, and 120-kDa SBDPs reflect changes in calpain and caspase activities (Brophy et al., 2009). The results strongly support the potential utility of SBDPs as important markers in the clinical monitoring of patients with severe TBI (Farkas et al., 2005). Spectrins are known to regulate surface chemistry and morphology of neural cells. Additional cytoskeletal substrates for calpains are tubulins, microtubule-associated proteins (MAP), and the neurofilament proteins. Marked decrease in MAP-2 (Posmantur et al., 1996) and neurofilament protein (Posmantur et al., 1994) is observed in experimental TBI. It is likely that their degradation may have pronounced and persistent effect on the synapse. This process may contribute to behavioral changes in TBI patients.

## 6.7 TBI-Mediated Alterations in Transcription Factors

Very little is known about the underlying mechanisms involved in the alterations of gene expression profiles modulating cell death and survival in TBI. The neurodegeneration is accelerated by the induction of pro-cell death gene expression profiles through an altered balance of pro- and anti-apoptotic transcription factors. Thus, alterations in regulation of these transcription factors may constitute one of the earliest events in TBI and may offer a therapeutic window of opportunity for intervention across a narrow time period prior to irreversible neuronal death (Kane and Citron, 2009) (Fig. 6.5). There has been considerable interest in the modulation of these cell death factors to prevent or mitigate damage to neurons with the goal of improving the lives of TBI patients. Following TBI, injured neurons degenerate while surviving neurons undergo neuritogenesis and synaptogenesis to establish neuronal connectivity disturbed and destroyed by the TBI.

### *6.7.1 Nuclear Factor Kappa B (NF-κB) in TBI*

Nuclear factor kappa B (NF-κB), an inducible transcription factor, acts as a master regulator of immune functions, inflammatory responses, secondary injury processes, and cell survival. NF-κB is rapidly activated in response to various stimuli, including trauma, infectious agents, and radiation-induced DNA double-strand breaks. Neuronal NF-κB is a mediator of trauma-triggered neuronal death, whereas astrocytic NF-κB is thought to be neuroprotective. Studies on the expression of NF-κB in human TBI indicate that a progressive upregulation of NF-κB activity occurs in the area surrounding the injured brain with the time from brain trauma to operation

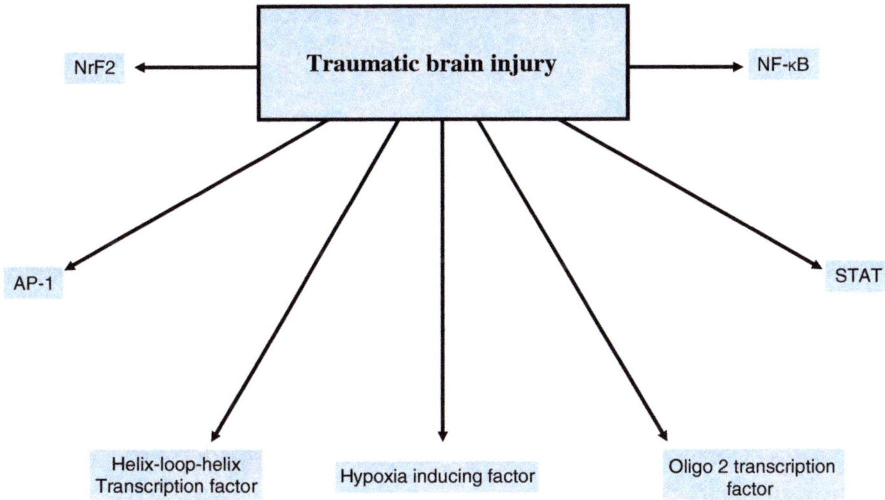


**Fig. 6.5** Transcription factors that are stimulated by TBI

(Hang et al., 2006). NF-κB consists of several subunits, including p65 (RelA), p50, p52, c-Rel, and RelB. NF-κB occurs in the cytoplasm of neural cells as a heteromeric protein consisting of a dimmer of two above-mentioned subunits complexed with an inhibitory subunit I-κB, which dissociates from the NF-κB dimer, allowing dimmer to translocate from cytoplasm to the nucleus and interact with target sequences in the genomẽ. Two protein kinases (IKKα and IKKβ) mediate phosphorylation of I-κB proteins and represent a convergence point for most signal transduction pathways leading to NF-κB activation. Most of the IKKα and IKKβ molecules in the cell are part of IKK complexes that also contain a regulatory subunit called IKKγ or NEMO. It is suggested that NF-κB activation represents a paradigm for controlling the function of a regulatory protein via ubiquitination-dependent proteolysis, as an integral part of a phosphorylation-based signaling cascade (Karin and Ben-Neriah, 2000).

In the brain, NF-κB regulates the expression of a large number of genes involved in immune responses, inflammation, cell survival, and apoptosis. NF-κB is rapidly activated in response to various stimuli, including cytokines, growth factors, and radiation-induced DNA double-strand breaks. TBI induces the expression of NF-κB p65, which is mainly found in glial and vascular endothelial cells. The expression of NF-κB p53 also occurs in glial cells. Very little expression of NF-κB occurs in neurons and vascular endothelial cells. TBI upregulates TLR2 and TLR4 mRNA and NF-κB binding activity at the injury site (Chen et al., 2008b). Levels of IL-β, TNF-α, IL-6, and ICAM-1 are significantly increased in the injured brain after brain contusion. Cortical levels of these mediators of TLRs/NF-κB signaling pathway are suppressed by treatment with progesterone. Progesterone administration also results in decreased number of TUNEL-positive apoptotic cells in the cortex surrounding the injured site suggesting that progesterone attenuates the TBI-induced TLRs/NF-κB signaling pathway and may inhibit development of secondary brain damage in experimental TBI (Chen et al., 2008b). In addition, within 24 h post-TBI, both NF-κB p65 and p53 immunoreactivities are mainly observed in the nucleus of damaged neural cells (Hang et al., 2006). Post-traumatic neurodegeneration (24 h) correlates with the increase in p53 levels and is significantly reduced by the selective p53 inhibitor pifithrin-α (PFT) (Plesnila, 2007). Importantly, neuroprotective effect is observed even when PFT treatment is delayed up to 6 h after TBI. Inhibition of p53 activity causes a concomitant increase in NF-κB transcriptional activity and upregulation of NF-κB-target proteins, for example, X-chromosomal-linked inhibitor of apoptosis (XIAP) (Plesnila et al., 2007). The XIAP-mediated inhibition blocks the neuroprotective effects of PFT in cultured neurons exposed to camptothecin, glutamate, or oxygen glucose deprivation. Based on these studies, it is concluded that delayed neuronal cell death after brain trauma is mediated by p53-dependent mechanisms that involve inhibition of NF-κB transcriptional activity (Plesnila et al., 2007).

### *6.7.2 Signal Transducers and Activators of Transcription (STATs) in TBI*

Signal transducers and activators of transcription (STATs) are latent cytoplasmic transcription factors that can be activated by a variety of tyrosine kinases in response to many different cytokines and growth factors. Accumulation of tyrosine phosphorylated STAT dimers in the nucleus is followed by DNA binding, activation of target gene transcription, dephosphorylation, and return to the cytoplasm (Levy and Darnell, 2002). In the brain, astrocytes and microglial cells respond to TBI. Double-labeling studies with Mac-1/CD11b and GFAP indicate that STAT2 is upregulated and phosphorylated following injury in astrocytes (Khorooshi et al., 2008). Both STAT2 upregulation and phosphorylation depend on NF-κB. These processes do not occur in the lesion-reactive hippocampus of transgenic mice with specific inhibition of NF-κB activation in astrocytes (Khorooshi et al., 2008). Collective evidence suggests that NF-κB signaling in astrocytes controls expression of both STAT2 and thus regulates infiltration of leukocytes into lesion-reactive hippocampus after axonal injury (Khorooshi et al., 2008).

### *6.7.3 Nuclear Factor E2-Related Factor 2 in TBI*

Nuclear factor E2-related factor 2 (Nrf2) is a basic leucine zipper redox-sensitive transcription factor that controls the basal and inducible expression of a battery of antioxidant genes. It induces expression and upregulation of cytoprotective and antioxidant/detoxifying genes that attenuate tissue injury (Lee and Johnson, 2004). Under physiological conditions, NrF2 is localized in the cytoplasm where it binds with the actin-binding protein, Kelch-like ECH-associating protein 1 (Keap1), and is rapidly degraded by ubiquitin-proteasome pathway. Keap1 acts as negative regulator of Nrf2. Oxidative stress liberates Nrf2 from Keap1 and allows Nrf2 translocation into nucleus, where it binds to stress or antioxidant response elements and facilitates expression of cytoprotective genes, numerous protective enzymes, and scavengers. Nrf2 protein levels are significantly increased following TBI (Yan et al., 2009). Studies on Wild-type $Nrf2^{+/+}$ and $Nrf2^{-/-}$-deficient mice indicate that $Nrf2^{-/-}$ mice have more NF-κB activation, inflammatory cytokines TNF-α, IL-1β and IL-6 production, and ICAM-1 expression in brain after TBI compared with their wild-type $Nrf2^{+/+}$ counterparts. These results suggest that Nrf2 plays an important protective role in limiting the cerebral upregulation of NF-κB activity, proinflammatory cytokine, and ICAM-1 after TBI. It is proposed that Nrf2 may play a protective role in the brain after TBI, possibly by reducing inflammation, oxidative stress, and brain edema (Jin et al., 2008).

### *6.7.4 AP-1 Transcription Factor in TBI*

The transcription factor activator protein-1 (AP-1) consists of a variety of dimers composed of members of the Jun and Fos families of proteins (Raivich and Behrens, 2006). However, it is the upregulation of c-jun that is a particularly common event in the adult as well as in injured nervous system that serves as a model of transcriptional control of brain function. It regulates genes expression in response to cytokines, neurotrophins, and oxidative stress. Depending on the AP-1 dimer combination, neuronal genes related to either apoptosis or survival is transcribed. A 35 kDa Fos-related antigen:JunD dimer is present in neurons that survive injury. Jun and JunD exist in neurons prior to undergoing apoptosis (Raivich and Behrens, 2006). Physiological and pathological stimuli induce the expression of Jun and Fos proteins. Involvement of AP-1 in neurodegeneration and neuroregeneration is associated with c-jun and its activation by JNKs. During excitotoxicity, apoptotic cell death involves the activation of c-jun, which affects hippocampal, nigral, and primary cultured neurons. The inhibition of JNKs exerts neuroprotective effects in neurons. Besides endogenous neuronal functions, the c-jun/AP-1 proteins can damage the nervous system by upregulation of harmful programs in non-neuronal cells (e.g., microglia) with release of neurodegenerative molecules. In contrast, the differentiation with neurite extension and maturation of neural cells in vitro indicates physiological and potentially neuroprotective functions of c-jun and JNKs, including sensoring for alterations in the cytoskeleton (Raivich and Behrens, 2006).

### *6.7.5 CCAAT/Enhancer-Binding Protein (C/EBP) in TBI*

CCAAT/enhancer-binding proteins (C/EBPs) are a family of transcription factors that contain a basic leucine zipper domain at the C-terminus that is involved in the dimerization and DNA binding. At least six members of the family have been isolated and characterized to date (C/EBP α[bond]C/EBP ς), with further diversity produced by the generation of different sized polypeptides, predominantly by differential use of translation initiation sites, and extensive protein–protein interactions both within the family and with other transcription factors (Ramji and Foka, 2002). They are encoded by an intronless gene, C/EBPβ, which is expressed as several distinct protein isoforms (LAP1, LAP2, LIP). These transcription factors regulate gene expression to control cellular proliferation, differentiation, inflammation, and metabolism. Upregulation of C/EBPβ is observed 1 day following injury in both the adult and the aged brain, but there were no major age-related differences in mRNA levels (Sandhir and Berman, 2009). C/EBP-β induces a variety of cytokines and thus may play a role in the induction of neuroinflammation. Differential expression of C/EBPβ, δ, and CCAAT/enhancer-binding protein homologous protein CHOP contributes to the hyper-inflammatory response. The molecular mechanism associated with C/EBPβ action is not clearly understood. However, interactions between activated NF-κB (Rel A) and C/EBP may aid to inflammatory response. RelA-C/EBP interactions are increased by phosphorylation of threonine at amino acid 75 and result in increased DNA binding compared with the wild-type nonphosphorylated C/EBP both in vitro and in vivo. It is suggested that interaction of the activated NF-κB pathway and C/EBP-ε may be important in selective activation of a subset of C/EBP-β-responsive genes (Chumakov et al., 2007).

## 6.8 TBI-Mediated Alterations in Gene Expression

TBI induces a complex sequence of putative autodestructive and neuroprotective cellular cascades. Genes involved in modulation of alteration in neuronal environment and apoptotic cell death are upregulated by TBI (Fig. 6.6). Glial cells, astrocytes and microglia, respond to neuronal death by transcribing genes to enhance the survival of remaining neurons and for regeneration and repair. Thus, TBI results in the expression of immediate early genes, heat shock proteins (Hsps), and cytokines (Raghupathi et al., 1995, 2000). The immediate early genes, c-fos, c-jun and junB are induced in the cortex and hippocampus as early as 5 min following lateral FPI rats. While levels of c-fos and junB mRNA come back to control levels by 2 h, c-jun mRNA remain elevated up to 6 h post-injury. These genes have been implicated not only in apoptotic cell death but also in repair and regeneration responses associated with TBI (McIntosh et al., 1998).

Increase in mRNA for the inducible heat shock protein (Hsp72) is observed up to 12 h following injury and is restricted to the cortex ipsilateral to the impact site (Raghupathi et al., 1995). The molecular mechanisms involved in TBI-mediated expression of Hsp are not fully understood. However, decrease in blood flow and decrease in ATP levels may be associated with the increased expression of stress protein following TBI (McIntosh et al., 1998). Mild induction of the glucose-regulated proteins (grp78 and grp94), which share sequence homology with Hsp72,

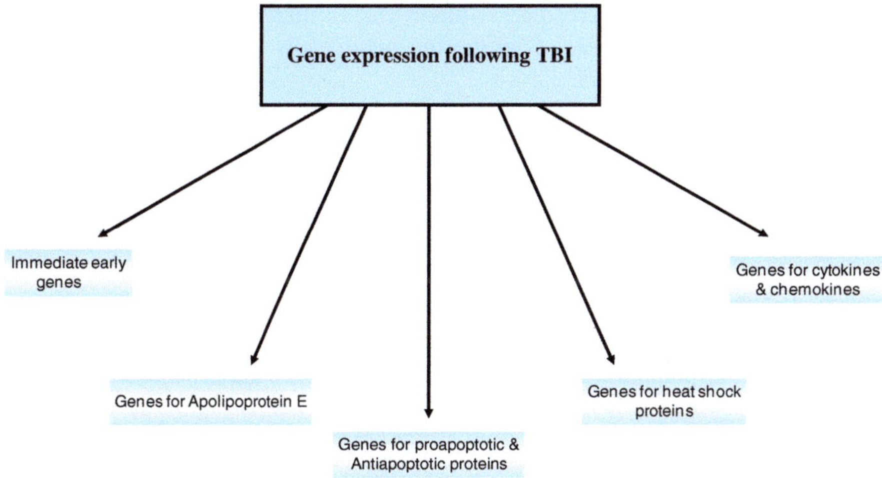


**Fig. 6.6** Increased expression of genes by TBI

is observed in the ipsilateral cortex. Induction of IL-1β and TNF-α is observed at 1 h following FPI. Expression of these cytokines remains elevated up to 6 h post-injury (Raghupathi et al., 1995). In addition, TBI modulates the expression of genes for death-inducing proteins such as Bax, c-jun N-terminal kinase, tumor-suppressor gene, p53, and calpains and caspases as well as neural cell survival proteins such as Bcl-2, Bcl-xL (Raghupathi et al., 1995; Raghupathi, 2004; Strauss et al., 2004). It is suggested that decrease in expression of Bcl-xL mRNA and increase in expression of bax mRNA coincides with apoptosis following TBI. The bcl-2 gene family is involved in neuronal apoptosis after TBI, and the changes of mRNA expression of the family members lead the neuronal cells to apoptosis. Among the upregulated genes 1 day post-TBI are transcription factors and genes involved in metabolism, e.g., STAT-3, C/EBP-δ, and cytochrome p450 (von Gertten et al., 2005). On $4^{th}$ day, TBI increases expression of inflammatory factors, proteases and their inhibitors, like cathepsins, α-2-macroglobulin and C1q. In addition, genes with biological function clustered to immune response are significantly upregulated 4 days after TBI, which is not observed following 1 day post-TBI. TBI also increases the mRNA and expression of osteopontin and one of its receptors, CD-44 in and around the injury site (von Gertten et al., 2005). Genes showing decreased expression both 1 and 4 days post-TBI include genes associated with transport, metabolism, signaling, and extra cellular matrix formation, e.g., vitronectin, neuroserpin, and angiotensinogen (von Gertten et al., 2005).

Time course analysis of gene expression levels using QRT-PCR in the acute phase of the penetrating ballistic brain injury between 3 and 6 h indicates an upregulation in the expression of cytokines TNF-α (eightfold to 11-fold), IL-1β (11-fold to 13-fold), and IL-6 (40-fold to 74-fold) as well as the cellular adhesion molecules VCAM (twofold to threefold), ICAM-1 (sevenfold to 15-fold), and E-selectin (11-fold to 13-fold). These processes are consistent with the upregulation of proinflammatory genes, peripheral blood cell infiltration is a prominent post-injury event with peak levels of infiltrating neutrophils (24 h) and macrophages (72 h) (Williams et al., 2007).

TBI also modulates the expression of different alleles of the apolipoprotein E gene (APOE gene, ApoE protein). Using the controlled cortical impact model of TBI and microarray technology, it is shown that gene expression profiles of APOE3 and APOE4 transgenic mice in cortex and hippocampus are significantly different from each other. It is suggested that the observed gene regulation predicts functional consequences, including effects on inflammatory processes, cell growth and proliferation, and cellular signaling (Crawford et al., 2009). In addition to above genes, genes encoding regulators of apoptosis, signal transduction, and metabolism are also altered following TBI. Collective evidence suggests that TBI is accompanied by different patterns of gene expression at different time with little overlap. The physiological relevant TBI-induced gene expression may explain molecular mechanisms associated with TBI-induced neurodegeneration.

## 6.9 TBI-Mediated Alterations in Adhesion Molecules

CNS responses to TBI include up- and downregulation of a vast number of proteins involved in the endogenous inflammatory responses and defense mechanisms developing post-injury. The neural cell adhesion molecule (NCAM), a member of the immunoglobulin superfamily, plays an important role during development and regeneration of the nervous system, mediating neuronal differentiation, survival, and plasticity. NCAM is found in three major forms. Two forms (NCAM-140 and NCAM-180) are transmembrane proteins, while the third form (NCAM-120) is attached to the membrane via a glycosylphosphatidyl inositol anchor. NCAM regulates cell adhesion, cell migration, and neurite outgrowth. NCAM also regulates learning and memory. The upregulation of cell adhesion molecules (NCAMs) is observed in animal model of TBI in rats (Klementiev et al., 2008; Pedersen et al., 2008). Alterations in the expression of junctional adhesion molecule cause BBB breakdown following TBI and this process promotes brain edema.

## 6.10 TBI-Mediated Alterations in Neurotrophic Factors

Significant neuronal degeneration and functional loss occurs following TBI. TBI also results in elevation in levels of basic fibroblast growth factor (FGF), brain-derived neurotrophic factor (BDNF), neurotrophin 4/5 (NT4/5), and insulin-like growth factor-1 (IGF-1) (Kizhakke Madathil et al., 2009). In vitro studies indicate that neurotrophic factors such as brain-derived neurotrophic factor (BDNF), ciliary neurotrophic factor (CNTF), and neurotrophin-4/5 (NT-4/5) can promote neuronal survival. Delivery of above neurotrophic factors to the injured brain is difficult due to their short half-lives, high molecular weight, and inability to cross the BBB efficiently. In addition, there is a possibility of potential immunogenicity and sequestration by binding proteins and other components of the blood and peripheral tissues. Studies on intranasal delivery of $^{125}$I-radiolabeled neurotrophic factors (BDNF, CNTF, NT-4, or erythropoietin, EPO) some neurotrophin can be delivered to brain parenchyma. These neurotrophic factors not only reach brain parenchyma but are present in sufficient concentrations to activate the prosurvival PtdIns 3-kinase/Akt pathway (Alcalá-Barraza et al., 2009). Neurochemical effects of neurotrophins are mediated through activation of TrkA, TrkB, and TrkC (Skaper, 2008). In addition, all neurotrophins activate the p75 neurotrophin receptor ($p75^{NTR}$), a member of the tumor necrosis factor receptor superfamily. Nerve growth factor (NGF), the best characterized member of the neurotrophin family, sends its survival signals through activation of TrkA and can induce death by binding to $p75^{NTR}$. Neurotrophin engagement of Trk receptors leads to activation of several signaling pathway, including Ras, PtdIns 3-kinase, PLC-$\gamma$1, and signaling pathways controlled through these proteins, including the mitogen-activated protein kinases. Neurotrophin availability is required for the modulation of synaptic function and plasticity and sustained neuronal cell survival, morphology, and differentiation (Skaper, 2008). The upregulation of nerve growth factor (NGF) and doublecortin (DCX) following TBI correlates with better neurologic outcome in children with severe TBI. Although the molecular mechanism of beneficial effects of neurotrophic factors is not fully understood, it is suggested that neurotrophic factors protect neurons (endogenous neuroprotection or repair) by promoting neuronal connection reorganization after TBI. This suggestion is supported by studies supporting the view that in the adult CNS, migrating neuroblasts can replace injured neurons after severe TBI (Chiaretti et al., 2008).

## 6.11 TBI-Mediated Alterations in Complement System

Inflammatory reactions in TBI induce activation of complement system, a biochemical cascade that not only facilitates phagocytosis of dead neural cells but also promotes the removal of immune complexes and induces adaptive immune responses. Under normal conditions, complement system plays a neuroprotective role and is a powerful and vital component of the innate immune system, but under pathological conditions, such as TBI-mediated oxidative stress, complement system induces the release of proinflammatory cytokines. These cytokines facilitate chronic neuroinflammation that promotes neural cell death. Activation of classical pathway involves the attachment of C1q to a target causing C1 dissociation. Amplification is facilitated through a cascade of proteases (C1r, C1s, C4, C2, and C3), and the hydrolyzed products C4b and C3b attach to the exposed sites close to the C1q binding site, opsonizing the target for phagocytosis. Following TBI, aggregated polypeptides can be potentially present their different charge patterns to C1q, which is a vital charge pattern of recognition molecule of the complement system. Consequently activation of complement leads to microglial activation, which in turn leads to defective clearance of the aggregated polypeptides by macrophages leading to chronic inflammation, especially in traumatized brain. As stated above, TBI is characterized, in part, by activation of the innate immune response, including the complement system. It is shown that mice devoid of a functional alternative pathway of complement activation (factor $B^{-/-}$ mice) are protected from complement-mediated neuroinflammation and neuropathology after TBI (Leinhase et al., 2006). In addition, inhibition of the alternative complement pathway by post-traumatic administration of a neutralizing anti-factor B antibody may represent a new promising avenue for pharmacological attenuation of the complement-mediated neuroinflammatory response after TBI (Leinhase et al., 2007).

Multiple organ injury results in a systemic inflammatory response syndrome (SIRS) due to the synthesis of proinflammatory cytokines and arachidonic acid metabolites, proteins of the contact phase and coagulation systems, complement factors and acute phase proteins, as well as hormonal mediators that may cause a single or multiple organ failure. As stated above, cytokines are integral components of immune response (McGreer et al., 2005). The local release of pro- and anti-inflammatory cytokines after severe trauma indicates their potential to induce systemic immunological alterations (Hildebrand et al., 2005). It appears that the balance or imbalance of these different cytokines partly controls the clinical course in multiple injury patients. Overproduction of proinflammatory cytokines (IL-6, TNF-α, IL-1β, KC, MIP-2, and MCP-1) and downregulation of anti-inflammatory cytokines (TNF-soluble receptors, IL-10, IL-1 receptor antagonist) may contribute to the development of multiple organ dysfunction syndrome (MODS) or multiple organ failure (MOF) (Hildebrand et al., 2005). Collective evidence suggests that inflammatory responses directly correlate not only to TBI but also to MODS and MOF.

## 6.12 TBI Mediators Alterations in Endocannabinoids

TBI also results in local and transient accumulation of 2-arachidonylglycerol (2-AG) at the site of injury, peaking at 4 h and sustained up to at least 24 h (Mechoulam and Shohami, 2007). It is suggested that 2-AG-mediated neuroprotection in TBI involves not only inhibition of NF-κB transactivation, inhibition of expression of cytokines (TNF-α, IL-6, and IL-1β), but also reduction in blood–brain barrier permeability. Moreover, the expression of $CB_1$, $CB_2$, and $TRVP_1$ receptors on microvascular endothelial cells, and their activation by 2-AG counteracts endothelin (ET-1)-mediated cerebral microvascular responses (namely, $Ca^{2+}$ mobilization and cytoskeleton rearrangement). This suggests the involvement of functional interaction among 2-AG, ET-1, and their receptors. The interplay between 2-AG and ET-1 may provide a potential alternative pathway for abrogating ET-1-inducible vasoconstriction after TBI (Mechoulam and Shohami, 2007).

## 6.13 TBI-Mediated Changes in Hydroxycholesterols

In humans, the brain represents only about 2% of the body's mass but contains about one-quarter of the body's free cholesterol. Most brain cholesterol is present in the myelin sheets (Pfrieger, 2003). The distribution of cholesterol in neural membranes is asymmetric across the plane of the membrane, with outer leaflet containing 25% and inner leaflet containing 75% cholesterol. Cholesterol condenses the packing of bilayer by positioning between these hydrocarbon chains below the large head groups of the sphingolipids. Such location of cholesterol in lipid bilayer not only controls exocytosis by modulating vesicle fusion and motion during synaptic transmission (Zhang et al., 2009) but also regulates activities of membrane-bound enzymes, receptors, and ion channels (Simons and Ikonen, 2000). Cholesterol is synthesized de novo in brain astrocytes and transported to neurons by apoE (Levi et al., 2005). It is removed from brain through metabolic conversion to oxysterols. 24S-Hydroxycholesterol represents the major metabolic product of cholesterol in brain, being formed via the cytochrome P450 (Cyp) enzyme Cyp46A1. Cyp46A1 is expressed exclusively in brain, normally by neurons. FPI increases Cyp46 levels at 7 days post-injury, and cell type-specific analysis at 3 days post-injury shows a significant increase in Cyp46 levels (84%) in microglia. FPI also increases apolipoprotein E and ATP-binding cassette transporter A1 at 7 days post-injury. This indicates that increased LXR activity coincides with increased Cyp46 levels. It is also reported that activation of primary rat microglia by LPS in vitro also results in increased Cyp46 levels, suggesting that increased microglial Cyp46 activity is part of a system for removal of damaged cell membranes post-injury (Cartagena et al., 2008). Levels of 24-hydroxycholesterol in brain, CSF, and plasma have not been determined. Recent studies on the determination of 24-hydroxycholesterol in closed TBI indicate that Plasma 24S-hydroxycholesterol levels do not change with severe closed head injury (Weiner et al., 2008). This suggests that more studies are required on the involvement of 24-hydroxycholesterol in TBI.

## 6.14 TBI and Apoptotic Cell Death

Morphologically, apoptotic cell death is characterized by nuclear chromatin condensation, DNA fragmentation, cell shrinkage, and bleb and apoptotic body formation. Plasma membrane and other subcellular organelles such as mitochondria and endoplasmic reticulum remain active during apoptosis (Mattson et al., 2000). Apoptotic cascades involve elevated levels of intracellular oxyradicals and calcium; upregulation of expression of proteins such as Par-4 (prostate apoptosis response-4), which act by promoting mitochondrial dysfunction and suppressing antiapoptotic mechanisms; mitochondrial membrane depolarization; calcium uptake; and release of cytochrome *c* that ultimately induces nuclear DNA condensation and fragmentation. In addition, activation of caspases along with transcription factor, AP-1 induces the expression of "killer genes" during apoptosis (Mattson et al., 2000). In contrast, necrotic cell death is characterized by massive $Na^{+}$ and $Ca^{2+}$ influxes, rapid ATP depletion, high levels of ROS, onset of rapid and prolonged mPT, activation of calpains, and other $Ca^{2+}$-dependent enzymes.

Apoptotic neuronal and glial cell death contributes to the overall pathology of TBI in both animals and humans (Raghupathi et al., 2000). In traumatized human brain and injured experimental animal brain, apoptotic cells have been observed alongside of cells undergoing necrosis. Neurons undergoing apoptosis have been identified within contusions in the acute post-traumatic period and in regions remote from the site of impact in the days and weeks after trauma. Apoptotic cell death in oligodendrocytes and astrocytes has been observed within injured white matter tracts.

In TBI, apoptotic cell death is triggered by interactions between excitotoxicity and oxidative stress. Apoptosis involves enhancement of glycerophospholipid, sphingolipid, and cholesterol metabolism not only due to changes in activities of phospholipases, sphingomyelinases, and cytochrome P450 oxygenases but also by alterations in levels of glycerophospholipid, sphingolipid, and cholesterol-derived lipid mediators. These processes along with abnormalities in signal transduction processes bring about neural cell demise through apoptosis (Farooqui et al., 2004). Neurochemical changes in apoptotic cell death occur in an orderly fashion due to sufficient levels of ATP that maintains normal ion homeostasis. The dead cells undergo phagocytosis without spilling cellular contents.

The clearance of debris after TBI is a critical step for restoration of the injured neural network. Although microglia contribute to the elimination of degenerating neurons and axons and facilitate the restoration of favorable environment after TBI, the mechanism underlying debris clearance remains elusive. It is recently suggested that activation of p38 mitogen-activated protein kinase (MAPK) in microglia promotes engulfment of cellular debris. This engulfment of axon debris can be blocked by the p38 MAPK inhibitor SB203580, indicating that p38 MAPK is required for phagocytic activity (Tanaka et al., 2009). In contrast, in necrosis rapid permeabilization of plasma membrane, rapid decrease in ATP, sudden loss of ion homeostasis, and activation of lysosomal enzymes result in a passive cell death through lysis (Farooqui et al., 2004; Farooqui and Horrocks, 2007). During necrosis release of cellular contents is accompanied by neuroinflammation and oxidative stress (Farooqui, 2009). In TBI, neurons die rapidly (hours to days) at the injury core by necrotic cell death, whereas in the surrounding area neurons undergo apoptotic cell death (several days to months) (McIntosh et al., 1998; Farooqui et al., 2004; Farooqui, 2009).

## 6.15 Molecular Mechanism of Neurodegeneration in TBI

As stated above, excitotoxicity, oxidative stress, and neuroinflammation are closely linked with the pathogenesis of TBI (Farooqui et al., 2004; Farooqui and Horrocks, 2007) (Fig. 6.7). Neurons are more susceptible to excitotoxic, oxidative, and inflammatory injuries than glial cells. In TBI, excitotoxicity and oxidative stress contribute to neuronal injury not only by intensifying the expression of inflammatory and stress-sensitive genes, including genes for cytokines and chemokines (Hayes et al., 2002; Farooqui et al., 2004), but also by activating mechanisms that result in a microglia and astrocytes-mediated secondary neuronal damage (Block and Hong, 2005; Farooqui and Horrocks, 2007). These activated glial cells are histopathological hallmarks of TBI and are closely associated with neurodegenerative processes (Farooqui et al., 2004). The direct contact of activated glia with neurons per se is not necessary for the commencement of neurodegenerative process. Immune mediators, e.g., NO, eicosanoids, ROS, and proinflammatory cytokines and chemokines, proteinases and complement proteins released by activated microglial cells and astrocytes may act as endogenous neurotoxins that promote and intensify neurodegenerative process in TBI. Neuritic beading may also

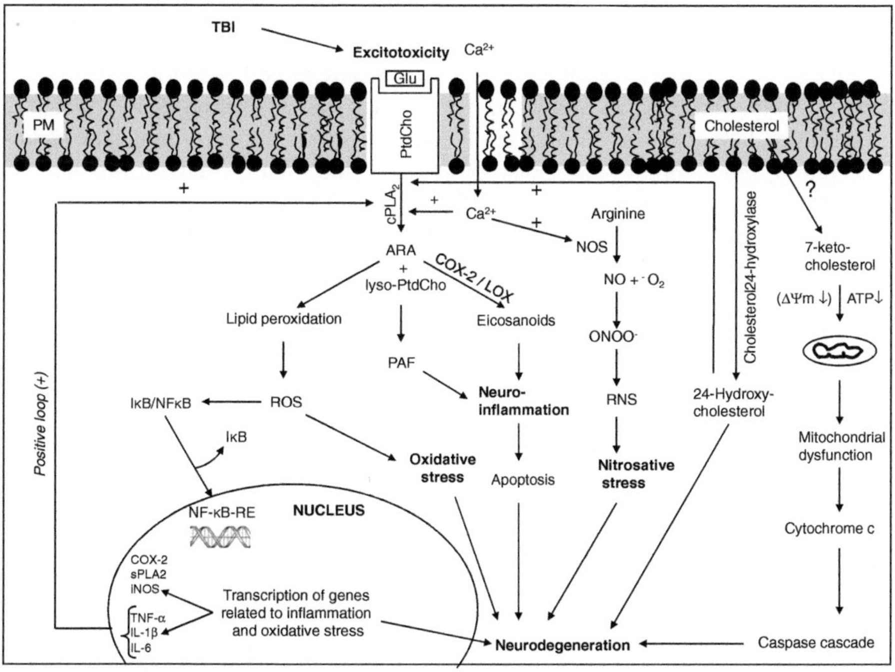


**Fig. 6.7** Interactions among excitotoxicity, neuroinflammation, and oxidative stress following TBI. Plasma membrane (PM); Glutamate (Glu); phosphatidylcholine (PtdCho); cytosolic phospholipase $A_2$ ($cPLA_2$); lyso-phosphatidylcholine (lyso-PtdCho); arachidonic acid (ARA); platelet-activating factor (PAF); secretory phospholipase $A_2$ ($sPLA_2$); reactive oxygen species (ROS); reactive nitrogen species (RNS); nitric oxide synthase (NOS); nuclear factor κB inhibited form (IκB/NFκB); nuclear factor κB-response element (NFκB-RE), inhibitory subunit of NFκB (IκB); tumor necrosis factor-α (TNF-α); interleukin-1β (IL-1β); interleukin-6 (IL-6); cyclooxygenase-2 (COX-2); lipoxygenase (LOX); peroxynitrite ($ONOO^-$); inducible nitric oxide synthase (iNOS). *Positive sign* (+) indicates stimulation

occur in TBI. It involves neuronal dysfunction due to abnormal signaling of NMDA receptor (Takeuchi et al., 2005). The molecular mechanism associated with neuritic beading that precedes neurodegeneration is not fully understood. However, a rapid drop in intracellular ATP levels is known to occur. Detailed investigations indicate that actual neurite beads contain collapsed cytoskeletal proteins and motor proteins arising from impaired neuronal transport secondary to cellular energy loss. TBI-mediated loss in intracellular ATP levels is caused by the inhibition of mitochondrial respiratory chain complex IV activity downstream of NMDA receptor signaling. Blockage of NMDA receptors nearly completely prevents mitochondrial dysfunction and neurotoxicity, suggesting that NMDA receptor antagonists may be an effective therapeutic approach for TBI (Takeuchi et al., 2005).

## 6.16 Conclusion

In TBI, the initial force of mechanical trauma causes distortion and destruction of brain tissue resulting in the release of glutamate from intracellular stores. This is followed by secondary injury leading to alterations in cell function and propagation of injury through processes such as depolarization, excitotoxicity, disruption of calcium homeostasis, activation of calcium dependent enzymes, free radical generation, blood–brain barrier disruption, ischemic injury, edema formation, and intracranial hypertension. The secondary injury also involves the initiation of an acute inflammatory response, including breakdown of the BBB, edema formation and swelling, infiltration of peripheral blood cells, and activation of resident immunocompetent cells. Neural cells at the injury site and infiltrating non-neural cells release chemokines, cytokines, and other intercellular signaling molecules. These molecules are involved in coordinating complex cellular responses, such as glial responses (release of GFAP and S100B), changes in neuronal survival and cellular repair events.

## References


Agoston DV, Gyorgy A, Eidelman O, Pollard HB (2009) Proteomic biomarkers for blast neurotrauma: targeting cerebral edema, inflammation, and neuronal death cascades. J Neurotrauma 26:901–911

Ahn MJ, Sherwood ER, Prough DS, Lin CY, DeWitt DS (2004) The effects of traumatic brain injury on cerebral blood flow and brain tissue nitric oxide levels and cytokine expression. J Neurotrauma 21:1431–1442

Alcalá-Barraza SR, Lee MS, Hanson LR, McDonald AA, Frey WH, McLoon LK (2009) J Drug Target 2009 Oct 6 [Epub ahead of print]

Amankulor NM, Hambardzumyan D, Pyonteck SM, Becher OJ, Joyce JA, Holland EC (2009) Sonic hedgehog pathway activation is induced by acute brain injury and regulated by injury-related inflammation. J Neurosci 29:10299–10308

Arundine M, Tymianski M (2004) Molecular mechanisms of glutamate-dependent neurodegeneration in ischemia and traumatic brain injury. Cell Mol Life Sci 61:657–668

Atkins CM, Oliva AA Jr, Alonso OF, Pearse DD, Bramlett HM, Dietrich WD (2007a) Modulation of the cAMP signaling pathway after traumatic brain injury. Exp Neurol 208:145–158

Atkins CM, Oliva AA Jr, Alonso OF, Chen S, Bramlett HM, Hu BR, Dietrich WD (2007b) Hypothermia treatment potentiates ERK1/2 activation after traumatic brain injury. Eur J Neurosci 26:810–819

Atkins CM, Chen S, Alonso OF, Dietrich WD, Hu BR (2009a) Activation of calcium/calmodulin-dependent protein kinases after traumatic brain injury. J Cereb Blood Flow Metab 26: 1507–1518

Atkins CM, Falo MC, Alonso OF, Bramlett HM, Dietrich WD (2009b) Deficits in ERK and CREB activation in the hippocampus after traumatic brain injury. Neurosci Lett 459:52–56

Auger C, Attwell D (2000) Fast removal of synaptic glutamate by postsynaptic transporters. Neuron 28:547–558

Bajetto A, Bonavia R, Barbero S, Florio T, Schettini G (2001) Chemokines and their receptors in the central nervous system. Front Neuroendocrinol 22:147–184

Bales JW, Ma X, Yan HQ, Jenkins LW, Dixon CE (2010) Expression of protein phosphatase 2B (calcineurin) subunit A isoforms in rat hippocampus after traumatic brain injury. J Neurotrauma 27:109–120

Beer R, Franz G, Srinivasan A, Hayes RL, Pike BR, Newcomb JK, Zhao X, Schmutzhard E, Poewe W, Kampfl A (2000) Temporal profile and cell subtype distribution of activated caspase-3 following experimental traumatic brain injury. J Neurochem 75:1264–1273

Bigford GE, Alonso OF, Dietrich D, Keane RW (2009) A novel protein complex in membrane rafts linking the NR2B glutamate receptor and autophagy is disrupted following traumatic brain injury. J Neurotrauma 26:703–720

Block ML, Hong JS (2005) Microglia and inflammation-mediated neurodegeneration: multiple triggers with a common mechanism. Prog Neurobiol 76:77–98

Brophy GM, Pineda JA, Papa L, Lewis SB, Valadka AB, Hannay HJ, Heaton SC, Demery JA, Liu MC, Tepas JJ 3rd, Gabrielli A, Robicsek S, Wang KK, Robertson CS, Hayes RL (2009) αII-Spectrin breakdown product cerebrospinal fluid exposure metrics suggest differences in cellular injury mechanisms after severe traumatic brain injury. J Neurotrauma 26:471–479

Buki A, Farkas O, Doczi T, Povlishock JT (2003) Preinjury administration of the calpain inhibitor MDL-28170 attenuates traumatically induced axonal injury. J Neurotrauma. 20:261–268

Callewaere C, Banisadr G, Rostene W, Parsadaniantz SM (2007) Chemokines and chemokine receptors in the brain: implication in neuroendocrine regulation. J Mol Endocrinol 38:358–363

Cardali S, Mangeri R (2006) Detection of αII-spectrin and breakdown products in humans after severe traumatic brain injury. J Neurosurg Sci 50:25–31

Carragher NO (2006) Calpain inhibition: a therapeutic strategy targeting multiple disease states. Curr Pharm Des 12:615–638

Cartagena CM, Ahmed F, Burns MP, Pajoohesh-Ganji A, Pak DT, Faden AI, Rebeck GW (2008) Cortical injury increases cholesterol 24S hydroxylase (Cyp46) levels in the rat brain. J Neurotrauma 25:1087–1098

Chen G, Shi J, Wei Jin W, Wang L, Xie W, Sun J, Hang C (2008b) Progesterone administration modulates TLRs/NF-κB signaling pathway in rat brain after cortical contusion. Ann Clin Lab Sci 38:65–74

Chen Y, Miles DK, Hoang T, Shi J, Hurlock E, Kernie SG, Lu QR (2008a) The basic helix-loop-helix transcription factor olig2 is critical for reactive astrocyte proliferation after cortical injury. J Neurosci 28:10983–10989

Cherian L, Hlatky R, Robertson CS (2004) Nitric oxide in traumatic brain injury. Brain Path 14:195–201

Chiaretti A, Antonelli A, Genovese O, Pezzotti P, Rocco CD, Viola L, Riccardi R (2008) Nerve growth factor and doublecortin expression correlates with improved outcome in children with severe traumatic brain injury. J Trauma 65:80–85

Chumakov AM, Silla A, Williamson EA, Koeffler HP (2007) Modulation of DNA binding properties of CCAAT/enhancer binding protein epsilon by heterodimer formation and interactions with NFkappaB pathway. Blood 109:4209–4219

Cohen GM (1997) Caspases: the executioners of apoptosis. Biochem J 326:1–16

Crawford F, Wood M, Ferguson S, Mathura V, Gupta P, Humphrey J, Mouzon B, Laporte V, Margenthaler E, O'Steen B, Hayes R, Roses A, Mullan M (2009) Apolipoprotein E-genotype dependent hippocampal and cortical responses to traumatic brain injury. Neuroscience 159:1349–1362

Creagh EM, Conroy H, Martin SJ (2003) Caspase-activation pathways in apoptosis and immunity. Immunol Rev 193:10–21

Demediuk P, Daly MP, Faden AI (1988) Free amino acid levels in laminectomized and traumatized rat spinal cord. Trans Am Soc Neurochem 19:176

de Rivero Vaccari JP, Lotocki G, Alonso OF, Bramlett HM, Dietrich WD, Keane RW (2009) Therapeutic neutralization of the NLRP1 inflammasome reduces the innate immune response and improves histopathology after traumatic brain injury. J Cereb Blood Flow Metab 29: 1251–1261

Dhillon HS, Donaldson D, Dempsey RJ, Prasad MR (1994) Regional levels of free fatty acids and Evans blue extravasation after experimental brain injury. J Neurotrauma 11:405–415

Dhillon HS, Dose JM, Prasad MR (1996) Regional generation of leukotriene C4 after experimental brain injury in anesthetized rats. J Neurotrauma 13:781–789

Ding JY, Kreipke CW, Schafer P, Schafer S, Speirs SL, Rafols JA (2009) Synapse loss regulated by matrix metalloproteinases in traumatic brain injury is associated with hypoxia inducible factor-1alpha expression. Brain Res 1268:125–134

Ellis RC, Earnhardt JN, Hayes RL, Wang KKW, Anderson DK (2004) Cathepsin B mRNA and protein expression following contusion spinal cord injury in rats. J Neurochem 88: 689–697

Farkas O, Polgar B, Szekeres-Bartho J, Doczi T, Povlishock JT, Buki A (2005) Spectrin breakdown products in the cerebrospinal fluid in severe head injury – Preliminary observations. Acta Neurochir (Wien) 147:855–861

Farooqui AA, Ong WY, Horrocks LA (2004) Biochemical aspects of neurodegeneration in human brain: involvement of neural membrane phospholipids and phospholipases $A_2$. Neurochem Res 29:1961–1977

Farooqui AA, Horrocks LA (2007) Glycerophospholipids in brain. Springer, New York, NY

Farooqui AA, Ong WY, Horrocks LA (2008) Neurochemical aspects of excitotoxicity. Springer, New York, NY

Farooqui AA (2009) Hot topics in neural membrane lipidology. Springer, New York, NY

Farooqui AA, Horrocks LA (2009) Glutamate and cytokine-mediated alterations of phospholipids head injury and spinal cord trauma. In: Banik NK (ed) Handbook of neurochemistry and molecular neurobiology, vol 24. Springer, New York, NY, pp 71–89

Fernandez AM, Fernandez S, Carrero P, Garcia-Garcia M, Torres-Aleman I (2007) Calcineurin in reactive astrocytes plays a key role in the interplay between proinflammatory and anti-inflammatory signals. J Neurosci 27:8745–8756

Fineman I, Hovda DA, Smith M, Yoshino A, Becker DP (1993) Concussive brain injury is associated with a prolonged accumulation of calcium: a 45Ca autoradiographic study. Brain Res 624:94–102

Ghirnikar RS, Lee YL, Eng LF (1998) Inflammation in traumatic brain injury: role of cytokines and chemokines. Neurochem Res 23:329–340

Girard J, Panizzon K, Wallis RA (1996) Azelastine protects against CA1 traumatic neuronal injury in the hippocampal slice. Eur J Pharmacol 300:43–49

Glantz SB, Cianci CD, Iyer R, Pradhan D, Wang KK, Morrow JS (2007) Sequential degradation of alphaII and betaII spectrin by calpain in glutamate or maitotoxin-stimulated cells. Biochemistry 46:502–513

Gomes-Leal W, Corkill DJ, Freire MA, Picanço-Diniz CW, Perry VH (2004) Astrocytosis, microglia activation, oligodendrocyte degeneration, and pyknosis following acute spinal cord injury. Exp Neurol 190:456–467

Gopez JJ, Yue H, Vasudevan R, Malik AS, Fogelsanger LN, Lewis S, Panikashvili D, Shohami E, Jansen SA, Narayan RK, Strauss KI (2005) Cyclooxygenase-2-specific inhibitor improves functional outcomes, provides neuroprotection, and reduces inflammation in a rat model of traumatic brain injury. Neurosurgery 56:590–604

Grossetete M, Phelps J, Arko L, Yonas H, Rosenberg GA (2009) Elevation of matrix metalloproteinases 3 and 9 in cerebrospinal fluid and blood in patients with severe traumatic brain injury. Neurosurgery 65:702–708

Hang CH, Chen G, Shi JX, Zhang X, Li JS (2006) Cortical expression of nuclear factor kappaB after human brain contusion. Brain Res 1109:14–21

Harris LK, Black RT, Golden KM, Reeves TM, Povlishack JT, Phillips LL (2001) Traumatic brain injury-induced changes in gene expression and functional activity of mitochondrial cytochrome C oxidase. J Neurotrauma 18:993–1009

Hayes KC, Hull TC, Delaney GA, Potter PJ, Sequeira KA, Campbell K, Popovich PG (2002) Elevated serum titers of proinflammatory cytokines and CNS autoantibodies in patients with chronic spinal cord injury. J Neurotrauma 19:753–761

Hickey RW, Adelson PD, Johnnides MJ, Davis DS, Yu Z, Rose ME, Chang YF, Graham SH (2007) Cyclooxygenase-2 activity following traumatic brain injury in the developing rat. Pediatr Res 62:271–276

Hildebrand F, Pape HC, Krettek C (2005) The importance of cytokines in the posttraumatic inflammatory reaction. Unfallchirurg 108:793–794

Homayoun P, Rodriguez de Turco EB, Parkins NE, Lane DC, Soblosky J, Carey ME, Bazan NG (1997) Delayed phospholipid degradation in rat brain after traumatic brain injury. J Neurochem 69:199–205

Homayoun P, Parkins NE, Soblosky J, Carey ME, Rodriguez de Turco EB, Bazan NG (2000) Cortical impact injury in rats promotes a rapid and sustained increase in polyunsaturated free fatty acids and diacylglycerols. Neurochem Res 25:269–276

Israelsson C, Bengtsson H, Kylberg A, Kullander K, Lewén A, Hillered L, Ebendal T (2008) Distinct cellular patterns of upregulated chemokine expression supporting a prominent inflammatory role in traumatic brain injury. J Neurotrauma 25:959–974

Jin W, Wang H, Yan W, Xu L, Wang X, Zhao X, Yang X, Chen G, Ji Y (2008) Disruption of Nrf2 enhances upregulation of nuclear factor-kappaB activity, proinflammatory cytokines, and intercellular adhesion molecule-1 in the brain after traumatic brain injury. Mediators Inflamm 2008:725174 Epub 2009 Jan 25

Jupp OJ, Vandenabeele P, MacEwan DJ (2003) Distinct regulation of cytosolic phospholipase $A_2$ phosphorylation, translocation, proteolysis and activation by tumour necrosis factor-receptor subtypes. Biochem J 374:453–461

Kadhim HJ, Duchateau J, Sebire G (2008) Cytokines and brain injury: invited review. J Intensive Care Med 23:236–249

Kane MJ, Citron BA (2009) Transcription factors as therapeutic targets in CNS disorders. Recent Pat Nanotechnol 4:190–199

Karin M, Ben-Neriah Y (2000) Phosphorylation meets ubiquitination: the control of NF-[kappa]B activity. Annu Rev Immunol 18:621–663

Khorooshi R, Babcock AA, Owens T (2008) NF-kappaB-driven STAT2 and CCL2 expression in astrocytes in response to brain injury. J Immunol 181:7284–7291

Kim GM, Xu J, Xu JM, Song SK, Yan P, Ku G, Xu XM, Hsu CY (2001) Tumor necrosis factor receptor deletion reduces nuclear factor-kappa B activation, cellular inhibitor of apoptosis protein 2 expression, and functional recovery after traumatic spinal cord injury. J Neurosci 21:6617–6625

Kizhakke Madathil S, Evans HN, Saatman KE (2009) Temporal and regional changes in IGF-1/IGF-1R signaling in the mouse brain following traumatic brain injury. J Neurotrauma 26:2269–2278

Klementiev B, Novikova T, Korshunova I, Berezin V, Bock E (2008) The NCAM-derived P2 peptide facilitates recovery of cognitive and motor function and ameliorates neuropathology following traumatic brain injury. Eur J Neurosci 27:2885–2896

Kunz T, Marklund N, Hillered L, Oliw EH (2002) Cyclooxygenase-2, prostaglandin synthases, and prostaglandin H2 metabolism in traumatic brain injury in the rat. J Neurotrauma 19:1051–1064

Kupina NC, Detloff MR, Bobrowski WF, Snyder BJ, Hall ED (2003) Cytoskeletal protein degradation and neurodegeneration evolves differently in males and females following experimental head injury. Exp Neurol 180:55–73

Kurz JE, Hamm RJ, Singleton RH, Povlishock JT, Churn SB (2005a) A persistent change in subcellular distribution of calcineurin following fluid percussion injury in the rat. Brain Res 1048:153–160

Kurz JE, Parsons JT, Rana A, Gibson CJ, Hamm RJ, Churn SB (2005b) A significant increase in both basal and maximal calcineurin activity following fluid percussion injury in the rat. J Neurotrauma 22:476–490

Lee JM, Johnson JA (2004) An important role of Nrf2-ARE pathway in the cellular defense mechanism, J. Biochem Mol Biol 37:139–143

Leinhase I, Schmidt OI, Thurman JM, Hossini AM, Rozanski M, Taha ME, Scheffler A, John T, Smith W, Holers VM, Stahel PF (2006) Pharmacological complement inhibition at the C3 convertase level promotes neuronal survival, neuroprotective intracerebral gene expression, and neurological outcome after traumatic brain injury. Exp Neurol 199: 454–464

Leinhase I, Rozanski M, Harhausen D, Thurman JM, Schmidt OI, Hossini AM, Taha ME, Rittirsch D, Ward PA, Holers VM, Ertel W, Stahel PF (2007) Inhibition of the alternative complement activation pathway in traumatic brain injury by a monoclonal anti-factor B antibody: a randomized placebo-controlled study in mice. J Neuroinflammation 4:13

Lenzlinger PM, Morganti-Kossmann MC, Laurer HL, McIntosh TK (2009) The duality of the inflammatory response to traumatic brain injury. Mol Neurobiol 24:169–181

Levi O, Lütjohann D, Devir A, von Bergmann K, Hartmann T, Michaelson DM (2005) Regulation of hippocampal cholesterol metabolism by apoE and environmental stimulation. J Neurochem 95:987–997

Levy DE, Darnell JE Jr. (2002) Stats: transcriptional control and biological impact. Nat Rev Mol Cell Biol 3:651–662

Li S, Stys PK (2000) Mechanisms of ionotropic glutamate receptor-mediated excitotoxicity in isolated spinal cord white matter. J Neurosci 20:1190–1198

Matute C, Domercq M, Sánchez-Gómez MV (2006) Glutamate-mediated glial injury: mechanisms and clinical importance. Glia 53:212–224

Mattson M, Culmsee C, Yu ZF (2000) Apoptotic and antiapoptotic mechanisms in stroke. Cell Tissue Res 301:173–187

McGeer EG, Klegeris A, McGeer PL (2005) Inflammation, the complement system and the diseases of aging. Neurobiol Aging 26(Suppl 1):94–97

McIntosh TK, Saatman KE, Raghupathi R, Graham DI, Smith DH, Lee VM, Trojanowski JQ (1998) The Dorothy Russell Memorial Lecture. The molecular and cellular sequelae of experimental traumatic brain injury: pathogenetic mechanisms. Neuropathol Appl Neurobiol 24:251–267

Mechoulam R, Shohami E (2007) Endocannabinoids and traumatic brain injury. Mol Neurobiol 36:68–74

Morganti-Kossmann MC, Satgunaseelan L, Bye N, Kossmann T (2007) Modulation of immune response by head injury. Injury 38:1392–1400

Morioka M, Hamada J, Ushio Y, Miyamoto E (1999) Potential role of calcineurin for brain ischemia and traumatic injury. Prog Neurobiol 58:1–30

Murphy TH, Miyamoto M, Sastre A, Schnaar RL, Coyle JT (1989) Glutamate toxicity in a neuronal cell line involves inhibition of cystine transport leading to oxidative stress. Neuron 2: 1547–1558

Oka A, Belliveau MJ, Rosenberg PA, Volpe JJ (1993) Vulnerability of oligodendroglia to glutamate: pharmacology, mechanisms, and prevention. J Neurosci 13: 1441–1453

Osuna E, Perez-Carceles MD, Luna A, Pounder DJ (1992) Efficacy of cerebro-spinal fluid biochemistry in the diagnosis of brain insult. Forensic Sci Int 52:193–198

Otani N, Nawashiro H, Fukui S, Ooigawa H, Ohsumi A, Toyooka T, Shima K (2007) Role of the activated extracellular signal-regulated kinase pathway on histological and behavioral outcome after traumatic brain injury in rats. J Clin Neurosci 14:42–48

Ottens AK, Golden EC, Bustamante L, Hayes RL, Denslow ND, Wang KK (2008) Proteolysis of multiple myelin basic protein isoforms after neurotrauma: characterization by mass spectrometry. J Neurochem 104:1404–1414

Panter SS, Yum SW, Faden AI (1990) Alteration in extracellular amino acids after traumatic spinal cord injury. Ann Neurol 27:96–99

Park E, Velumian AA, Fehlings MG (2004) The role of excitotoxicity in secondary mechanisms of spinal cord injury: a review with an emphasis on the implications for white matter degeneration. J Neurotrauma 21:754–774

Pavel J, Lukácová N, Marsala J, Marsala M (2001) The regional changes of the catalytic NOS activity in the spinal cord of the rabbit after repeated sublethal ischemia. Neurochem Res 26:833–839

Pedersen MV, Helweg-Larsen RB, Nielsen FC, Berezin V, Bock E, Penkowa M (2008) The synthetic NCAM-derived peptide, FGL, modulates the transcriptional response to traumatic brain injury. Neurosci Lett 437:148–153

Pelinka LE, Kroepfl A, Leixnering M, Buchinger W, Raabe A, Redl H (2004) GFAP versus S100B in serum after traumatic brain injury: relationship to brain damage and outcome. J Neurotrauma 21:1553–1561

Pereire CFM, Resende de Oliveira C (2000) Oxidative glutamate toxicity involves mitochondrial dysfunction and perturbation of intracellular $Ca^{2+}$ homeostasis. Neurosci Res 83:2758–2762

Pfrieger FW (2003) Outsourcing in the brain: do neurons depend on cholesterol delivery by astrocytes? Bioessays 25:72–78

Phillis JW, Horrocks LA, Farooqui AA (2006) Cyclooxygenases, lipoxygenases, and epoxygenases in CNS: their role and involvement in neurological disorders. Brain Res Rev 52:201–243

Pilitsis JG, Coplin WM, O'Regan MH, Wellwood JM, Diaz FG, Fairfax MR, Michael DB, Phillis JW (2003) Free fatty acids in cerebrospinal fluids from patients with traumatic brain injury. Neurosci Lett 349:136–138

Pike BR, Zhao X, Newcomb JK, Posmantur RM, Wang KK, Hayes RL (1998) Regional calpain and caspase-3 proteolysis of alpha-spectrin after traumatic brain injury. Neuroreport 9:2437–2442

Pineda JA, Lewis SB, Valadka AB, Papa L, Hannay HJ, Heaton SC, Demery JA, Liu MC, Aikman JM, Akle V, Brophy GM, Tepas JJ, Wang KK, Robertson CS, Hayes RL (2007) Clinical significance of alphaII-spectrin breakdown products in cerebrospinal fluid after severe traumatic brain injury. J Neurotrauma 24:354–366

Plesnila N (2007) Decompression craniectomy after traumatic brain injury: recent experimental results. Prog Brain Res 161:393–400

Plesnila N, von Baumgarten L, Retiounskaia M, Engel D, Ardeshiri A, Zimmermann R, Hoffmann F, Landshamer S, Wagner E, Culmsee C (2007) Delayed neuronal death after brain trauma involves p53-dependent inhibition of NF-kappaB transcriptional activity. Cell Death Differ 14:1529–1541

Posmantur R, Hayes RL, Dixon CE, Taft WC (1994) Neurofilament 68 and neurofilament 200 protein levels decrease after traumatic brain injury. J Neurotrauma 11:533–545

Posmantur RM, Kampfl A, Liu S, Heck K, Taft WC, Clifton GL, Hayes RL (1996) Cytoskeletal derangements of cortical neuronal processes three hours after traumatic brain injury in rats: an immunofluorescence study. J Neuropathol Exp Neurol 55:68–80

Praticò D, Reiss P, Tang LX, Sung S, Rokach J, McIntosh TK (2002) Local and systemic increase in lipid peroxidation after moderate experimental traumatic brain injury. J Neurochem 80: 894–898

Raghupathi R, McIntosh TK, Smith DH (1995) Cellular responses to experimental brain injury. Brain Path 5:437–442

Raghupathi R, Graham DI, McIntosh TK (2000) Apoptosis after traumatic brain injury. J Neurotrauma 17:927–938

Raghupathi R (2004) Cell death mechanisms following traumatic brain injury. Brain Path 14: 215–222
Raivich G, Behrens A (2006) Role of the AP-1 transcription factor c-Jun in developing, adult and injured brain. Prog Neurobiol 78:347–363
Ramji DP, Foka P (2002) CCAAT/enhancer-binding proteins: structure, function and regulation. Biochem J 365:561–575
Ray SK, Banik NL (2003) Calpain and its involvement in the pathophysiology of CNS injuries and diseases: therapeutic potential of calpain inhibitors for prevention of neurodegeneration. Curr Drug Targets CNS Neurol Disord 2:173–189
Ray SK, Hogan EL, Banik NL (2003) Calpain in the pathophysiology of spinal cord injury: neuroprotection with calpain inhibitors. Brain Res Rev 42:169–185
Ray SK (2006) Currently evaluated calpain and caspase inhibitors for neuroprotection in experimental brain ischemia. Curr Med Chem 13:3425–3440
Rhodes JK, Sharkey J, Andrews PJ (2009) The temporal expression, cellular localization, and inhibition of the chemokines MIP-2 and MCP-1 after traumatic brain injury in the rat. J Neurotrauma 26:507–525
Sandhir R, Puri V, Klein RM, Berman NE (2004) Differential expression of cytokines and chemokines during secondary neuron death following brain injury in old and young mice. Neurosci Lett 369:28–32
Sandhir R, Berman NE (2009) Age-dependent response of CCAAT/enhancer binding proteins following traumatic brain injury in mice. Neurochem Int 2009 Oct 12 [Epub ahead of print]
Schwab JM, Seid K, Schluesener HJ (2001) Traumatic brain injury induces prolonged accumulation of cyclooxygenase-1 expressing microglia/brain macrophages in rats. J Neurotrauma 18:881–890
Schwab JM, Beschorner R, Meyermann R, Gözalan F, Schluesener HJ (2002) Persistent accumulation of cyclooxygenase-1-expressing microglial cells and macrophages and transient upregulation by endothelium in human brain injury. J Neurosurg 96:892–899
Schuhmann MU, Mokhtarzadeh M, Stichtenoth DO, Skardelly M, Klinge PM, Gutzki FM, Samii M, Brinker T (2003) Temporal profiles of cerebrospinal fluid leukotrienes, brain edema and inflammatory response following experimental brain injury. Neurol Res 25:481–491
Shapira M, Licht A, Milman A, Pick CG, Shohami E, Eldar-Finkelman H (2007) Role of glycogen synthase kinase-3beta in early depressive behavior induced by mild traumatic brain injury. Mol Cell Neurosci 34:571–577
Sharma S, Zhuang Y, Ying Z, Wu A, Gomez-Pinilla F (2009) Dietary curcumin supplementation counteracts reduction in levels of molecules involved in energy homeostasis after brain trauma. Neuroscience 161:1037–1044
Shohami E, Shapira Y, Sidi A, Cotev S (1987) Head injury induces increased prostaglandin synthesis in rat brain. J Cereb Blood Flow Metab 7:58–63
Shohami E, Shapira Y, Yadid G, Reisfeld N, Yedgar S (1989) Brain phospholipase A2 is activated after experimental closed head injury in the rat. J Neurochem 53:1541–1546
Sifringer M, Stefovska V, Zentner I, Hansen B, Stepula K, Knaute C, Marzahn J, Ikonomidou C (2007) The role of matrix metalloproteinases in infant traumatic brain injury. Neurobiol Dis 25:526–535
Siman R, Toraskar N, Dang A, McNeil E, McGarvey M, Plaum J, Maloney E, Grady MS (2009) A panel of neuron-enriched proteins as markers for traumatic brain injury in humans. J Neurotrauma 26:1867–1877
Simons K, Ikonen E (2000) How cells handle cholesterol? Science 290:1721–1726
Skaper SD (2008) The biology of neurotrophins, signalling pathways, and functional peptide mimetics of neurotrophins and their receptors. CNS Neurol Disord Drug Targets 7:46–62
Strauss KI, Narayan RK, Raghupathi R (2004) Common patterns of bcl-2 family gene expression in two traumatic brain injury models. Neurotox Res 6:333–342
Sun D, Newman TA, Perry VH, Weller RO (2004) Cytokine-induced enhancement of autoimmune inflammation in the brain and spinal cord: implications for multiple sclerosis. Neuropathol Appl Neurobiol 30:374–384

Sundström E, Mo LL (2002) Mechanisms of glutamate release in the rat spinal cord slices during metabolic inhibition. J Neurotrauma 19:257–266

Svetlov SI, Larner SF, Kirk DR, Atkinson J, Hayes RL, Wang KK (2009) Biomarkers of blast-induced neurotrauma: profiling molecular and cellular mechanisms of blast brain injury. J Neurotrauma 26:913–921

Szmydynger-Chodobska J, Strazielle N, Zink BJ, Ghersi-Egea JF, Chodobski A (2009) The role of the choroid plexus in neutrophil invasion after traumatic brain injury. J Cereb Blood Flow Metab 29:1503–1516

Takadera T, Yumoto H, Tozuka Y, Ohyashiki T (2002) Prostaglandin E2 induces caspase-dependent apoptosis in rat cortical cells. Neurosci Lett 317:61–64

Takeuchi H, Mizuno T, Zhang GQ, Wang JY, Kawanokuchi J, Kuno R, Suzumura A (2005) Neuritic beading induced by activated microglia is an early feature of neuronal dysfunction toward neuronal death by inhibition of mitochondrial respiration and axonal transport. J Biol Chem 280:10444–10454

Tanaka T, Ueno M, Yamashita T (2009) Engulfment of axon debris by microglia requires p38 MAPK activity. J Biol Chem 284:21626–21636

Tatsumi K, Takebayashi H, Manabe T, Tanaka KF, Makinodan M, Yamauchi T, Makinodan E, Matsuyoshi H, Okuda H, Ikenaka K, Wanaka A (2008) Genetic fate mapping of Olig2 progenitors in the injured adult cerebral cortex reveals preferential differentiation into astrocytes. J Neurosci Res 86:3494–3502

Van Santbrink H, Schouten JW, Steyerberg EW, Avezaat CJ, Maas AL (2002) Serial transcranial Doppler measurements in traumatic brain injury with special focus on the early posttraumatic period. Acta Neurochir (Wien) 144:1141–1149

Varma S, Janesko KL, Wisniewski SR, Bayir H, Adelson PD, Thomas NJ, Kochanek PM (2003) F2-isoprostane and neuron-specific enolase in cerebrospinal fluid after severe traumatic brain injury in infants and children. J Neurotrauma 20:781–786

Vázquez MD, Sánchez-Rodriguez F, Osuna E, Diaz J, Cox DE, Pérez-Cárceles MD, Martinez P, Luna A, Pounder DJ (1995) Creatine kinase BB and neuron-specific enolase in cerebrospinal fluid in the diagnosis of brain insult. Am J Forensic Med Pathol 16:210–214

Von Gertten C, Flores Morales A, Holmir S, Mathiesen T, Nordgvist AC (2005) Genomic responses in rat cerebral cortex after traumatic brain injury. BMC Neurosci 6:69–73

Wada K, Chatzipanteli K, Kraydieh S, Busto R, Dietrich WD (1998) Inducible nitric oxide synthase expression after traumatic brain injury and neuroprotection with aminoguanidine treatment in rats. Neurosurg 43:1427–1436

Wei EP, Lamb RG, Kontos HA (1982) Increased phospholipase C activity after experimental brain injury. J Neurosurg 56:695–698

Weiner MF, Vega GL, Diaz-Arrastia R, Moore C, Madden C, Hudak A, Lütjohann D (2008) Plasma 24S-hydroxycholesterol and other oxysterols in acute closed head injury. Brain Inj 22: 611–615

Wiesmann M, Steinmeier E, Magerkurth O, Linn J, Gottmann D, Missler U (2010) Outcome prediction in traumatic brain injury: comparison of neurological status, CT findings, and blood levels of S100B and GFAP. Acta Neurol Scand 121:178–185

Wilson CJ, Finch CE, Cohen HJ (2002) Cytokines and cognition - the case for a head-to-toe inflammatory paradigm. J Am Geriatr Soc 50:2041–2056

Williams AJ, Wei HH, Dave JR, Tortella FC (2007) Acute and delayed neuroinflammatory response following experimental penetrating ballistic brain injury in the rat. J Neuroinflammation 4:17–19

Wissing D, Nouritzen H, Egeblad M, Poirier GG, Jaattela M (1997) Involvement of caspase-dependent activation of cytosolic phospholipase A2 in tumor necrosis factor-induced apoptosis. Proc Natl Acad Sci USA 94:5073–5077

Xing G, Ren M, Watson WA, O'Neil JT, Verma A (2009) Traumatic brain injury-induced expression and phosphorylation of pyruvate dehydrogenase: a mechanism of dysregulated glucose metabolism. Neurosci Lett 454:38–42

Xiong Y, Rabchevsky AG, Hall ED (2007) Role of peroxynitrite in secondary oxidative damage after spinal cord injury. J Neurochem 100:639–649

Yan W, Wang HD, Feng XM, Ding YS, Jin W, Tang K (2009) The expression of NF-E2-related factor 2 in the rat brain after traumatic brain injury. J Trauma 66:1431–1435

Zhang J, Xue R, Ong WY, Chen P (2009) Roles of cholesterol in vesicle fusion and motion. Biophys J 97:1371–1380

Zhivotovsky B, Samali A, Gahm A, Orrenius S (1999) Caspases: their intracellular localization and translocation during apoptosis. Cell Death Differ 6:644–651

# Chapter 7
# Potential Neuroprotective Strategies for Traumatic Brain Injury

## 7.1 Introduction

Traumatic brain injury (TBI) is caused by physical trauma to the brain tissue that temporarily or permanently impairs brain function. According to Centers for Disease Control and Prevention about 2 million people sustain a TBI in the USA each year, of which approximately 70,000–90,000 suffer from long-term disability (Nolan, 2005). Symptoms and severity of a TBI can be mild, moderate, or severe depending on the intensity of impact and extent of the damage to the brain. Some TBI symptoms appear immediately, while others do not appear until several days or weeks. Mild TBI symptoms include headache, confusion, lightheadedness, dizziness, blurred vision, fatigue, and trouble with memory (Bahraini et al., 2009). Moderate TBI produces a headache that gets worse with time, seizures, inability to awaken from sleep, dilation of one or both pupils of the eyes, slurred speech, loss of coordination, increased confusion. Severe TBI causes loss of consciousness and repeated very severe seizures. Elevated troponin, post-traumatic cerebral infarction, and coagulopathy are frequently observed after severe TBI. The level of troponin correlates with the severity of head injury and is an independent predictor of adverse outcomes (Salim et al., 2008). Mild brain injuries usually do not cause lasting effects; however, severe brain injuries can cause devastating consequences, including coma and death. Diagnosis is suspected clinically and confirmed by neuroimaging (primarily CT). CT can rapidly detect intracranial hematoma, intraparenchymal contusion, skull fracture, and cerebral edema, as well as transependymal flow and obliteration of the basal cisterns, which are concerns for increased intracranial pressure (Chun et al., 2009).

As stated in Chapter 6, TBI is accompanied by primary and secondary injuries. Primary injury is irreversible and is caused by the direct mechanical damage to neurons, axons, glial cells, and blood vessels. Focal traumas such as contusions and hematomas are caused by contact, linear forces when the head is struck by a moving object. Inertial, angular forces generated by acceleration–deceleration may result in immediate physical shearing or tearing and stretching of axons and blood vessels. These processes may also result in contusions, hemorrhage, and laceration, with immediate clinical effects. In contrast, secondary injury involves cellular, neurochemical, and metabolic alterations initiated by the primary injury

A.A. Farooqui, *Neurochemical Aspects of Neurotraumatic and Neurodegenerative Diseases*, DOI 10.1007/978-1-4419-6652-0_7,

that continue to develop over time (Povlishock and Christman, 1995). Secondary injury includes two events. The first event of secondary brain injury is accompanied by hypoxemia, hypotension, intracranial hypertension, hypercarbia, hyper- or hypoglycemia, electrolyte abnormalities, enlarging hematomas, coagulopathy, seizures, and hyperthermia which are potentially avoidable or treatable (Kochanek et al., 2008; Huh and Raghupathi, 2009). Initial treatment of severe TBI patients consists of ensuring a reliable airway and maintaining adequate ventilation, oxygenation, and blood pressure. The second event of secondary brain injury involves an endogenous cascade of cellular and neurochemical events in the brain that occurs within minutes and continues for months after the primary brain injury, leading to ongoing traumatic axonal injury and neuronal cell damage (delayed brain injury), and ultimately, neuronal cell death (Lenzlinger et al., 2001).

## 7.2 Regeneration and Neuritogenesis in TBI

Functional recovery in TBI victims is very limited because injured axons in brain do not regenerate spontaneously and do not respond to therapeutic strategies. This is because of the induction of opposing permissive (growth factors) and hostile signals (repulsive cues) resulting in growth cone collapsing and concomitant inhibition of restructuring of the cytoskeleton (Hou et al., 2008). Repulsive cues, such as semaphorins, ephrins, slits, netrins, Wnts, and myelin-secreted inhibitory glycoproteins (MAG and Nogo), act through their respective receptors to initiate the collapsing of growth cones through the participation of Rho GTPases-mediated signaling associated with microtubule changes in cytoskeleton remodeling. A brain-specific protein called collapsin response mediator protein (CRMP) has been reported to modulate microtubules (Hou et al., 2008). It is shown that cleavage of CRMPs in response to injury-activated proteases, such as calpain, signals axonal retraction and neuronal death in adult post-mitotic neurons, while inhibiting this signal transduction retards axonal retraction and death following excitotoxic insult and cerebral ischemia (Hou et al., 2008). Although the molecular mechanism associated with the obstruction of axonal regeneration is not fully understood, receptor complex comprising of the Nogo receptor (NgR1), the $p75^{NTR}$ receptor, and LINGO-1 (a nervous system-specific transmembrane protein that binds to NgR1-p75 complex and impedes the axonal regeneration) transduces the signals from all of these inhibitors. Downstream of these inhibitors, activation of small GTPase RhoA and its effector Rho-kinase has been shown to be an important element for neurite growth inhibition and growth cone collapse elicited by these inhibitors (Skaper et al., 2001). In addition to direct effects on axonal regeneration, many axonal guidance molecules have effects on glial, meningeal, or immune system cells, which also modulate the responses of the brain tissue to injury (Hou et al., 2008). Astrocytes also contribute to the inhibition of axonal regeneration by synthesizing multiple inhibitory proteoglycans, such as chondroitin sulfate proteoglycans (CSPGs) and facilitating the formation of a glial scar, a major obstacle to axonal growth after injury to the adult CNS. Inhibition of regeneration results in significant functional deficits and, depending on the severity of injury, may contribute to permanent paralysis or loss of senses distal to the site of injury (Yamashita, 2007). Collective evidence suggests that brain tissue contains multiple axon growth inhibitors that contribute to inability of the injured axons to regenerate. However, some regeneration in adult injured brain (neurogenesis) does occur in limited areas where synaptic plasticity is prevalent. In the absence of axonal regeneration, there is not only an inevitable loss-of-functional connections but also a loss of neurons. It is stated that a detailed understanding of the molecular mechanisms that limit neuronal growth in the injured brain will be an important step toward the development of specific strategies aimed at restoring functional connectivity lost as a consequence of injury (Skaper et al., 2001; Yamashita, 2007; Hou et al., 2008).

## 7.3 Potential Neuroprotective Strategies for TBI

TBI is a devastating and complex clinical condition involving release of glutamate, generation of reactive oxygen species, release of proinflammatory cytokines, and production of nitric oxide. These processes result in a progressive injury entailing neuronal loss, axonal destruction, and demyelination not only at the site of impact but also in the surrounding area. Many drugs such as NMDA receptor antagonists, opioid receptor antagonists, calcium channel blockers, platelet-activating factor antagonists, gangliosides, thyrotropin hormone analogs, aminosteroids (tirilazad mesylate), and antioxidants have been used for the treatment of TBI in animal models, but these drugs do not produce beneficial effects (Faden and Salzman, 1994; Faden, 2002). The failure of above drugs in TBI may be due to the fact that pathophysiology of TBI involves not only a number of mechanisms including excitotoxicity, oxidative stress, and inflammation but also two different modes of cell death, necrosis and apoptosis (McIntosh et al., 1998; Farooqui et al., 2004). In addition, brain injury also triggers auto-protective mechanisms, including the upregulation of anti-inflammatory cytokines and endogenous antioxidants. Introduction of "omics" technology (lipidomics, proteomics, and genomics) has not only resulted in better understanding of genes, enzymes, lipid mediator associated with the interplay among excitotoxicity, oxidative stress, and inflammation but also provided information that can be used for characterizing biomarkers and developing of new drugs for TBI treatment. It should be recognized that there are methodological differences between animal and human studies. The therapeutic window and treatment optimization in human may be very different from animal models. Pharmacokinetics related factors, such as the rate of drug penetration in the brain may modulate the efficacy of therapeutic agent. Safety and tolerability of the drug may also contribute to differences between human and experimental models of TBI. Since drugs such as NMDA receptor antagonists, opioid receptor antagonists, calcium channel blockers, platelet-activating factor antagonists, gangliosides, thyrotropin hormone analogs, aminosteroids (tirilazad mesylate), and antioxidants do not produce beneficial effects, the following sections will describe effect of new drugs for the treatment of TBI.

## *7.3.1 Statins and TBI*

Statins are drugs that inhibit HMG-CoA (3-hydroxy-3-methylglutaryl coenzyme A) reductase and significantly reduce risk for cardiovascular and cerebrovascular diseases (Endres, 2005, 2006; Vaughan, 2003). Beneficial effects of statins in cardiovascular and cerebrovascular systems are due to their antiexcitotoxic, antioxidant, and anti-inflammatory properties (Table 7.1). Statins are commercially available and include lovastatin and pravastatin (naturally occurring statins); simvastatin (semisynthetic statins); and atorvastatin, fluvastatin, cerivastatin, rosuvastatin, and pitavastatin (synthetic statins). Structural differences among statins (Figs. 7.1 and 7.2) determine their lipophilicity, half-lives, and potency in mammalian tissues. Because of their lipophilicity simvastatin, lovastatin, and cerivastatin pass the blood–brain barrier. In contrast, pravastatin, fluvastatin, and atorvastatin do not pass the blood–brain barrier (Vuletic et al., 2006). In addition to inhibiting HMG-CoA

**Table 7.1** Statins, their commercial names, and pleiotropic effects

| Generic name | Trademark | Pleiotropic effects | $IC_{50}$ (nM) | References |
|---|---|---|---|---|
| Atorvastatin | Lipitor | Antioxidant/anti-inflammatory | 8.0 | Amarenco (2005), Endres (2005) |
| Lovastatin | Mevacor, Altocor | Antioxidant/anti-inflammatory, antithrombotic | – | Amarenco (2005), Endres (2005) |
| Cerivastatin | Lipobay, Baycol | Antioxidant/anti-inflammatory, antithrombotic | 10.0 | Rajanikant et al. (2007), Vaughan (2003) |
| Fluvastatin | Lescol RXL | Antioxidant/anti-inflammatory | 28.0 | Endres (2005), Vaughan (2003) |
| Mevastatin | Compactin | Antioxidant/anti-inflammatory, antithrombotic | 23.0 | Amarenco (2005), Vaughan (2003) |
| Pitavastatin | Livalo, Pitava | Antioxidant/anti-inflammatory, antithrombotic | – | Amarenco (2005), Vaughan (2003) |
| Pravastatin | Pravachol | Antioxidant/anti-inflammatory, antithrombotic | – | Rajanikant et al. (2007), Vaughan (2003) |
| Rosuvastatin | Crestor | Antioxidant/anti-inflammatory, antithrombotic | 5.0 | Endres (2005), Vaughan (2003) |
| Simvastatin | Zocor, Lipex | Antioxidant/anti-inflammatory, antithrombotic | 11.0 | Endres (2005), Vaughan (2003) |

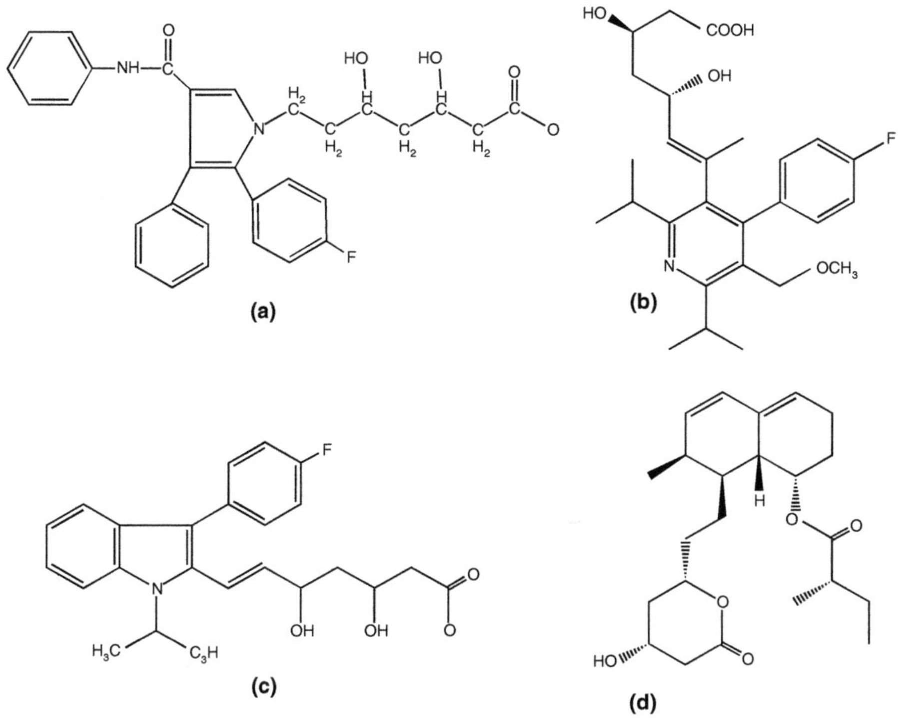


**Fig. 7.1** Chemical structures of statins. Atorvastatin (Lipitor) (**a**); cerivastatin (Baycol) (**b**); fluvastatin (lescol RXL) (**c**); and mevastatin (compactin) (**d**)

reductase, statins not only modulate activities of other enzymes (nitric oxide synthases, PtdIns 3-kinases, and metalloproteinases) but also modulate gene expression and have a variety of "pleiotropic effects" in visceral and brain tissues (Johnson-Anuna et al., 2005, 2007; Kirsch et al., 2003). The pleiotropic effects include modification of endothelial cell function, immunoinflammatory responses, smooth muscle cell activation, proliferation, and stabilization of atherosclerotic plaques.

Treatment of rats with atorvastatin and simvastatin 1 day after TBI and daily for 14 days not only improves spatial learning on days 31–35 after onset of TBI but also reduces the neuronal loss in hippocampal CA3 region, lowers microglial activation, and decreases TBI-mediated increases in β-amyloid (Aβ) (Lu et al., 2007; Abrahamson et al., 2009). In addition, statin treatment enhances neurogenesis in the dentate gyrus, augments TBI-induced angiogenesis, and reduces cortical apoptosis (Lu et al., 2007). Although the molecular mechanism associated with statin-induced effects is not known, recent studies have indicated that simvastatin modulates neural activities in several ways. It (a) stimulates phosphorylation of v-akt murine thymoma viral oncogene homolog (Akt), glycogen synthase kinase-3β (GSK-3β), and cAMP response element-binding proteins (CREB); (b) upregulates the expression of BDNF and VEGF in the dentate gyrus; (c) increases in the cell proliferation and

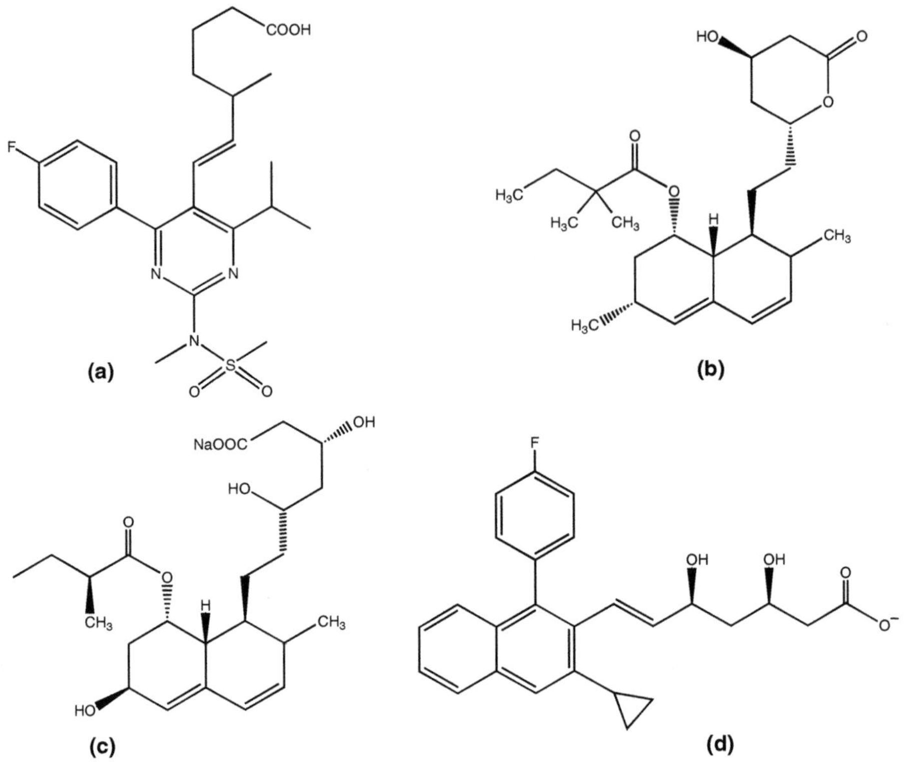


**Fig. 7.2** Chemical structures of statins. Rosuvastatin (Crestor) (**a**); simvastatin (Zocor) (**b**); pravastatin (Pravachol) (**c**); and pitavastatin (**d**)

differentiation in the dentate gyrus; and (d) enhances the recovery of spatial learning (Wu et al., 2008). Furthermore, statins may also modulate apoptosis. This possibility is supported by the observation that the ratio of Bax/Bcl-2 is significantly reduced in simvastatin-treated animals, favoring an antiapoptotic state (Lu et al., 2007). Similar beneficial effects on locomotor outcome have also been described in spinal cord injury (SCI), with authors attributing the neuroprotection to effects on endothelial dysfunction (Maas, 2001). In injured animals, simvastatin administration also attenuates TLR4/NF-κB-mediated inflammatory response in the injured rat brain (Chen et al., 2009). Collective evidence suggests that the neurorestorative and neuroprotective effects of statin may be mediated through activation of the Akt-mediated signaling pathway, upregulation of growth factor expression (BDNF), and induction of neurogenesis in the dentate gyrus. Statins produce antiexcitotoxic, anti-inflammatory, and antioxidant effects in brain. In addition, statins also affect microvasculature by increasing nitric oxide bioavailability, which regulates cerebral perfusion and improves endothelial function. These processes may lead to restoration of cognitive function and improved outcome after TBI in rats (Bösel et al., 2005; Nakazawa et al., 2007; Wu et al., 2008; Mahmood et al., 2009a).

### *7.3.2 Progesterone and TBI*

Progesterone is a C-21 steroid hormone associated with the female menstrual cycle, pregnancy, and embryogenesis of humans and mammals. It occurs in small amounts in both male and female brains, where it is synthesized de novo by glial cells. Human brain tissue is loaded with progesterone receptors (PRs). Progesterone is not only critical for the normal development of neurons but also associated with the regulation of cognition, mood, inflammation, mitochondrial function, neurogenesis, regeneration, and myelination (Brinton et al., 2008). Progesterone action is mediated by a progesterone receptor (PR), which occurs in two isoforms, namely PR-A and PR-B. In the absence of progesterone, PRs form complex with several chaperone molecules, such as heat shock protein (Hsp) 90, Hsp70, and Hsp40. The interaction of PRs with the chaperones is a prerequisite for hormone binding (Pratt, 1998). The binding of PR with chaperone links it with protein trafficking systems. In classical response, the binding of progesterone produces conformational changes in PR resulting in the dissociation of PR with the chaperone proteins. Chaperone protein-free PR dimerizes and directly interacts with specific response elements (PREs) in the promoters of target genes (Leonhardt et al., 2003; Brinton et al., 2008). PREs-bound PRs interact with components of the basal transcription machinery through steroid receptor co-activators. These co-activators bind to PR via a conserved LXXLL amphipathic helix or nuclear receptor box motifs, which make initial contacts with several helices in the AF-2 (activation function) region of the PR ligand-binding domain (McKenna and O'Malley, 2002; Brinton et al., 2008). Detailed investigations indicate that human PR has a polyproline motif in the amino-terminal domain that interacts with the SH3 domain of Src and mediates rapid stimulation of c-Src and downstream MAPK (Erk-1/-2) independent of the transcriptional activity of PR (Boonyaratanakomkit et al., 2008). The activation of neural PRs is much more complex and diverse than previously realized. Four distinct classes of molecules, neurotransmitters, peptide growth factors, cyclic nucleotides, and neurosteroids activate the PRs via cross talk and pathway convergence. In addition, rapid signaling events associated with membrane receptors and/or subpopulations of cytoplasmic PRs, via activation of protein kinase cascades, regulate PR gene expression in the cytoplasm independent of PR nuclear action. The increasing in vitro and in vivo evidence of differential transcriptional activities and coregulator interactions between PR-A and PR-B predicts that these isoforms could have distinct roles in mediating additional and/or alternate signaling pathways within steroid-sensitive neurons (Mani, 2008). The non-classical PR responses occur through alternative genomic mechanisms, such as PR tethering to the SP1 transcription factor (Owen et al., 1998) or nongenomic mechanisms, such as activation of second messenger signaling cascades (Nilsen and Brinton, 2002; Brinton et al., 2008). Progesterone and its metabolites also modulate neuronal excitability by interacting with the inhibitory GABA receptors, as well as by modulating other neurotransmitter receptors, including serotonin, glycine, nicotinic acetylcholine, and kainate receptors (Rupprecht and Holsboer, 1999).

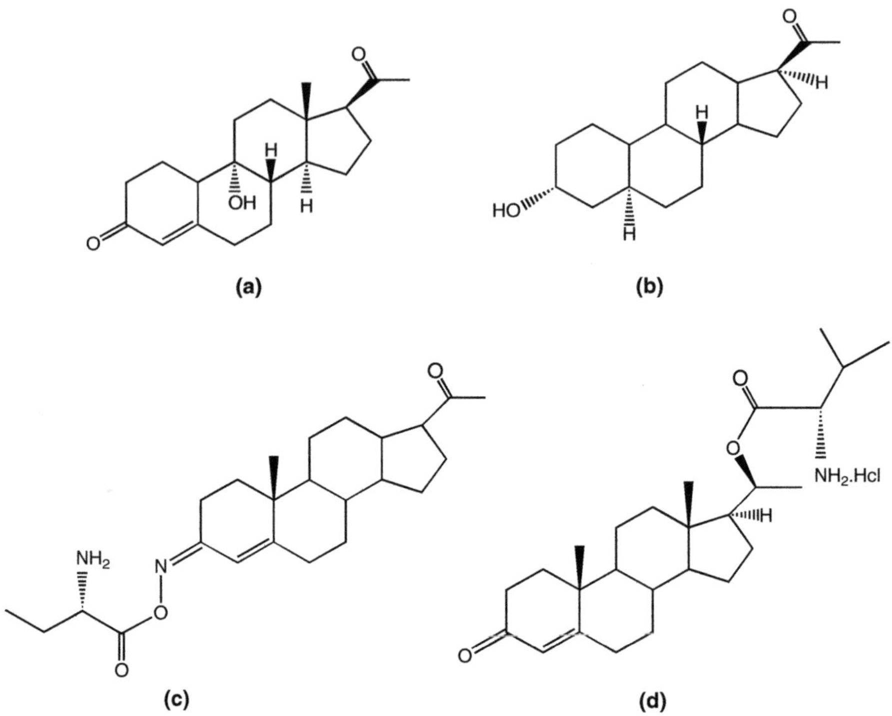


**Fig. 7.3** Chemical structures of progesterone, allopregnanolone, oxime derivative of progesterone, and valine tethered progesterone analogs. Progesterone (**a**); allopregnanolone (**b**); oxime derivative of progesterone (**c**); and valine tethered progesterone analog (**d**)

Progesterone and its metabolite allopregnanolone (Fig. 7.3) attenuate pathophysiological events associated with TBI in young adult rats (Sayeed and Stein, 2009). Thus, administration of progesterone to TBI patients is safe and the rate of mortality among severely injured patients treated with progesterone has been reported to be reduced by over 60% relative to the placebo group. In addition, patients in the moderate group show significantly better functional outcomes at 30 days post-TBI. Amino acid tethering has been used for greatly enhancing the solubility of progesterone (Fig. 7.3) and other related steroidal compounds (He et al., 2004; MacNevin et al., 2009). Progesterone and allopregnanolone act by attenuating the production of proinflammatory cytokines early after TBI, and this may be one mechanism by which progesterone and allopregnanolone reduce cerebral edema and promote functional recovery from TBI (He et al., 2004; De Nicola et al., 2009). These steroids not only reduce the size of glial fibrillary acid protein (GFAP)-positive astrocytes at the lesion site after TBI but also facilitate improved performance in a spatial learning task compared to injured rats given only the vehicle (Djebaili et al., 2005). These results support the view that anti-apoptotic and anti-astrogliotic effects of progesterone and allopregnanolone (Fig. 7.4) may be associated with better cognitive

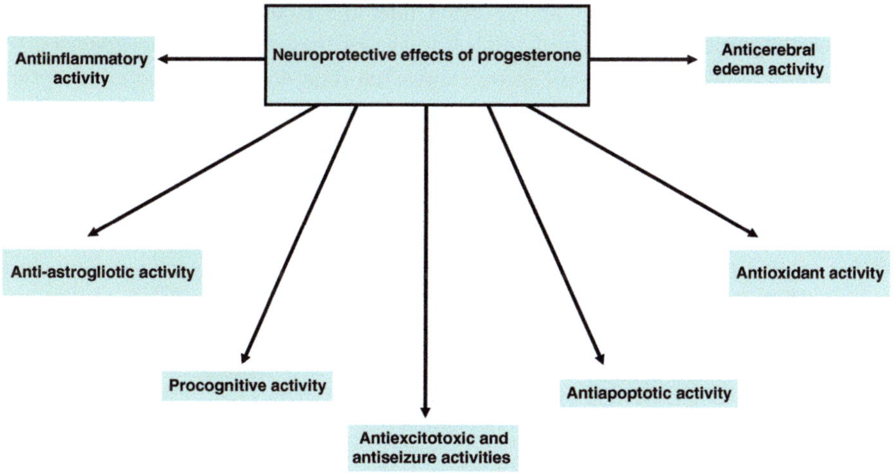


**Fig. 7.4** Progesterone-mediated neuroprotective mechanisms in TBI

performance following TBI. It is suggested that effects of progesterone may be due to the regulation of myelin synthesis in glial cells and also due to direct actions on neuronal function (Hu et al., 2009; Atif et al., 2009; Cekic et al., 2009a). In addition, neurotrophins have also been proposed as possible mediators of hormone action. Progesterone treatment increases the expression of brain-derived neurotrophic factor (BDNF) at both the mRNA and protein levels in ventral horn motor neurons from rats with spinal cord injury (SCI) (Gonzalez et al., 2004). It is also stated that progesterone exerts its neuroprotective effects by protecting or rebuilding the blood–brain barrier, downregulating the inflammatory cascade, and limiting cellular necrosis and apoptosis (He et al., 2004; MacNevin et al., 2009; Sayeed and Stein, 2009; Hu et al., 2009; Atif et al., 2009; Cekic et al., 2009a). These effects are mediated through classical nuclear receptors, extra nuclear receptors, and membrane receptors. Progesterone also reduces membrane lipid peroxidation after TBI, indicating that it acts as an antioxidant and reduces oxidative stress (Roof et al., 1992, 1997). This effect on oxidative stress has been confirmed in tissue culture as well as in an in vitro stretch model of TBI. In both cases, progesterone retards oxidative stress as reflected by 2-thiobarbituric acid, cytochrome oxidase, or manganese superoxide dismutase levels. Inhibition of inflammation by progesterone after TBI (Pettus et al., 2005) may also contribute to the widely observed beneficial effects of the hormone on edema. Thus, administration of progesterone after brain injury attenuates edema in both female and male animals (Pettus et al., 2005; O'Connor et al., 2005) irrespective of estrogen levels. Progesterone treatment also modulates gene expression in injured animals. Thus, progesterone treatment not only downregulates bax and bad mRNA as well as Bax and Bad protein levels in the cerebral cortex of injured rats, but also upregulates expression of bcl-2 and bcl-x(L) mRNA and protein levels injured cortex tissue (Yao et al., 2005). Under the sham-treated condition, progesterone significantly increased mRNA levels of the anti-apoptotic gene, bcl-2, but downregulated pro-apoptotic gene expression (bax and bad) in cerebral cortex (Yao et al., 2005). Collective evidence suggests that progesterone and its analogs promote neuroregeneration not only by reducing inflammation, swelling, and apoptosis but also by increasing the BDNF-mediated survival of neurons, rebuilding of blood–brain barrier, improving vascular tone and by promoting the formation of new myelin sheaths. Progesterone upregulates GABA, reduces excitotoxicity and seizure activity, and modulates hemostatic proteins. In addition, progesterone suppresses TLRs/NF-kB signaling pathway, which is remarkably upregulated following TBI (Chen et al., 2008a). These observations support the view that progesterone and its analogs can be used as potential therapeutic agents in experimental and human TBI (He et al., 2004; Schumacher et al., 2008; MacNevin et al., 2009; Sayeed and Stein, 2009).

1,25-Dihydroxyvitamin $D_3$ hormone (VDH) (Fig. 7.5) is another steroid ring containing compound that in combination with progesterone show better neuroprotection than progesterone alone following excitotoxic neuronal injury in vitro. Vitamin D-deficient animals have increased levels of proinflammatory cytokines (TNF-α, IL-1β, IL-6, and NF-κB p65) in the brain even without injury. Vitamin D-deficient rats with TBI show increased neuroinflammation and greater open-field

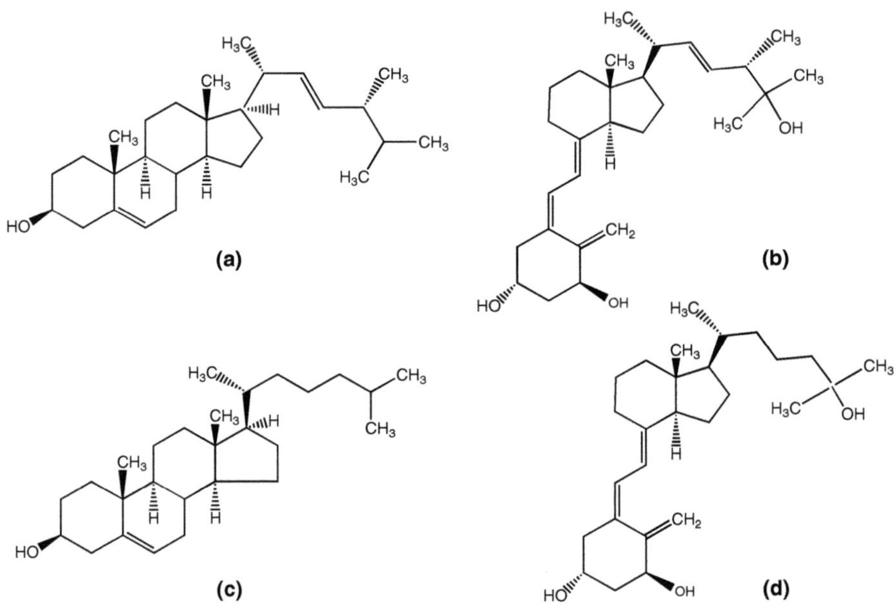


**Fig. 7.5** Chemical structures of vitamin D-related metabolites. Ergosterol (**a**); 1,25-dihydroxyvitamin $D_3$ (**b**); 7-dehydrocholesterol (**c**); and 25-hydroxyvitamin $D_3$

behavioral deficits compared to vitamin D-normal animals after progesterone treatment (Cekic et al., 2009b). Progesterone administration is beneficial for injured vitamin D-normal animals, but in vitamin D-deficient animals, progesterone treatment confers no improvement over vehicle. Supplemental dose of VDH with the first progesterone treatment dramatically improves results in vitamin D-deficient rats, but treatment with VDH alone has no effect. It is suggested that vitamin D-deficiency increases baseline brain inflammation, exacerbates the effects of TBI, and attenuates the benefits of progesterone treatment. These effects may be reversed if the deficiency is corrected (Cekic et al., 2009b). Although the molecular mechanism associated with beneficial effects of VDH is not fully understood, it is shown that VDH confers neuroprotection in parallel with downregulation of L-type calcium channel (VSCC) expression in hippocampal neurons (Brewer et al., 2001). This suggestion is supported by electrophysiological and real-time PCR studies, which indicate that VDH monotonically downregulate mRNA expression for the $alpha_{1C}$ and $alpha_{1D}$ pore-forming subunits of L-VSCCs (Brewer et al., 2001).

## *7.3.3 Erythropoietin and TBI*

Erythropoietin (Epo) is a glycoprotein hormone synthesized by the kidneys in response to hypoxia. It has a molecular mass of 30.4 kDa and is considered as a hematopoietic growth factor that stimulates the production of red cells in bone marrow and has been used for the treatment of anemia in humans (Eckartdt and Kurtz, 2005). Epo is an essential growth and survival factor for erythroid progenitor cells. It acts through a specific erythropoietin receptor (Epo-R) on the surface of red cell precursors in the bone marrow and facilitates their transformation into mature red blood cells. Epo production is inversely proportional to oxygen availability, so that an effective feedback loop is established, which controls erythropoiesis. As a result, the oxygen level in blood reaching the kidney rises and the amount of Epo generation decreases. The mechanisms modulating the expression of EPO encoding gene are exemplary for oxygen-regulated gene expression. In mammals, hypoxia modulates Epo levels by increasing expression of the Epo gene. Recent studies have led to the identification of a widespread cellular oxygen-sensing mechanism. Central to oxygen-sensing mechanism is the transcription factor complex hypoxia-inducible factor (HIF)-1. The abundance and activity of HIF-1, a heterodimer of an α- and β-subunit, is predominantly regulated by oxygen-dependent post-translational hydroxylation of the α-subunit (Eckartdt and Kurtz, 2005; Stockmann and Fandrey, 2006; De Spiegelaere et al., 2009). Non-heme ferrous iron containing hydroxylases uses dioxygen and 2-oxoglutarate to specifically target proline and an asparagine residue in HIF-1α. Three prolyl hydroxylases (PHD1, PHD2, and PHD3) and asparagyl hydroxylase (factor inhibiting HIF (FIH)-1) have been reported to act as cellular oxygen sensors. In addition to erythropoiesis, HIF-1 regulates a broad range of physiologically relevant genes involved in angiogenesis, apoptosis, vasomotor control, and energy metabolism (Eckartdt and Kurtz, 2005; De Spiegelaere et al., 2009).

Epo and erythropoietin receptors (Epo-R) are expressed in the nervous system. Epo interacts through Epo-R and induces non-hematopoietic effects. Neuronal expression of Epo and Epo-R peaks during brain development and is upregulated in the adult brain after injury. Peripherally administered Epo, and at least some of its variants, crosses the blood–brain barrier, stimulates neurogenesis and neuronal differentiation, and activates brain neurotrophic, anti-apoptotic, anti-oxidant and anti-inflammatory signaling (Siren et al., 2009). Delayed post-traumatic administration of Epo significantly improves histological and long-term functional outcomes compared with saline treatment in rats after TBI. The triple doses of delayed Epo treatment induces better histological and functional outcomes in rats, although a single dose provided substantial benefits (Xiong et al., 2009). Studies on the treatment of injured rat with recombinant Epo, carbamylated erythropoietin (CEpo), and asialoerythropoietin (ASEpo) indicate that Epo and CEpo are equally effective in enhancing spatial learning and promoting neural plasticity after TBI (Mahmood et al., 2009b).

Although the molecular mechanisms associated with therapeutic action of Epo remain unclear, it is well known that cerebral inflammation, excitotoxicity, oxidative stress play an important role in the pathogenesis of secondary brain injury after TBI (Farooqui and Horrocks, 2009) and levels of NF-κB, proinflammatory cytokines, and ICAM-1 are markedly increased in all injured animals (Chen et al., 2009). Rats receiving recombinant human erythropoietin (rhEpo) post-TBI show considerable decrease in NF-κB, IL-1β, TNF-α, and ICAM-1 levels compared to vehicle-treated animals. No changes are observed in IL-6 levels after rhEpo treatment. Furthermore, administration of rhEpo also reduces brain edema, blood–brain barrier permeability, and apoptotic cells in the injured brain. Accumulating evidence suggests that post-TBI rhEpo administration may attenuate inflammatory response in the injured rat brain, and this may be one mechanism by which rhEpo improves outcome following TBI (Chen et al., 2009). Another mechanism of neuroprotection by Epo may involve prevention of $Zn^{2+}$-mediated toxicity. It is well known that $Zn^{2+}$ plays a key role in excitotoxicity-mediated neural cell injury. Injections of recombinant human Epo (rhEpo) 30 min after TBI in rats dramatically protect neuronal death, suggesting that rhEpo can significantly reduce the pathological $Zn^{2+}$ accumulation in rat hippocampus after TBI as well as zinc-induced cell death in cultured cells (Zhu et al., 2009). In a cryogenic model of cortical brain injury (Grasso et al., 2007), Epo administration significantly decreases vasogenic brain edema, attenuate blood–brain barrier breakdown, reduces lesion volume, and ameliorate motor dysfunction. Similarly, following TBI, Epo administration increases the neuronal density in the CA1 and CA3 region of the hippocampus and significantly reduces the total contusion volume when administered within 6 h of injury (Cherian et al., 2007). Mice lacking Epo or Epo-R exhibit increased neural cell apoptosis during development before embryonic death due to severe anemia (Noguchi et al., 2007). Collectively, these studies suggest that Epo facilitates neurorestoration not only by enhancing cytoprotection and neurogenesis through activation of multiple signaling pathways but also by blocking apoptosis, reducing inflammation, restoring vascular integrity, and subsequently improving sensorimotor and spatial learning function (Xiong et al., 2008; Matis and Birbillis, 2009).

### 7.3.4 Minocycline and TBI

As stated in Chapter 5, minocycline is a lipophilic second-generation tetracycline analog that crosses blood–brain barrier and produces neuroprotective in animal models of acute neural trauma and neurodegenerative diseases. Although molecular mechanisms associated with beneficial effects of minocycline are not fully understood, it is reported that minocycline may block mitochondrial permeability transition-mediated cytochrome *c* release from mitochondria, inhibit caspase-1 and -3 expressions, upregulate iNOS, inhibit NADH-cytochrome *c* reductase and cytochrome *c* oxidase activities, and then suppress microglial activation (Garcia-Martinez et al., 2010). In addition, minocycline also blocks expression and activities of phospholipase $A_2$ ($PLA_2$), cyclooxygenase-2 (COX-2), 5-lipoxygenase (LOX), MMP-2, MMP-9, and p38 mitogen-activated protein kinase (MARK) and decreases the expression of c-fos in brain (Hua et al., 2005; Machado et al., 2006; Marchand et al., 2009). These enzymes are involved in nociception, neuroinflammation, and apoptotic cell death. Minocycline reduces the number of reactive astrocytes and augment survival of oligodendrocytes in the spared white matter. Thus, minocycline is a multifaceted therapeutic agent that has proven clinical safety and efficacy during a clinically relevant therapeutic window. It can be effective in treating acute SCI. Because of the high tolerance and the excellent penetration through blood–brain barrier, minocycline has been used for the treatment of many neurological disorders, including stroke, multiple sclerosis, TBI, SCI, amyotrophic lateral sclerosis, Huntington disease, and Parkinson disease (Kim and Suh, 2009).

Cerebral edema, microglial activation, and thrombin formation are important complications of TBI. They contribute to brain injury after intracerebral hemorrhage and should be treated to prevent further brain damage. Minocycline administration not only reduces cerebral edema but also downregulates inflammatory markers at 6 h post-TBI without effecting TBI-induced oxidized glutathione increases. The anti-edematous effect of minocycline persists up to 24 h and is accompanied by a neurological recovery (Homsi et al., 2009). Minocycline decreases thrombin-mediated increase in TNF-$\alpha$ and IL-1$\beta$ levels. In vivo, minocycline reduces neurological deficits and brain atrophy (Wu et al., 2009). Studies on gene expression patterns of sham TBI and minocycline-treated brain TBI indicate that many genes are modulated by minocycline treatment and significant differences are observed in genes modulating chemokines, proinflammatory cytokines, and genes involved in cell surface receptor-linked signal transduction. Expression levels of some key genes are validated by real-time quantitative PCR and it is suggested that multiple regulatory pathways are affected following brain injury and these genes are affected by minocycline following brain injury (Crack et al., 2009). Collective evidence suggests that minocycline is safe and effectively penetrates the blood–brain barrier and provides neuroprotection in TBI through its anti-inflammatory and anti-apoptotic effects and protease inhibition properties.

## *7.3.5 PPARα Agonist and TBI*

Fenofibrate (propan-2-yl 2-[4-(4-chlorobenzoyl)phenoxy]-2-methylpropanoate), a fibric acid derivative (Fig. 7.6), mainly exerts its effect via the activation of specific nuclear receptor called peroxisome proliferator-activated receptor alpha (PPARα). This PPARα agonist is primarily used to decrease the cholesterol levels in cardiovascular diseases patients. Like statins, fenofibrate also reduces triglycerides and low- and very low density protein levels. It also increases high-density lipoprotein levels in the body. Fenofibrate also has nonlipid (i.e., pleiotropic) effects (reduction in fibrinogen, C-reactive protein, and uric acid levels and improvement in the flow-mediated dilatation).

Fenofibrate also reduces TBI-mediated neurological deficits, the edema, and the cerebral lesion (Chen et al., 2007). Fenofibrate promotes neurological recovery by exerting anti-inflammatory effect as evidenced by downregulation in

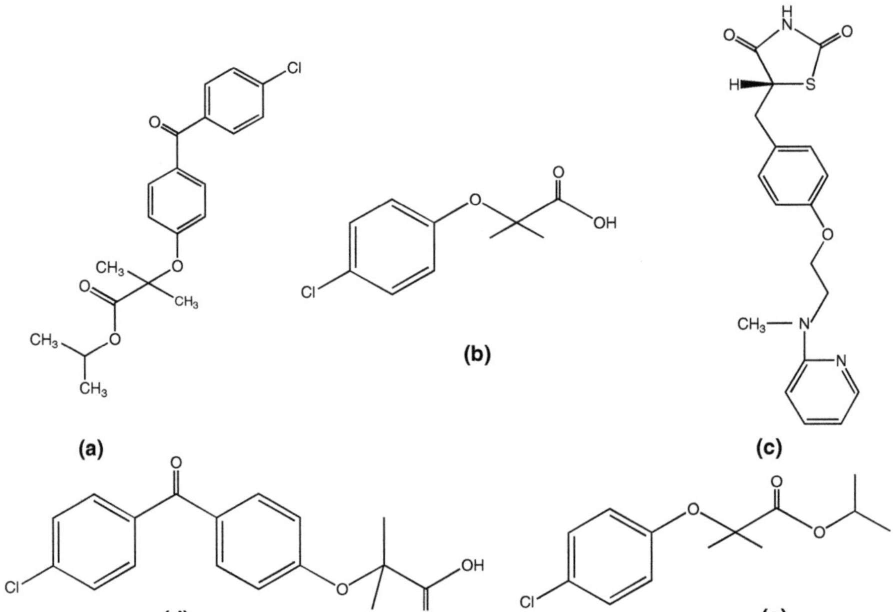


**Fig. 7.6** Chemical structures of fibric acid and its derivatives. Fenofibrate (**a**); clofibric acid (**b**); rosiglitazone (**c**); fenofibric acid (**d**); and clofibrate (**e**)

expression of iNOS, COX-2, and MMP9 activities. In addition, fenofibrate also shows antioxidant effect as demonstrated by decrease in markers of oxidative stress, such as loss of glutathione, glutathione oxidation ratio, 3NT, and 4HNE staining. Fibrates modulate many cellular activities (Fig. 7.7). Thus, they induce the expression of I-κB, the cytoplasmic inhibitory protein of NF-κB (Delerive et al., 2002). This process involves a PPARα-dependent mechanism. Fibrates downgrade the expression of the transcription factor YY1 (CCATT), an enhancer-binding protein β that facilitates the expression of the interleukin 6 gene (Gervois et al., 2004). Fibrates also decrease the activation of the transcription factor c-jun by preventing its phosphorylation and finally fibrates block the activity of the promoter of adhesion molecule-encoding gene transrepression (without binding to the DNA) (Marx et al., 1999) (Fig. 7.7). These observations suggest that PPARα activation may mediate many pleiotropic effects that may be responsible for better recovery from experimental TBI (Chen et al., 2007). The combinations of simvastatin and fibrate synergistically enhance PPARα activation as well as prolong beneficial effects of improving functional outcome in experimental TBI than each alone (Chen et al., 2008b).

Rosiglitazone (RSG) (Fig. 7.6) is another PPARγ agonist that has been used to reduce inflammation and provide neuroprotection in experimental models of ischemia, intracerebral hemorrhage, and surgical brain injury (SBI) (Hyong et al., 2008). SBI can cause postoperative complications such as brain edema after

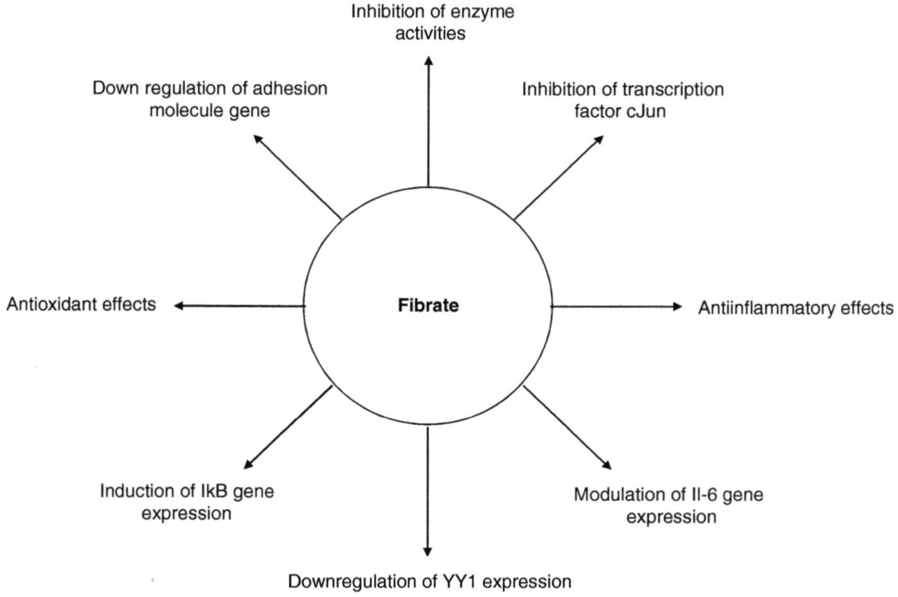


**Fig. 7.7** Modulation of neurochemical activities by fibrate

blood–brain barrier (BBB) disruption and inflammation localized along the periphery of the site of surgical resection. Although RSG attenuates inflammatory changes, it has no effect on brain edema, BBB disruption, and neurological outcomes after SBI (Hyong et al., 2008). RSG also blocks the expression of CD40, TNF-α, and microglial activation in different regions of hippocampus. RSG prevents neuronal loss in the CA1 area after lithium pilocarpine-induced status epilepticus (SE). The protective effects of RSG are significantly reversed by the cotreatment with T0070907, a selective antagonist of the PPARγ, supporting the involvement of a PPARγ-dependent mechanism. Based on these results, it is suggested that RSG attenuates inflammatory responses after SE by suppressing CD40 expression and microglial activation (Sun et al., 2008).

## 7.3.6 Endocannabinoids and TBI

Endocannabinoids include arachidonylethanolamine, noladin, arachidonyldopamine, 2-arachidonylglycerol (2-AG), and arachidonylethanolamide (anandamide) (Fig. 7.8). 2-AG and anandamide are derived from the non-oxidative metabolism of arachidonic acid (ARA). 2-AG and anandamide are synthesized through

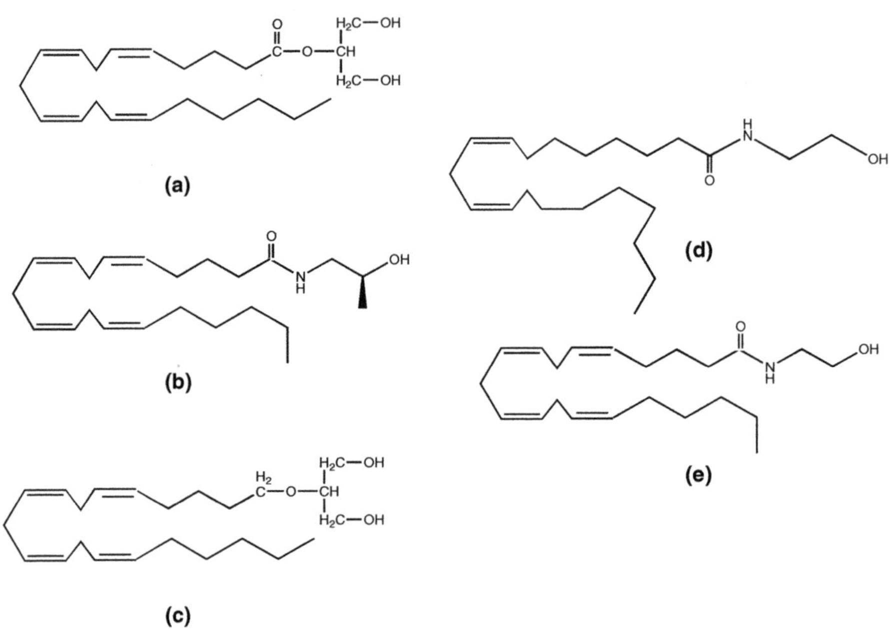


**Fig. 7.8** Structures of some cannabinoid receptor agonists. 2-Arachidonylglycerol (**a**); anandamide (**b**); noladin ether (**c**); homo-γ-linolenylethanolamide (**d**); and docosatetraenoyl ethanolamide (**e**)

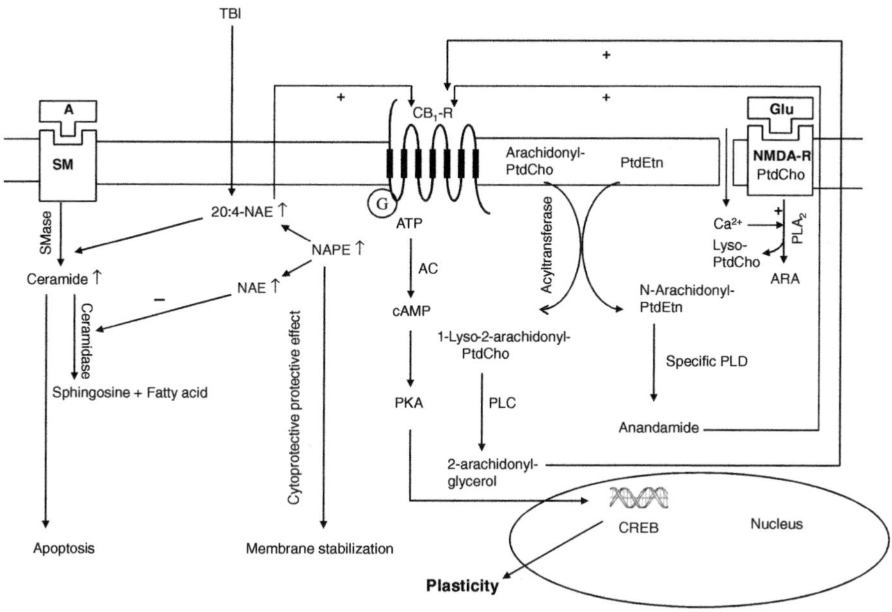


**Fig. 7.9** Generation of *N*-acylethanolamine (NAE) and *N*-acylphosphatidylethanolamine (NAPE) in brain. Phosphatidylcholine (PtdCho); phosphatidylethanolamine (PtdEtn); cannabinoid receptor1 ($CB_1$-R); *N*-methyl-D-aspartate receptor (NMDA-R); anandamide and 2-arachidonyl-gltcerol not only stimulate $CB_1$-R but also have stabilizing effects on neural membranes. TBI increases the formation of NAE and NAPE. *N*-arachidonylethanolamine stimulate ceramide formation, *N*-acylethanolamine inhibit ceramidase. Ceramide induces apoptosis. *Plus sign* indicate stimulation and *minus sign* indicates inhibition. (↑) Indicate increase in levels

two distinct pathways (Fig. 7.9). Transfer of ARA from sn-1 position of 1,2-arachidonyl-PtdCho to the N-position of PtdEtn results in the generation of 1-lyso-2-arachidonyl-PtdCho and *N*-arachidonyl-PtdEtn. This reaction is catalyzed by a $Ca^{2+}$-dependent, membrane-associated *N*-acyltransferase. 1-Lyso-arachidonyl-PtdCho is converted to 2-AG by PLC and *N*-arachidonyl-PtdEtn is transformed into anandamide by *N*-acylphosphatidylethanolamine-specific PLD (NAPE-PLD), a member of the metallo-β-lactamase family, which specifically hydrolyzes *N*-acylphosphatidylethanolamine among glycerophospholipids, and appears to be constitutively active (Di Marzo et al., 1996; Ueda et al., 2005) (Fig. 7.9). An alternative pathway for the synthesis of 2-AG involves the hydrolysis of 1,2-arachidonyl-PtdCho by PLC, followed by the action of DAG-lipase on 1-acyl-2-arachidonylglycerol. Two types of cannabinoid receptor ($CB_1$ and $CB_2$) have been reported to occur in mammalian tissues. The $CB_1$ receptors are abundantly expressed in the brain, whereas $CB_2$ receptors are limited to lymphoid organs. 2-AG and anandamide nonselectively bind to both $CB_1$ and $CB_2$ receptors and act as neurotransmitter or neuromodulators in the brain, immune, and cardiovascular systems. Endocannabinoids modulate brain function through cannabinoid receptor-dependent and cannabinoid receptor-independent mechanisms. Receptor-dependent mechanisms include modulation of protein kinases (Childers and Breivogel, 1998), whereas receptor-independent mechanisms involve modulation of ion channels. Thus, $CB_1$ and $CB_2$ receptors are coupled to adenylyl cyclase through heterotrimeric Gi/o-proteins. Generation of cAMP initiates CREB phosphorylation at serine 133 that is located at the upstream element TGACGTCA of gene encoding c-fos protein. In pharmacologically relevant concentrations, endocannabinoids modulate the functional properties of voltage-gated ion channels, including P/Q-type $Ca^{2+}$ channels, $Na^+$ channels, and inwardly rectifying $K^+$ channels, and ligand-gated ion channels such as 5-$HT_3$ and nicotinic ACh receptors (Oz, 2006).

As stated in Chapter 6, significant decrease in phospholipids occurs following TBI (Homayoun et al., 1997, 2000) except in *N*-acylethanolamine phospholipids (NAPE) and *N*-acylethanolamine (NAE), which are markedly increased following TBI (Hansen et al., 2001a, b). The generation of NAPEs and NAEs from PtdEtn may be endogenous neuroprotective mechanism. This process induces membrane stabilizing effects resulting in endocannabinoid receptor-mediated decrease in pain and increase in neuroprotection (Fig. 7.8). Several NAE are synthesized in brain tissue. They produce neuroprotective effects by (a) inhibiting necrosis, (b) enhancing apoptosis, and (c) blocking the release of mediators that promote necrosis and inflammation (Hansen et al., 2002).

Dexanabinol (also known as HU-211) is a nonpsychotropic analog of tetrahydrocannabinol and cannabinoid NMDA receptor antagonist that has a number of neuroprotective properties. It produces beneficial effects in severe closed head injury, ischemia, and nerve crush injury (Shohami et al., 1993; Feigenbaum et al., 1989). It not only interacts with cannabinoid receptor but also acts as a week NMDA receptor antagonist (Feigenbaum et al., 1989). It is capable of scavenging free radicals and inhibiting cytokine TNF-α (Eshhar et al., 1995). Dexanabinol can cross the blood–brain barrier rapidly and weakly blocks NMDA receptors by interacting with a site close to, but distinct from, that of uncompetitive NMDA antagonists (Eshhar et al., 1995). The beneficial effects of dexanabinol are due to uncompetitive NMDA receptor antagonistic activity. By inhibiting the NMDA receptor, it blocks calcium influx, which results in inhibiting calcium-induced proteolysis and lipolysis. As stated above, dexanabinol is an antioxidant that has ability to block the synthesis of TNF-α and other inflammatory cytokines both in vitro and in vivo settings (Shohami et al., 1997). This property may contribute to the attenuation of blood–brain barrier permeability after injury, with a consequent reduction in edema formation. Collective evidence suggests that dexanabinol is a multifactorial drug that has been used for phase II and III trials in human TBI (Knoller et al., 2002; Maas et al., 2006). Phase II trials indicate that dexanabinol is a safe drug that can be well tolerated by severe TBI patients. Treatment not only results in better control of intracranial pressure/cerebral perfusion pressure without jeopardizing blood pressure but also faster and better neurologic outcome (Knoller et al., 2002). Unfortunately in larger phase III trials on dexanabinol have failed, but it is conclusively demonstrated that dexanabinol is a safe drug (Maas et al., 2006).

### 7.3.7 Thyrotropin-Releasing Hormone (TRH) and TBI

Most of above drugs are directed toward a single pathophysiological mechanism of TBI. Since pathogenesis of TBI involves multiple pathogenic processes, it is suggested that multifunctional drugs that target multiple injury mechanisms, particularly those that occur later after the insult may be useful for the treatment of TBI (Stoica et al., 2009). Thyrotropin-releasing hormone (TRH) is one of multifunctional hormone that stimulates the release of thyroid-stimulating hormone and prolactin from the anterior pituitary. It acts through two G protein-coupled receptors for TRH (namely, TRH-R1 and TRH-R2) (Monga et al., 2008), which are distributed differently in the brain and peripheral tissues, but exhibit indistinguishable binding affinities for TRH and TRH analogs. TRH inhibits multiple secondary injury processes, including declines of blood flow and bioenergetics, lipid degradation products, such as peptidyl leukotriene and platelet-activating factor, ionic dyshomeostasis ($Na^+$, $K^+$, $Ca^{2+}$, and $Mg^{2+}$), endogenous opioids, and excitotoxins (Faden et al., 1999; Stoica et al., 2009). In addition, it is shown that TRH analogs that modify either the N-terminal or the middle amino acid of the tripeptide hormone have longer half-lives and are more effective in neuroprotection than TRH. These analogs are highly effective in improving functional recovery and reducing lesion volume after experimental SCI or TBI (Faden et al., 1999; Stoica et al., 2009). Diketopiperazines are structurally related to the TRH metabolites that reduce neuronal cell death primary cell cultures (Faden et al., 2005). These cyclic dipeptides not only protect against glutamate toxicity and Aβ-induced injury but also strongly block glutamate-mediated increase in intracellular calcium. Injections of cyclic peptide produce highly significant improvement in motor and cognitive recovery after controlled cortical impact CCI and markedly reduce lesion volumes as shown by high field magnetic resonance imaging. DNA microarray studies in rat model of TBI show that treatment with one of these dipeptides after injury significantly downregulates expression of mRNAs for cell cycle proteins, aquaporins, cathepsins, and calpain in ipsilateral cortex and/or hippocampus, while upregulating expression of brain-derived neurotrophic factor, hypoxia-inducible factor, and several heat shock proteins (Faden et al., 2005). Collective evidence suggests that small cyclic peptides provide neuroprotection and neural cell survival by modulating multiple mechanisms as well as their ability to improve functional outcome and reduce post-traumatic lesion size (Faden et al., 2005).

### 7.3.8 Citicoline (CDP-Choline) and TBI

Citicoline (CDP-choline) is an intermediate in PtdCho biosynthesis. It has been used for the treatment of ischemic and head injuries (Andersen et al., 1999; Dempsey and Rao, 2003). It not only restores the concentration of PtdCho following ischemic injury by increasing PtdCho synthesis from diacylglycerol but also blocks the activation of $cPLA_2$ activity (Adibhatla et al., 2002; Adibhatla and Hatcher, 2003). The decrease in $cPLA_2$ activity may lead to a reduction in levels of arachidonic acid and reactive oxygen species, with stabilization of neural membranes. CDP-choline also protects cerebellar granule neurons from glutamate-mediated neurotoxicity (Mir et al., 2003), suggesting that CDP-choline may protect neurons from excitotoxicity. Citicoline brain injury treatment (COBRIT) is a randomized, double-blind, placebo-controlled, multi-center trial of the effects of 90 days of citicoline on functional outcome in patients with complicated mild, moderate, and severe TBI (Zafonte et al., 2009). Citicoline (1000 mg bid) or placebo (bid), administered enterally or orally and functional outcomes have been assessed at 30, 90, and 180 days after the day of randomization. Results of these trials have not been published. Similarly, citicoline has been used in phase III clinical trials for stroke and is being evaluated for the treatment of AD and PD.

### *7.3.9 ω-3 Fatty Acids and TBI*

ω-3 fatty acids have been used for the treatment of experimental TBI (Wu et al., 2003, 2004a, b). Thus, ω-3 fatty acids supplementation inhibits increase in oxidative stress and reduces impairment in learning ability in the Morris water maze test following FPI. Although the molecular mechanism of ω-3 fatty acid action is not clear, it is well known that dietary ω-3 fatty acids not only normalizes levels of BDNF, synapsin I, and CREB but also reduces oxidative damage and restores learning and memory disability (Wu et al., 2003, 2004a; Farooqui, 2009a, b). In contrast, consumption of high saturated fat diet reduces levels of BDNF, compromises neuroplasticity, impairs cognitive function, and aggravates the outcome of TBI (Wu et al., 2004b, 2005). Supplementation of the high-fat diet with vitamin E dramatically retards oxidative damage, normalizes levels of BDNF, synapsin I, and transcription factor, CREB (cAMP response element binding), induced by the consumption of high-fat diet. The molecular mechanism associated with ω-3 fatty acid-mediated modulation of BDNF may involve binding to the cell surface Trk receptors, a family of three receptor tyrosine kinases, each of which can be activated by neurotrophins, such as nerve growth factor (NGF), BDNF, and neurotrophins 3 and 4 (NT3 and NT4) (Huang and Reichardt, 2003; Rao et al., 2007). The cytoplasmic domains of Trk receptors contain several sites of tyrosine phosphorylation that recruit intermediates in intracellular signaling cascades. Trk receptor signaling activates Ras, Rap-1, and the Cdc-42-Rac-Rho family, as well as pathways regulated by MAP kinase, PtdIns 3-kinase, and phospholipase-C-γ-PKC cascade (Huang and Reichardt, 2003). It is likely that ω-3 fatty acid supports neural cell survival and maintenance of neuroplasticity through modulating MAP kinase, PtdIns 3-kinase, and phospholipase-C-γ-PKC cascade. In addition, ω-3 fatty acids are metabolized to docosanoids (resolvins and neuroprotectins), which produce anti-inflammatory, antioxidant, and anti-apoptotic effects in the injured brain. Generation of these metabolites may increase neuronal survival not only by BDNF-mediated neuroplasticity but also by anti-inflammatory, antioxidant, and anti-apoptotic effects of ω-3 fatty acid-derived lipid mediators (Wu et al., 2003, 2004a; Wu et al., 2004b; Vaynman et al., 2004; Wu et al., 2005; Bazan, 2006, 2007; Serhan, 2005a, b; Farooqui, 2009a, b).

### *7.3.10 Hypothermia and TBI*

In warm-blooded animals, core body temperature is maintained constant (35°C) through the process of thermoregulation. Hypothermia is a condition in which body's temperature drops below that required for normal metabolism and body functions. Although hyperthermia is common following TBI and is associated with poor neurological outcomes, hypothermia has emerged as a potentially effective therapy for TBI. The molecular mechanisms associated with hypothermic effects are not clear. However, it is shown that hypothermia decreases endogenous antioxidant consumption and lipid peroxidation after TBI (Sahuguillo and Vilalta, 2009). In CSF, glutathione levels are inversely associated with patient temperature (Bayir et al., 2009). Although $F_2$-isoprostane levels in CSF are approximately threefold lower in patients randomized to hypothermia vs. normothermia, this difference was not statistically significant. It is stated that hypothermic therapy improves survival and the neurologic outcome in animal models of TBI (Sahuguillo and Vilalta, 2009). Hypothermia reduces brain edema and intracranial pressure in TBI patients. In TBI patients, therapeutic hypothermia is performed by cooling of the body to less than 36°C. Therapeutic hypothermia decreases mortality and morbidity and improves long-term outcomes by protecting the brain from secondary brain injury. The most commonly seen benefits of hypothermic temperatures of 32°C to 33°C are a significant reduction in intracranial hypertension and improved cerebral perfusion and oxygenation. However, hypothermic therapy among TBI patients has been very controversial and results have been inconsistent (Hutchison et al., 2008; Grände et al., 2009). In hypothermic therapy, the main problem has been the lack of a systematic methodology to induce and maintain hypothermic conditions. In addition, optimal duration of hypothermic therapy, methodology, and timing for bringing the body at the normal temperature have not been determined. It is also essential to establish velocity and other important parameters needed for rewarming the body. In the rewarming phase, condition of many successfully controlled patients deteriorates and they die (Sahuguillo et al., 2001; Sahuguillo and Vilalta, 2009). The molecular mechanism by which hypothermia provides neuroprotection is multifactorial and includes (a) reduction in brain metabolic rate, (b) modulation in cerebral blood flow, (c) reduction of the critical threshold for oxygen delivery, (d) blockade of excitotoxic mechanisms and inhibition of calcium influx, (e) preservation of protein synthesis, and (f) reduction of brain thermopooling (Sahuguillo and Vilalta, 2009). Following TBI, alterations in intracellular signaling cascades are of great importance because they are associated with the regulation of cellular repair, plasticity, and homeostatic functions. It is reported that the MAPK pathways for ERK and JNK, but not p38, are stimulated soon after TBI in astrocytes in vitro, and that temperature markedly modulates these responses. Hypothermia decreases JNK activation, which is involved in the reduction of caspase-3 expression (Huang et al., 2009). In contrast, hyperthermia activates both ERK and JNK and upregulates expression of cleaved caspase-3. These observations support the involvement of JNK activation in apoptosis after TBI. In addition, hypothermia protects against TNF-α-induced endothelial barrier dysfunction and apoptosis through a MAPK phosphatase-1 (MKP-1)-dependent mechanism (Yang et al., 2009). It is also shown that temperature-dependent modulation of excitotoxic neuronal death is mediated in part by temperature-dependent changes in the synaptic release/translocation of $Zn^{2+}$. Neurodegeneration under hypoglycemic conditions is temperature dependent and is mediated by increased $Zn^{2+}$ release (Shin et al., 2009). Toxic $Zn^{2+}$ accumulation may result from either trans-synaptic $Zn^{2+}$movement and/or cation mobilization from intracellular sites. To gain entry to the cytosol, $Zn^{2+}$ can flux through glutamate receptor-associated channels, voltage-sensitive calcium channels, or $Zn^{2+}$-sensitive membrane transporters, while metallothioneins and mitochondria provide sites of intracellular $Zn^{2+}$ release. $Zn^{2+}$-mediated neurotoxicity involves many signaling pathways, including mitochondrial and extra-mitochondrial generation of ROS, disruption of metabolic enzymic activities, and microglial activation, which ultimately result in the induction of apoptotic and/or necrotic cell death-related processes. It is likely that similar to $Ca^{2+}$ homeostasis, neuronal mitochondria take up $Zn^{2+}$ to maintain cellular $Zn^{2+}$ homeostasis. However, excessive mitochondrial $Zn^{2+}$ sequestration may lead to a marked dysfunction of these organelles, characterized by prolonged ROS generation (Sensi and Jeng, 2004).

## 7.4 Cell Therapy and TBI

Because the adult brain cells have a limited capacity to regenerate at sites of injury, stem cell transplantations may provide enormous potential to replace the lost cells following TBI. Several types of cell lines such as immortalized progenitors cells, embryonic rodent, and human stem cells and bone marrow-derived cells have been successfully transplanted in experimental models of SCI and TBI, resulting in reduced neurobehavioral deficits and attenuation of histological damage (Longhi et al., 2005). For example, transplantation of human neuroteratocarcinoma-derived neuronal (NT2N) cells results in integration and survival of these cells at the injury, but no changes in behavior and histopathological damage (Philips et al., 1999). Among above cell types, neural stem cells are multipotent. Their differentiating progeny give rise to neurons, astrocytes, and oligodendrocytes. The restoration of brain damage and function after TBI will require more than cellular replacement. In the ideal scenario, stem cells implanted in the damaged brain area will differentiate in situ into those cells that have died, integrating properly into functioning brain circuitries and/or will contribute to the repair of axonal damage. This suggests that more studied are required not only on effective following transplantation of stem cells into the injured brain, but also on factors that promote trophic support and manipulate of the local environment to stimulate endogenous neuroprotective/neuroregenerative mechanisms (Longhi et al., 2005; Maegele and Schaefer, 2008). Thus, use of embryonic stem cells can provide repair and regeneration of damaged tissues through the prolonged release of neuroprotective substances in animal model (Jain, 2009), but there are serious safety concerns about the use of such cells in human.

## 7.5 Conclusion

TBI survivors suffer from long-lasting disability, which is mainly related to cognitive deficits. Such deficits include slow information processing, deficits of learning and memory, attention, working memory, and executive functions, associated with behavioral and personality modifications. Earlier studies on the treatment of TBI in patients using NMDA receptor antagonists, opioid receptor antagonists, calcium channel blockers, platelet-activating factor antagonists, gangliosides, aminosteroids (tirilazad mesylate), and antioxidants have failed. Investigators are developing and using new drugs, such as statins, progesterone, erythropoietin, minocycline, PPAR$\alpha$ agonists, thyrotropin-releasing hormone analogs, citicoline, and hypothermia for the treatment of TBI in experimental models. These drugs provide neuroprotection by facilitating and promoting angiogenesis, neurogenesis, and synaptogenesis. Clinical trials of these drugs have been planned. It is expected that the next decade will witness an increasing number of clinical trials that seek to translate preclinical research discoveries to the clinic.

## References

Abrahamson EE, Ikonomovic MD, Dixon CE, DeKosky ST (2009) Simvastatin therapy prevents brain trauma-induced increases in beta-amyloid peptide levels. Ann Neurol 66: 407–414

Adibhatla RM, Hatcher JF, Dempsey RJ (2002) Citicoline: neuroprotective mechanisms in cerebral ischemia. J Neurochem 80:12–23

Adibhatla RM, Hatcher JF (2003) Citicoline decreases phospholipase $A_2$ stimulation and hydroxyl radical generation in transient cerebral ischemia. J Neurosci Res 73:308–315

Amarenco P (2005) Effect of statins in stroke prevention. Curr Opin Lipidol 16:614–618

Andersen M, Overgaard K, Meden P, Boysen G (1999) Effects of citicoline combined with thrombolytic therapy in a rat embolic stroke model. Stroke 30:1464–1470

Atif F, Sayeed I, Ishrat T, Stein DG (2009) Progesterone with vitamin D affords better neuroprotection against excitotoxicity in cultured cortical neurons than progesterone alone. Mol Med 15:328–336

Bahraini NH, Brenner LA, Harwood JE, Homaifar BY, Ladley-O'Brien SE, Filley CM, Kelly JP, Adler LE (2009) Utility of the trauma symptom inventory for the assessment of post-traumatic stress symptoms in veterans with a history of psychological trauma and/or brain injury. Mil Med 174:1005–1009

Bayir H, Andelson PD, Wisniewski SR, Shore P, Lai Y, Brown D, Janesko KL, Kagan VE, Kochanek PM (2009) Therapeutic hypothermia preserves antioxidant defenses after severe traumatic brain injury in infants and children. Crit Care Med 37:689–695

Bazan NG (2006) The onset of brain injury and neurodegeneration triggers the synthesis of docosanoid neuroprotective signaling. Cell Mol Neurobiol 26:901–913

Bazan NG (2007) Omega-3 fatty acids, pro-inflammatory signaling and neuroprotection. Curr Opin Clin Nutr Metab Care 10:136–141

Boonyaratanakornkit V, Bi Y, Rudd M, Edwards DP (2008) The role and mechanism of progesterone receptor activation of extra-nuclear signaling pathways in regulating gene transcription and cell cycle progression. Steroids 73:922–928

Bösel J, Gandor F, Harms C, Synowitz M, Harms U, Djoufack PC, Megow D, Dirnagl U, Hörtnagl H, Fink KB, Endres M (2005) Neuroprotective effects of atorvastatin against glutamate-induced excitotoxicity in primary cortical neurones. J Neurochem 92:1386–1398

Brewer LD, Thibault V, Chen KC, Langub MC, Landfield PW, Porter NM (2001) Vitamin D hormone confers neuroprotection in parallel with downregulation of L-type calcium channel expression in hippocampal neurons. J Neurosci 21:98–108

Brinton RD, Thompson RF, Foy MR, Baudry M, Wang J, Finch CE, Morgan TE, Pike CJ, Mack WJ, Stanczyk FZ, Nilsen J (2008) Progesterone receptors: form and function in brain. Front Neuroendocrinol 29:313–339

Cekic M, Sayeed I, Stein DG (2009a) Combination treatment with progesterone and vitamin D hormone may be more effective than monotherapy for nervous system injury and disease. Front Neuroendocrinol 30:158–172

Cekic M, Culter SM, Vanlandingham JW, Stein DG (2009b) Vitamin D deficiency reduces the benefits of progesterone treatment after brain injury in aged rats. Neurobiol Aging DOI: 10.1016/j.neurobiolaging.2009.04.017, May 29. [Epub ahead of print]

Chen XR, Besson VC, Palmier B, Garcia Y, Plotkine M, Marchand-Leroux C (2007) Neurological recovery-promoting, anti-inflammatory, and anti-oxidative effects afforded by fenofibrate, a PPAR alpha agonist, in traumatic brain injury. J Neurotrauma 24:1119–1131

Chen G, Shi J, Wei Jin W, Wang L, Xie W, Sun J, Hang C (2008a) Progesterone administration modulates TLRs/NF-κB signaling pathway in rat brain after cortical contusion. Ann Clin Lab Sci 38:65–74

Chen XR, Besson VC, Beziand T, Plotkine M, Marchand-Leroux C (2008b) Combination therapy with fenofibrate, a peroxisome proliferator-activated receptor alpha agonist, and simvastatin, a 3-hydroxy-3-methylglutaryl-coenzyme A reductase inhibitor, on experimental traumatic brain injury. J Pharmacol Exp Ther 326:966–974

Chen G, Zhang S, Shi J, Ai J, Qi M, Hang C (2009) Simvastatin reduces secondary brain injury caused by cortical contusion in rats: possible involvement of TLR4/NF-kappaB pathway. Exp Neurol 216:398–406

Cherian L, Goodman JC, Robertson C (2007) Neuroprotection with erythropoietin administration following controlled cortical impact injury in rats. J Pharmacol Exp Ther 322:789–794

Childers SR, Breivogel CS (1998) Cannabis and endogenous cannabinoid systems. Drug Alcohol Dependence 51:173–187

Chun KA, Manley GT, Stiver SI, Aiken AH, Phan N, Wang V, Meeker M, Cheng SC, Gean AD, Wintermark M (2009) Interobserver Variability in the Assessment of CT Imaging Features of Traumatic Brain Injury. J Neurotrauma 2009 Nov 6 [Epub ahead of print]

Crack PJ, Gould J, Bye N, Ross S, Ali U, Habgood MD, Morganti-Kossman C, Saunders NR, Hertzog PJ Victorian Neurotrauma Research Group (2009) The genomic profile of the cerebral cortex after closed head injury in mice: effects of minocycline. J Neural Transm 116:1–12

Dempsey RJ, Rao VLR (2003) Cytidinediphosphocholine treatment to decrease traumatic brain injury-induced hippocampal neuronal death, cortical contusion volume, and neurological dysfunction in rats. J Neurosurg 98:867–873

Delerive P, De Bosscher K, Vanden Berghe W, Fruchart JC, Haegeman G, Staels B (2002) DNA binding-independent induction of IkappaBalpha gene transcription by PPARalpha. Mol Endocrinol 16:1029–1039

De Nicola AF, Labombarada F, Deniselle MC, Gonzalez SL, Garay L, Meyer M, Gargiulo G, Guennoun R, Schumacher M (2009) Progesterone neuroprotection in traumatic CNS injury and motoneuron degeneration. Front Neuroendocrinol 30:173–187

De Spiegelaere W, Cornillie P, Van den Broeck W (2009) Localization of erythropoietin in and around growing cartilage. Mol Cell Biochem 2009 Nov 12 [Epub ahead of print]

Di Marzo V, De Petrocellis L, Sugiura T, Waku K (1996) Potential biosynthetic connections between the two cannabimimetic eicosanoids, anandamide and 2-arachidonoyl-glycerol, in mouse neuroblastoma cells. Biochem and Biophys Res Commun 227:281–288

Djebaili M, Guo Q, Pettus EH, Hoffman SW, Stein DG (2005) The neurosteroids progesterone and allopregnanolone reduce cell death, gliosis, and functional deficits after traumatic brain injury in rats. J Neurotrauma 22:106–118

Eckartdt KU, Kurtz A (2005) Regulation of erythropoietin production. Eur J Clin Invest 35(Suppl 3):13–19

Endres M (2005) Statins and stroke. J Cereb Blood Flow Metab 25:1093–1110

Endres M (2006) Statins: potential new indications in inflammatory conditions. Atheroscler Suppl 7:31–35

Eshhar N, Striem S, Kohen R, Tirosh O, Biegon A (1995) Neuroprotectant and antioxidant activities of HU-211, a novel NMDA receptor antagonist. Eur J Pharmacol 283:19–29

Faden AI, Salzman SK (1994) Experimental pharmacology. In: Salzman SK, Faden AI (eds) The neurobiology of central nervous system trauma. Oxford University Press, New York, Oxford, pp 227–244

Faden AI, Fox GB, Fan L, Araldi GL, Qiao L, Wang S, Kozikowski AP (1999) Novel TRH analog improves motor and cognitive recovery after traumatic brain injury in rodents. Am J Physiol 277:R1196–R1204

Faden AI (2002) Neuroprotection and traumatic brain injury:theoretical option or realistic proposition. Curr Opin Neurol 15:707–712

Faden AI, Movsesyan VA, Knoblach SM, Ahmed F, Cernak I (2005) Neuroprotective effects of novel small peptides in vitro and after brain injury. Neuropharmacology 49:410–424

Farooqui AA, Ong WY, Horrocks LA (2004) Biochemical aspects of neurodegeneration in human brain: involvement of neural membrane phospholipids and phospholipases $A_2$. Neurochem Res 29:1961–1977

Farooqui AA, Horrocks, LA (2009) Glutamate and cytokine-mediated alterations of phospholipids in head injury and spinal cord trauma. In: Banik NK, Ray SK (eds) Handbook of Neurochemistry and Molecular Neurobiology vol 24, 3rd edn. Springer, New York, NY, pp 71–89

Farooqui AA (2009a) Hot topics in neural membrane lipidology. Springer, New York, NY

Farooqui AA (2009b) Beneficial effects of fish oil on human brain. Springer, New York, NY

Feigenbaum JJ, Bergmann F, Richmond SA, Mechoulam R, Nadler V, Kloog Y, Sokolovsky M (1989) Nonpsychotropic cannabinoid acts as a functional *N*-methyl-D-aspartate receptor blocker. Proc Natl Acad Sci USA 86:9584–9587

Garcia-Martinez EM, Sanz-Blasco S, Karachitos A, Bandez MJ, Fernandez-Gomez FJ, Perez-Alvarez S, de Mera RM, Jordan MJ, Aguirre N, Galindo MF, Villalobos C, Navarro A, Kmita H, Jordán J (2010) Mitochondria and calcium flux as targets of neuroprotection caused by minocycline in cerebellar granule cells. Biochem Pharmacol 79:239–2350

Gervois P, Kleemann R, Pilon A, Percevault F, Koenig W, Staels B, Kooistra T (2004) Global suppression of IL-6-induced acute phase response gene expression after chronic in vivo treatment with the peroxisome proliferator-activated receptor-alpha activator fenofibrate. J Biol Chem 279:16154–16160

Gonzalez SL, Labombarda F, Gonzalez Deniselle MC, Guennoun R, Schumacher M, De Nicola AF (2004) Progesterone up-regulates neuronal brain-derived neurotrophic factor expression in the injured spinal cord. Neuroscience 125:605–614

Grände PO, Reinstrup P, Romner B (2009) Active cooling in traumatic brain-injured patients: a questionable therapy? Acta Anaesthesiol Scand 53:1233–1238 Epub 2009 Aug 13

Grasso G, Sfacteria A, Meli F, Fodale V, Buemi M, Iacopino DG (2007) Neuroprotection by erythropoietin administration after experimental traumatic brain injury. Brain Res 1182: 99–105

Hansen HH, Schmid PC, Bittigau P, Lastres-Becker I, Berrendero F, Manzanares J, Ikonomidou C, Schmid HH, Fernández-Ruiz JJ, Hansen HS (2001a) Anandamide, but not 2-arachidonoylglycerol, accumulates during in vivo neurodegeneration. J Neurochem 78: 1415–1427

Hansen HH, Ikonomidou C, Bittigau P, Hansen SH, Hansen HS (2001b) Accumulation of the anandamide precursor and other N-acylethanolamine phospholipids in infant rat models of in vivo necrotic and apoptotic neuronal death. J Neurochem 76:39–46

Hansen HS, Moesgaard B, Petersen G, Hansen HH (2002) Putative neuroprotective actions of N-acyl-ethanolamines. Pharmacol Ther 95:119–126

He J, Evans CO, Hoffman SW, Oyesiku MN, Stein DG (2004) Progesterone and allopregnanolone reduce inflammatory cytokines after traumatic brain injury. Exp Neurol 189:404–412

Homayoun P, Parkins NE, Soblosky J, Carey ME, Rodriguez de Turco EB, Bazan NG (2000) Cortical impact injury in rats promotes a rapid and sustained increase in polyunsaturated free fatty acids and diacylglycerols. Neurochem Res 25:269–276

Homayoun P, Rodriguez de Turco EB, Parkins NE, Lane DC, Soblosky J, Carey ME, Bazan NG (1997) Delayed phospholipid degradation in rat brain after traumatic brain injury. J Neurochem 69:199–205

Homsi S, Federico F, Croci N, Palmier B, Plotkine M, Marchand-Leroux C, Jafarian-Tehrani M (2009) Minocycline effects on cerebral edema: relations with inflammatory and oxidative stress markers following traumatic brain injury in mice. Brain Res 1291:122–132

Hou ST, Jiang SX, Smith RA (2008) Permissive and repulsive cues and signalling pathways of axonal outgrowth and regeneration. Int Rev Cell Mol Biol 267:125–181

Hu Z, Li Y, Fang M, Wai MS, Yew DT (2009) Exogenous progesterone: a potential therapeutic candidate in CNS injury and neurodegeneration. Curr Med Chem 16:1418–1425

Hua XY, Svensson CI, Matsui T, Fitzsimmons B, Yaksh TL, Webb M (2005) Intrathecal minocycline attenuates peripheral inflammation-induced hyperalgesia by inhibiting p38 MAPK in spinal microglia. Eur J Neurosci 22:2431–2440

Huang EJ, Reichardt LF (2003) Trk receptors: roles in neuronal signal transduction. Ann Rev Biochem 72:609–642

Huang T, Solano J, He D, Loutfi M, Dietrich WD, Kuluz JW (2009) Traumatic injury activates MAP kinases in astrocytes: mechanisms of hypothermia and hyperthermia. J Neurotrauma 26:1535–1545

Huh JW, And Raghupathi R (2009) New concepts in treatment of pediatric traumatic brain injury. Anesthesiol Clin 27:213–240

Hutchison JS, Ward RE, Lacroix J, Hébert PC, Barnes MA, Bohn DJ, Dirks PB, Doucette S, Fergusson D, Gottesman R et al (2008) Hypothermia therapy after traumatic brain injury in children. N Engl J Med 358:2447–2456

Hyong A, Jadhav V, Lee S, Tong W, Rowe J, Zhang JH, Tang J (2008) Rosiglitazone, a PPAR gamma agonist, attenuates inflammation after surgical brain injury in rodents. Brain Res 1215:218–224

Jain KK (2009) Cell therapy for CNS trauma. Mol Biotechnol 2009 Mar 28 [Epub ahead of print]

Johnson-Anuna LN, Eckert GP, Franke C, Igbavboa U, Müller WE, Wood WG (2007) Simvastatin protects neurons from cytotoxicity by up-regulating Bcl-2 mRNA and protein. J Neurochem 101:77–86

Johnson-Anuna LN, Eckert GP, Keller JH, Igbavboa U, Franke C, Fechner T, Schubert-Zsilavecz M, Karas M, Müller WE, Wood WG (2005) Chronic administration of statins alters multiple gene expression patterns in mouse cerebral cortex. J Pharmacol Exp Ther 312:786–793

Kochanek PM, Bayir H, Jenkins LW (2008) Molecular biology of brain injury. In: Nichols D (ed) Textbook of pediatric intensive care, 4th edn. Lippincott Williams & Wilkins, Pennsylvania, PA, pp 826–845

Kim HS, Suh YH (2009) Minocycline and neurodegenerative diseases. Behav Brain Res 196: 168–179

Kirsch C, Eckert GP, Muller EE (2003) Brain cholesterol, statins and Alzheimer's Disease. Pharmacopsychiatry 36(Suppl 2):S113–S119

Knoller N, Levi L, Shoshan I, Reichenthal E, Razon N, Rappaport ZH, Biegon A (2002) Dexanabinol (HU-211) in the treatment of severe closed head injury: a randomized, placebo-controlled, phase II clinical trial. Crit Care Med 30:548–554

Lenzlinger PM, Saatman K, Raghupathi R (2001) Overview of basic mechanisms underlying neuropathological consequences of head trauma. In: Miller G, Hayes R (eds) Head trauma – basic, preclinical, and clinical directions. Wiley-Liss, Hoboken, NJ, pp 3–36

Leonhardt SA, Boonyaratanakornkit V, Edwards DP (2003) Progesterone receptor transcription and non-transcription signaling mechanisms. Steroids 68:761–770

Longhi L, Zanier ER, Royo N, Stocchetti N, McIntosh TK (2005) Stem cell transplantation as a therapeutic strategy for traumatic brain injury. Transpl Immunol 15:143–148

Lu D, Qu C, Goussev A, Jiang H, Lu C, Schallert T, Mahmood A, Chen J, Li Y, Chopp M (2007) Statins increase neurogenesis in the dentate gyrus, reduce delayed neuronal death in the hippocampal CA3 region, and improve spatial learning in rat after traumatic brain injury. J Neurotrauma 24:1132–1146

Maas AI (2001) Neuroprotective agents in traumatic brain injury. Expert Opin Investig Drugs 10:753–767

Maas AI, Murray G, Henney H 3rd, Kassem N, Legrand V, Mangelus M, Muizelaar JP, Stocchetti N, Knoller N Pharmos TBI Investigators (2006) Efficacy and safety of dexanabinol in severe traumatic brain injury: results of a phase III randomised, placebo-controlled, clinical trial. Lancet Neurol 5:38–45

Machado LS, Kozak A, Erqul A, Hess DC, Borlougan CV, Fagan SC (2006) Delayed minocycline inhibits ischemia-activated matrix metalloproteinases 2 and 9 after experimental stroke. BMC 7:56

Maegele M, Schaefer U (2008) Stem cell-based cellular replacement strategies following traumatic brain injury (TBI). Minim Invasive Ther Allied Technol 17:119–131

Marchand F, Tsantoulas C, Singh D, Grist J, Clark AK, Bradbury EJ, McMahon SB (2009) Effects of Etanercept and Minocycline in a rat model of spinal cord injury. Eur J Pain 13: 673–681

MacNevin CJ, Atif F, Sayeed I, Stein DG, Liotta DC (2009) Development and screening of water-soluble analogues of progesterone and allopregnanolone in models of brain injury. J Med Chem 52:6012–6023

Mahmood A, Gousser A, Kazmi H, Qu C, Lu D, Chopp M (2009a) Long-term benefits after treatment of traumatic brain injury with simvastatin in rats. Neurosurg 65:187–191

Mahmood A, Lu D, Qu C, Goussev A, Zhang Y, Chopp M (2009b) Treatment of traumatic brain injury in rats with erythropoietin and carbamylated erythropoietin. J Neurosurg 107: 392–397

Mani S (2008) Progestin receptor subtypes in the brain: the known and the unknown. Endocrinology 149:2750–2756

Marx N, Sukhova GK, Collins T, Libby P, Plutzky J (1999) PPARalpha activators inhibit cytokine-induced vascular cell adhesion molecule-1 expression in human endothelial cells. Circulation 99:3125–3131

Matis GK, Birbillis TA (2009) Erythropoietin in spinal cord injury. Eur Spine J 18:314–323

McIntosh TK, Saatman KE, Raghupathi R, Graham DI, Smith DH, Lee VM, Trojanowski JQ (1998) The Dorothy Russell Memorial Lecture. The molecular and cellular sequelae of experimental traumatic brain injury: pathogenetic mechanisms. Neuropathol Appl Neurobiol 24:251–267

McKenna NJ, O'Malley BW (2002) Minireview: nuclear receptor coactivators—an update. Endocrinology 143:2461–2465

Mir C, Clotet J, Aledo R, Durany N, Argemi J, Lozano R, Cervos-Navarro J, Casals N (2003) CDP-choline prevents glutamate-mediated cell death in cerebellar granule neurons. J Mol Neurosci 20:53–59

Monga V, Meena CL, Kaura N, Jain R (2008) Chemistry and biology of thyrotropin-releasing hormone (TRH) and its analogs. Curr Med Chem 15:2718–2733

Nakazawa T, Takahashi H, Nishijima K, Shimura M, Fuse N, Tamai M, Hafezi-Moghadam A, Nishida K (2007) Pitavastatin prevents NMDA-induced retinal ganglion cell death by suppressing leukocyte recruitment. J Neurochem 100:1018–1031

Nilsen J, Brinton RD (2002) Impact of progestins on estrogen-induced neuroprotection: synergy by progesterone and 19-norprogesterone and antagonism by medroxyprogesterone acetate. Endocrinology 143:205–212

Noguchi CT, Asavaritikrai P, Teng R, Jia Y (2007) Role of erythropoietin in the brain. Crit Rev Oncol Hematol 64:159–171

Nolan S (2005) Traumatic brain injury: a review. Ctrt Care Nurse 28:188–194

O'Connor CA, Cernak I, Vink R (2005) Both estrogen and progesterone attenuate edema formation following diffuse traumatic brain injury in rats. Brain Res 1062:171–174

Owen GI, Richer JK, Tung L, Takimoto G, Horwitz KB (1998) Progesterone regulates transcription of the p21(WAF1) cyclindependent kinase inhibitor gene through Sp1 and CBP/p300. J Biol Chem 273(1998):10696–10701

Oz M (2006) Receptor-independent effects of endocannabinoids on ion channels. Curr Pharm Design 12:227–239

Pettus EH, Wright DW, Stein DG, Hoffman SW (2005) Progesterone treatment inhibits the inflammatory agents that accompany traumatic brain injury. Brain Res 1049:112–119

Philips MF, Muir JK, Saatman KE, Raghupathi R, Lee VM, Trojanowski JQ, McIntosh TK (1999) Survival and integration of transplanted postmitotic human neurons following experimental brain injury in immunocompetent rats. J Neurosurg 90:116–124

Povlishock JT, Christman CW (1995) The pathobiology of traumatically induced axonal injury in animals and humans: a review of current thoughts. J Neurotrauma 12:555–564

Pratt WB (1998) The hsp90-based chaperone system: involvement in signal transduction from a variety of hormone and growth factor receptors, Proc. Soc Exp Biol Med 217:420–434

Rajanikant GK, Zemke D, Kassab M, Majid A (2007) The therapeutic potential of statins in neurological disorders. Curr Med Chem 14:103–112

Rao JS, Ertley RN, Lee H-J, DeMar JC Jr, Arnold JT, Repoport SI, Bazinet RP (2007) N-3 Polyunsaturated fatty acid deprivation in rats decreases frontal cortex BDNF via a p38 MARK-dependent mechanism. Mol Psychiatry 12:36–46

Roof RL, Duvdevani R, Stein DG (1992) Progesterone treatment attenuates brain edema following contusion injury in male and female rats. Restor Neurol Neurosci 4:425–427

Roof RL, Hoffman SW, Stein DG (1997) Progesterone protects against lipid peroxidation following traumatic brain injury in rats. Mol Chem Neuropathol 31:1–11

Rupprecht R, Holsboer F (1999) Neuroactive steroids: mechanisms of action and neuropsychopharmacological perspectives. Trends Neurosci 22:410–416

Sahuguillo J, Bietro A, Amoros S, Poca MA, Baguena M, Ibanez J, Noguer M, Garnacho A (2001) The use of moderate hypothermia in the treatment of patients with severe craniocerebral trauma. Neurocirgia (Astur) 12:23–35

Sahuguillo J, Vilalta A (2009) Cooling the injured brain: how does moderate hypothermia influence the pathophysiology of traumatic brain injury. Curr Pharm Des 13:2310–2322

Salim A, Hadjizacharea P, Brown C, Inaba K, Teixeira PG, Chan L, Rhee P, Demetriades D (2008) Significance of troponin elevation after severe traumatic brain injury. J Trauma 64:46–57

Sayeed I, Stein DG (2009) Progesterone as a neuroprotective factor in traumatic and ischemic brain injury. Prog Brain Res 175:219–237

Schumacher M, Sitruk-ware R, De Nicola AF (2008) Progesterone and progestins: neuroprotection and myelin repair. Curr Opin Pharmacol 8:740–746

Sensi SL, Jeng JM (2004) Rethinking the excitotoxic ionic milieu: the emerging role of $Zn^{2+}$ in ischemic neuronal injury. Curr Mol Med 4:87–111

Serhan CN (2005a) Novel eicosanoid and docosanoid mediators: resolvins, docosatrienes, and neuroprotectins. Curr Opin Clin Nutr Metab Care 8:115–121

Serhan CN (2005b) Novel ω-3-derived local mediators in anti-inflammation and resolution. Pharmacol Ther 105:7–21

Shin BS, Won SJ, Yoo BH, Kauppinen TM, Suh SW (2009) Prevention of hypoglycemia-induced neuronal death by hypothermia. J Cereb Blood Flow Metab 2009 Oct 28 [Epub ahead of print]

Shohami E, Novikov M, Mechoulam R (1993) A nonpsychotropic cannabinoid, HU-211, has cerebrovascular effects after closed head injury in the rat. J Neurotrauma 10:109–119

Shohami E, Beit-Yannai E, Horowitz M, Kohen R (1997) Oxidative stress in closed-head injury: brain antioxidant capacity as an indicator of functional outcome. J Cereb Blood Flow Metab 17:1007–1019

Siren AL, Fasshauer T, Bartels C, Ehrenreich H (2009) Therapeutic potential of erythropoietin and its structural or functional variants in the nervous system. Neurotherapeutics 6:108–127

Skaper SD, Moore SE, Walsh FS (2001) Cell signalling cascades regulating neuronal growth-promoting and inhibitory cues. Prog Neurobiol 65:593–608

Stockmann C, Fandrey J (2006) Hypoxia-induced erythropoietin production: a paradigm for oxygen-regulated gene expression. Clin Exp Pharmacol Physiol 33:968–979

Stoica BA, Byrnes KR, Faden AI (2009) Cell cycle activation and CNS injury. Neurotox Res 16:221–237

Sun H, Huang Y, Yu X, Li Y, Yang J, Li R, Deng Y, Zhao G (2008) Peroxisome proliferator-activated receptor gamma agonist, rosiglitazone, suppresses CD40 expression and attenuates inflammatory responses after lithium pilocarpine-induced status epilepticus in rats. Int J Neurosci 26:505–515

Ueda N, Okamoto Y, Tsuboi K (2005) Endocannabinoid-related enzymes as drug targets with special reference to N-acylphosphatidylethanolamine-hydrolyzing phospholipase D. Curr Med Chem 2005(12):1413–1422

Vaughan CJ (2003) Prevention of stroke and dementia with statins: effects beyond lipid lowering. Am J Cardiol 91:23B–29B

Vaynman S, Ying Z, Gomez-Pinilla F (2004) Hippocampal BDNF mediates the efficacy of exercise on synaptic plasticity and cognition. Eur J Neurosci 20:2580–2590

Vuletic S, Riekse RG, Marcovina SM, Peskind ER, Hazzard WR, Albers JJ (2006) Statins of different brain penetrability differentially affect CSFPLTP activity. Dement Geriatr Cogn Disord 22:392–398

Wu A, Molteni R, Ying Z, Gomez-Pinilla F (2003) A saturated-fat diet aggravates the outcome of traumatic brain injury on hippocampal plasticity and cognitive function by reducing brain-derived neurotrophic factor. Neuroscience 119:365–375

Wu A, Ying Z, Gomez-Pinilla F (2004a) Dietary omega-3 fatty acids normalize BDNF levels, reduce oxidative damage, and counteract learning disability after traumatic brain injury in rats. J Neurotrauma 21:1457–1467

Wu A, Ying Z, Gomez-Pinilla F (2004b) The interplay between oxidative stress and brain-derived neurotrophic factor modulates the outcome of a saturated fat diet on synaptic plasticity and cognition. Eur J Neurosci 19:1699–1707

Wu A, Ying Z, Gomez-Pinilla F (2005) Omega-3 fatty acids supplementation restores homeostatic mechanisms disrupted by traumatic brain injury. J Neurotrauma 22:1212

Wu H, Jiang H, Xiong Y, Qu C, Li B, Mahmood A, Zhou D, Chopp M (2008) Simvastatin-mediated upregulation of VEGF and BDNF, activation of the PI3K/Akt pathway, and increase of neurogenesis are associated with therapeutic improvement after traumatic brain injury. J Neurotrauma 25:130–139

Wu J, Yang S, Xi G, Fu G, Keep RF, Hua Y (2009) Minocycline reduces intracerebral hemorrhage-induced brain injury. Neurol Res 31:183–188

Xiong Y, Lu D, Qu C (2008) Effects of erythropoietin on reducing brain damage and improving functional outcome after traumatic brain injury in mice. J Neurosurg 109:510–521

Xiong Y, Mahmood A, Meng Y, Zhang Y, Qu C, Schallert T, Chopp M (2009) Delayed administration of erythropoietin reducing hippocampal cell loss, enhancing angiogenesis and neurogenesis, and improving functional outcome following traumatic brain injury in rats: comparison of treatment with single and triple dose. J Neurosurg 2009 Oct 9 [Epub ahead of print]

Yamashita T (2007) Molecular mechanism and regulation of axon growth inhibition. Brain Nerve 59:1347–1353

Yang D, Xie P, Guo S, Li H (2009) Induction of MAPK phosphatase-1 by hypothermia inhibits TNF-{alpha}-induced endothelial barrier dysfunction and apoptosis. Cardiovasc Res Oct 22 [Epub ahead of print]

Yao XL, Liu J, Lee E, Ling GS, McCabe JT (2005) Progesterone differentially regulates pro- and anti-apoptotic gene expression in cerebral cortex following traumatic brain injury in rats. J Neurotrauma 22:656–668

Zafonte R, Friedewald WT, Lee SM, Levin B, Diaz-Arrastia R, Ansel B, Eisenberg H, Timmons SD, Temkin N, Novack T, Ricker J, Merchant R, Jallo J (2009) The citicoline brain injury treatment (COBRIT) trial: design and methods. J Neurotrauma 26:2207–2216

Zhu L, Wang HD, Yu XG, Jin W, Qiao L, Lu TJ, Hu ZL, Zhou J (2009) Erythropoietin prevents zinc accumulation and neuronal death after traumatic brain injury in rat hippocampus: in vitro and in vivo studies. Brain Res 1289:96–105

# Chapter 8
# Neurochemical Aspects of Neurodegenerative Diseases

## 8.1 Introduction

Neurodegenerative diseases are a debilitating group of diseases associated with site-specific premature and slow death of specific neuronal populations and synapses in brain and spinal cord that modulate thinking, skilled movements, decision making, cognition, and memory (Graeber and Moran, 2002; Soto and Estrada, 2008). These diseases include Alzheimer disease (AD), Parkinson disease (PD), Huntington disease (HD), amyotrophic lateral sclerosis (ALS), and prion diseases. In AD, neurons die in the nucleus basalis; in PD, neurodegeneration occurs in the substantia nigra; degeneration of striatal medium spiny neurons is involved in the pathogenesis of HD; and ALS is characterized by damage to motor neurons in the brain and spinal cord. It is not clear when does a neurodegenerative disease actually start and how long does it take for neuropathological changes to appear. As stated in Chapter 1, the most important risk factors for neurodegenerative diseases are old age, positive family history, unhealthy lifestyle, and exposure to toxic environment (Fig. 8.1) (Farooqui and Farooqui, 2009). Normal aging is accompanied by alterations in structural organization and functioning of brain tissue. Aging also causes an increase in inflammatory signaling in the nervous system as well as dysfunction of the immune system elsewhere in the body. Chronic neuroinflammation is characterized not only by long-standing chronic activation of microglia but also by sustained release of inflammatory mediators. The sustained release of inflammatory mediators causes an imbalance in the inflammatory cycle homeostasis by activating additional microglia, promoting their proliferation, and leading to further release of inflammatory factors (Farooqui, 2010a). Collectively, these studies suggest that there are many age-related changes that contribute to the modulation of brain function in aged brain resulting in decline of brain activities and increase in brain frailty, which may singly and collectively affect neuronal viability and vulnerability (Farooqui and Horrocks, 2007). Due to premature and slow death of specific neuronal populations, neurodegenerative diseases are accompanied by the loss of modulation of structural organization and functioning of the brain tissue. Despite the important differences in clinical manifestation and progressive cell loss of specific neuronal populations in a specific region, neurodegenerative diseases share some common features such

A.A. Farooqui, *Neurochemical Aspects of Neurotraumatic and Neurodegenerative Diseases*, DOI 10.1007/978-1-4419-6652-0_8,

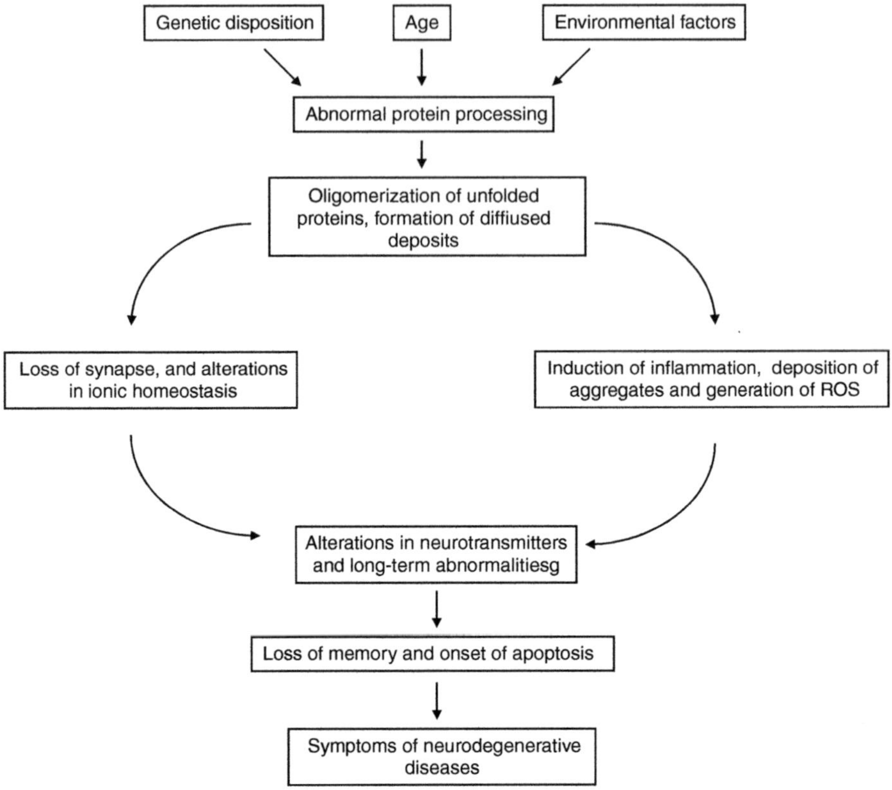


**Fig. 8.1** Factors associated with the pathogenesis of Neurodegenerative diseases

as appearance in aged brain, the extensive neurodegeneration, synaptic dysfunction, and the accumulation of intracellular or extracellular cerebral deposits of misfolded protein aggregates with a β-sheet conformation, such as β-amyloid (Aβ) in AD, α-synuclein in PD, mutated huntingtin in HD, elevation in membrane-associated oxidative stress in ALS resulting in genetic abnormalities (Cu/Zn superoxide dismutase mutations) (Table 8.1) (Farooqui and Horrocks, 2007), and accumulation of advanced glycation end products (AGEs) (Farooqui, 2009a; Miranda and Outerio, 2009). Post-translational modifications (glycation) facilitate misfolding, aggregation, and accumulation of Aβ, tau (τ), prions, and transthyretin proteins in patients with neurodegenerative diseases (Chen et al., 2009, 2010). AGEs through their receptor, RAGE, may cause an increase in oxidative stress and inflammation through the formation of ROS and the induction of NF-κB (Miranda and Outerio, 2009).

In AD, misfolded Aβ peptide 1–42 accumulates in the neuronal endoplasmic reticulum extracellularly as plaques. In contrast, in PD and dementia with Lewy bodies (DLB) abnormal accumulation of α-synuclein occurs in neuronal cell bodies, axons, and synapses. Furthermore, in DLB, Aβ 1–42 has been reported to promote α-synuclein accumulation and neurodegeneration (Hashimoto et al., 2003).

**Table 8.1** Accumulation of various types of protein aggregates and their location in neurodegenerative diseases

| Neurodegenerative disease | Protein | Type of aggregate | Location | References |
|---|---|---|---|---|
| AD | β-Amyloid | Amyloid | Extracellular | Haass and Selkoe (2007) |
| PD | α-Synuclein | Fibrillar non-amyloid | Intracellular | Beyer (2007), Burke (2004) |
| HD | Huntingtin | Fibrillar non-amyloid | Intracellular | Bonilla (2000), Gil and Rego (2008) |
| ALS | Superoxide dismutase I | Fibrillar non-amyloid | Intracellular | Jellinger (2009), Kucic and Kiernan (2009) |
| CJD | Prion protein | Amyloid | Extracellular | DeArmond and Prusiner (2003), Behrens (2003) |
| Other prion diseases | Prion protein | Amyloid | Extracellular | DeArmond and Prusiner (2003), Behrens (2003) |

Interactions between fragments of α-synuclein and Aβ peptide promote the aggregation of α-synuclein in vivo. In addition under pathlogical condition, interactions between Aβ and α-synuclein may initiate the formation of toxic oligomers and nanopores that increase intracellular calcium leading to induction of oxidative stress, leakage of lysosomal membranes, and mitochondrial dysfunction (Crews et al., 2009).

## 8.2 Factors and Molecular Mechanisms that Modulate Neurodegeneration in Neurodegenerative Diseases

Molecular mechanisms associated with the pathogenesis of neurodegenerative diseases remain unknown. Causes of neuronal death in neurodegenerative diseases include decline in cellular antioxidant defenses (activities of superoxide dismutase, glutathione peroxidase, catalase, and glutathione reductase); generation of ROS; and accumulation of peroxidized lipids, proteins, and DNA oxidative products along with genetic and environmental factors (Farooqui, 2009a) (Fig. 8.1), supporting the view that neural cell death in neurodegenerative diseases is a multifactorial process involving genetic, environmental, and endogenous factors. Endogenous factors that contribute to neurodegenerative diseases include neuroinflammation, abnormal protein dynamics with defective protein degradation, and aggregation related to the ubiquitin-proteasomal system resulting in generation and accumulation of misfolded proteins, autoimmunity, and mitochondrial dysfunction resulting in impaired energy metabolism (Farooqui et al., 2007a; Farooqui and Horrocks, 2007; Farooqui, 2009a; Lahiri et al., 2007). Disease-specific proteins (τ, Aβ, α-synuclein, huntingtin), which accumulate in neurodegenerative diseases, are substrates for transglutaminase 2, a calcium-dependent cross-linking enzyme involved in the post-translational modification of intra- and extracellular proteins. It generates isopeptide bonds, which stabilize polymeric aggregates of accumulated proteins. This indicates the importance of transglutaminase 2-mediated cross-linking reactions in neurodegenerative processes (Hartley et al., 2008; Caccamo et al., 2009; Wilhelmus et al., 2009). The abnormal protein aggregates cannot be degraded by cytosolic proteases, ubiquitin-protesome system, and autophagy, therefore, accumulate in cells and extracellular compartments as residual debris. In addition, blood–brain barrier (BBB) dysfunction and hypertension may also contribute to the pathogenesis of neurodegenerative diseases (Rao and Balachandran, 2002; Farooqui, 2009a). The dysfunction of BBB is accompanied by the disruption of tight junctions, alterations in transport of molecules (plasma proteins) between blood and brain and brain and blood, aberrant angiogenesis, vessel regression, brain hypoperfusion, and changes in inflammatory responses. These processes may contribute to a "vicious circle" that leads to progressive synaptic loss and neurodegeneration in disorders of neurodegenerative diseases (Zlokovic, 2008). Most of the above mechanisms are interrelated in vicious circles finally leading to programmed cell death. A common feature of neurodegenerative diseases is a long course until sufficient protein accumulates, followed by a cascade of symptoms over many years with increasing disability leading to death (Jellinger, 2009).

The sources of increased oxidative damage are not entirely clear. Occurrence of increased localization of redox-active transition metals (copper and iron) in the brain regions most affected by neurodegenerative diseases is consistent with the hypothesis that redox-active transition metals may contribute to oxidative stress (Bolognin et al., 2009). The redox state is regulated by oxidative and antioxidative processes, and changes in redox state stimulate or inhibit activities of various signal proteins, resulting in modulation of cell fate. Furthermore, high concentration of ROS generated by the oxidative phosphorylation pathway in mitochondria exposes mitochondrial genome to oxidative stress leading to mitochondrial DNA injury. Mitochondrial dysfunction induces abatement in ATP production, alterations in calcium homeostasis, oxidative damage, and induction of apoptotic cell death. All these processes are closely associated with the pathogenesis of neurodegenerative diseases. In neurodegenerative diseases, oxidative stress initially occurs at the disease-specific site, for example Aβ-mediated oxidative stress in the cerebral cortex and hippocampal region of AD patients, α-synuclein-induced oxidative stress in the brain stem of PD patients, and glutamate receptor-mediated oxidative stress in the motor system of ALS spinal cord. In addition, oxidation of $K^+$ channels by ROS has been reported to be a major mechanism underlying the loss of neuronal function in neurodegenerative diseases (Sesti et al., 2009).

Loss of synapses is another feature that plays an important role in loss of skilled movements, decision making, cognition, and memory-related processes in neurodegenerative diseases (Wishart et al., 2006). As stated above, neurodegenerative diseases also involve the accumulation of ubiquitinated proteins in neuronal inclusions along with signs of inflammation. These abnormal protein aggregates may trigger the expression of inflammatory mediator generating enzymes, such as phospholipase $A_2$ ($PLA_2$), cyclooxygenase-2 (COX-2), and lipoxygenase (LOX), indicating that impairment of the ubiquitin-proteasome pathway may contribute to this neurodegenerative process (Farooqui and Horrocks, 2007; Farooqui, 2009a). In addition to the generation of ROS, pathophysiology of neurodegenerative diseases may also share many common terminal neurochemical processes, such as inflammation, and excitotoxicity (Farooqui and Horrocks, 2007; Forman et al., 2004). Excitotoxicity increases cytosolic $Ca^{2+}$ levels, resulting in activation of $Ca^{2+}$-dependent enzymes, including NADPH oxidase, cytosolic phospholipase $A_2$, xanthine oxidase, and neuronal nitric oxide synthase (NOS), in the neurons. Activation of these enzymes is common to many neurodegenerative diseases. This activation generates ROS, nitric oxide, and peroxynitrite, which oxidatively modify nucleic acid, lipid, sugar, and protein, leading to nuclear damage, mitochondrial damage, proteasome inhibition, and endoplasmic reticulum (ER) stress (Shibata and Kobayashi, 2008). NO and peroxynitrite not only depelete glutathione but also S-nitrosylate many proteins. S-Nitrosylation also contributes to protein misfolding (Lipton et al., 2007). One such enzyme protein is protein disulfide isomerase (PDI). This enzyme is responsible for normal protein folding in the endoplasmic reticulum (ER). S-Nitrosylation of PDI compromises its function and induces misfolding (Lipton et al., 2007).

Oxidative stress also stimulates astrocytes and microglia to facilitate the generation and secretion of cytokines such as TNF-α and FasL that not only cause neuronal caspase-8 activation but also induce glial inflammatory response through induction of nuclear factor-κB-mediated generation and secretion of IL-1, TNF-α, NO, $PGE_2$ (Shibata and Kobayashi, 2008; Farooqui, 2009a). The sustained release of above mediators works to perpetuate the inflammatory cycle, activating additional microglia, promoting their proliferation, and resulting in further release of inflammatory factors. High concentrations of these metabolites are not only toxic to neurons but also propagate neuronal injury (Dheen et al., 2007). Moreover, oxidative DNA damage mediates the release of mitochondrial apoptosis-inducing kinase, which triggers apoptosis-like programmed cell death via cyclophilin A.

Normal aging is accompanied by a moderate upregulation of interplay among excitotoxicity, oxidative stress and neuroinflammation (Facheris et al., 2004; Farooqui and Horrocks, 2007; Farooqui, 2010a). The high intensity of interplay among exicitotoxicity, oxidative stress, and neuroinflammation in neurodegenerative diseases turns on specific genes that affect only a specific neuronal population in a particular region where neuronal degeneration occurs (Dwyer et al., 2005; Migliore et al., 2005). This proposal is supported by the hypothesis that the nature of neuron–neuron connections as well as interactions between neurons and glial cells is essential for determining the selective neuronal vulnerability of neurons in neurodegenerative diseases (Wilde et al., 1997; Farooqui et al., 2007a, b). Although it remains elusive whether exicitotoxicity, oxidative stress, and neuroinflammation are the cause or consequence of neural cell death in neurodegenerative diseases (Andersen, 2004; Juranek and Bezek, 2005; Farooqui, 2009a), oxidative stress initially occurs in specific neuronal population of a specific region. Even if the interplay among exicitotoxicity, oxidative stress, and neuroinflammation is not the primary triggering mechanism that initiates and maintains the pathogenetic cascade in neurodegenerative diseases, it is likely that this cascade may promote and maintain factors that aid the progression of AD, PD, ALS, and HD (Farooqui and Horrocks, 2007; Farooqui and Farooqui, 2009). Collective evidence suggests that excitotoxicity, oxidative stress, and neuroinflammation is closely associated with pathomechanisms of apoptotic cell death in neurodegenerative diseases (Farooqui et al., 2007a, b; Shibata and Kobayashi, 2008; Farooqui, 2009a).

Another important finding is that many neurodegenerative diseases are characterized by aberrant protein phosphorylation and ubiquitination (Thomas et al., 2009). Thus, disruption of the phosphorylation of neurotransmitter receptors and hyperphosphorylation of τ-protein has been implicated in impaired memory function in AD. Similarly, AD also involves aberrant accumulation of proteins that are normally degraded by the ubiquitin-proteasome system. It is suggested that phosphorylation and ubiquitination of proteins can serve biomarkers for neurodegenerative diseases (Thomas et al., 2009).

## 8.3 Neurochemical Aspects of Alzheimer Disease

Changes in glutamate homeostasis have been reported to occur in AD (Bi and Sze, 2002). These changes are not due to increased release of glutamate, but significantly lower expression of NR2A and NR2B transcripts in susceptible regions of AD brain supporting the view that NR2 subunit composition may modulate NMDA receptor-mediated excitotoxicity (Hynd et al., 2004). In AD, NMDA receptors are overactivated by glutamate in a tonic rather than a phasic manner (Parson et al., 2007). This continuous mild activation may lead to neuronal damage through the stimulation of $Ca^{2+}$-dependent enzymes related to lipid, protein, and nucleic acid metabolism. $Ca^{2+}$-dependent enzymes associated with glycerophospholipid, sphingolipid, and cholesterol metabolism modulate levels of glycerophospholipid-, sphingolipid-, and cholesterol-derived lipid mediators. The function of glycerophospholipid, sphingolipid, and cholesterol-derived lipid mediator network is to convey extracellular signals from the cell surface to the nucleus to induce a biological response at the gene level. In neural cells, glycerophospholipid-, sphingolipid-, and cholesterol-derived lipid mediators are involved in signal transduction, adhesion, sorting, and trafficking (Simons and Ikonen, 2000; Farooqui and Horrocks, 2007). The intensity of interactions among glycerophospholipids, sphingolipids, and cholesterol-derived lipid mediators not only modulates cellular function through signal transduction processes but also adaptive responses (Farooqui, 2009a). Alterations in composition and levels of lipid mediators are

**Table 8.2** Neurochemical alterations in Alzheimer disease

| Neurochemical parameter | Effect | References |
|---|---|---|
| Glycerophospholipid metabolism | Altered | Farooqui (2009a, b) |
| Free fatty acid composition | Altered | Farooqui (2009a, b) |
| $PLA_2$ activity | Increased | Farooqui (2009a, b) |
| Eicosanoids | Increased | Farooqui (2009a, b) |
| Lipid peroxidation | Increased | Farooqui (2009a, b) |
| 4-Hydroxynonenal | Increased | Farooqui (2009a, b) |
| Cholesterol | Increased | Farooqui (2009a, b) |
| 8-OHdGua | Increased | Farooqui (2009a, b) |
| APP processing | Abnormal | Nathlie and Jean-Noel (2008) |
| BACE and γ-secretase | Increased | Siman and Salidas (2004) |
| NF-κB | Upregulated | Farooqui (2009a, b) |
| Synapse integrity | Lost | Farooqui (2009a, b) |
| Excitotoxicity | Increased | Farooqui (2009a, b) |
| Oxidative stress | Increased | Farooqui (2009a, b) |
| Neuroinflammation | Increased | Farooqui (2009a, b) |
| Neurodegeneration | Increased | Farooqui (2009a, b) |

associated with progression of AD (Table 8.2) (Farooqui and Horrocks, 2007). In addition, there are cumulative metabolic alterations that impair neuronal function and decrease neuron viability. In vivo, mitochondria form dynamic networks that undergo frequent morphologic changes through fission and fusion. In neurons, the imbalance of mitochondrial fission/fusion influences neuronal physiology, such as synaptic transmission and plasticity, and affects neuronal survival. In AD, major changes occur in mitochondria and synapses. Mitochondrial changes include decrease in complex IV of the respiratory chain, damage to complex V, and alterations in voltage-dependent anion channel VDAC, a mitochondrial porin involved in redox homeostasis and apoptosis (Ferrer, 2009). In AD, neurochemical changes in the synapse include activation of PlsEtn-$PLA_2$, loss of plasmalogens, and reduction (25%) in the presynaptic vesicle protein synaptophysin (Farooqui et al., 2003; Farooqui and Horrocks, 2007; Masliah et al., 2001). These changes may cause aberrant sprouting and synaptic loss. Although the mechanisms that trigger above neurochemical changes resulting in loss of synapse are not understood, they may be related to alterations in intensity of cross talk among various lipid mediators in cytoplasmic and nuclear compartments (Farooqui, 2009a; Farooqui et al., 2010c) and loss of normal function of misfolded or aggregated of synaptic proteins (Masliah et al., 2001). Aging itself causes synaptic loss in the dentate region of the hippocampus, but in advancing AD, synapses are disproportionately lost relative to neurons, and this loss can be correlated with dementia (Terry et al., 1991). Although the molecular mechanism of synaptic loss is not fully understood, it is proposed that soluble Aβ oligomers, also referred to as Aβ-derived diffusible ligands (ADDLs), act as highly specific pathogenic ligands, binding to sites localized at particular synapses (De Felice et al., 2009). This binding may not only stimulate PlsEtn-$PLA_2$

(Farooqui et al., 2003) but also trigger oxidative stress through the oxidation of arachidonic acid, loss of synaptic spines, and ectopic redistribution of receptors critical to plasticity and memory (De Felice et al., 2009). Studies in incipient AD cases have shown that this alteration occurs very early in the progression of the disease preceding tangle formation and neuronal loss. Collectively, these studies indicate that reduction in energy production and loss of synapse in AD may cause impairment in neuronal function, alteration in cognitive function, reduction in molecular turnover, and enhanced cell death.

### *8.3.1 Lipids in AD*

Levels of glycerophospholipids are decreased in neural membranes from different regions of AD patients compared to age-matched control human brain (Stokes and Howthrone, 1987; Söderberg et al., 1991; Wells et al., 1995; Guan et al., 1999; Han et al., 2001; Pettegrew et al., 2001). This is due to the stimulation of isoforms of $PLA_2$ activities (Farooqui et al., 1997; Stephenson et al., 1999; Farooqui et al., 2003; Farooqui and Horrocks, 2007). Stimulation of $PLA_2$ isoforms is accompanied by elevation in glycerophospholipid degradation metabolites which include phosphodiesters, phosphomonoesters, fatty acids, prostaglandins, isoprostanes, 4-hydroxynonenals, and other lipid mediators (Table 8.2) (Farooqui and Horrocks, 2006, 2007). Physicochemical and pathological consequences of enhanced glycerophospholipid metabolism in neural membranes include alterations in membrane fluidity and permeability; alterations in ion homeostasis; and changes in activities of membrane-bound enzymes, receptors, and ion channels and in oxidative stress. Many of these lipid mediators are proinflammatory. Their effects are accompanied by the activation of astrocytes and microglia and the release of inflammatory cytokines. These cytokines in turn propagate and intensify neuroinflammation by a number of mechanisms including further upregulation of $PLA_2$ isoforms, generation of platelet-activating factor, and stimulation of nitric oxide synthases (Farooqui and Horrocks, 2007; Farooqui, 2009a).

The cause of increased activities of $PLA_2$ isoforms in AD brain is not fully understood. However, there are several possibilities. Aβ, which accumulates in AD, has been reported to activate $cPLA_2$ activity (Kanfer et al., 1998). Thus, the treatment of cortical cultures with Aβ stimulates $cPLA_2$ activity in a dose-dependent manner and this stimulation is blocked by $cPLA_2$ antisense oligonucleotides (ODN), strongly suggesting the involvement of $cPLA_2$ in the pathogenesis of AD (Kriem et al., 2005). The second possibility is that the activation of astrocytes and microglia in AD may result in expression of the cytokines, TNF-α, IL-1β, and IL-6, that are known to stimulate $cPLA_2$ activity (Sun et al., 2004). Another mechanism of $cPLA_2$ activation may involve the proteolytic cleavage of $cPLA_2$ by caspase-3 (Wissing et al., 1997). A specific tetrapeptide inhibitor of caspase-3 (acetyl-Asp-Glu-Val-Asp-aldehyde) prevents the activation of $cPLA_2$ supporting the view that caspase-mediated proteolysis of $cPLA_2$ retards cell injury and death. Finally, ceramide, a metabolite of sphingolipid metabolism, which accumulates in AD brain, stimulates isoforms of $PLA_2$ (Farooqui, 2010b). It is proposed that in AD, PlsEtn-$PLA_2$ may be the first $PLA_2$ that initiates neural injury. Its stimulation may alter neural membrane permeability due to the loss of plasmalogens, allowing slow $Ca^{2+}$ influx. This slow $Ca^{2+}$ influx and generation of ceramide may facilitate translocation of c$PLA_2$ from cytosol to neural membranes and its activation resulting in the hydrolysis of neural membrane PtdCho. As concentration of $Ca^{2+}$ reaches in millimolar level, the s$PLA_2$ may be activated promoting neural cell injury and death. Thus in injury process sequence, PlsEtn-$PLA_2$ is situated at the proximal end, c$PLA_2$ in the middle, and s$PLA_2$ at the distal end (Farooqui, 2010b). AD patients show a significant decrease in plasma and hippocampal levels of DHA compared to age-matched control (Conquer et al., 2000; Tully et al., 2003; Söderberg et al., 1991) (Fig. 8.5). This decrease correlates not only with upregulation of PlsEtn-$PLA_2$ (Farooqui et al., 2006) but also with significant reduction in plasmalogen levels in AD patient (Söderberg et al., 1991; Guan et al., 1999; Han et al., 2001).

Alterations in sphingolipid metabolism also play an important role in the etiology of AD. Increase in ceramide and decrease in sulfatide levels have been detected in the brains of patients with AD (Han et al., 2002; Cutler et al., 2004; Satoi et al., 2005). It is suggested that apoE is involved in sulfatide transport and mediates sulfatide homeostasis in the nervous system through lipoprotein metabolism pathways, and these alterations in apoE-mediated sulfatide trafficking are associated with sulfatide depletion in the brain (Han, 2007). Increase in ceramide may be due to elevation in activities of acid sphingomyelinase (ASM) and acid ceramidase in AD (He et al., 2010; Huang et al., 2004). Microarray studies on AD brain also indicate that there is an upregulation of gene expression of the enzymes associated with de novo synthesis of ceramide and the downregulation of the enzymes involved in glycosphingolipid synthesis in early AD progression (Katsel et al., 2007). It is suggested that reduction in sphingosine-1-phosphate levels in the AD brain, together with elevated ceramide, may contribute to the pathogenesis of AD. These studies are supported by results on accumulation of ceramide in the cortex of $APP^{SL}$ mice, but not in PS1Ki mice, whereas all other major sphingolipids (except galactosylceramides) are not altered in comparison with those from age-matched wild-type mice (Barrier et al., 2008). Increase in ceramide levels may produce changes in multiple enzymes and cell signaling components. The early inhibition of the neuronal survival pathway regulated by phosphatidylinositol-3-kinase/protein kinase B or AKT mediated by ceramide may be a relevant early event in the decision of neuronal survival/death (Arboleda et al., 2009). Ceramide may also perturb several molecular and metabolic functions. In particular it might decrease glycolysis through rapid modulation of hexokinase activity. This would in turn generate limited amounts of mitochondrial substrates leading to mitochondrial dysfunction and neuronal apoptosis. Subtle and early metabolic alterations caused by inhibition of the PtdIns3K/AKT pathway mediated by ceramide may potentially work with genes associated with neurodegeneration in AD (Arboleda et al., 2009).

Brain is the richest source of cholesterol in the body accounting approximately 23% of total body cholesterol. Most brain cholesterol is present in myelin, neural membranes, and small amounts of cholesterol are associated with the nucleus, which contain activities of cholesterol-metabolizing enzymes (Pfrieger, 2003; Farooqui, 2009a). In neural membranes, cholesterol modulates not only the physicochemical properties and endocytosis but also the antigen expression, exocytosis, synaptic transmission, and activities of membrane-bound enzymes, receptors, and ion channels (Simons and Ikonen, 2000; Farooqui, 2009a). Both neurons and glial cells can synthesize cholesterol. In brain, cholesterol is metabolized by cytochrome P450-dependent oxygenases, cholesterol oxidases, and acyl-CoA: cholesterol acyltransferase. These enzymes transform cholesterol into hydroxycholesterols (24-hydroxycholesterol, 25-hydroxycholesterol, and 27-hydroxycholesterol), cholesterol oxides, and cholesterol esters, respectively (Mast et al., 2003). Cholesterol-metabolizing enzymes are expressed almost exclusively in neurons in the normal brain (Russell et al., 2009). 24-Hydroxycholesterol is the major brain cholesterol metabolite and responsible for maintaining cholesterol homeostasis and the removal of excess cholesterol from the brain into plasma. It exerts a unique modulatory effect on APP processing and increases the α-secretase activity as well as the α/β-secretase activity ratio. 22-Hydroxycholesterol and 27-hydroxycholesterol are minor hydroxycholesterols. Recent studies indicate that significant net uptake of 27-hydroxycholesterol occurs from the circulation to the brain tissue, and patients with AD have increased brain levels of 27-hydroxycholesterol, which may affect the production of β-amyloid (Farooqui, 2009a; Ong et al., 2010). Cholesterol contents regulate compartmentation of the amyloid precursor protein (APP) molecule within the neural cell membrane bilayer. The amyloid precursor protein molecule is found inside or outside the rafts. Processes altering the compartmentation of the APP molecule by transferring it to the neural membrane rafts, favor its cleavage by secretases and are closely associated with amyloidogenic processing (see below). Intact blood–brain barrier retards lipoprotein uptake into the brain. Instead, neurons and glial cells synthesize their own cholesterol through de novo synthesis. This process is controlled by 3-hydroxy-3-methylglutaryl coenzyme A. The decreased CYP46A1 activity in AD brain patients may increase membrane cholesterol levels, and as a consequence the APP is shifted and deposited in the cholesterol-rich lipid rafts leading to amyloidogenic β-amyloid peptide generation.

Among the polymorphic variants of the apolipoprotein E gene (ApoE), the E4 allele is considered as a major risk factor for AD. ApoE is also a risk factor for coronary artery disease (CAD) (Martins et al., 2009). Lipidation status of apoE influences the metabolism of Aβ peptides that accumulate as amyloid deposits in the neural parenchyma and cerebrovasculature. ApoE not only inhibits the transport of Aβ across the blood–brain barrier (BBB) but also facilitates the proteolytic degradation of Aβ by neprilysin and insulin degrading enzyme (IDE), which are enhanced when apoE is lipidated. It is suggested that AD and CAD share other risk factors, such as altered cholesterol levels, particularly high levels of low-density lipoproteins together with low levels of high-density lipoproteins (Martins et al., 2009). Statins, the inhibitors of HMG-CoA reductase lower cholesterol levels in CAD, have been shown to protect against AD. Although the molecular mechanisms associated with neuroprotective and cardioprotective effects are still elusive, recent studies indicate the downregulation of ABCA1 expression in human macrophages through the inhibition of LXR ligand 24(*S*), 25-epoxycholesterol synthesis (Wong et al., 2006; Velazquez et al., 2006). Collective evidence suggests that cholesterol metabolism and transport may be linked to the pathogenesis of AD.

In AD brain, stimulation of PlsEtn-$PLA_2$ may produce a deficiency of docosahexaenoic acid, a ω-3 long-chain polyunsaturated fatty acid, which increases viscosity and augments energy consumption (Farooqui, 2009b; Ferrer, 2009). It is proposed that abnormal neural membrane composition may modify the activity of key enzymes that modulate the cleavage of the amyloid precursor protein to form toxic Aβ (see below).

## 8.3.2 Protein in AD

The two classical pathological hallmarks of AD are deposits of aggregated Aβ peptide and neurofibrillary tangles composed of hyperphosphorylated τ-protein. Pathophysiologic hypotheses are centered on the role of Aβ peptide and τ-protein hyperphosphorylation and mechanisms of their production in AD brain (Fig. 8.2). Experimental evidence indicates that Aβ accumulation precedes and drives τ

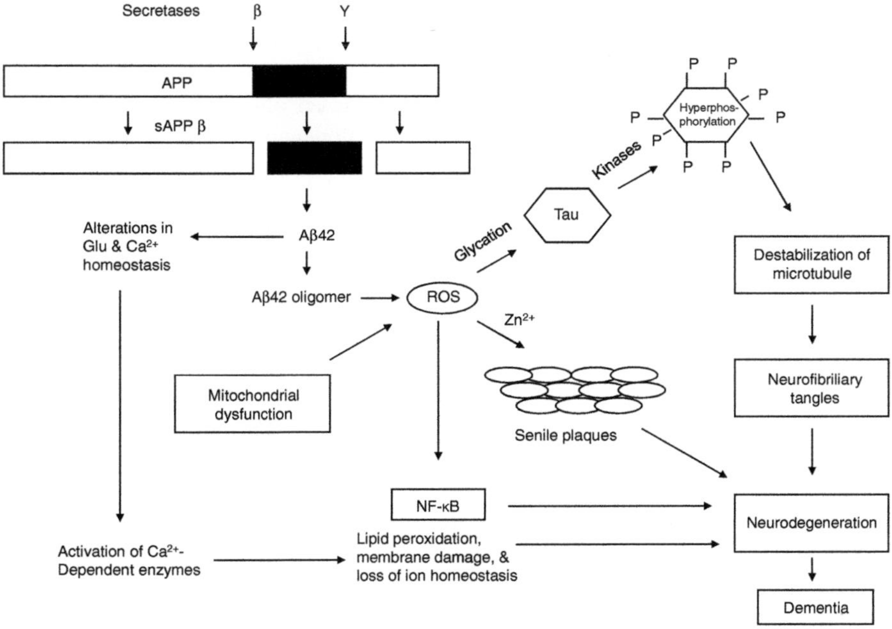


**Fig. 8.2** Hypothetical diagram showing pathogenesis of AD. Amyloid precursor protein (APP); C-terminal membrane-spanning fragment amyloid precursor proteinβ (sAPPβ); amyloid Aβ (Aβ); Glutamate (Glu); $Ca^{2+}$-dependent enzymes include phospholipase $A_2$; nitric oxide synthase, and calpains; other protein kinases include protein kinase C, ERK2, and cck5/p25

aggregation (Oddo et al., 2003). Moreover, Aβ-induced degeneration of cultured neurons and cognitive deficits in mice with an AD-like disease requires the presence of endogenous τ (Roberson et al., 2007).

Aβ peptide is generated by cleavage of the amyloid precursor protein (APP), an integral membrane protein with a large extracellular domain, by transmembrane proteases. APP is catabolized by two pathways. A non-amyloidogenic pathway involves the cleavage of APP by β-secretase within the sequence of the amyloid peptide. This cleavage precludes the formation of the full-length Aβ42 found in the amyloid core of senile plaques. A second catabolic pathway of APP leads to the production of Aβ42 from its precursor. In this amyloidogenic pathway, APP is cleaved by β-secretase at the N-terminus of Aβ. The C-terminal fragment of APP thus formed is in turn cleaved by γ-secretase to release the full-length amyloid peptide (Octave, 2005; Nathlie and Jean-Noel, 2008). The γ-secretase is identical to the presenilin proteins, PS1 and PS2, where as the β-secretase is a novel transmembrane aspartic protease called β-site APP cleaving enzyme 1 (BACE1; also called Asp2 and memapsin2). Another protease, BACE2, is homologous to BACE1 also occurs in the brain tissue.

The most abundant 40 amino acid species (Aβ40) is rather benign, whereas the less abundant 42 amino acid variant (Aβ42) aggregates much faster and may therefore be directly related to the pathogenesis of AD (Haass and Selkoe, 2007). It is not known how the addition of the two amino acids at the C-terminus of Aβ changes the biophysical properties of the peptide in a way that it aggregates faster than other species including Aβ40̃. Although all Aβ42 species are secreted from healthy neurons throughout life, why does Aβ42 species tend to form soluble oligomers (Haass and Selkoe, 2007) remains unknown. In vivo secreted oligomeric assemblies can be as small as dimers or trimers or as large as dodecamers. Prefibrillar, soluble oligomers of Aβ42 have been recognized to be early and key intermediates in AD-related synaptic dysfunction. BACE1 and 2 contribute to the formation of neuritic plaques in AD. At nanomolar concentrations, soluble oligomers of Aβ42 block hippocampal long-term potentiation, cause dendritic spine retraction from pyramidal cells, and impair rodent spatial memory (Lacor et al., 2004). Long before the onset of widespread synaptic loss and neurodegeneration, mild cognitive impairment in early AD may be due to synaptic dysfunction caused by the accumulation of non-fibrillar, oligomeric Aβ42. Soluble Aβ42 oligomers can rapidly disrupt synaptic memory mechanisms at very low concentrations via stress-activated kinases and oxidative/nitrosative stress mediators (Farooqui, 2009a). Accumulating evidence suggests that Aβ42 plays a central role in the pathogenesis of AD, and τ acts downstream of Aβ42 as a modulator of the disease progression.

Cellular prion protein ($PrP^C$) is a receptor for Aβ oligomer (Fig. 8.3). Aβ oligomers bind to $PrP^C$ with nanomolar affinity and the interaction does not require the infectious $PrP^{Sc}$ conformation (Lauren et al., 2009). Synaptic responsiveness in hippocampal slices from young adult PrP null mice is normal, but the blockade of long-term potentiation by amyloid oligomer is not observed. Anti-PrP antibodies prevent Aβ oligomer binding to $PrP^C$ and rescue synaptic plasticity in hippocampal slices from oligomeric Aβ. Thus, $PrP^C$ is a mediator of Aβ oligomer-induced

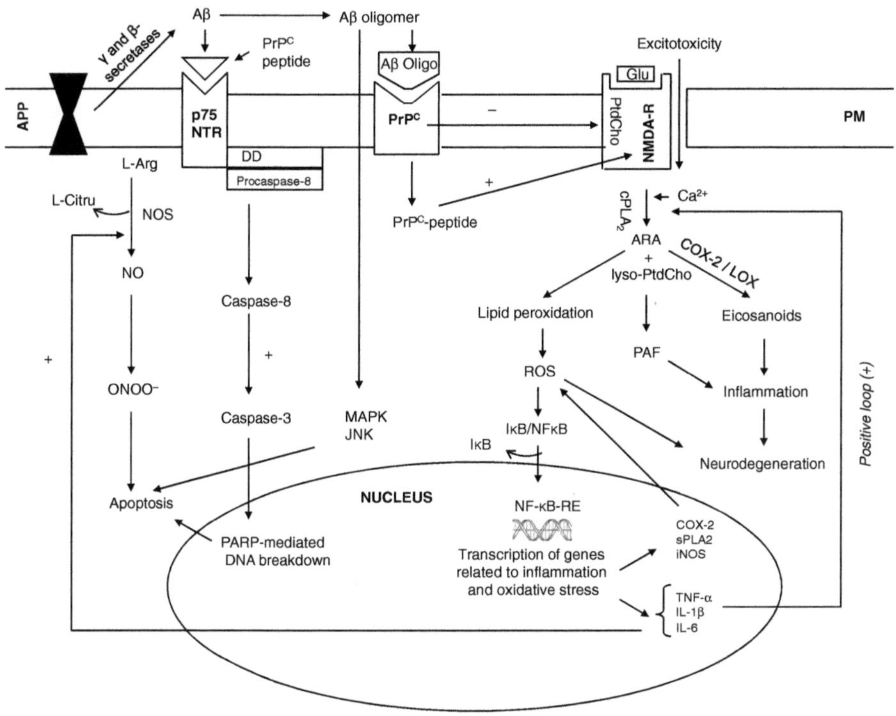


**Fig. 8.3** Interactions of Aβ42 oligomer with $PrP^C$ protein and Aβ and $p75^{NTR}$ in AD and prion diseases. The amyloid precursor protein (APP) is cleaved by β-secretase and γ-secretase to produce monomeric Aβ peptides that is transformed into toxic Aβ oligomers. β-Amyloid oligomers bind to cellular prion protein ($PrP^C$) and suppress LTP by altering neurotransmission through *N*-methyl-D-aspartate receptor (NMDA-R). Generation of prion peptide initiates downstream signal transduction processes that involve $cPLA_2$ and result in generation of lipid mediators closely associated with neuroinflammation and oxidative stress. Aβ also interacts with p75 NTR and initiates apoptosis. Amyloid precursor protein (APP); β-amyloid (Aβ); cellular prion protein ($PrP^C$); glutamate (Glu); NMDA receptor (NMDA-R); phosphatidylcholine (PtdCho); cytosolic phospholipase $A_2$ ($cPLA_2$); lyso-phosphatidylcholine (lyso-PtdCho); cyclooxygenase (COX); lipoxygenase (LOX); arachidonic acid (ARA); platelet-activating factor (PAF); reactive oxygen species (ROS); nuclear factor-κB (NF-κB); nuclear factor-κB-response element (NF-κB-RE); inhibitory subunit of NF-κB (I-κB); tumor necrosis factor-α (TNF-α); interleukin-1β (IL-1β); interleukin-6 (IL-6); inducible nitric oxide synthase (iNOS); secretory phospholipase $A_2$ ($sPLA_2$); death domain (DD); nitric oxide (NO); poly(ADP)ribose polymerase (PARP). Positive sign indicates stimulation

synaptic dysfunction. Aβ hypothesis of AD pathogenesis is based on the induction of oxidative stress (Lauren et al., 2009; Nygaard and Strittmatter, 2009). Oxidative modification of the protein results in structural modifications of proteins. This may lead to functional impairment of modified proteins. A number of oxidatively modified brain proteins have been identified using redox proteomics in AD, mild cognitive impairment (MCI), and Aβ models of AD. These findings support a role of Aβ in the alteration of a number of biochemical and cellular processes such as energy metabolism, protein degradation, synaptic function, neuritic growth, neurotransmission, cellular defense system, and long-term potentiation involved in formation of memory (Sultana et al., 2009).

Another cell surface target for Aβ is the p75 neurotrophin receptor ($p75^{NTR}$) (Chiarini et al., 2006). By using SK-N-BE neuroblastoma cells with and without neurotrophin receptors $p75^{NTR}$, it is shown that $p75^{NTR}$ mediates the Aβ-induced cell death via intracellular death domain (DD). This signaling involves activation of caspase-8, which then activates caspase-3, resulting into apoptogenesis. The binding of Aβ with p75 activates downstream signaling pathways, such as JNK, NF-κB, and PtdIns3 kinase. Several studies suggest that overexpression of $p75^{NTR}$ in a variety of cell lines confers more sensitivity to Aβ-mediated neurotoxicity (Perini et al., 2002), whereas p75-deficient mouse hippocampal neurons are resistant to Aβ-mediated neurotoxicity (Sotthibundhu et al., 2008). In addition, $p75^{NTR}$ and Aβ-mediated signaling not only involves strong cytocidal interactions with proinflammatory cytokines but also facilitates the determination of selective vulnerability of cholinergic neuronal population in basal forebrain (Fombonne et al., 2009) (Fig. 8.3). It is proposed that in AD these proinflammatory cytokines amplify neuronal death through astrocytes, which flood neurons with NO and its lethal metabolite, $ONOO^-$. Furthermore, $p75^{NTR}$ and its DD also interact with prion protein fragment PrP106–126 and induce the degeneration of SK-N-BE human neuroblastoma cells. Collective evidence suggests that neurons expressing $p75^{NTR}$ as well as proinflammatory cytokine receptors are preferential targets of Aβ and prions toxicity in AD as well as prion diseases (Chiarini et al., 2006; Bai et al., 2008).

Tau, a neuronal microtubule-bound protein, is a component of intracellular neurofibrillary tangles (NFT). Hyperphosphorylation of τ-protein is one of the critical steps in the formation of neurofibrillary tangles. Hyperphosphorylation results in an imbalance between protein kinases and protein phosphatases which are tightly regulated by the process of phosphorylation/dephosphorylation. Two main protein kinases are associated with anomalous τ phosphorylations: the cyclin-dependent kinase Cdk5 and glycogen synthase kinase GSK3β. Cdk5 plays a critical role in brain development and is involved in neurogenesis (Maccioni et al., 2001; Churcher, 2006). Deregulation of this protein kinase as induced by extracellular amyloid loading results in τ-protein hyperphosphorylations, thus triggering a sequence of molecular events that lead not only to collapse of the microtubule network and disturbances of axoplasmic transports but also to loss of synapses, neuritic atrophy, impairment in learning and memory, and neuronal death. Administration of calyculin A, a potent and specific inhibitor of protein phosphatase (PP) 2A and PP1, into rat hippocampus bilaterally induces AD-like deficiency in dephosphorylation system, resulting not only decline in memory retention ability in rats undergoing Morris water maze test, but also mediating hyperphosphorylation of τ at Ser396/Ser404 (PHF-1) and Ser-262/Ser-356 (12E8). Hyperphosphorylation of τ may be a crucial step in mediating alterations in spatial memory formation in AD and its animal model (Sun et al., 2003; Chen, 2005). Saturated free fatty acids (FFAs) such as palmitic and stearic acids promote amyloidogenesis and τ hyperphosphorylation in primary rat cortical neurons (Patil et al., 2007). These FFA-induced effects in neurons are supported and mediated by astroglial FFA metabolism. Thus, palmitic acid significantly increases de novo synthesis of ceramide in astroglia, which in turn regulates induction and upregulation of Aβ protein production and hyperphosphorylation of the τ-protein. Increased amyloidogenesis and hyperphoshorylation of τ lead to the formation of senile plaques and neurofibrillary tangles, respectively (Patil et al., 2007). In addition to above pathophysiological changes, AD is also characterized by abnormal cerebral glucose metabolism. In this context, it is shown that palmitic acid significantly decreases the levels of astroglial glucose transporter (GLUT1) and downregulates glucose uptake and lactate release by astroglia. Collective evidence suggests that saturated fatty acids may contribute to the pathophysiology of AD (Patil et al., 2007).

In contrast to the above view, it is recently proposed that generation and aggregation of Aβ, τ phosphorylation, cytoskeleton rearrangement, oxidative stress, and lipid peroxidation in AD, and a number of other neurodegenerative diseases, are secondary pathological pathophysiological processes, which represent natural compensatory mechanisms for impaired primary neurodegeneration, membrane dynamic deterioration, and/or associated failures of neurotransmission, synaptic function, and neuroplasticity (Koudinov et al., 2009). In the initial stage of AD, Aβ deposition and hyperphosphorylation of τ-protein not only upregulate the antioxidant enzymes and activate stress-activated protein kinases as compensatory responses but also modulate downstream adaptations to ensure that neuronal cells do not succumb to oxidative damage (Su et al., 2008; Petersen et al., 2007). These observations support the view that pathogenesis of AD may involve a novel balance in oxidant/antioxidant homeostasis. It is suggested that Aβ, lipid peroxidation, and τ-protein may function to sense changes in activity-dependent membrane properties, therefore, biochemically modulate membrane lipid homeostasis for more efficient synaptic action.

Although the levels of glutamate are not altered in AD, a marked reduction in the expression of NR2A and NR2B subunit mRNA in the hippocampus and entorhinal cortex in brain of AD patients and alteration in glutamate transporters have been observed in AD (Bi and Sze, 2002). This may induce changes in glutamate homeostasis in AD causing a major disturbance in $Ca^{2+}$ homeostasis and activation of $Ca^{2+}$-dependent enzymes including $PLA_2$, NOS, calpains, and downstream enzymes of arachidonic acid cascade (Farooqui and Horrocks, 2007). The aldehydic products of arachidonic acid, 4-hydroxynonenal (4-HNE), which accumulate in AD brain, co-localize with intraneuronal neurofibrillary tangles and may contribute to the cytoskeletal derangement found in AD.

In general, 4-HNE reacts with lysine, cysteine, and histidine residues in proteins (Farooqui and Horrocks, 2007). 4-HNE also modifies neprilysin (NEP), a major protease that plays a crucial role in maintaining a physiologic balance between Aβ production and catabolism (Wang et al., 2009). In addition, many proteins are targeted by ROS. In AD brain, they are generated at high concentrations due to mitochondrial dysfunction. These proteins target components of the glycolysis, lipid metabolism, and cycle of the citric acid that fuels oxidative phosphorylation, mitochondrial respiration, and energy production.

Cellular molecular chaperones (heat shock protein 70 and 90 (Hsp70 and 90)) are ubiquitous stress-induced proteins involved in preventing misfolding of different disease-associated proteins. Cellular molecular chaperones reduce the severity of several neurodegenerative disorders by providing structural integrity and proper regulation to a subset of cytosolic proteins. Thus, chaperone proteins play a neuroprotective role because of their ability to modulate the earliest aberrant protein interactions that trigger pathogenic cascades (Muchowski and Wacker, 2005; Chaudhuri and Paul, 2006). These proteins fulfill a housekeeping function in contributing to the folding, maintenance of structural integrity, and proper regulation of a subset of cytosolic proteins (King et al., 2009). Levels of small heat shock protein Hsp27 are increased in AD brains and accumulate in plaques from AD patients, but whether this represents a potentially protective response to stress or is part of the disease process is not known. Based on various studies, it is hypothesized that increased expression of Hsp27 in neurons can promote neuronal survival and stabilize the cytoskeleton in the face of Aβ exposure (King et al., 2009).

## *8.3.3 Nucleic Acid in AD*

ROS interact with nucleic acid and ROS mediate damage to nucleic acid, resulting in the production of RNA/DNA oxidative products. ROS attack on DNA bases that result in the hydroxylation, ring opening, and fragmentation. These reactions generate 8-hydroxy-2′-deoxyguanosine and 2, 6-diamino-4-hydroxy-5-formamidopyrimidine (Jenkinson et al., 1999). In addition, increased levels of 4-HNE in AD inhibit DNA synthesis (Farooqui, 2009a). RNA is more susceptible to oxidative damage than DNA because RNA is largely single stranded and its bases are not protected by hydrogen bonding and specific RNA-binding proteins (Nunomura et al., 2009). Also, in cytoplasm, cellular RNA is located close to mitochondria, which are the primary generator of ROS. In AD significant amounts of polyA$^+$ mRNAs are oxidized. The oxidation of RNA oxidation is not random but highly selective. Quantitative analysis in AD brain indicates that some mRNA species are more susceptible to oxidative damage (Shan et al., 2003). Oxidative modification can occur not only in protein-coding RNAs but also in non-coding RNAs. Damage to coding and non-coding RNAs may induce errors in proteins and alter the regulation of gene expression in AD (Nunomura et al., 2009).

Another important factor is the change(s) in microRNAs (miRNAs), which represent a family of small ribonucleic acids (21–24 nucleotide (nt) duplex RNAs) that post-transcriptionally regulate the messenger RNA (mRNA) complexity in the brain and other tissues (Yokota, 2009). Brain cells maintain distinct populations of miRNAs, which support physiologically not only normal patterns of expression but also CNS-specific gene expression during development, plasticity, aging, and diseases. Upregulation of miRNA-9, miRNA-125b, and miRNA-146a has been reported to occur in temporal cortex of AD patient at short postmortem interval (Sethi and Lukiw, 2009). This suggests unless specifically stabilized, certain brain-enriched miRNAs represent a rapidly executed signaling system employing highly transient effectors of CNS gene expression. It remains an open question whether this upregulation correlates with neuropathological changes in AD.

### *8.3.4 Transcription Factors in AD*

In general transcription factors are defined as proteins, which interact with specific DNA sequences and thereby control the transfer of genetic information from DNA to mRNA. Transcription factors perform their function alone or with other proteins in a complex, by activating (activator) or inhibiting (repressor) the recruitment of RNA polymerase to specific genes. Accumulation of Aβ protein induces ROS-mediated neurodegeneration. Redox factor-1 (Ref-1), also known as HAP1, APE or APEX, is a multifunctional protein associated with the regulation of gene transcription as well as the response to oxidative stress (Tell et al., 2005). By interacting with transcription factors, such as AP-1, NF-κB, CREB, and p53, and directly participating in the cleavage of apurininic/apyrimidinic DNA lesions, Ref-1 plays crucial roles in both cell death signaling pathways and DNA repair. Immunocytochemical studies indicate that an increased expression of APE1/Ref-1 in AD cerebral cortex compared to normal age-matched subject supports the view that the cellular adaptive response to the oxidative stress condition is involved in the pathogenesis of AD (Marcon et al., 2009).

Hypoxia-inducible transcription factor (HIF) is a transcription factor central to oxygen homeostasis. Active HIFs are heterodimers (HIF-α/β) that regulate a cassette of genes that can provide compensation for hypoxia, metabolic compromise, and oxidative stress including erythropoietin, vascular endothelial growth factor, or glycolytic enzymes (Siddiq et al., 2007). Hypoxic insult has been implicated in AD pathogenesis (Carvalho et al., 2009). Acute hypoxic injury increases the expression and the enzymic activity of BACE1 by upregulating the level of BACE1 mRNA, resulting in significant increase in the APP C-terminal fragment-β (βCTF) and Aβ (Zhang et al., 2007). In AD, the accumulation of Aβ peptide-dependent astrocyte activation causes a long-term decrease in hypoxia-inducible factor (HIF)-1α expression and a reduction in the rate of glycolysis (Schubert et al., 2009). Glial activation and the glycolytic alterations are reversed by the maintenance of HIF-1α levels with conditions that prevent the proteolysis of HIF-1α. Aβ stimulates long-term ROS production through the activation of NADP oxidase and reduces the amount of HIF-1α via the activation of the proteasome. Collectively, these studies not only suggest the importance of HIF-1α-mediated transcription in maintaining the metabolic integrity of the AD brain but also identify the probable cause of lower energy metabolism in afflicted areas (Schubert et al., 2009; Carvalho et al., 2009).

The transcription factor NF-κB controls the expression of numerous genes that modulate the immune and stress responses, onset, and the resolution of inflammation, cell adhesion, calcium homeostasis, maintenance of intercellular communications, and regulation of cellular proliferation, and protection against apoptosis (Mattson and Meffert, 2006). Activation of NF-κB in neurons promotes their survival through the expression of genes encoding anti-apoptotic proteins such as Bcl-2 and the antioxidant enzyme Mn-superoxide dismutase. In contrast, activation of NF-κB in glial and immune cells mediates pathological processes through the induction of inflammatory cytokines release and production of ROS. Chronic inflammation constitutes a risk factor for AD. The inflammatory mediators activate a signaling cascade involving NF-κB translocation to the nucleus and a beneficial or detrimental transcriptional response in neuronal and glial cells. In glial and immune cells, an inflammatory response is typically accompanied by activation of $PLA_2$, induction of arachidonic acid cascade, generation of free radicals, stimulation of cytokines, chemokines, and growth factors (Granic et al., 2009). In contrast, anti-apoptotic signaling involves the participation of Bcl-2 and the antioxidant enzyme Mn-superoxide dismutase (Mattson and Meffert, 2006).

### *8.3.5 Gene Expression in AD*

AD is a complex multifactorial disease that involves many biological processes that are controlled by many genes. Three "causative" AD genes for early-onset familial AD and one "susceptibility" gene that modulates onset of AD in familial and sporadic late-onset AD have been identified (Levy-Lahad et al., 1998; Priller et al., 2007). Two genes, namely amyloid precursor protein gene (APP gene) and the presenilin-1 and -2 genes (PS-1 and PS-2), located on chromosomes 21, 14, and 1, respectively. The third susceptibility gene, apolipoprotein E (APOE) gene, is located on chromosome 19 (Levy-Lahad et al., 1998). Mutations in genes associated with familial AD (amyloid β protein precursor, presenilin-1, or presenilin-2 gene) lead to intensification of oxidative stress. In addition, exposure to metals or pesticides may promote sporadic AD (Nunomura et al., 2007). Mutations in presenilin proteins cause the most aggressive form of familial AD. These mutations lead to altered intramembranous cleavage of the β-amyloid precursor protein by the protease called γ-secretase. γ-Secretase is a multiprotein complex composed of presenilin, nicastrin (NCT), APH-1, and PEN-2. These subunits are expressed predominantly in neurons and to some extent in axons. Their distributions and levels of expression are not affected by mutant presenilin-1. In a presenilin-1/amyloid precursor protein double knock-in mouse, γ-secretase subunits are associated with plaques. In familial AD, PS1 mutations not only result in enhanced calcium responses and increased sensitivity of cells to undergo apoptosis but also upregulate γ-secretase activity (Selkoe, 2001; Siman and Salidas, 2004; Urano et al., 2005; Popescu et al., 2004; Cedazo-Minguez et al., 2002; Priller et al., 2007). APOE ε4 gene is another important risk factor for AD. The presence of one copy of the APOE ε4 gene increases the risk of AD two to three times, and among subjects with two copies the risk is increased by 12–15 times compared to those without the ε4 allele (Petot and Friedland, 2004). Involvement of APOE ε4 gene links lipid metabolism to the pathogenesis of AD. APOE is associated with lipid transport in the blood, brain, and cerebrospinal fluid.

Using subtractive transcription-based amplification of mRNA (STAR) technology, over 800 genes have been identified in AD brain. These genes are upregulated and downregulated compared to age-matched controls. Over 55% of the sequences represent genes of unknown function and roughly half of them are novel with unknown function in the human brain. The STAR technology can be used to identify new gene sequences associated with subtle changes in gene expression that potentially contribute to the development and/or progression of AD (Liu et al., 2006). Other studies have indicated that in the late stages of AD, proinflammatory and pro-apoptotic gene expression spreads into the primary visual sensory cortex. This upregulation of pathological gene expression may be involved in the visual disturbances associated with AD (Cui et al., 2007).

### *8.3.6 Neurotrophins in AD*

Neurotrophins promote proliferation, differentiation, and survival of neurons and glia, and they mediate learning, memory, and behavior. The normally high levels of neurotrophin receptors in cholinergic neurons in the basal forebrain are severely reduced in late-stage AD. Neurotrophins play an important role in maintaining neuronal homeostasis by modulating proliferation, differentiation, and survival of neurons and glia as well as synaptic plasticity, learning, memory, and behavior (Fumagalli et al., 2008). Normal levels of neurotrophin and high density of their receptors in cholinergic neurons of the basal forebrain are severely reduced in late-stage AD. Thus, marked reduction in the levels of brain-derived neurotrophic factor (BDNF), TGF-β1, and precursor form of the nerve growth factor (proNGF) have been reported to occur in AD (Murer et al., 2001; Cotman, 2005). Reduction in levels of BDNF and its receptor, tropomyosin receptor kinase B (TrkB), is accompanied by a decrease in BDNF-mediated signaling related to synaptic dysfunction, neurodegeneration, and cognitive deficits (Murer et al., 2001; Cotman, 2005). In AD, the accumulation of Aβ aggregates and increase in TNF-α and IL-1β signaling interfere with BDNF signaling by impairing the axonal transport of BDNF in neurons of AD transgenic mice (Tg2576) (Poon et al., 2009).

In brain, BDNF also interacts with other neurotrophins such as TGF-β1, which is an anti-inflammatory cytokine. It regulates the balance between T helper-1 and T helper-2 cytokines, but it can also act as a neurotrophic factor in the CNS protecting neurons against a diverse number of insults, including excitotoxicity, hypoxia, ischemia, and most importantly β-amyloid (Caraci et al., 2008). TGF-β1 can also increase synaptic plasticity by enhancing the expression of BDNF and TrkB (Sometani et al., 2001). Significant decrease in TGF-β1 expression and signaling has been reported to occur in very early stages of AD. This impairment TGF-β1 signaling may facilitate and support to the pathogenesis of AD not only by decreasing BDNF and increasing the accumulation of Aβ but also by promoting Aβ-induced neurodegeneration in different models of AD (Wyss-Coray, 2006; Caraci et al., 2008).

The expression profiling of single cholinergic nucleus basalis (NB) neurons indicates that TrkA but not $p75^{NTR}$ mRNA is reduced in mild cognitive impairment (MCI), suggesting that reduction in neurotrophin responsiveness may be an early biomarker for AD (Mufson et al., 2007). Infact, levels of proneurotrophins, which bind to $p75^{NTR}$ to promote neuronal death, are increased in postmortem brains of AD patients. Upregulation and ligand activation of $p75^{NTR}$ have been shown to be involved in neuronal cell death in cultured cells and animal models of neurodegenerative diseases (Fujii and Kunugi, 2009). In addition, NGF precursor molecule, proNGF, is upregulated in the cortex of MCI and AD patients. Accumulation of proNGF in the presence of reduced cortical TrkA and sustained levels of $p75^{NTR}$ causes a shift in the balance between cell survival and death molecules may occur in AD. Alterations in BDNF and its precursor molecule, pro-BDNF, also coincide with changes in proNGF/NGF system. In addition, gene expression studies indicate that there is a shift in the ratio of 3-repeat τ to 4-repeat τ gene expression, whereas total τ message remains stable in NB neurons during the disease process (Mufson et al., 2007). Collective evidence suggests that alterations and interplay among BDNF, TGF-β1, and proNGF/NGF system may modulate the onset and progression of AD.

### *8.3.7 Insulin and Insulin-Like Growth Factor in AD*

Brain imaging studies show that alterations in glucose metabolism are an early sign of cognitive decline in AD. Elevation in peripheral insulin is associated with reduction in AD-related brain atrophy, cognitive dysfunction, and dementia severity, suggesting that insulin signaling may play a role in the pathophysiology of AD (Burns et al., 2007). Insulin signaling involves insulin receptor, which belongs to a subfamily of receptor tyrosine kinases that includes the IGF (insulin-like growth factor) receptor and the IRR (insulin receptor-related receptor). These receptors contain two α and two β subunits (tetrameric proteins) that function as allosteric enzymes in which the α subunit inhibits the tyrosine kinase activity of the β subunit. Decrease in expression of insulin and insulin-like growth factor type I and II (IGF-I and IGF-II) signaling in AD brains is closely associated not only with reduction in levels of insulin receptor substrate (IRS) mRNA, τ mRNA, IRS-associated phosphotidylinositol 3-kinase, and phospho-Akt (activated) but also with increased glycogen synthase kinase-3β activity, amyloid precursor protein mRNA expression, and clearance of Aβ from brain. Decrease in IGF signaling is also related to ATP levels and choline acetyltransferase (ChAT) expression (Rivera et al., 2005). Insulin and IGF also modulate the activity of excitatory and inhibitory receptors, including the glutamate and γ-aminobutyric acid receptors, and activate two biochemical pathways: the shc-ras-mitogen-activated protein kinase pathway and the PtdIns3K/PKC pathway. Both pathways are involved in memory processing. A marked decrease in CNS expression of genes encoding insulin, IGF-I, and IGF-II as well as the insulin and IGF-I receptors suggests that AD may be called as a Type 3 diabetes (Steen et al., 2005; Nelson and Alkon, 2005; Rivera et al., 2005). In addition, Aβ binds to cholesterol and catalyzes its oxidation to 7β-hydroxycholesterol, which potently inhibits isoforms of PKC, an enzyme critical in memory consolidation and synaptic plasticity, and has been implicated in pathogenesis of AD (Tanimukai et al., 2002). Oxidized cholesterol may also act as a second messenger for insulin. Oxidized low-density lipoprotein inhibits insulin-dependent phosphorylation of the signaling kinases ERK (extracellular signal-regulated kinase) and PKB/Akt. In sporadic AD patients, insulin levels are decreased, supporting the view that there is a link between AD and diabetes. Collective evidence suggests that loss of insulin function may be closely related with AD because insulin is not only a key regulator of cellular carbohydrate metabolism but also involved in other brain functions, including cognition, learning and memory, and inhibition of neuronal apoptosis (Craft and Watson, 2004).

## 8.4 Neurochemical Aspects of Parkinson Disease

PD is caused by the gradual and selective loss of dopaminergic neurons in the substantia nigra pars compacta (Beal, 1998; Jenner and Olanow, 2006). Loss of these neurons causes pathological changes in neurotransmission in the basal ganglia motor circuit. The vulnerability of dopaminergic neurons in the substantia nigra pars compacta to oxidative stress is due to monoamine oxidase-mediated abnormal dopamine metabolism and hydrogen peroxide generation. This enzyme catalyzes the oxidative deamination of dietary amines and monoamine neurotransmitters, such as serotonin, norepinephrine, dopamine, β-phenylethylamine, and other trace amines. The rapid degradation of these molecules ensures the proper functioning of synaptic neurotransmission and is critically important for the regulation of emotional behaviors and other brain functions. Since dopaminergic neurons in the substantia nigra pars compacta regulate body movement, their loss in PD result in resting tremor, rigidity, bradykinesia, postural instability, and gait disturbance in the patients with PD. The neuropathologic hallmark of PD is the presence of Lewy bodies composed mostly of α-synuclein and ubiquitin. To a minor extent other non-dopaminergic systems, such as norepinephrinergic neurons in the locus coeruleus and serotoninergic neurons in the raphe nuclei, are also affected by the pathological processes in PD.

In addition, in vivo brain imaging studies show significant increase of iron levels in the substantia nigra pars compacta in PD (Gerlach et al., 2006). This increase in iron, however, occurs only in the advanced stages of PD, suggesting that this phenomenon may be a secondary rather than a primary initiating event in the disease process. The major pathways associated with pathophysiology of sporadic and familial PD involve mitochondrial dysfunction, free radical generation, oxidative and nitrosative stress, glutamate receptor-mediated excitotoxicity, inflammation, oligodendrocytic interaction and neurotrophic factors, accumulation of aberrant or misfolded proteins, and ubiquitin-proteasome system dysfunction (Fig. 8.4) (Beal, 1998; Jenner and Olanow, 2006).

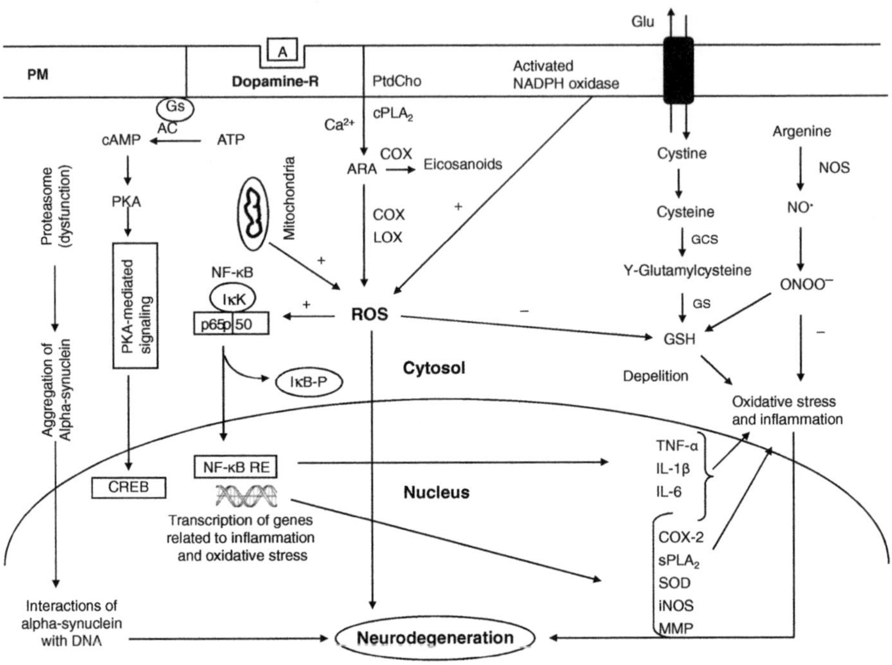


**Fig. 8.4** Hypothetical diagram showing involvement of ROS and peroxynitrite in pathogenesis of PD. Cytosolic phospholipase $A_2$ ($cPLA_2$); cyclooxygenase (COX); lipoxygenase (LOX); reactive oxygen species (ROS); arachidonic acid (ARA); agonist (A); dopamine receptor (dopamine-R); nuclear factor-κB (NF-κB); nuclear factor κB-response element (NF-κB-RE); inhibitory subunit of NF-κB (I-κB); tumor necrosis factor-α (TNF-α); interleukin-1β (IL-1β); interleukin-6 (IL-6); inducible nitric oxide synthase (iNOS); nitric oxide (NO$^{\cdot}$); glutathione (GSH); glutathione synthase (GS); γ-glutamylcysteine synthase (GCS); superoxide dismutase (SOD1); secretory phospholipase $A_2$ ($sPLA_2$). Positive sign indicates stimulation

## *8.4.1 Lipids in PD*

Very little information is available on the neural membrane glycerophospholipid composition of PD patients (Farooqui and Horrocks, 1998). Studies on determination of glycerophospholipid composition in brains from wild-type and α-synuclein –/– mice indicate that total brain glycerophospholipid mass is not altered, but cardiolipin and phosphatidylglycerol masses are decreased by 16% and 27%, respectively. No changes are observed in plasmalogen and polyphosphoinositide. In ethanolamine glycerophospholipids and phosphatidylserine, DHA is decreased 7%, while palmitic acid is increased 1.1-fold and 1.4-fold (Barcelo-Coblijn et al., 2007). Although the exact mechanism of α-synuclein-mediated changes in fatty acid metabolism is not known, it is suggested that α-synuclein facilitates the incorporation of fatty acid in glycerophospholipids (Barcelo-coblijn et al., 2007; Golovko et al., 2006).

**Table 8.3** Neurochemical alterations in Parkinson disease

| Neurochemical parameter | Effect | References |
|---|---|---|
| Glycerophospholipid metabolism | Altered | Farooqui and Horrocks (2007) |
| Free fatty acid composition | Altered | Farooqui and Horrocks (2007) |
| $PLA_2$ activity | Increased | Yoshinaga et al. (2000), Lee et al. (2009b) |
| Eicosanoids | Increased | Farooqui and Horrocks (2007) |
| Lipid peroxidation | Increased | Farooqui and Horrocks (2007) |
| 4-Hydroxynonenal | Increased | Farooqui and Horrocks (2007) |
| Hydroxycholesterol | Increased | Seet et al. (2009) |
| 8-OHdGua | Increased | Seet et al. (2009) |
| Parkin | Abnormal | Burler (2009) |
| PINK | Increased | Bueler (2009) |
| Aggregated α-synuclein | Increased | Beyer (2007) |
| NF-κB | Upregulated | Ghosh et al. (2007) |
| Synapse integrity | Lost | Farooqui (2009a) |
| Excitotoxicity | Increased | Farooqui (2009a) |
| Oxidative stress | Increased | Farooqui (2009a) |
| Neuroinflammation | Increased | Farooqui (2009a) |
| Neurodegeneration | Increased | Farooqui (2009a) |

Studies on *N*-methyl,4-phenyl-1,2,3,6 tetrahydropyridine (MPTP)-induced model of PD indicate that activities of $PLA_2$, COX-2, and rate of lipid peroxidation are increased (Table 8.3), and $PLA_2$, COX inhibitors, and some antioxidants protect neural cells from MPTP-mediated neurodegeneration (Yoshinaga et al., 2000; Teismann et al., 2003; Farooqui et al., 2006). Similarly, arachidonic acid signaling is upregulated in the caudate-putamen and frontal cortex of unilaterally 6-hydroxydopamine lesioned rats, a model for asymmetrical PD. Stimulation of $D_2$-like dopamine receptor initiates arachidonic acid release from glycerophospholipids by $cPLA_2$ and subsequent metabolism by COX-2 (Lee et al., 2009a). Generation of $PLA_2$ and COX-2-derived lipid mediators (4-HNE, $F_2$-IsoP, HETEs) in brain and plasma of PD patients may contribute to neuroinflammation and oxidative stress, which promotes the progressive loss of dopaminergic nigral neurons (Table 8.3) (Beal, 1998; Jenner and Olanow, 2006; Farooqui and Horrocks, 2007; Seet et al., 2009). In addition, levels of plasma 27-hydroxycholesterol, 7-ketocholesterol, $F_4$-NPs, and urinary 8-hydroxy-2′-deoxyguanosine (8-OHdG) are also increased in the earlier stages of PD (Seet et al., 2009). Recent studies indicate that 27-hydroxycholesterol can cross the blood–brain barrier and may facilitate neurodegeneration (Rantham Prabhakara et al., 2008; Heverin et al., 2005). Incubation of the human neuroblastoma SH-SY5Y cells with 24-hydroxycholesterol, 27-hydroxycholesterol, or a mixture of 24-hydroxycholesterol plus 27-hydroxycholesterol for 24 h indicate that 24-hydroxycholesterol increases the levels of tyrosine hydroxylase and 27-hydroxycholesterol increases levels of α-synuclein and induces apoptosis (Rantham Prabhakara et al., 2008). Collectively, these studies indicate that oxysterols can trigger changes in levels of proteins that are associated with the pathogenesis of PD. In normal and PD brain, hydroxysterols and cholesterol oxides may be involved in the modulation of sphingolipid metabolism, platelet aggregation, apoptosis, and protein prenylation. Increase in levels of above metabolites is an indication of oxidative damage in PD.

### *8.4.2 Proteins in PD*

The synucleins (α-, β-, and γ-synucleins) are a small, soluble, highly conserved group of neuronal proteins that have been implicated in both neurodegenerative diseases and cancer (Ahmad et al., 2007). A typical structural feature of synucleins is the presence of a repetitive, degenerative AA motif KTKEGV throughout the first 87 residues and acidic stretches within the C-terminal region. α-, β-, and γ-synucleins share sequence homologies and structural properties. Although roles of the synucleins in neural and non-neural tissues are still unclear at the present time, their involvement in the pathogenesis of PD and cancer may provide insights into the pathological processes. Recently, elevated levels of γ-synuclein proteins have been detected in various types of cancer, especially in advanced stages of the disease (Ahmad et al., 2007).

Dominant mutations in the gene that encodes α-synuclein, a small protein containing 140 amino acids, is widely distributed throughout the brain, may be closely associated with pathophysiology of PD. α-Synuclein has been identified in the presynaptic terminals and in the synaptosomal preparations. It occurs as a monomer in an aqueous solution. Self-aggregation leads to a variety of β-structures, while membrane association may result in the formation of an amphipathic helical structure (Beyer, 2007). Accumulating evidence suggests that α-synuclein becomes toxic to vulnerable neurons as a result of its tendency to aggregate. Under in vitro conditions conversion from monomer to aggregate is complex, and aggregation rates are sensitive to changes in amino acid sequence and environmental conditions. α-Synuclein aggregates faster at low pH than at neutral pH. In vivo, several aggregation mechanisms have been described. Purified tissue transglutaminase (tTGase) catalyzes α-synuclein cross-linking that leads to the formation of high molecular weight aggregates in vitro, and overexpression of tTGase produces detergent-insoluble α-synuclein aggregates in the cellular models (Junn et al., 2003). Immunocytochemical studies indicate the presence of α-synuclein-positive cytoplasmic inclusions in 8% of tTGase-expressing cells. The formation of α-synuclein aggregates is significantly inhibited by the calcium ionophore and abolished by the inhibitor cystamine (Junn et al., 2003). Immunohistochemical studies in PD brain tissue confirm the presence of transglutaminase-catalyzed epsilon (γ-glutamyl)lysine cross-links in the halo of Lewy bodies in PD and dementia with Lewy bodies, colocalizing with α-synuclein. These observations support the view that tTGase activity leads to α-synuclein aggregation to form Lewy bodies and perhaps contributes to neurodegeneration (Junn et al., 2003). Another mechanism of α-synuclein toxicity indicates that oligomers of α-synuclein consist of spheres, chains, and rings (Rochet et al., 2004). α-Synuclein protofibrils permeabilize synthetic vesicles and produce pore-like assemblies on the surface of brain-derived vesicles. Dopamine reacts with α-synuclein to form a covalent adduct that slows the conversion of protofibrils to fibrils (oligomers and insoluble fibrils with increased ss-sheet configuration) indicating that cytosolic dopamine in dopaminergic neurons promotes the accumulation of toxic α-synuclein protofibrils (Rochet et al., 2004). In the third mechanism α-synuclein forms a triple complex with anionic lipids (such as cardiolipin) and cytochrome *c*, which exerts a peroxidase activity (Bayir et al., 2009). The latter catalyzes covalent hetero-oligomerization of α-synuclein with cytochrome *c* into high molecular weight aggregates. α-Synuclein is a preferred substrate of this reaction and is oxidized more readily than cardiolipin, dopamine, and other phenolic substrates. α-Synuclein-cardiolipin complex protects against cytochrome *c*-mediated caspase-3 activation in a cell-free system, particularly in the presence of $H_2O_2$ (Bayir et al., 2009) Direct delivery of α-synuclein into mouse embryonic cells induces resistance to pro-apoptotic caspase-3 activation, but small interfering RNA-mediated depletion of α-synuclein in HeLa cells makes them more sensitive to dopamine-mediated apoptosis (Bayir et al., 2009). Thus, α-synuclein aggregates are toxic and major components of Lewy bodies found in PD (Moore et al., 2005). Although the precise nature of in vivo α-synuclein function remains elusive, there are evidences indicating its involvement in the regulation of vesicular release and/or turnover and synaptic function in the central nervous system. It is also suggested that this protein not only plays a role in neuronal plasticity responses, binds fatty acids, but also may be involved in the regulation of certain enzymes, transporters, neurotransmitter vesicles, and neuronal survival or even act as a molecular chaperone (Uversky, 2008). Although the molecular mechanism involved in α-synuclein-mediated neurodegeneration in PD is not known, induction of oxidative stress may be responsible for the neurodegeneration in PD (Kumar et al., 2005). Genetic studies indicate that point mutations or genetic alteration (duplications or triplications) that increases the number of copies of the α-synuclein gene can cause PD or the related disorder dementia with Lewy bodies (Uversky, 2008). Although there is a substantial evidence supporting the toxic nature of α-synuclein inclusions, other modes of toxicity such as oligomers have also been suggested (Hegde et al., 2010).

In vitro studies on the interactions between several brain sphingolipids and α-synuclein indicate that α-synuclein specifically binds to ganglioside $GM_1$-containing small unilamellar vesicles (SUVs) (Martinez et al., 2007). This results in the induction of substantial α-helical structure and inhibition or elimination of α-synuclein fibril formation, depending on the amount of $GM_1$ present. SUVs containing total brain gangliosides, gangliosides $GM_2$ or $GM_3$, or asialo-$GM_1$ produce weak inhibitory effects on α-synuclein fibrillation and induce some α-helical structure, while all other sphingolipids studied show negligible interaction with α-synuclein. α-Synuclein binding to $GM_1$-containing SUVs is accompanied by the formation of oligomers of α-synuclein (Martinez et al., 2007). In the familial mutant A53T, α-synuclein binds with $GM_1$-containing SUVs in an analogous manner to wild-type, whereas the A30P mutant shows minimal interaction, indicating that interactions between $GM_1$ and α-synuclein may be attributed to both the sialic acid and the carbohydrate moieties of $GM_1$. The recruitment of α-synuclein by $GM_1$ to caveolae and lipid raft regions in membranes may explain α-synuclein's localization to presynaptic membranes and raises the possibility that perturbation of $GM_1$/raft association may induce changes in α-synuclein that contributes to the pathogenesis of PD (Martinez et al., 2007). Parkin, an $E_3$ ubiquitin-protein ligase, involved in the degradation of cellular proteins by the proteasomal pathway has been recently shown to protect cells against α-synuclein toxicity (Baptista et al., 2004; Burke, 2004). Deletions or point mutations in the gene for parkin also cause an autosomal recessive, early-onset form of PD. It is possible that mutations and interactions between α-synuclein and parkin genes may play important roles in the pathophysiology of idiopathic PD.

Generation of excessive nitric oxide (NO) facilitates protein misfolding (Nakamura and Lipton, 2008). S-Nitrosylation, which is a covalent reaction of a NO group with a cysteine thiol, represents one such mechanism. NO contributes to degenerative conditions by *S*-nitrosylating protein disulfide isomerase (PDI) (forming SNO-PDI) and the ubiquitin-protein ligase, parkin (forming SNO-parkin). It is reported that addition of memantine, an uncompetitive inhibitor of NMDA receptor, ameliorates excessive production of NO, protein misfolding, and neurodegeneration (Nakamura and Lipton, 2008).

### 8.4.3 Nucleic Acids in PD

Interactions between DNA and α-synuclein have been recently observed. Thus, double-stranded oligos induce partial folding in α-synuclein and promote its aggregation, whereas single-stranded circular DNA and supercoiled plasmid DNA produce a helix-rich conformation and protect the protein from fibrillation. In turn, α-synuclein induces DNA conformation from B- to an altered B-form, which may modulate DNA transactions (Fig. 8.4) (Hegde et al., 2010). Studies on the effect of osmolytes on DNA-induced folding/aggregation of α-synuclein as a model system indicate that glycerol, trimethylamine-*N*-oxide, betaine, and taurine induce partially folded conformation and in turn enhance the aggregation of α-synuclein. The ability of DNA and osmolytes in inducing conformational transition in α-synuclein indicates that two factors are closely associated with α-synuclein folding: (a) electrostatic interaction as in the case of DNA and (b) hydrophobic interactions as in the case of osmolytes. DNA-induced changes in α-helical conformation of α-synuclein and inhibition of the fibrillation may be important in developing engineering DNA chip-based therapy of PD (Hegde et al., 2010).

In addition, marked oxidative damage to nucleic acids (DNA and RNA) has been reported to occur in PD. Oxidative insults are more pronounced in RNA than DNA because RNA is mostly single stranded and its bases are not protected by hydrogen bonding and specific proteins (Nunomura et al., 2007). As stated above, the oxidative damage to RNA may result in errors in proteins expression or dysregulation of gene expression. Studies on analysis of oxidized RNA species have revealed that both messenger RNA (mRNA) and ribosomal RNA (rRNA) are damaged not only in PD but also in AD and ALS (Kong et al., 2008; Nunomura et al., 2007). The magnitude of the RNA oxidation, at least in mRNA, is significantly high at the early stage of these neurodegenerative diseases. Oxidative damage to mRNA is not random but selective, and many oxidized mRNAs are related to the pathogenesis of the disease. It is suggested that oxidative damage to RNA may cause alterations in the translational process and resulting in the expression of less protein and/or defective protein. Thus, RNA damage in PD may contribute to neurochemical alterations related to the onset or development of this disease in aged brain. Although the molecular sequence associated with the effect of oxidative RNA damage to protein synthesis is attenuated, at least in part, by the existence of mechanisms that avoid the incorporation of the damaged ribonucleotides into the translational machinery, studies on consequences and processing mechanisms are beginning to emerge (Kong et al., 2008; Nunomura et al., 2007).

### *8.4.4 Transcription Factors in PD*

Alterations in transcription factors associated with oxidative stress and neuroinflammation have been reported to occur in PD (Fig. 8.4). Thus, activation of transcription factor, NF-κB occurs in the substantia nigra pars compacta of PD patients as well as in MPTP-intoxicated mice (Ghosh et al., 2007). Injections (i.p.) of wild-type NF-κB essential modifier-binding domain (NBD) peptide not only inhibits nigral activation of NF-κB and suppresses nigral microglial activation but also protects both the nigrostriatal axis and neurotransmitters and improves motor functions in MPTP-injected mice (Ghosh et al., 2007). The mutated NBD peptide has no effect on NF-κB. Another transcription factor called as Nurr1 is critical in the development and maintenance of the dopaminergic system and may be involved in the pathogenesis of PD. Human Nurr1 gene has been mapped to chromosome 2q22–23. It regulates the expression of tyrosine hydroxylase (TH), dopamine transporter (DAT), vesicular monoamine transporter 2 (VMAT2), and l-aromatic amino acid decarboxylase (AADC), all of which are important in the synthesis and storage of dopamine (Jankovic et al., 2005). Studies on Nurr1 knockout mice indicate that Nurr1 deficiency results in impaired dopaminergic function and increased vulnerability of those midbrain dopaminergic neurons that degenerate in PD. Decreased Nurr1 expression has been reported in the PD midbrains autopsies, particularly in neurons containing Lewy bodies, as well as in peripheral lymphocytes of patients with PD (Backman et al., 1999; Jankovic et al., 2005). Nurr1-mediated signaling involves retinoid X receptor (RXR). Heteromerization of Nurr-1 with RXR (Nurr1-RXR heterodimers) facilitates the survival of DA neurons (Perlmann and Wallen-Mackenzie, 2004). Collective evidence suggests that not only is Nurr1 essential for the development of mensencephalic dopaminergic neurons and maintenance of their functions but it may also play a role in the pathogenesis of PD.

As stated above, HIF is a transcription factor central to oxygen homeostasis. It mediates complex adaptations to reduce oxygen supply (Siddiq et al., 2007). Modulation of HIF activity occurs mainly through oxygen-dependent destruction of its alpha subunit. In the presence of oxygen, two HIFα prolyl residues undergo enzymic hydroxylation, which is required for its proteasomal degradation. Under hypoxic conditions, the $O_2$-labile a-subunit of HIF is translocated to the nucleus, where it targets genes, such as enolase1 and vascular endothelial growth factor. The translational products of these genes (erythropietin and glycolytic enzymes) increase $O_2$ delivery to hypoxic tissues (Bruegge et al., 2007). HIF prolyl 4 hydroxylase, inhibitor (3,4-dihydroxybenzoate) protects nigral dopaminergic neurons from 1-methyl-4-phenyl-1,2,3,6-tetrahydropyridine (MPTP)-induced neurotoxicity by upregulating HIF-1α. Furthermore, elevations in mRNA and protein levels of HIF-dependent genes heme oxygenase-1 (HO-1) and manganese super oxide dismutase (Mn-SOD) following 3,4-dihydroxybenzoate pretreatment alone are also maintained in the presence of MPTP (Lee et al., 2009b). Collective evidence suggests that HIF-1α plays an important role in cell survival by regulating iron, antioxidant defense, and mitochondrial function (Bruegge et al., 2007; Lee et al., 2009b).

### *8.4.5 Gene Expression in PD*

Two forms of PD, namely sporadic and familial, occur in humans. Although the cause of sporadic PD is unknown, familial form of PD has been linked to mutations in genes for α-synuclein, Parkin (PARK2), DJ-1, PTEN-induced kinase 1 (PINK1), ubiquitin-C-terminal hydrolase-L1 (UCH-L1), and leucine-rich repeat kinase 2 (LRRK2). The discovery of these genes provides new avenues to study PD pathogenesis and the mechanisms underlying the selective dopaminergic neuron death in PD (Bueler, 2009). As stated above, studies in humans, as well as molecular studies in toxin-mediated and genetic animal models of PD show that mitochondrial dysfunction is a defect occurring early in the pathogenesis of both sporadic and familial PD. PINK1 and Parkin play crucial roles in the regulation of mitochondrial dynamics and function (Bueler, 2009). The PINK1/Parkin pathway promotes mitochondrial fission and that the loss of mitochondrial and tissue integrity in PINK1 and parkin mutants derives from reduced mitochondrial fission (Poole et al., 2008). Thus, mutations in Parkin render animals more susceptible to oxidative stress and mitochondrial toxins implicated in sporadic PD, supporting the hypothesis that some PD cases may be caused by gene–environmental factor interactions. In addition, mutations in the gene encoding LRRK2 have also been linked to autosomal dominant, late-onset PD that is clinically indistinguishable from typical, idiopathic disease (Gandhi et al., 2009). LRRK2 protein contains two functional domains, namely MAPKKK-like kinase and Rab-like GTPase domains. Emerging evidence shows that LRRK2 contains kinase and GTPase activities, which are enhanced in several PD-associated mutants of LRRK2. PD-associated mutations are found throughout the multidomain structure of the protein. LRRK2, however, is unique among the PD-causing genes, because a missense mutation, G2019S, is a frequent determinant of not only familial but also sporadic PD (Gandhi et al., 2009). Disease-associated mutations in LRRK2 also promote and facilitate the formation of cytoplasmic inclusions and induce neuronal toxicity in cultured cells in a kinase-dependent manner.

DJ-1 interacts with mRNA in an oxidation-dependent manner. The oxidation of DJ-1 occurs more in cortex from cases of sporadic PD compared to controls (Blackinton et al., 2009). These observations suggest that in PD post-transcriptional modification of many proteins level may involve translational regulation by DJ-1. Measurement of protein and RNA expression for four DJ-1 target genes GPx4, MAPK8IP1, ND2, and ND5 indicates an increase in GPx4 and MAPK8IP1 protein expression in PD cases. Furthermore, same patients show a decrease in mRNA and protein levels of two mitochondrial targets, ND2 and ND5, suggesting that these proteins may undergo regulation at the post-transcriptional level that may involve translational regulation by DJ-1 (Blackinton et al., 2009). Collective evidence suggests that compromising cellular energy production, mitochondrial dysfunction, aberrant or misfolded protein deposition, oxidative stress, and induction of apoptosis may be closely associated with the pathogenesis of PD (Bueler, 2009).

### *8.4.6 Neurotrophins in PD*

Many animal studies indicate that the glial cell line-derived neurotrophic factor (GDNF) has strong neuroprotective and neurorestorative effects on dopaminergic neurons. Continuous intraputaminal infusion of GDNF in animal models of PD indicates that GDNF not only produces beneficial effects (Eslamboli, 2005) but also boosts the functional outcome of widespread intrastriatal dopaminergic grafts in intrastriatal transplantation experiments (Winkler et al., 2006). Positive results in monkeys have encouraged the use of GDNF in human trials. These trials have shown mixed results, which may be due to the influence of parameters related to administration procedures on the clinical outcome (Eslamboli, 2005; Yasuhara et al., 2007). GDNF has tolerance with few side effects and clinical benefits following 3 months of the treatment. The clinical improvement is sustained and progressive, and by 24-months patients show a 57 and 63% improvement in their off-medication motor activities of daily living along with better UPDRS subscores with clear benefit in dyskinesias (Patel and Gill, 2007). For GDNF treatment to become a clinical reality, appropriate delivery techniques will have to be developed. Studies on the potential of encapsulated cells and viral vectors to locally release neurotrophic factors in experimental models of PD are at the present time in progress.

In addition, $p75^{NTR}$, the low-affinity NGF receptor, acts as a "molecular signal switch" that determines cell death or survival through several mechanisms (Chen et al., 2008a). First, proNGF triggers neural cell death by its high-affinity binding to $p75^{NTR}$, while NGF induces neuronal survival with low-affinity binding. Second, $p75^{NTR}$ induces cell death by combining with co-receptor sortilin, whereas it promotes neuronal survival through combination with proNGF. Third, release of the intracellular domain chopper or cleavaged "short $p75^{NTR}$" can independently initiate neuronal apoptosis. Thus through these cell self-destructive proNGF-$p75^{NTR}$-sortilin signaling apparatus, dopaminergic neurons in the substantia nigra pars compacta may die via $p75^{NTR}$ signaling in PD (Chen et al., 2008a).

## 8.5 Neurochemical Aspects of Amyotropic Lateral Sclerosis

ALS is a major motor neuronal disorder that causes progressive loss of neurons leading to muscle loss, paralysis, and death from respiratory failure. An important pathological hallmark of ALS is the presence of axonal spheroids and perikaryal accumulations/aggregations comprised of the neuronal intermediate filament proteins, neurofilaments, and peripherin (Beaulieu and Julien, 2003). Although the exact cause of neurodegeneration in ALS is not known, multiple pathophysiological mechanisms that trigger the loss of motor neuron have been proposed. These mechanisms include oxidative stress, mitochondrial impairment, protein aggregation, axonal dysfunction, reactive astrocytosis, mutant superoxide dismutase expression, peroxynitrite toxicity, cytoskeletal disorganization, glutamate cytotoxicity, transcription dysfunction, inflammation, and apoptotic cell death (Table 8.4) (Farooqui and Horrocks, 2007). It is suggested that synergistic interactions among excitotoxicity, oxidative stress, and neuroinflammation may play a major role in pathogenesis of ALS (Fig. 8.5) (Shaw and Ince, 1997; Rao and Weiss, 2004; Farooqui and Horrocks,

**Table 8.4** Neurochemical alterations in amyotrophic lateral sclerosis

| Neurochemical parameter | Effect | References |
|---|---|---|
| Glycerophospholipid metabolism | Altered | Farooqui and Horrocks (2007) |
| Free fatty acid composition | Altered | Farooqui and Horrocks (2007) |
| $PLA_2$ activity | Not known | – |
| Eicosanoids | Increased | Kivenyi et al. (2004) |
| Lipid peroxidation | Increased | Kivenyi et al. (2004) |
| 4-Hydroxynonenal | Increased | Kim et al. (2009) |
| Cholesterol ester | Increased | Cutler et al. (2004) |
| 8-OHdGua | Increased | Bogdanov et al. (2000) |
| SOD1 processing | Abnormal | Beaulieu and Julien (2003) |
| SOD1 activity | Abnormal | Beaulieu and Julien (2003) |
| E-selectin | Increased | Sathasivam (2010) |
| NF-κB | upregulated | Migheli et al. (1997) |
| Synapse integrity | Lost | Farooqui (2009a) |
| Excitotoxicity | Increased | Farooqui (2009a) |
| Oxidative stress | Increased | Drachman et al. (2002) |
| Neuroinflammation | Increased | Rao and Weiss (2004) |
| Neurodegeneration | Increased | Kabashi et al. (2007) |

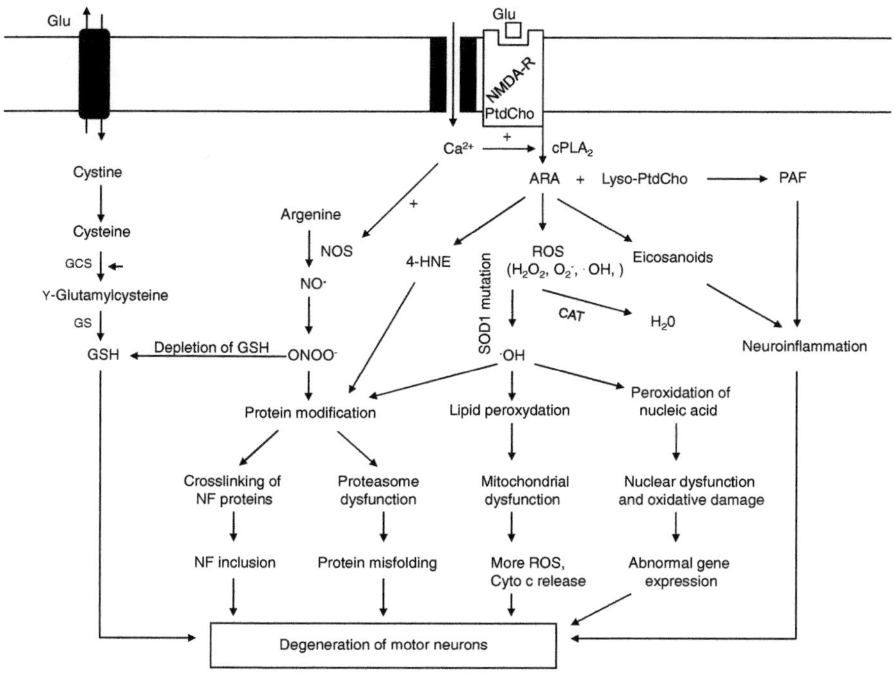


**Fig. 8.5** Hypothetical schematic model showing degeneration of motor neurons in ALS. Glutamate (Glu); NMDA receptor (NMDA-R); phosphatidylcholine (PtdCho); cytosolic phospholipase $A_2$ ($cPLA_2$); lyso-phosphatidylcholine (lyso-PtdCho); arachidonic acid (ARA); platelet-activating factor (PAF); superoxide dismutase (SOD1); reactive oxygen species (ROS); hydrogen peroxide ($H_2O_2$); superoxide (); hydroxyl radical (·OH); catalase (CAT); nitric oxide (NO·); peoxynitrite ($ONOO^-$); nitric oxide synthase (NOS); neurofilament (NF); glutathione (GSH); glutathione synthase (GS); γ-glutamylcysteine synthase (GCS); and glutamate (Glu)

2007). In addition, there is evidence for the involvement of immune system in the ALS, and activation of components of the classical complement pathway have been observed in the serum, cerebrospinal fluid, and neuronal tissue of diseased individuals (Woodruff et al., 2008). Thus, some patients of ALS have antibodies against ganglioside complexes including $GM_2$ and $GD_2$ gangliosides and GalNAc-$GD_1$a (Mizutani et al., 2003; Yamazaki et al., 2008) along with other components as cholesterol which are known to form lipid rafts in which the carbohydrate portions of above gangliosides may form a new conformational epitope. Within the rafts, gangliosides interact with important receptors or signal transducers. The antibodies against ganglioside complexes may therefore directly cause nerve conduction failure and severe disability, which ultimately may contribute to the degeneration of motor neurons in ALS (Mizutani et al., 2003; Yamazaki et al., 2008). Occurrence of antibodies to sulfoglucuronyl paragloboside (SGPG) has also been reported in ALS, although the pathogenic significance of the antibodies is still unknown (Ikeda et al., 2000). Levels of sE-selectin are significantly increased in patients with ALS with other neurological diseases. It is proposed that anti-SGPG antibodies may be responsible for the activation of endothelial cells in ALS and the increased expression of E-selectin may be related to immunological disturbances in some ALS patients (Ikeda et al., 2000).

ALS occurs in sporadic and familial forms. The pathogenesis of neuronal degeneration in both sporadic and familial ALS may involve mutations in copper/zinc superoxide dismutase, mitochondrial dysfunction (alterations in respiratory complexes I and III), protein aggregation, and neuroinflammation (Almer et al., 2001; Liu et al., 2002). Cytosolic Cu/Zn superoxide dismutase (SOD1) is a ubiquitous small cytosolic metalloenzyme that catalyzes the conversion of superoxide anion to hydrogen peroxide. The mutant copper/zinc superoxide dismutase exhibits a toxic gain of function that adversely affects the function of neurons in the spinal cord, brain stem, and motor cortex.

Oxidation of wild-type SOD1 results in its misfolding, causing it to gain many of the same toxic properties as mutant SOD1 (Kabashi et al., 2007). In vitro studies of oxidized/misfolded SOD1 and in vivo studies of misfolded SOD1 indicate that these protein species are selectively toxic to motor neurons, supporting the view that oxidized/misfolded SOD1 may lead to ALS even in individuals who do not carry an SOD1 mutation. It is also shown that glial cells secrete oxidized/misfolded mutant SOD1 to the extracellular environment, where it can trigger the selective death of motor neurons, offering a possible explanation for the noncell autonomous nature of mutant SOD1 toxicity and the rapid progression of disease once the first symptoms develop (Kabashi et al., 2007). The mechanism by which mutant SOD1s cause ALS is not understood. Transgenic mice expressing multiple copies of fALS-mutant SOD1s develop an ALS-like motoneuron disease resembling ALS. The sporadic form of ALS is characterized by a prominent neuroinflammatory component, upregulation of COX-2 (Fig. 8.5) mRNA, and oxidative stress along with abnormalities in glutamate homeostasis (Drachman and Rothstein, 2000; Yasojima et al., 2001; Drachman et al., 2002). Oral administration of either celecoxib or rofecoxib, inhibitors of COX-2 enzyme not only significantly improve motor performance, attenuate weight loss, and extend survival but also significantly reduce prostaglandin $E_2$ levels at 110 days of age. The combination of creatine with COX-2 inhibitors causes additive neuroprotective effects and extends survival by approximately 30% (Kivenyi et al., 2004).

### *8.5.1 Lipids in ALS*

Studies on spinal cord myelin lipid composition in mouse model of ALS indicate that levels of lipids, phospholipids, cholesterol, and cerebrosides are decreased compared to wild-type mice (Table 8.4) (Niebroj-Dobos et al., 2007). Although a progressive decrease in proteolipid, DM-20, and Wolfgram proteins occurs in this ALS model, myelin basic proteins I and II are not affected. Electron microscopy indicates massive myelin disorganization (Niebroj-Dobos et al., 2007). Production and high concentrations of ROS in ALS may contribute to neural membrane damage, which may cause an uncontrolled sustained increase in calcium ion influx resulting into increased membrane permeability and stimulation of many calcium-dependent enzymes associated with lipolysis, proteolysis, and disaggregation of microtubules with a disruption of cytoskeleton and membrane structure. Arachidonic acid is metabolized to 4-hydroxynonenal (4-HNE), an α, β unsaturated aldehyde, which reacts with nucleophilic sites of proteins on lysine, cysteine, and histidine residues. In G93A SOD1 Tg mice, levels of 4-HNE are increased around zinc-accumulating cells and mSOD1-positive cells, suggesting a link between 4-HNE, SOD1 mutation, and zinc accumulation. The exposure of G93A SOD1 Tg mice cultured spinal neurons and astrocytes to 4-HNE increases labile zinc levels and decreases glutamate transporter (Kim et al., 2009; Yao, 2009). Administration of the zinc chelator TPEN increases the survival of G93A SOD1 Tg mice suggesting that zinc dyshomeostasis occurs in the spinal cords of Tg mice, and that this dyshomeostasis may contribute to motoneuron degeneration (Kim et al., 2009; Yao, 2009). Levels of 4-HNE are also elevated in spinal cord motor neurons in the cerebrospinal fluid of patients with ALS (Vigh et al., 2005; Perluigi et al., 2005). Furthermore, inhibition of $Na^+.K^+$-ATPase by 4-HNE can result in the depolarization of neuronal membranes leading to the opening of NMDA receptor channels and influx of additional $Ca^{2+}$ into the cell. In cortical neurons, 4-HNE disrupts G protein-linked muscarinic cholinergic receptors (mAChR) and metabotropic glutamate receptors (mGluRs). This may alter the activity of phospholipase C and phospholipase $A_2$, indicating that 4-HNE modulates signal transduction processes in brain tissue.

Elevations in levels of sphingomyelin, ceramides, and cholesterol esters have been reported in the spinal cords of ALS patients and in a transgenic mouse model (Cu/ZnSOD mutant mice) (Cutler et al., 2002). Increase in sphingomyelin, ceramides, and cholesterol esters; generation of their lipid mediators; and interplay among phospholipid, sphingolipid, and cholesterol-derived lipid mediators intensify neuroinflammation and oxidative stress in ALS (Farooqui, 2009a).

### *8.5.2 Proteins in ALS*

A proteomic analysis of protein expression in mouse model of ALS indicates differences in protein expression in the spinal cords of mice expressing a mutant protein with the G93A mutation found in human ALS (Lukas et al., 2006). Alterations in the expression of proteins associated with mitochondria are particularly prevalent in spinal cord proteins from both mutant G93A-SOD1 and wild-type SOD1 transgenic mice. G93A-SOD1 mouse spinal cord also shows differences in proteins associated with metabolism, protein kinase regulation, antioxidant activity, and lysosomes (Lukas et al., 2006). There is an overlap of changes in mRNA expression in presymptomatic mice in three different gene categories, which include selected protein kinase signaling systems, ATP-driven ion transport, and neurotransmission. Therefore, alterations in selected cellular processes are detectable before symptomatic onset in ALS mouse models. However, in late-stage ALS, mRNA expression analysis does not reveal significant changes in mitochondrial gene expression, but shows concordant changes in lipid metabolism, lysosomes, and the regulation of neurotransmission (Lukas et al., 2006). Similarly, proteomic analysis of spinal cord in ALS mouse models (mice overexpressing wild-type (WT) and G93A mutant SOD1) indicates that lipid rafts contain 413 and 421 proteins, respectively. Functional classification of 67 altered proteins shows that three most affected subsets of proteins are associated with vesicular transport and neurotransmitter synthesis and release, cytoskeletal organization and linkage to the plasma membrane, and metabolism. Other protein changes correlate with alterations in microglia activation and inflammation, astrocyte and oligodendrocyte function, cell signaling, cellular stress response and apoptosis, and neuronal ion channels and neurotransmitter receptor functions (Zhai et al., 2009).

Transgenic mice overexpressing the mutant human SOD1 gene also show nitrated and oxidized proteins in the motor cortex, the cerebellar cortex, and the nucleus of hypoglossal nerves (regions related with movement). Significantly elevated protein nitration and nitric oxide synthesis have also been observed in brain tissues and CSF of mutant SOD1 mice. This study correlates mutation of the SOD1 gene to increased nitric oxide, nitration, and oxidation of proteins in ALS (Liu et al., 2007).

### 8.5.3 Nucleic Acids in ALS

Oxidative stress and generation of ROS in ALS result in oxidation of DNA and RNA. Thus, levels of DNA 8-hydroxy-2′-deoxyguanosine (8-OHdG) are increased not only in sporadic ALS motor cortex and spinal cord but also in plasma, urine, and CSF of ALS patients (Table 8.4) (Bogdanov et al., 2000; Aguirre et al., 2005). Similarly, increase in 8-OHdG levels is also observed in the spinal cord, frontal cortex, and striatum from G93A SOD1 transgenic mice (Sasaki et al., 2005). In addition, aberrant RNA metabolism has been reported to occur in ALS (Strong, 2010). Perturbed expression of RNA-binding proteins is causally related to the selective suppression of the low molecular weight subunit protein (NFL) steady-state mRNA levels in degenerating motor neurons in ALS. The occurrence of mtSOD1, TDP-43, and 14-3-3 proteins and their cytosolic aggregates in ALS can each modulate the stability of NFL mRNA (Strong, 2010). It is suggested that a fundamental alteration in ALS may be due to the interaction of mRNA species with key trans-acting binding factors. Furthermore, the oxidation of messenger RNA (mRNA) has also been reported to occur in ALS patients as well as in many different transgenic mice expressing familial ALS-linked mutant copper-zinc superoxide dismutase (SOD1). The analysis of oxidized RNA species indicates that oxidation of mRNA occurs at early stages of the disease. Oxidation of mRNA in ALS is not random but selective and many oxidized mRNAs are related to the pathogenesis of the disease (Kong et al., 2008). Oxidative modification of RNA in ALS and other neurodegenerative diseases modulates the translational process that results in the production of defective protein. In mutant SOD1 mice, increased oxidation of mRNA primarily occurs in the motor neurons and oligodendrocytes of the spinal cord at an early, presymptomatic stage, indicating that mRNA oxidation is an early event associated with motor neuron deterioration in ALS (Chang et al., 2008).

### *8.5.4 Transcription Factors in ALS*

Several transcription factors are stimulated in ALS. They maintain and intensify neuroinflammation. Thus in ALS patients, STAT3 (a proinflammatory transcription factor) is translocated from cytosol to the nucleus, where it remains persistently activated but no upregulation of STAT3 is observed in ALS spinal cord microglia (Shibata et al., 2009). Proliferator-activated receptorγ (PPARγ) agonist, pioglitazone, protects motor neurons against p38-mediated neuronal death and NF-κB-mediated glial inflammation via a PPARγ-independent mechanism (Shibata et al., 2009). Nuclear erythroid 2-related factor 2 (Nrf2) is a basic region leucine-zipper transcription factor that binds to the antioxidant response element and modulates the expression of many genes that are associated with cellular antioxidant and anti-inflammatory defenses. Under normal conditions, Nrf2 activation is blocked by Kelch-like ECH-associated protein 1 (Keap1). In ALS samples, there is a reduction of Nrf2 mRNA and protein expression in neurons, whereas Keap1 mRNA expression is increased in the motor cortex (Sarlette et al., 2008). Thus, alterations in signaling transduction pathways occur in motor neurons in ALS. Studies on over-expression of Nrf2 in astrocytes in chronic model indicate the activation of Nrf2 not only provides neuroprotection, but also delays ALS symptoms and extends neuronal survival (Vargas et al., 2008).

High threshold for stress-induced activation of the heat shock transcription factor, Hsf1, is known to contribute to the vulnerability of motor neurons to disease and limit efficacy of agents promoting expression of neuroprotective heat shock proteins (Hsps) through this transcription factor (Batulan et al., 2006; Taylor et al., 2007). Plasmid encoding a constitutively active form of Hsf1 (Hsf1act) is shown to activate Hsf1 in a primary culture model of familial ALS. Hsf1 induces high expression of multiple Hsps in cultured motor neurons and confers dramatic neuroprotection against SOD1G93A in comparison to Hsp70 or Hsp25 alone (Batulan et al., 2006). Thus, Hsf1 is the primary transcription factor responsible for the transcriptional response to heat stress in mammalian cells. It is tightly regulated by a series of inhibitory checkpoints that include sequestration in multichaperone complexes governed by Hsp90 (Taylor et al., 2007).

### *8.5.5 Gene Expression in ALS*

Familial ALS is characterized by mutations in the SOD1 gene (*SOD-1*). At least, 135 mutations have been reported in the *SOD-1* gene, accounting for approximately

20% of familial ALS cases. Mutations are widely distributed throughout the gene with preponderance for exons 4 and 5 (Vucic and Kiernan, 2009). Although mutations result in a toxic gain of function of the SOD1 enzyme, which normally functions as a free radical scavenger, the mechanisms underlying motor neuron degeneration have not been clearly elucidated. Studies in G93A-SOD1 mice and rats indicate that oxidative damage is part of an unmitigated neuroinflammatory reaction, arising in combination from mitochondrial dysfunction plus pathophysiologic activation of both astrocytes and microglia. Lesions to redox signal-transduction pathways in mutant $SOD1^{+}$ glial cells may stimulate broad-spectrum upregulation of proinflammatory genes, including genes for enzymes of arachidonic acid cascade ($sPLA_2$; COX-2, 5-LOX) and nitric oxide synthase (NOS); as well as cytokines, chemokines and immunoglobulin Fc receptors. The integration of these processes creates a paracrine milieu consistent with situation that arises from interplay among excitotoxicity, oxidative stress, and neuroinflammation (Farooqui and Horrocks, 2007). Complex interactions between genetic and above molecular events may account for neurodegeneration and damage of critical target proteins and organelles within the motor neuron (Vucic and Kiernan, 2009). Gene expression profiles of degenerating spinal motor neurons isolated from ALS patients obtained using microarray procedure and laser-captured microdissection technique indicate that some genes are downregulated, while others are upregulated in motor neurons (Jiang et al., 2005). Downregulated genes include genes associated with cytoskeleton/axonal transport, transcription, and cell surface antigens/receptors, such as dynactin, microtubule-associated proteins, and early growth response 3. In contrast, cell death-associated genes are mostly upregulated. Promoters for cell death pathway, death receptor 5, cyclins A1 and C, and caspases-1, -3, and -9, are upregulated. In addition, cell death inhibitors, acetyl-CoA transporter, and NF-κB as well as neuroprotective neurotrophic factors such as ciliary neurotrophic factor, hepatocyte growth factor, and glial cell line-derived neurotrophic factor are also upregulated (Jiang et al., 2005).

It is reported that homozygous SMN1 (survival motor neuron) gene deletion causes spinal muscular atrophy, and SMN2 gene deletions are possible risk factors in lower motor neuron disease. A study of SMN1 and SMN2 gene copy numbers in 167 ALS patients and in 167 matched controls indicates that 16% of ALS patients had an abnormal copy number of the SMN1 gene (1 or 3 copies), compared with 4% of controls. It is suggested that an abnormal SMN1 gene locus may be a susceptibility factor for amyotrophic lateral sclerosis (Corcia et al., 2002).

### *8.5.6 Neurotrophins in ALS*

Alterations in serum levels of insulin, IGF-I, and their binding proteins (IGFBPs) have been reported to occur in ALS, cerebellar ataxia (CA), ataxia-telangiectasia (AT) and Charcot-Marie-Tooth 1A disease (Busiguina et al., 2000). Transplantation of human neural progenitor cells (hNPs) or hNPs expressing growth factor into the brain of mutant Cu/Zn superoxide dismutase (SOD1(G93A)) transgenic mice indicate that hNPs expressing BDNF IGF-1, VEGF, neurotrophin-3 (NT-3), or GDNF are not only engrafted and migrated into the spinal cord or brain of ALS mice but also differentiated into neurons, oligodendrocytes, or glutamate transporter-1 (GLT1)-expressing astrocytes while some cells retained immature markers (Park et al. 2009a). Although transplantation of GDNF- or IGF-1-expressing hNPs attenuates the loss of motor neurons and mediates trophic factor-mediated changes in motor neurons of the spinal cord, it produces no improvement in motor performance and extension of lifespan suggesting that trophic support for degenerating neurons was inadequate, and more studies are required on this important topic in ALS mice (Park et al., 2009a; Lunn et al., 2009). Similarly, isolation and seeding of hNPs along with modification using lentivirus to secrete GDNF (hNPs(GDNF)) in culture result in cells survival up to 11 weeks following transplantation into the lumbar spinal cord of rats overexpressing the G93A SOD1 mutation (SOD1 (G93A)) (Klein et al., 2005). Integration of cells is observed into both gray and white matter without any adverse behavioral effects. All transplants secreted GDNF within the region of cell survival, but not outside this area. Upregulation of cholinergic markers also occurs in response to GDNF, indicating that developing fibers are physiologically active (Klein et al., 2005).

## 8.6 Neurochemical Aspects of Huntington Disease

HD is a progressive neurological genetic disease characterized by midlife onset causing involuntary movements, cognitive, physical and emotional deterioration, personality changes, dementia, and premature death. HD is characterized by neuronal dysfunction and death in the basal ganglia and cortex (Cepeda et al., 2001). The genetic defect underlying HD has been identified as an unstable CAG trinucleotide repeat in exon 1 of the HD gene, which encodes for a polyglutamine expansion near the N-terminal end of a large protein called huntingtin. Insoluble aggregates containing huntingtin occur in cytosol and nuclei of HD patients, transgenic animal, and cell culture models of HD. The molecular mechanism involved in aggregate formation is not fully understood. However, it is proposed that interactions of huntingtin with other proteins may promote its own polymerization to form insoluble aggregates. These aggregates have been shown to trigger neurodegeneration (Scherzinger et al., 1997). The intraneuronal aggregates of huntingtin may induce neurodegeneration by modulating gene transcription, protein interactions, protein transport inside the nucleus and cytoplasm as well as vesicular transport (Bonilla, 2000). Despite its widespread expression in the brain and body, mutant huntingtin causes selective neurodegeneration in striatal medium-sized spiny GABAergic projection neurons (MSNs), resulting in the appearance of generalized involuntary movements, the main phenotypic alteration in HD. Although the molecular mechanism associated with selective neurodegeneration is not known, the selective nuclear localization of mutant huntingtin in striatal nuclei may contribute to the region-specific atrophy in transgenic models of HD (Van Raamsdonk et al., 2007). Selective phosphorylation of mutant huntingtin on serine 421 results in phosphorylation of mutant huntingtin and reduction in its toxicity in the striatum compared to other regions of the brain. MSNs constitute 95% of all striatal neurons. Since MDN neurons are innervated by glutamatergic axons, they prone to be subjected to excitotoxicity. In addition, gliosis (reactive astrocytosis) has also been reported to occur in the striatum and cerebral cortex (Cepeda et al., 2001) from HD patient's brains. In contrast, HD mouse models expressing mutant huntingtin do not have obvious neurodegeneration despite significant neurological symptoms (Li and Li, 2004). Most HD mouse models display the accumulation of toxic N-terminal mutant huntingtin fragments in both the nucleus and neuronal processes, suggesting that these subcellular sites may be target sites for the early neuropathology of HD (Li and Li, 2004). Other neurochemical changes include decrease in levels of GABA, dynorphin, and substance P and increase in somatostatin and neuropeptide Y. In addition, hippocalcin (a neuronal calcium sensor protein) is highly expressed in the medium spiny striatal output neurons that degenerate selectively in HD. Decrease in hippocalcin expression occurs in parallel with the onset of disease in mouse models of HD. In situ hybridization histochemistry studies have indicated that hippocalcin RNA is diminished by 63% in human HD brain (Rudinskiy et al., 2009). It is proposed that degeneration of neurons starts at the distal part of axons, leading to defective neuronal interaction, abnormal synaptic transmission, and impaired supply of growth factors to the cell body (Li and Li, 2004). Axonal degeneration in the form of dying back then results in the loss of neuronal body where apoptosis and other cellular pathological pathways are activated. These pathways include impairment of energy metabolism, sensitivity to oxidative stress, cytotoxic effects of glutamate, and aggregation of huntingtin.

### *8.6.1 Lipids in HD*

Although earlier studies have indicated that levels of phospholipid degradation products (phosphodiesters) are increased in HD (Abood and Butler, 1979; Pettegrew et al., 1987), recent studies suggests that phospholipid composition of the synaptic membranes is not affected in HD (Table 8.5) (Suopanki et al., 2006). In vitro studies show that large unilamellar vesicles of brain lipids readily bind with soluble N-terminal huntingtin exon 1 fragments and promote fibrillogenesis of mutant huntingtin aggregates. Moreover, binding of both mutant and wild-type huntingtin exon 1 fragment with brain lipids induces bilayer perturbation mediated by a proline-rich region adjacent to the polyglutamines. It is proposed that lipid interactions in vivo may influence misfolding of huntingtin and initiate early HD pathogenesis. Huntingtin interacts with various phospholipids. Thus, in vitro studies indicate that huntingtin from normal (Hdh(7Q/7Q)) mouse brain and mutant huntingtin from Hdh(140Q/140Q) mouse brain binds with large unilamellar vesicles containing PtdIns, PtdIns 3,4-$P_2$, PtdIns 3,5-$P_2$, and PtdIns 3,4,5-$P_3$. Mutant huntingtin binds more tightly with PtdEtn and PtdIns 3,4,5-$P_3$ than wild-type huntingtin. The recruitment of endogenous huntingtin to the plasma membrane is facilitated by

**Table 8.5** Neurochemical alterations in Huntington disease

| Neurochemical parameter | Effect | References |
|---|---|---|
| Glycerophospholipid metabolism | No effect | Suopanki et al. (2006) |
| Lipid peroxidation | Increased | Coyle and Schwarcz (1976) |
| Hydroxycholesterol | Increased | Leoni et al. (2008) |
| Huntingtin processing | Abnormal | Scherzing et al. (1997) |
| Caspase activity | Increased | Hermel et al. (2004) |
| NF-κB | upregulated | Napolitano et al. (2008) |
| Excitotoxicity | Increased | Coyle and Schwarcz (1976) |
| Oxidative stress | Increased | Thomas (2006) |
| Neuroinflammation | Increased | Thomas (2006) |
| Neurodegeneration | Increased | Thomas (2006) |

exogenous PtdIns 3,4-$P_2$ and PtdIns 3,4,5-$P_3$ and is stimulated by platelet-derived growth factor or insulin growth factor 1. It is proposed that huntingtin interacts with membranes through specific phospholipid associations and that mutant huntingtin may disrupt membrane trafficking and signaling at membranes (Kegel et al., 2009).

In transgenic HD mice, affymetrix microarray studies using a custom-designed GLYCOv2 chip indicate that an abnormal expression levels of genes encoding glycosyltransferases in the striatum of R6/1 transgenic mice, as well as in postmortem caudate from human HD subjects (Desplats et al., 2007). The disrupted patterns of glycolipids (acidic and neutral lipids) and/or ganglioside levels in both the forebrain of the R6/1 transgenic mice and caudate samples from human HD subjects indicate a disruption in glycolipid/ganglioside metabolic pathways in the pathology of HD and suggest that the development of new targets to restore glycosphingolipid balance may control some symptoms of HD (Desplats et al., 2007).

Cholesterol metabolism alterations have been reported in murine HD models and HD patients (Leoni et al., 2008). 24*S*-Hydroxycholesterol (24OHC) is closely associated with neurodegeneration. Plasma levels of 24OHC were significantly higher in controls than in HD patients at all disease stages (Table 8.5) (Leoni et al., 2008). Changes in 24OHC levels parallel the large decrease in caudate volumes, a process that occurs in HD stage 1 and is associated with neuronal loss (Leoni et al., 2008). It is likely that dysregulation in cholesterol metabolism is linked to specific actions of the mutant huntingtin on sterol regulatory element-binding proteins leading to lower cholesterol levels in caudate areas of the brain. Collective evidence suggests that dysregulation in lipid metabolism may contribute to the pathogenesis of HD.

### *8.6.2 Proteins in HD*

Although the normal function of huntingtin in brain is not known, recent studies indicate that huntingtin interacts with many proteins, including heme activator protein1 (HAP1), huntingtin interacting protein1 (HIP1), microtubules, glyceraldehyde-3-phosphate dehydrogenase (GADPH), calmodulin, and an ubiquitin-conjugating enzyme (Walling et al., 1998). Polyglutamine expansion alters many of these interactions and causes huntingtin to aggregate and form neuronal nuclear inclusions that ultimately facilitate cell death. Mutant huntingtin not only binds to cAMP response element-binding protein and modulates gene expression but also affects axonal transport and facilitates mitochondrial dysfunction in HD. It is not known whether mitochondrial dysfunction occurs in early HD brain or is specifically induced by N-terminal mutant huntingtin (Guidetti et al., 2001). Caspases have been shown to hydrolyze huntingtin, but it is not known which caspase cleaves huntingtin in vivo or whether regional expression of caspases contributes to selective neuronal cells loss. Caspase-2 cleaves huntingtin selectively at amino acid 552. Furthermore, huntingtin recruits caspase-2 into an apoptosome-like complex. Binding of caspase-2 to huntingtin depends on the length of polyglutamine repeat, therefore may serve as a critical initiation step in HD cell death. This hypothesis is supported by the upregulation of caspase-2, which correlates directly with decrease in levels of BDNF in the cortex and striatum of 3-month YAC72 transgenic mice, supporting the view that upregulation of caspase-2 may be an early event in HD pathogenesis (Hermel et al., 2004). It is also shown that mutant huntingtin activates caspase cascades (Li et al., 2000). Caspase antagonists have been reported to delay neurological symptoms in HD mouse models (Chen et al., 2000). Huntingtin undergoes proteolysis by calpains and caspases within an N-terminal region between amino acids 460 and 600. Generation of shorter N-terminal fragments, which are termed as cp-1 and cp-2 (distinct from previously described cp-A/cp-B), has also been reported (Ratovitski et al., 2009). cp-1 cleavage occurs between residues 81 and 129 of huntingtin, whereas the cp-2 fragment is generated by cleavage of huntingtin at position Arg(167). Based on structural studies, it is suggested that cp-2 mediates mutant huntingtin toxicity in HD (Ratovitski et al., 2009). In addition, the involvement of excitotoxicity in HD is supported by the observation that administration of NMDA receptor agonists to the striatum of animals produces a selective degeneration of MSNs with neurological symptoms similar to those seen in HD patients (Coyle and Schwarcz, 1976). A decrease in expression of glutamate transporters along with NMDA receptor alterations in transgenic models and HD patients also supports the presence of excitotoxic damage. Thus, excitotoxicity, dopamine toxicity, metabolic impairment, mitochondrial dysfunction, oxidative stress, apoptosis, and autophagy may promote progressive degeneration observed in HD (Table 8.5). It is also speculated that huntingtin may play a role in protein trafficking, vesicle transport, postsynaptic signaling, transcriptional regulation, and apoptosis (Gil and Rego, 2008).

At the molecular level, huntingtin is phosphorylated by the inflammatory kinase known as I-κB kinase (IKK), increasing normal clearance of huntingtin by the proteasomal and lysosomal pathways. Phosphorylation of huntingtin not only modulates ubiquitination of huntingtin but also facilitates SUMOylation (small ubiquitin-like modifier-mediated process) and acetylation and increases huntingtin nuclear localization, cleavage, and clearance mediated by lysosomal-associated membrane protein 2A and Hsc70 (Thompson et al., 2009). IKK enhances mutant huntingtin clearance until an age-related loss of proteasome/lysosome function and promotes accumulation of toxic post-translationally modified mutant huntingtin.

Thus, IKK activation may modulate mutant huntingtin neurotoxicity depending on the cell's ability to degrade the modified species (Thompson et al., 2009). Increased expression of several key inflammatory mediators, such as CCL2 and IL-10, specifically in the striatum occurs in HD patients. In addition, an upregulation of IL-6, IL-8, and MMP9 is found in the cortex and notably in the cerebellum. These observations suggest that neuroinflammation may be a prominent feature associated with HD.

In addition, two-dimensional gel/mass spectrometry-based proteomics studies in mouse model of HD indicate an upregulation of proteins associated with glycolysis/gluconeogenesis and downregulation of cytoskeleton proteins such as actins (Zabel et al., 2009) in early stages of HD. Although the upregulation of glycolysis/gluconeogenesis-related protein remains dominant during HD progression, late stages of HD also show an upregulation of proteins involved in proteasomal function, supporting the view that HD is accompanied by a highly dynamic pathology not represented by linear protein concentration alterations (Zabel et al., 2009).

### *8.6.3 Nucleic Acids in HD*

Nucleic acid alterations have been described in brain tissue from HD patients as well as in rodent models of HD. Studies on isolation and determination of total RNA from the cortex and striatum of HD patients and control subjects indicate that the mechanism of disease expression does not occur during transcription or in the stability of the RNA, but rather occurs during translation or posttranslational stages (Stine et al., 1995).

Alterations in DNA are particularly important in the mitochondrial DNA (mtDNA) and nuclear DNA, which play important roles in the pathogenesis of the respiratory chain complex activities and oxidative stress in HD (Banoei et al., 2007). Determination of mtDNA damage in 60 HD patients and 70 healthy controls indicate that HD patients have higher frequencies of mtDNA deletions in lymphocytes than control subjects. Although the molecular mechanism associated with mtDNA damage is not known, it is proposed that CAG repeats instability and mutant huntingtin may be causative factors in mtDNA damage (Banoei et al., 2007).

### *8.6.4 Transcription Factors in HD*

Several transcription factors have been implicated in the pathogenesis of HD. These transcription factors include NF-κB, Bcl11b, and RE1/NRSE. As stated earlier, NF-kB is a family of DNA-binding proteins that play important roles in modulation of immune and inflammatory responses, as well as in cell survival and apoptosis. In 3-nitroproprionic acid-induced neurotoxicity, an experimental model of HD, NF-kB is translocated from cytosol to the nucleus where it modulates transcriptional activation of NOS genes (Napolitano et al., 2008). Expression of Bcl11b (also known as CTIP2), a transcription factor that has highly enriched localization in adult striatum, is significantly decreased in HD cells including mouse models and human subjects (Desplats et al., 2008). The overexpression of Bcl11b attenuates toxic effects of mutant huntingtin in cultured striatal neurons. Bcl11b directly activates the proximal promoter regions of striatal-enriched genes and can increase mRNA levels of striatal-expressing genes (Desplats et al., 2008). It is proposed that decreased expression of Bcl11b in HD, at least in part, may be responsible for the dysregulation of striatal gene expression seen in HD. Repressor element-1 (RE1) silencing transcription/neuron-restrictive silencer factor (REST/NRSF) is a transcriptional repressor that can block transcription of a battery of neuronal differentiation genes by binding to a specific consensus DNA sequence present in their regulatory region. In neurons, the REST protein is sequestered in the cytoplasm in part through binding to huntingtin. Mutant huntingtin abrogates REST-huntingtin binding. Consequently, REST translocates to the nucleus, occupies RE1 repressor sequences, and decreases neuronal gene expression. Increase in binding of the RE1 silencing transcription factor/neuron-restrictive silencer factor (REST/NRSF) repressor occurs at multiple genomic RE1/NRSE loci in HD cells not only in animal models but also in postmortem brains, resulting in a decrease of RE1/NRSE-mediated gene transcription. Restoration of BDNF through attenuation of REST/NRSF binding may result in repression of aberrant neuronal gene transcription in HD (Zuccato et al., 2007).

Transcriptional dysregulation and aberrant chromatin remodeling are central features of HD pathogenesis. Studies on histone profiles and associated gene changes in transgenic N171-82Q (82Q) and R6/2 HD mice indicate that significant chromatin modifications take place due to reduction in histone acetylation with concomitant upregulation of histone methylation in above transgenic models of HD (Stack et al., 2007). It is suggested that mutant huntingtin alters histone acetyltransferase activity, and aberrant activity of this enzyme may be an underlying mechanism of transcriptional dysregulation in HD. Alterations in nucleosomal dynamics are accompanied by significant improvement in the behavioral and neuropathological phenotype observed in HD mice (Stack et al., 2007). In the nucleus, huntingtin interacts with the transcriptional activator Sp1 and coactivator TAFII130 (Dunah et al., 2002). Coexpression of Sp1 and TAFII130 in cultured striatal cells from wild-type and HD transgenic mice reverses the transcriptional blockage of the dopamine $D_2$ receptor gene induced by mutant huntingtin, as well as protects neurons from huntingtin-induced neurotoxicity. It is also reported that soluble mutant huntingtin retards Sp1 binding to DNA in postmortem brain tissues of both presymptomatic and affected HD patients (Dunah et al., 2002).

### *8.6.5 Gene Expression in HD*

Mutant huntingtin has been reported to interfere with the function of widely expressed transcription factors, suggesting that gene expression may be altered in a variety of tissues in HD. Thus, mutant huntingtin reduces levels of brain-derived neurotrophic factor (BDNF) in the striatum by inhibiting cortical BDNF gene expression and anterograde transport of BDNF from cortex to striatum (Gharami et al., 2006). Based on above observation, it is suggested that mutation in huntingtin reduces BDNF gene transcription. One mechanism involved in this process is the activation of repressor element 1/neuron-restrictive silencer element (RE1/NRSE) located within the BDNF promoter (Zuccato et al., 2007). The effectiveness of BDNF therapy in HD also depends on the proper expression of its receptor TrkB. A specific reduction in expression of TrkB receptors also has been observed in transgenic exon-1 and full-length knock-in HD mouse models and also in the motor cortex and caudate nucleus of HD brains (Gines et al., 2006). This suggests that abnormal expression of BDNF and TrkB due to accumulation of mutant huntingtin may contribute to the altered neurotrophic support in HD.

Involvement of transcriptional gene dysregulation in the pathophysiology of HD has attracted considerable attention in recent years, but this hypothesis does not explain the specificity of dysfunction and neurodegeneration in HD (Thomas, 2006). Microarray studies in mouse model of HD indicate alterations in hundreds of gene expression as well as in postmortem brain samples from HD subjects. Alterations in "striatal-enriched" genes are associated with disturbances in transcriptional processes, behavioral disturbances, calcium homeostasis, abnormal vesicle trafficking, and defective mitochondrial bioenergetics.

At least two potential mechanisms explaining gene alterations in HD have been described (Johnson and Buckley, 2009). They include (a) involvement of REST (RE1-silencing transcription factor), a master regulator of neuronal genes and (b) dysregulation of post-transcriptional gene regulation by microRNAs in neurons of the forebrain. Such dysregulation may cause aberrant nuclear localization of the transcriptional repressor, REST. Furthermore, expression of key neuronal microRNAs including mir-9/9*, mir-124, and mir-132-is repressed in the brains of human HD patients and mouse models. These changes occur downstream of REST and are likely to result in major disruption of mRNA regulation and neuronal function (Johnson and Buckley, 2009). Thus, both transcriptional and post-transcriptional mechanisms may induce a loss of neuronal identity associated with molecular etiology of HD.

### *8.6.6 Neurotrophins in HD*

Pathogenesis of HD involves decrease in mRNA and protein levels of BDNF in the brains of several HD rodent models and in striatum of human HD patients (Zuccato and Cattaneo, 2007; Conforti et al., 2008), suggesting that decrease in the expression of BDNF in HD may be related with cognition, learning impairment, and other clinical manifestation of HD progression (Giratt et al., 2009). Studies on R6/1 and R6/1:BDNF$^{+/-}$ mice indicate that R6/1:BDNF$^{+/-}$ mice show earlier and more accentuated cognitive impairment than R6/1 mice at 5 weeks of age in discrimination learning; at 5 weeks of age in procedural learning, and at 9 weeks of age in learning alternations. Studies on BDNF-trkB signaling and glutamate receptor expression in the hippocampus of these mice indicate that reduction in phospholipaseC$\gamma$ activity in R61, BDNF$^{+/-}$, and R6/1:BDNF$^{+/-}$ mice hippocampus is accompanied by alterations in LTP (Giratt et al., 2009). However, a specific decrease in the expression of glutamate receptors NR1, NR2A, and GluR1 is observed only in hippocampus from R6/1:BDNF$^{+/-}$ mice. This suggests that BDNF modulates the learning and memory alterations in mice with huntingtin mutation.

## 8.7 Neurochemical Aspects of Prion Diseases

Prion diseases are fatal neurodegenerative disorders characterized by the accumulation of abnormal isoforms of a host protein known as cellular prion protein ($PrP^C$), motor dysfunctions, dementia, and neuropathological changes such as spongiosis, astroglyosis, and neuronal loss. The cellular prion protein ($PrP^C$), a membrane-bound glycoprotein, is abundantly expressed in neurons and glial cells within the brain tissue. $PrP^C$ associates with cholesterol- and glycosphingolipid-rich lipid rafts through association of its glycosylphosphatidylinositol (GPI) anchor with saturated raft lipids and through interaction of its N-terminal region with an as yet unidentified raft associated molecule. $PrP^C$ contains two Asn-linked glycosylation sites. The function of $PrP^C$ remains elusive. Its amino-terminal region contains a repeated five octapeptide domain that binds copper. $PrP^C$ binds as many as six $Cu^{2+}$ ions with submicromolar affinity (Viles et al., 2008; Singh et al., 2009). The affinity and number of $Cu^{2+}$-binding sites support the view that $PrP^C$ may act as an antioxidant by binding potentially harmful $Cu^{2+}$ ions and sacrificially quenching of free radicals generated as a result of copper redox cycling. In addition, $PrP^C$ displays a superoxide dismutase-like activity and hence a possible protective function against oxidative stress (Rachidi et al., 2005). It is present in the cells in three different glycoforms, including an unglycosylated form. Although bound to the membrane by means of GPI, $PrP^C$ on neurons is rapidly and constitutively endocytosed by means of coated pits, a property dependent upon basic amino acids at its N-terminus (Parkyn et al., 2008). Low-density lipoprotein receptor-related protein 1 (LRP1), which interacts with multiple ligands through basic motifs, binds to $PrP^C$ during its endocytosis, and may be closely associated with $Cu^{2+}$-mediated endocytosis and trafficking of $PrP^C$ in neurons (Parkyn et al., 2008).

Prion diseases differ from other amyloid-associated protein misfolding diseases, such as AD, in that they are naturally transmitted between individuals and involve spread of protein aggregation between tissues (Speare et al., 2010). Very little is known about the factors modulating above characteristics of prion diseases. In addition, among protein misfolding disorders, only prion diseases involve the misfolding of a glycosylphosphatidylinositol (GPI)-anchored protein (Speare et al., 2010). Recent studies with live cell imaging indicate that GPI anchoring facilitates the propagation and spread of protein aggregation and thus may enhance the transmissibility and pathogenesis of prion diseases relative to other protein misfolding diseases (Speare et al., 2010). Thus, Prion diseases are unique in that they can be inherited and can also occur sporadically through prion infection.

In addition, Aβ oligomers bind to $PrP^C$ with nanomolar affinity (Lauren et al., 2009). Anti-$PrP^C$ antibodies inhibit Aβoligomer binding to $PrP^C$ and rescue synaptic plasticity in hippocampal slices from oligomeric Aβ toxicity. Based on these results, it is proposed that $PrP^C$ mediates Aβ-oligomer-induced synaptic dysfunction. These results also support the view that there are mechanistic similarities between AD and prion disease (CJD) (Fig. 8.3) (Lauren et al., 2009; Nygaard and Strittmatter, 2009; Gunther and Strittmater, 2010).

The scrapie prion protein ($PrP^{Sc}$) is a misfolded and altered β-sheet-rich isoform of $PrP^C$ formed by post-translational modification of the $PrP^C$. Molecular mechanisms, which lead to the conformational changes in $PrP^C$ are still unknown, but heparan sulfate stimulates conversion of purified $PrP^C$ into $PrP^{Sc}$ in vitro, and heparan sulfate proteoglycan molecules are required for efficient $PrP^{Sc}$ formation in prion-infected cells (Supattapone, 2004). In addition, the expression of $PrP^C$ in neuronal cells is required to mediate neurotoxic effects of $PrP^{Sc}$ (Chesebro et al., 2005). The generation of $PrP^{Sc}$ is followed by its aggregation and possibly fragmentation of aggregates, which replicates in the body in the absence of nucleic acids. The neurotoxicity of $PrP^{Sc}$ is linked to its propagation in neuronal cells, or $PrP^{Sc}$ may elicit a deadly signal through a $PrP^C$-dependent signaling pathway. Although there is no formal proof of the correctness of this model, a wealth of information is available on properties and replication of pathogen. $PrP^{Sc}$ is relatively resistant to proteinase K digestion. $PrP^{Sc}$ causes prion diseases (transmissible spongiform encephalopathies, TSE), a group of incurable neurodegenerative disorders that affect a wide variety of mammalian species. Prion diseases include scrapie found in goats and sheep, bovine spongiform encephalopathy (mad cow disease) in cattle, and fetal familial insomnia, Creutzfeldt–Jakob disease (CJD), kuru, and Gerstmann–Sträussler–Scheinker syndrome in humans (Prusiner, 2001; Grossman et al., 2003). Neuronal loss, spongiform degeneration, and glial cell proliferation are pathological characteristics of prion diseases. Human prion protein ($PrP^C$) contains 209 amino acids, a disulfide bridge between residues 179 and 214, and two sites of non-obligatory N-linked glycosylation at amino acids 181 and 197 (DeArmond and Prusiner, 2003). The accumulation of $PrP^{Sc}$ in the cytoplasm and in secondary lysosomes as well as in the neuronal plasmalemma and synaptic regions may be responsible for the loss of cognitive function in prion diseases (Jeffrey et al., 1992). The pathogenesis and molecular basis of the neurodegeneration are not fully understood. Limited structural information is available on aggregate formation by this protein as the possible cause of prion diseases and on its toxicity. The region comprising the residues 106–126 of human $PrP^C$ plays a key role in this conformational conversion between $PrP^C$ and $PrP^{Sc}$ because a synthetic peptide homologous with this sequence (PrP106–126) adopts different secondary structures in different environments. This peptide has been largely used to explore the neurotoxic mechanisms underlying the prion diseases. PrP106–126 peptide replicates the fundamental properties of full-length $PrP^{Sc}$, including the destabilization of neural membranes, dysregulation of intracellular calcium homeostasis; increase in oxidative stress, and enhancement of pro-apoptotic signaling pathways (proteasome dysfunction and mitochondrial and endoplasmic reticulum stress).

### 8.7.1 Lipids in Prion Diseases

The determination of neutral and phospholipid composition of mouse brain infected with scrapie prions indicates that during the later stages of prion disease the level of dolichol is decreased by 30%, whereas the level of dolichyl phosphate is increased by 30% (Guan et al., 1996). The terminally ill mice show a 2.5-fold increase in both total ubiquinone and its reduced form. Although levels of α-tocopherol are increased at this stage by 50%, no changes are observed in phospholipid amount, in phospholipid composition, and in phosphatidylethanolamine plasmalogen content during the entire disease process (Guan et al., 1996). Furthermore, no changes are observed in fatty acid and aldehyde composition of individual phospholipids and cholesterol contents. Thus, no changes are observed in neutral lipids and phospholipids in scrapie-infected mouse brain. In contrast, other studies indicate that prion infection alters the membrane composition and significant increase in total cholesterol levels in two neuronal cell lines (ScGT1 and ScN2a cells) (Bate et al., 2008). There is a good correlation between the concentration of free cholesterol in ScGT1 cells and the amounts of PrPSc. This elevation is entirely a result of increased amounts of free cholesterol as prion infection reduces the amounts of cholesterol esters in cells (Bate et al., 2008). Treatment of cerebellar granule neuronal culture with $PrP^{Sc}$ and PrP106–126 (a neurotoxic prion peptide) results in the stimulation of NMDA receptor (Fig. 8.3), and this stimulation is blocked by MK-801, memantine, and flupirtine (Muller et al., 1993; Perovic et al., 1995). PrP106–126 peptide-induced stimulation of the NMDA receptor generates of arachidonic acid in cerebellar granule neurons. This observation implicates $PLA_2$ in the pathogenesis of prion diseases (Stewart et al., 2001) (Table 8.6). Released arachidonic acid is converted into eicosanoids

**Table 8.6** Neurochemical alterations in prion diseases

| Neurochemical parameter | Effect | References |
|---|---|---|
| Glycerophospholipid metabolism | Altered | Bate et al. (2008) |
| $PLA_2$ activity | Increased | Bate et al. (2008) |
| Eicosanoids | Increased | Bate et al. (2004) |
| Lipid peroxidation | Increased | Freixes et al. (2006) |
| Cholesterol | No effect | Lobesto et al. (2005) |
| PrP processing | Abnormal | Supattapone (2004) |
| NF-κB | Upregulated | Kim et al. (1999) |
| Synapse integrity | Lost | Jellinger (2009) |
| Excitotoxicity | Increased | Perovic et al. (1995) |
| Oxidative stress | Increased | Bate et al. (2008) |
| Neuroinflammation | Increased | Bate et al. (2008) |
| Neurodegeneration | Increased | Bate et al. (2008) |

and 4-HNE through enzymic and non-enzymic oxidation, respectively. The association of $PLA_2$ with the pathogenesis of prion diseases is also supported by recent neuronal cell culture studies (Bate et al., 2004). In a tissue culture model of prion disease, neuronal $PLA_2$ is activated by GPI isolated from $PrP^C$ or $PrP^{Sc}$. The ability of GPI to activate $PLA_2$ is lost by either the removal of acyl chains or the cleavage of the phosphatidylinositolglycan linkage and inhibited by a monoclonal antibody that recognizes phosphatidylinositol (Bate et al., 2004, 2008). Immunoprecipitation studies show that $cPLA_2$ co-precipitates with $PrP^{Sc}$ in ScGT1 cells. Furthermore, prion infection not only increases the phosphorylation of $cPLA_2$ but also enhances prostaglandin $E_2$ production (Table 8.6). The treatment of neuronal cultures with inositol monophosphate or sialic acid provides resistance to the toxic effects of prion neurotoxic peptides (Bate et al., 2008, 2004). Intensity of oxidative stress is studied in a mouse model of scrapie in the brain at various stages of disease progression. A significant increase in concentration of lipid peroxidation markers, malondialdehyde and 4-HNE, and mRNA level of an oxidative stress-response enzyme, heme oxygenase-1, is observed at early preclinical stages of scrapie (Yun et al., 2006). The changes precede dramatic synaptic loss as demonstrated by decrease in synaptophysin immunostaining. These findings indicate that brain undergoes oxidative stress even from an early stage of prion invasion. Given the well-known deleterious effects of ROS-mediated damage in the brain, it is considered that the oxidative stress occurs at the preclinical stage of prion diseases (Yun et al., 2006).

Studies on composition of subcellular structures in primary cultured rat cerebellar neurons indicate that about 45% of total cellular prion protein is associated with a low-density, sphingolipid- and cholesterol-enriched membrane fraction. Compositional analysis indicates that prion protein-enriched membrane domains contain non-receptor tyrosine kinases Lyn and Fyn, caveolin-1, and the neuronal glycosylphosphatidylinositol-anchored protein Thy-1 (Loberto et al., 2005). In addition, prion protein-rich membrane domains also contain 50% of the sphingolipids, cholesterol, and phosphatidylcholine. All main sphingolipids, including sphingomyelin, neutral glycosphingolipids, and gangliosides, are also enriched in the prion protein-rich membrane domains (Loberto et al., 2005). Studies on the induction of apoptotic cell death in primary cultured rat cerebellar neurons indicate that levels of ceramide are increased and sphingomyelin levels are decreased, while cholesterol and ganglioside contents are not affected during apoptosis. Changes in ceramide and sphingomyelin composition are exclusively restricted to a detergent-resistant membrane fraction (Rivaroli et al., 2007). Sphingolipids metabolism in PrP-infected ScN2a cells indicates that ceramide synthase inhibitor fumonisin $B_1$ ($FB_1$) decreases both sphingomyelin and ganglioside $GM_1$ in cells by upto 50%, whereas $PrP^{Sc}$ is increased by three to four-fold (Naslavsky et al., 1999). Metabolic radiolabeling shows that $PrP^C$ production is either unchanged or slightly decreased in $FB_1$-treated cells, whereas $PrP^{Sc}$ formation is augmented by three to four-fold. Incubation of cells with sphingomyelinase for 3 days decreases sphingomyelin levels, but has no affect on $GM_1$, and $PrP^{Sc}$ is increased by three to four-fold. In contrast, treatment of ScN2a cells with glycosphingolipid inhibitor PDMP reduces

$PrP^{Sc}$, but increases sphingomyelin levels. Thus, generation of $PrP^{Sc}$ seems to correlate inversely with sphingomyelin levels (Naslavsky et al., 1999).

## *8.7.2 Proteins in Prion Diseases*

A large number of studies support the view that transmission of prion diseases does not require nucleic acids and that $PrP^{Sc}$ alone can act as an infectious agent. The hypothesis that misfolded proteins can be infectious is also supported by recent findings regarding prion phenomena in yeast and other fungi. One of the most characteristic of prions is their ability to form different strains, causing distinct phenotypes of prion diseases (Cobb and Surewicz, 2009). Prion infection causes alteration in many proteins. Changes have been observed in activities of $PLA_2$, PLC, caspases, and protein kinases. An increase in $PLA_2$ (Bate et al., 2004, 2008) and decrease in metabotropic glutamate receptor/phospholipase C signaling is observed in the cerebral cortex in Creutzfeldt–Jakob disease (CJD), suggesting that this important neuromodulatory and neuroprotective pathway is attenuated in CJD (Rodriguez et al., 2004, 2006). Intra-cerebrally inoculated hamsters or C57BL mice brain with scrapie agents 263 K or 139A show increase in casein kinase 2 (CK2) activities (Chen et al., 2008a).

In N2a cells, synthetic prion peptide PrP106–126 interacts with $p75^{NTR}$ and induces apoptosis through NF-κB signaling pathway (Bai et al., 2008). PrP106–126-mediated activation of NF-κB in N2a cells is due to PrP106–126-induced upregulation of $p75^{NTR}$ mRNA expression and protein levels. Pretreatment with $p75^{NTR}$ polyclonal antibody (sc-6189) or pretreatment with inhibitor (NF-κB SN50) blocks the activation of NF-κB and attenuates the apoptotic death by PrP106–126 (Bai et al., 2008). Furthermore, recombinant protease-resistant domain of the prion protein (PrP90–231) induces the secretion of several cytokines, chemokines, and nitric oxide (NO) release, in both type I astrocytes and microglial cells (Thellung et al., 2007). In both type of cells, PrP90–231 mediates the activation of ERK1/2 MAP kinase and this process modulates proliferative and secretive responses of astrocytes and functional activation of microglia, both dependent on MAP kinase activation (Corosaro et al., 2003; Thellung et al., 2007; Chen et al., 2008b). Microglial cell activation is accompanied by expression of TNF-α, interleukin (IL)-6, IL-1β, inducible nitric oxide synthase (iNOS), and cyclooxygenase-2 (COX-2). The induction of mRNAs of the inflammatory cytokines, IL-1α, IL-1β, and TNF-α is detected only in the brains of scrapie-infected mice (Kim et al., 1999). This is accompanied by increased activity of NF-κB in the nuclear extracts from brains of the scrapie-infected group, and the immunoreactivity of NF-κB is increased in the hippocampus and thalamus in the brains of scrapie-infected mice (Kim et al., 1999). In addition, generation of ROS is significantly increased in the brain mitochondrial fractions of scrapie-infected mice. Collective evidence suggests that prion diseases are characterized by a significant inflammatory component, which is supported by increased production of cytokines and chemokines and stimulation of $PLA_2$, COX-2, iNOS, and protein kinases.

### 8.7.3 Nucleic Acids in Prion Diseases

Although many investigators believe that infectious agent in prion diseases constitute a single host protein, more rigorous evaluations in scrapie have shown reasonably abundant nucleic acids (Akowitz et al., 1994). In addition, treatment of highly purified 120S CJD preparations with nucleases generates RNA fragments with 6,000 bases. RNA analysis of CJD microglial cells with relevant cDNA arrays results in identification of approximately 30 transcripts not previously examined in any prion disease. This CJD expression profile is different from uninfected microglia exposed to prototypic inflammatory stimuli such as lipopolysaccharide and IFN-$\gamma$ as well as PrP-amyloid (Baker and Manuelidis, 2003), supporting the view that prion contains nucleic acids and alterations in prion nucleic acid metabolism may contribute to the pathogenesis of prion diseases.

### 8.7.4 Transcription Factors in Prion Diseases

As stated above, prion diseases are accompanied by neuroinflammation. The activity of NF-$\kappa$B is significantly elevated in the nuclear extracts from brains of the scrapie-infected group, and the immunoreactivity of NF-$\kappa$B is increased in the hippocampus and thalamus in the brains of scrapie-infected mice (Kim et al., 1999). The NF-$\kappa$B immunoreactivity is observed mainly in GFAP-positive astrocytes and also detected in the PrP-amyloid plaques in the brains of 87 V scrapie-infected mice. These results suggest that prion accumulation in astrocytes might activate NF-$\kappa$B through the increase of ROS generation (Kim et al., 1999).

### 8.7.5 Gene Expression in Prion Diseases

The prion gene family currently consists of three members: Prnp which encodes $PrP^{C}$, the precursor to prion disease-associated isoforms such as $PrP^{Sc}$; Prnd which encodes Doppel (Dpl), a testis-specific protein involved in the male reproductive system; and Sprn which encodes the newest PrP-like protein, Shadoo, which is expressed in the CNS (Watt and Westaway, 2007). Overexpression of Dpl is neurotoxic and causes a neurological disease (Behrens, 2003). In contrast to its homologue $PrP^{C}$, Dpl does not participate in prion disease progression or to achieve an abnormal $PrP^{Sc}$-like state. Interestingly, Dpl neurotoxicity can be retarded by $PrP^{C}$. In contrast to its homolog $PrP^{C}$, Dpl is dispensable for prion disease progression and for the generation of $PrP^{Sc}$, but Dpl appears to have an essential function in male spermatogenesis (Behrens, 2003).

Quantitative RT-PCR studies on brain tissues from scrapie-infected mice and age-matched, mock-inoculated controls before inoculation and at different time points of post-inoculation indicate that a total of 449 probe sets representing 430 genes exhibit differential expression between scrapie- and mock-inoculated mice over the time course of the disease (Xiang et al., 2007; Skinner et al., 2006). These genes are grouped into two clusters according to expression patterns: the genes in cluster 1 demonstrate lower mRNA levels in scrapie-infected brains when compared with mock-inoculated brains, whereas genes in cluster 2 show higher mRNA levels in scrapie-infected brains (Xiang et al., 2007). Functional analysis of differentially expressed genes reveals the most severely affected biological process: cholesterol metabolism. The expression patterns of the cholesterol-related genes indicate an inhibited cholesterol synthesis in the diseased brains. Conspicuously, a number of cluster 1 genes, including some of cholesterol-related genes, show not only decreasing mRNA levels in scrapie-infected brains but also increasing mRNA levels in mock-inoculated brains with increasing age. Quantitative RT-PCR analysis of some cholesterol-related genes in untreated mice suggests that changes of the examined genes observed in mock-inoculated brains are mainly age related (Xiang et al., 2007). This finding suggests a link between age-related genes and scrapie-associated neurodegeneration. The microarray analysis of control and sCJD subjects indicates that 79 genes are upregulated and 275 genes are downregulated in sCJD frontal cortex. In sCJD brains upregulated genes not only include genes encoding immune and stress-response factors but also include genes associated with cell death and cell cycle. The prominent downregulated genes encode for synaptic proteins (Xiang et al., 2005). The range of the upregulated genes and the degree of the increased expression correlates with the degree of the neuropathological alterations in particular subtypes. Overall the gene array studies demonstrate the presence of a strong inflammatory component, oxidative stress response, and gene expression patterns in prion diseases. The genes that are downregulated in prion diseases include genes associated with synapse function, calcium signaling, long-term potentiation, and ERK/MAPK signaling and also genes coding for the transcription regulators, EGR1, and CREB1 (Sorensen et al., 2008).

As stated above, heparin sulfate stimulates the conversion of $PrP^{C}$ into $PrP^{Sc}$. Comparative analysis of 200 glycosylation-related genes on prion-infected and prion-uninfected hypothalamus-derived GT1 cells indicates that some genes, such as (ChGn1), are upregulated, while others (such as Chst8) are downregulated in prion-infected cells (Barret et al., 2005). ChGn1 and Chst8 are involved in the initiation of the synthesis of chondroitin sulfate and in the 4-*O*-sulfation of non-reducing *N*-acetylgalactosamine residues, respectively. It is suggested that hyposulfated chondroitin plays an important role in $PrP^{Sc}$ accumulation. Treatment of Sc-GT1 cells with a heparan mimetic (HM2602) results in a reduction of the amount of $PrP^{Sc}$, associated with a total reversion of the transcription pattern of the *N*-acetylgalactosamine-4-*O*-sulfotransferase 8. These observations suggest a link between the genetic control of 4-*O*-sulfation and $PrP^{Sc}$ accumulation (Barret et al., 2005).

### 8.7.6 Neurotrophins in Prion Diseases

Very little is known about neurotrophin alterations in prion diseases. NGF has been reported to increase PrP mRNA levels in brains of neonatal hamsters (Mobley et al.,

1988). The expression of the BDNF gene is markedly decreased in cerebrum, cerebellum, and brainstem regions of zitter rat with genetic spongiform encephalopathy than normal mice. Changes in BDNF are accompanied by significant decrease in mitogen-activated protein kinase (MAPK) Erk2 activity but not in MAPK protein expression. These observations suggest that alterations in MAPK pathway may be related with BDNF mRNA reduction in the zitter rat brain (Muto et al., 1999).

## 8.8 Complement System Changes and Neurodegenerative Diseases

The complement system is a component of the innate immune system (Lu et al., 2008), which includes phagocytosis, the generation of cytokines, chemokines, and adhesion molecules, and killing of abnormal or infected cells by natural killer cells along with cytokine-dependent resistance of leukocytes to viral infection. The complement system plays an important role in neurodegenerative diseases. When present at an optimum level in the normal brain, the complement system not only provides neuroprotection but is also involved in clearing apoptotic cells, ingestion, and destruction of pathogens (opsonization). Complement system is modulated by age and oxidative stress, which are closely associated with neurodegeneration. Neurodegenerative diseases are accompanied by upregulation of C1q, a recognition molecule of the complement system (Fraser et al., 2009). This multimeric protein triggers an enhancement of phagocytosis of suboptimally opsonized targets by microglia, the phagocytic cells of the CNS, similar to other phagocytes, enhances the uptake of apoptotic cells in peripheral phagocytes, and suppresses inflammatory cytokine production in human monocytes, macrophages, and dendritic cells in the absence of activation of the entire complement cascade (Lu et al., 2008; Fraser et al., 2009). In neurodegenerative diseases, aggregated polypeptides may potentially activate complement system leading to microglial cell activation which in turn leads to defective clearance of the aggregated polypeptides by macrophages leading to chronic inflammation. Occurrence of activated microglia has been reported in various brain regions, such as the hippocampus, substantia nigra, and cortex in PD, AD, and HD (Dheen et al., 2007). Several lines of evidence support the contribution of microglial cells in neurodegenerative diseases: (a) microglial activation precedes the neurodegenerative changes, (b) activated microglia surround the region that undergo neurodegeneration and phagocytose the degenerating cells, (c) activated microglia release neurotoxic molecules such as IL-1β, IL-6, TNF-α, nitric oxide, and ROS, (d) inhibition of microglial activation causes the amelioration of neurodegeneration, and (e) microglia derived from aged animal induce more toxicity to neurons in an age-dependent fashion in the same way as occurs in neurodegenerative diseases (Sugama et al., 2009). In AD, plaques and dystrophic neuritis are surrounded by activated microglia, which express histocompatibility class II antigens and complement receptor. Aggregated Aβ peptides are potentially capable of mediating the secretion and release of proinflammatory cytokines (TNF-α, IL-1β, and IL6), ROS, complement factors, and chemokines from microglia (Heneka and O'Banion, 2007). Proinflammatory cytokines induce abnormal processing and hyperphosphorylation of the τ-protein through the downregulation of the cdk5/p35 pathway (Quintanilla et al., 2004). Collective evidence suggests that the complement system has a Jenus face with dual contrasting properties (neuroprotection and neurodegeneration). When the complement levels are normal, it acts as a boon to the immune system through aiding in various processes, including recognition of pathogens, opsonization, and clearance of apoptotic cells (Lu et al., 2008). However, factors such as oxidative stress due to the presence of excess free radicals and aging can reverse this protective role and hence bring about the destructive aspect of complement, i.e., lead to neurodegeneration (Fraser et al., 2009). This takes place especially in the presence of aggregated polypeptides that can present themselves to vital charge pattern recognition molecules of the complement system, especially C1q. This aggregation leads to augmentation of the microglial activity and hence leads to microglial activation, initiated by C1q. This defect in the efficiency of the complement system leads to defective clearance of the aggregated polypeptides by macrophages which in turn lead to chronic inflammation, a process closely associated with the pathogenesis of chronic neurodegenerative diseases.

## 8.9 Apoptotic and Necrotic Cell Death and Autophagy in Neurodegenerative Diseases

In neurodegenerative diseases, neural cell in brain and spinal cord dies by two major mechanisms, namely apoptosis and necrosis. Apoptotic and necrotic cell deaths are triggered by a variety of stimuli including developmental signals, disruption of cell cycle, withdrawal of neurotrophic factors, release of excitatory amino acids, accumulation of protein aggregates, inflammatory reactions, and oxidative stress (Sastry and Subba Rao, 2000; Farooqui et al., 2004). During apoptosis and necrosis, neural cells undergo through events that are controlled by intricate interplay among lipid mediators, intracellular enzymes, changes in integrity of subcellular organelles especially mitochondria, and levels of ATP (Sastry and Subba Rao, 2000; Farooqui et al., 2004). In addition to neuronal cell death, apoptosis also contributes to synaptic dysfunction and breakdown of neural circuitry (Mattson et al., 2000). Morphologically, apoptotic cell death is characterized by nuclear chromatin condensation, DNA fragmentation, cell shrinkage, and bleb and apoptotic body formation. Plasma membrane and other subcellular organelles such as mitochondria and endoplasmic reticulum remain active during apoptosis. During the execution phase of apoptosis, the asymmetric distribution of glycerophospholipid in lipid bilayer is lost due to the externalization of PtdSer from inner leaflet to the outer leaflet where it functions as a tag on the dying cell for recognition (eat-me signal) and removal by phagocytosis (Farooqui, 2009a). In addition, apoptosis is accompanied by an increase in activities of phospholipases, sphingomyelinases, ceramidases, kinases, and caspases, alterations in levels of glycerophospholipid, sphingolipid, and cholesterol-derived lipid mediators, and abnormal signal transduction processes (Farooqui, 2009a). During apoptosis, these changes occur in an orderly fashion due to sufficient levels of ATP that maintains normal ion homeostasis. The dead cells are removed from the tissue through apoptotic body formation and phagocytosis.

In contrast, necrosis is accompanied by rapid permeabilization of plasma membrane, rapid decrease in ATP, sudden loss of ion homeostasis, glutathione depletion, and activation of lysosomal enzymes resulting in a passive cell death through lysis (Nicotera et al., 1999; Farooqui et al., 2004). The release of cellular contents is accompanied by neuroinflammation and oxidative stress. Treatment of PC12 cells with NO donors or specific inhibitors of mitochondrial respiration (myxothiazol, rotenone, or azide) in the absence of glucose produces total ATP depletion and results in 80–100% necrosis (Bal-Price and Brown, 2000). The presence of glucose almost completely prevents the decrease in ATP level and the increase in necrosis mediated by the NO donors or mitochondrial inhibitors, suggesting that the NO-induced necrosis in the absence of glucose may be due to the inhibition of mitochondrial respiration and subsequent ATP depletion. However, in the presence of glucose, NO donors and mitochondrial inhibitors induce apoptosis of PC12 cells as determined by nuclear morphology (Bal-Price and Brown, 2000). It is now becoming increasingly evident that apoptosis and necrosis are interrelated processes, which are induced by common stimuli (cytokines, ischemia, heat, irradiation, and pathogens) and abnormal signaling pathways (Nicotera and Lipton, 1999; Farooqui et al., 2004).

Many factors, such as alteration in expression of p53 and Bcl-2, free radicals, insufficient levels of nerve growth factors, accumulation of self-aggregating proteins (Aβ, τ protein, α-synuclein, and Huntingtin), and mitochondrial dysfunction, cause impaired calcium buffering, generation of free radicals, and activation of the mitochondrial permeability transition. All these factors have been implicated in apoptotic cell death in neurodegenerative diseases. In experimental model of AD, PD, ALS, and HD, both extracellular amyloid, parkin, and huntingtin deposits activate caspases and induce apoptosis. P53, a nuclear phosphoprotein transcription factor, is critical for activating the expression of genes involved in cell cycle arrest and stress-induced apoptosis. In neurodegenerative diseases, the expression of p53 is significantly increased in glial cells, and microglial numbers are decreased. These processes contribute to apoptotic cell death in neurodegenerative diseases (Davenport et al., 2009). Permeabilization of the outer mitochondrial membrane and subsequent release of intermembrane space proteins is closely associated with apoptosis and necrosis. Alterations in mitochondrial permeability transition are associated mainly with necrosis, whereas, Bcl-2 family of protein-mediated release of cytochrome *c* and activation of caspase is involved in apoptotic cell death (Orrenius et al., 2007). In brain, mitochondrial mechanisms involved in the release of mitochondrial proteins depend not only on the type of neural cell but also on the nature of stimuli. A relationship between apoptotic and necrotic signaling cascades, disruption of mitochondrial energy metabolism, balance of cross talk between apoptotic and anti-apoptotic pathways, and duration of stimulus dictates the feasibility of mode of cell death in brain tissue (Soane et al., 2007).

Autophagy "self-eating" is a cell survival mechanism in which portions of the cellular cytoplasm and organelles are sequestered in a double membrane-bound vesicle called an autophagosome. Fusion of autophagosomes with lysosomes facilitates the formation of autolysosomes where the cytoplasmic long-lived proteins and organelles are degraded (Rajawat and Bossis, 2008). Thus, autophagy allows the removal of damaged proteins without disturbing nearby functional ones. Although the molecular mechanism of autophagy is not fully understood, PtdIns3K has been reported to stimulate autophagy in concert with the autophagy-regulatory protein beclin 1/Atg6. Beclin 1 is an essential role in autophagosome formation and Atg5 is associated with early stages of autophagosome formation (Gorman, 2008). PtdIns3K inhibitors and RNA interference knockdown of beclin 1 effectively block autophagy elicited by amino acid deprivation (Zhu et al., 2007). In addition, autophagic death may also require the activation of c-jun NH2-terminal kinase, phosphorylation of Bcl-2 and p53, and inhibition of caspase-8 (Park et al., 2009a). An insufficient autophagic response may make neural cells more susceptible to stress, whereas prolonged overactivation of autophagy may lead to a complete self digestion of the cell. The extent of autophagy modulates cross talk or interplay between apoptosis and autophagy, which may represent a master switch between cell survival and cell death (Fig. 8.6). Inclusion bodies in neurodegenerative

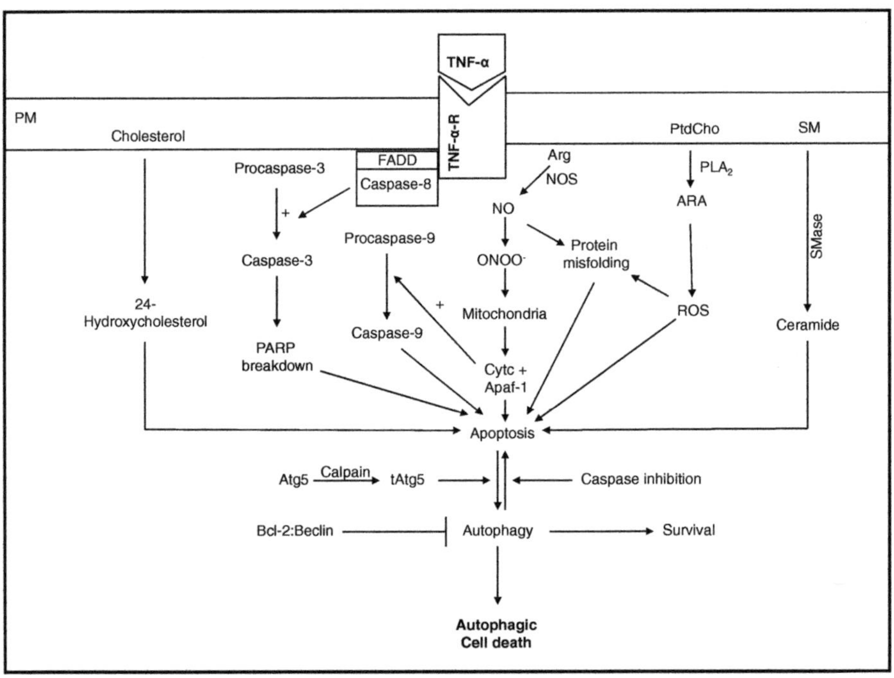


**Fig. 8.6** Interplay between apoptosis, autophagy, and autophagic cell death. Plasma membrane (PM); reactive oxygen species (ROS); arginine (Arg); nitric oxide synthase (NOS); nitric oxide (NO); peroxynitrite ($ONOO^-$); arachidonic acid (ARA); cytochrome *c* (Cytc); apoptosome complex with apoptosis-activating factor-1 (Apaf-1); and poly(ADP)ribose polymerase (PARP). Modified from Gorman (2008)

diseases consist of insoluble, unfolded proteins, which are tagged with ubiquitin. This reaction is catalyzed by ubiquitin-protein ligases (E3 s). Covalent tagging of proteins with chains of ubiquitin generally targets them for degradation through the ubiquitin-proteasome system (UPS), a major route through which intracellular proteolysis is regulated. Because ubiquitin tags proteins that must be eliminated from cells, it is hypothesized that the ubiquitin-proteasome system (UPS) is inactivated or malfunctions are due to overload of aggregated and unfolded proteins in neurodegenerative diseases (Matsuda and Tanaka, 2010). This inactivation may result in accumulation of ubiquitylated proteins with their concomitant aggregation into inclusion bodies and subsequent neurodegeneration. Although autophagy prevents neurons from undergoing protein aggregation-induced neurodegeneration, excessive or imbalanced induction of autophagy and disturbance in cross talk between autophagy and apoptosis can actively contribute to neuronal atrophy, neurite degeneration, and cell death (Fig. 8.6). The regulation of autophagy is a very complex process. It overlaps with the regulation of cell growth, proliferation, cell survival, and death. Collective evidence suggests that many signal transduction pathways, including target of rapamycin (TOR) or mammalian target of rapamycin (mTOR), PtdIns3K-I/PKB, c-jun NH2-terminal kinase, GTPases, calcium, p53, and protein synthesis along with autophagy-regulatory protein beclin 1/Atg6, are closely associated with induction, maintenance, and regulation of autophagy (Gorman, 2008; Bitomsky and Hofmann, 2009).

## 8.10 Mechanisms of Neurodegeneration in Neurodegenerative Diseases

The precise molecular mechanism of neurodegeneration in neurodegenerative diseases is a complex process, which still remains illusive (Farooqui and Horrocks, 2007; Farooqui, 2009a, b). The complex interplay among aging, genetic, and environmental factors may result in induction of pathologic processes, such as aggregation of proteins, excessive inflammation, the production of ROS, and depletion of glutathione may contribute to the onset and progression of neurodegenerative diseases. It is proposed that in neurodegenerative diseases protein misfolding and the overload of toxic products derived from the free radical oxidation of polyunsaturated fatty acids, cholesterol, and sphingolipid generated through the action of $PLA_2$, sphingomyelinase, and cholesterol hydroxylases may contribute to the disruption of the cellular redox and ion homeostasis (Fig. 8.7). Most common hypotheses of neurodegeneration include interactions among neuroinflammation, oxidative stress, excitotoxicity, mitochondrial dysfunction, alterations in calcium homeostasis, proteasomal dysfunction, protein aggregation, decrease in blood flow, alterations in blood brain barrier, and neuronal cell cycle induction (Farooqui and Horrocks, 2007; Golde, 2009). However, placing these pathways in the proper relationship to the onset, time course, and progress of neurodegeneration and their relationship to the cytoskeletal pathology are challenging issues that are not fully understood (Golde, 2009).

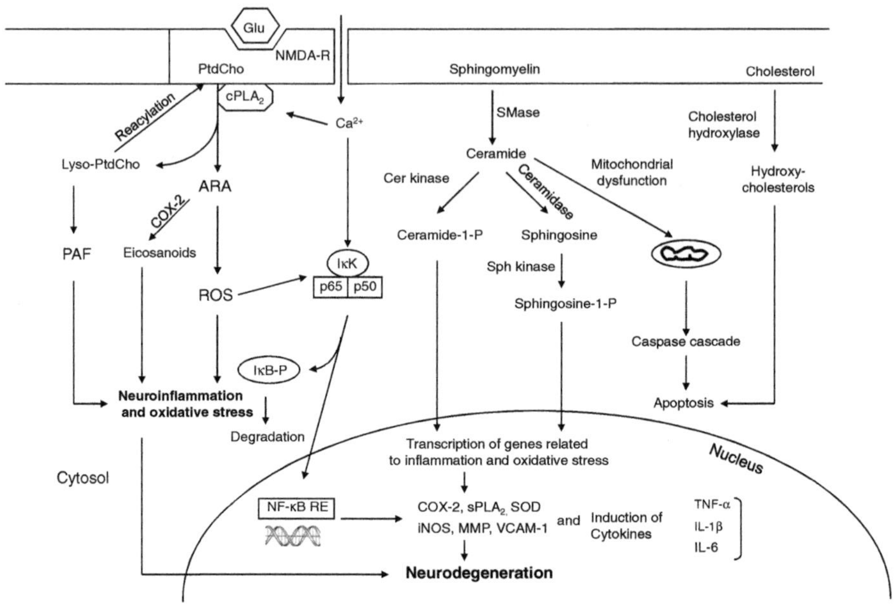


**Fig. 8.7** Interactions among signal transduction pathways associated with excitotoxicity, oxidative stress, and neuroinflammation result in neurodegeneration in neurodegenerative diseases. Glutamate and its analogs (A); NMDA receptor (NMDA-R); phosphatidylcholine (PtdCho); lyso-phosphatidylcholine (lyso-PtdCho); arachidonic acid (ARA); platelet-activating factor (PAF); cytosolic phospholipase $A_2$ ($cPLA_2$); cyclooxygenase-2 (COX-2); sphingomyelinase (SMase); ceramide kinase (Cer kinase); sphingosine kinase (Sph kinase); reactive oxygen species (ROS); nuclear factor-κB (NF-κB); nuclear factor-κB-response element (NF-κB-RE); inhibitory subunit of NF-κB (I-κB); phosphorylated I-κB (I-κB-P); tumor necrosis factor-α (TNF-α); interleukin-1β (IL-1β); interleukin-6 (IL-6); inducible nitric oxide synthase (iNOS); matrix metalloproteinases (MMPs); superoxide dismutase (SOD); vascular adhesion molecule-1 (VCAM-1); and secretory phospholipase $A_2$ ($sPLA_2$). Alterations in glutamate homeostasis contribute to inflammation and oxidative stress-mediated neural cell injury

Above described neurodegenerative diseases are associated with protein misfolding and the formation of distinct aggregates, resulting in a putative pathological protein load on the nervous system. Thus, Aβ aggregates are associated with the pathogenesis of AD, α-synuclein aggregates are involved in PD, aggregates comprising neuronal intermediate filament proteins, neurofilaments, and peripherin have been implicated in ALS, insoluble aggregates containing huntingtin are associated with HD, and misfolded PrPsc polymerized amyloid fibril is involved in neurodegeneration in prion diseases (Fig. 8.8). A causative link between protein aggregate formation and neurodegenerative diseases has not yet been established, but it is suggested that the toxic action of soluble oligomers and protofibrillar derivatives of misfolded proteins may play a pathogenic role (Jellinger, 2009; Farooqui, 2009a, b). This suggestion is supported by the observation that a single-domain antibody can

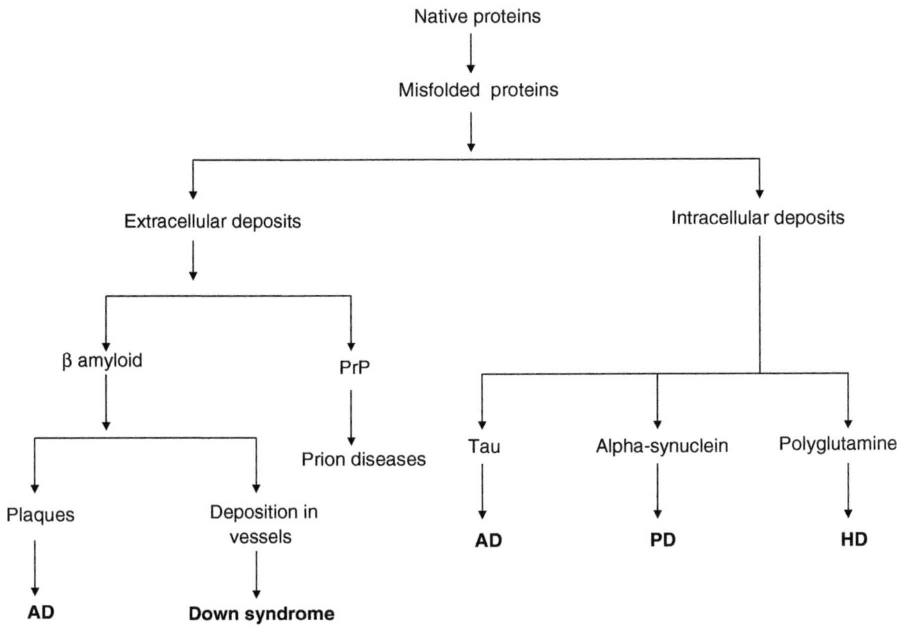


**Fig. 8.8** Protein aggregation and classification of neurodegenerative diseases. Modified from Jellinger (2009)

recognize a common conformational epitope that is displayed by several disease-associated proteins, including Aβ, α-synuclein, τ-protein, prions, and polyglutamine (polyQ)-containing peptides (Jellinger, 2009; Farooqui, 2009a). The transformation of native proteins into pathological aggregates results not only in loss of protein functions but also in neurotoxic effects of accumulated aggregates resulting in neural cell death through oxidative stress, apoptosis, loss of synapses, abnormalities in axonal transport, and defects in neuronal development (Fig. 8.9). Many factors mediate and modulate protein aggregation in neurodegenerative diseases. They include aggregation-prone sequences, specific mutations, environmental factors, protein modifications, and also dysregulation of the protein degradation machinery (Bandopadhyay and de Belleroche, 2009). To get rid of misfolded proteins, neuronal cells contain a large number of intracellular proteases, which, together with the chaperones (Hsp72), comprise the cellular protein quality control systems in the endoplasmic reticulum (ER) (Scheper and Hoozemans, 2009). Chaperones promote refolding or degradation of misfolded polypeptides, inhibit protein aggregation, and facilitate the formation of aggresome. Some molecular chaperones and chaperone-related proteases, such as the proteasome, can also hydrolyze ATP to forcefully convert stable harmful protein aggregates into harmless natively refoldable, or protease-degradable, polypeptides. The ubiquitin-proteasome proteolytic system (UPS) participates in reducing the levels of soluble abnormal proteins, while autophagy clears cells containing protein aggregates (Fig. 8.9). Accumulation of the

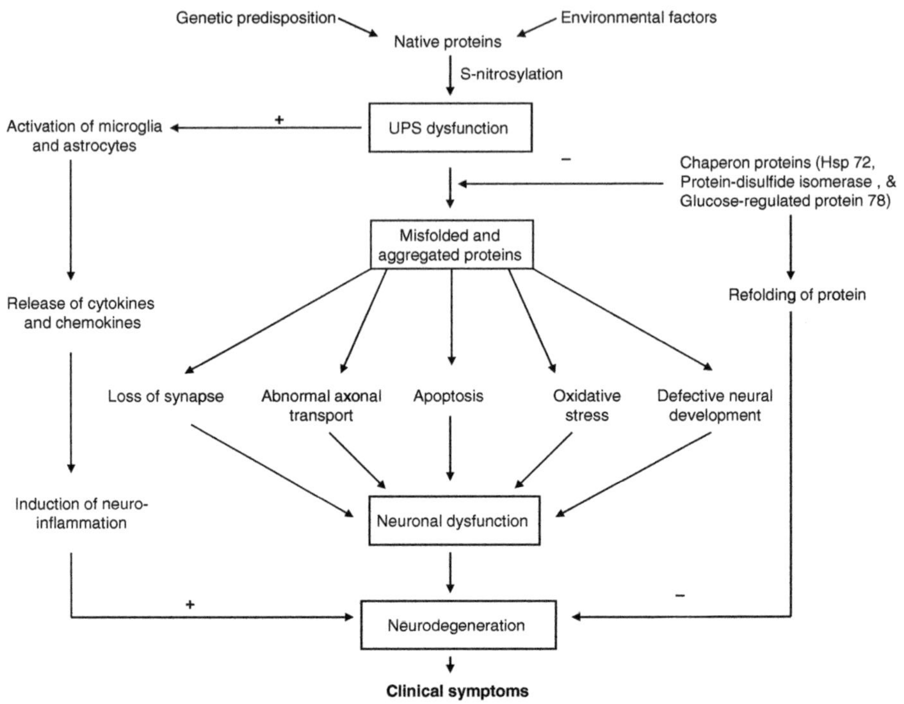


**Fig. 8.9** Neurotoxic effects of misfolded protein aggregates and their association with neurodegeneration

aggregation-prone proteins activates signal transduction pathways that control cell death, including JNK pathway controlling viability of a cell in various models of neurodegenerative diseases (Scheper and Hoozemans, 2009). The major chaperone Hsp72 interferes with this signaling pathway, thus promoting survival. In a mouse model of HD, the deletion of the molecular chaperones Hsp70.1 and Hsp70.3 significantly exacerbates many physical and behavioral parameters and neuropathological outcome measures, such as survival, body weight, tremor, limb clasping, and open-field activities (Wacker et al., 2009). Although the deletion of Hsp70.1 and Hsp70.3 significantly increases the size of inclusion bodies generated by mutant htt exon 1, has no effect on the levels of fibrillar aggregates. In addition, the deficiency of Hsp70s significantly downregulates levels of c-fos, a marker for neuronal activity. In contrast, deletion of Hsp70s does not enhance prion-mediated neurodegeneration, ruling out the possibility that the Hsp70.1/70.3 mice are non-specifically sensitized to all protein misfolding disorders (Wacker et al., 2009). Collective evidence suggests that endogenous Hsp70s are a critical component of the cellular defense against the toxic effects of misfolded htt protein in neurons, and its mechanism of action does not involve the deposition of fibrillar aggregates.

Accumulation of misfolded aggregated proteins impairs UPS and suppresses heat shock protein response. Such an inhibition of major cell defense systems may play a

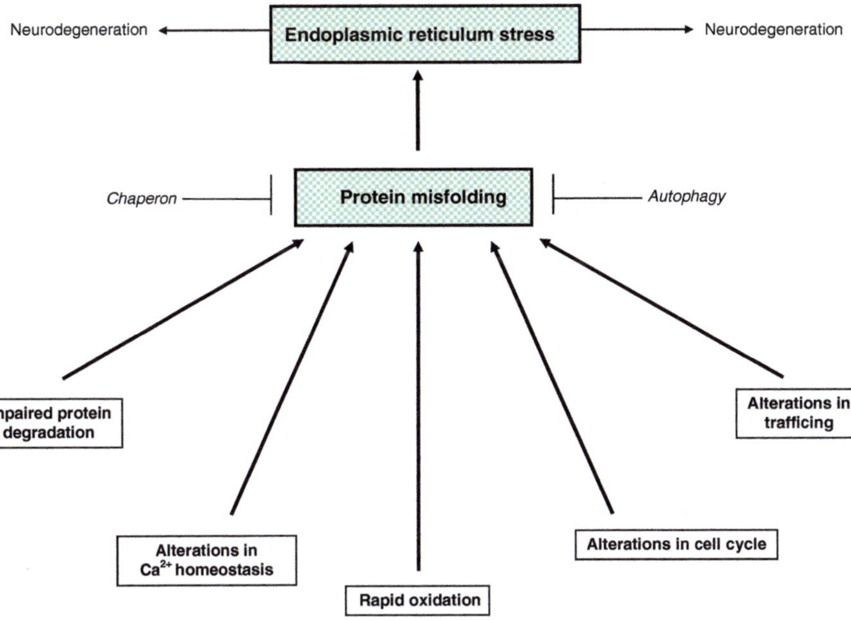


**Fig. 8.10** Factors modulating protein misfolding in neurodegenerative diseases

critical role in cell death in neurodegenerative diseases. Accumulation of misfolded proteins in the endoplasmic reticulum triggers a cellular stress response called the unfolded protein response (UPR) that protects the cell against the toxic buildup of misfolded proteins (Figs. 8.9 and 8.10). The UPR include translational attenuation, induction of ER-resident chaperones, and degradation of unfolded proteins through the ER-associated degradation. In the case of severe and/or prolonged ER stress (Fig. 8.10), cellular signals leading to cell death are activated. Thus, UPR results not only in inhibition of global protein synthesis but also in activation of expression of genes coding for ER-resident proteins that are involved in the folding and processing reactions. Neurodegenerative diseases are accompanied by the activation of UPR. How does UPR relate to the pathological hallmarks of neurodegenerative diseases is still elusive (Scheper and Hoozemans, 2009).

## 8.11 Conclusion

A majority of neurodegenerative diseases are accompanied by excessive oxidative stress, chronic inflammation, and alterations in glutamate homeostasis. Excessive oxidative stress and chronic neuroinflammation are long-standing and self-perpetuating processes that persist long after insult. Chronic neuroinflammation includes not only long-standing activation of microglia and astrocytes but also subsequent sustained release of inflammatory mediators that increase in oxidative stress. The sustained release of inflammatory mediators, such as cytokines, adhesion molecules, and proteases, works to perpetuate the inflammatory cycle, activating additional microglia and astrocytes promoting their proliferation, and resulting in further release of inflammatory factors. In neurodegenerative diseases inflammation and oxidative stress are supported by the generation of excess of glycerophospholipid-, sphingolipid-, and cholesterol-derived lipid mediators and their interactions with each other, along with disruption of cellular calcium homeostasis and alterations in redox status. Increased intensity of interplay among glycerophospholipid-, sphingolipid-, and cholesterol-derived lipid mediators triggers neuronal apoptosis. Phospholipases $A_2$, sphingomyelinases, and cholesterol hydroxylases are families of enzymes associated with the generation of above lipid mediators. In addition mitogen-activated protein kinases (MAPKs) are a family of serine/threonine protein kinases responsible for most cellular responses to cytokines and external stress signals and crucial for regulation of the production of inflammation mediators.

Genetic mutations, lifestyle, and environmental factors modulate the risk of neurodegenerative diseases. In addition, neurodegenerative diseases also involve abnormal protein aggregates generated by aberrant post-translational modifications, solubility, aggregation, and fibril formation of selected proteins, which cannot be degraded by cytosolic proteases, ubiquitin-protesome system, and autophagy. These aggregated proteins include Aβ, τ-protein, α-synuclein, huntingtin, and prion protein. Interactions of α-synuclein, β-amyloid peptides, and prion proteins with DNA suggest that DNA-binding activity may be a common property of many amyloidogenic proteins associated with various neurodegenerative disorders. The binding of $PrP^C$ with Aβ oligomer is another important finding that has been recently reported. Collective evidence suggests that there are many mechanistic similarities among neurodegenerative diseases. Counteracting neurodegenerative processes, brain tissue contains many mechanisms, such as neurotrophic factor signaling, antioxidant enzymes, protein chaperones, and anti-apoptotic proteins that facilitate endogenous neuroprotective processes.

## References


Abood ME, Butler M (1979) Membrane fluidity and fatty acid composition of phospholipids in erythrocyte membranes of patients with Huntington disease. J Neurosci Res 4:183–187

Aguirre N, Beal MF, Matson WR, Bogdanov MB (2005) Increased oxidative damage to DNA in an animal model of amyotrophic lateral sclerosis. Free Radic Res 39:383–388

Ahmad M, Attoub S, Singh MN, Martin GL, El-Agnaf M (2007) Gamma-synuclein and the progression of cancer. FASEB J 21:3419–3430

Akowitz A, Sklaviadis T, Manuelidis L (1994) Endogenous viral complexes with long RNA cosediment with the agent of Creutzfeldt-Jakob disease. Nucleic Acid Res 22:1101–1107

Almer G, Guegan C, Teismann P, Naini A, Rosoklija G, Hays AP, Chen CP, Przedborski S (2001) Increased expression of the pro-inflammatory enzyme cyclooxygenase-2 in amyotrophic lateral sclerosis. Ann Neurol 49:176–185

Andersen JK (2004) Oxidative stress in neurodegeneration: cause or consequence? Nat Rev Neurosci 5:S18–S25

Arboleda G, Morales LC, Benítez B, Arboleda H (2009) Regulation of ceramide-induced neuronal death: cell metabolism meets neurodegeneration. Brain Res Rev 59:333–346

Baker CA, Manuelidis L (2003) Unique inflammatory RNA profiles of microglia in Creutzfeldt-Jakob disease. Proc Natl Acad Sci USA 100:675–679

Backman C, Perlmann T, Wallen A, Hoffer BJ, Morales M (1999) Selective group of dopaminergic neurons express Nurr1 in the adult mouse brain. Brain Res 851:125–132

Bai Y, Li Q, Yang J, Zhou X, Yin X, Zhao D (2008) p75(NTR) activation of NF-kappaB is involved in PrP106–126-induced apoptosis in mouse neuroblastoma cells. Neurosci Res 62:9–14

Bal-Price A, Brown GC (2000) Nitric-oxide-induced necrosis and apoptosis in PC12 cells mediated by mitochondria. J Neurochem 75:1455–1464

Blackinton J, Kumaran R, van der Brug MP, Ahmad R, Olson L, Galter D, Lees A, Bandopadhyay R, Cookson MR (2009) Post-transcriptional regulation of mRNA associated with DJ-1 in sporadic Parkinson disease. Neurosci Lett 452:8–11

Bandopadhyay R, de Belleroche J (2010) Pathogenesis of Parkinson's disease: emerging role of molecular chaperones. Trends Mol Med 16:27–36

Banoei MM, Houshmand M, Panathi MS, Shariati P (2007) Huntington's disease and mitochondrial DNA deletions: event or regular mechanism for mutant huntingtin protein and CAG repeats expansion? Cell Mol Neurobiol 27:867–875

Baptista M, Cookson MR, Miller DW (2004) Parkin and α-synuclein: opponent actions in the pathogenesis of Parkinson's disease. Neuroscientist 10:63–72

Barcelo-Coblijn G, Golovko MY, Weinhofer I, Berger J, Murphy EJ (2007) Brain neutral lipids mass is increased in alpha-synuclein gene-ablated mice. J Neurochem 101:132–141

Barret A, Forestier L, Deslys JP, Julien R, Gallet PF (2005) Glycosylation-related gene expression in prion diseases: PrPSc accumulation in scrapie infected GT1 cells depends on beta-1,4-linked GalNAc-4-SO4 hyposulfation. J Biol Chem 280:10516–10523

Barrier L, Ingrand S, Fauconneau B, Page G (2008) Gender-dependent accumulation of ceramides in the cerebral cortex of the APP(SL)/PS1Ki mouse model of Alzheimer's disease. Neurobiol Aging 24 Nov 2008 [Epub ahead of print]

Bate C, Reid S, Williams A (2004) Phospholipase $A_2$ inhibitors or platelet-activating factor antagonists prevent prion replication. J Biol Chem 279:36405–36411

Bate C, Tayebi M, Williams A (2008) Sequestration of free cholesterol in cell membranes by prions correlates with cytoplasmic phospholipase $A_2$ activation. BMC 6:8

Batulan Z, Taylor DM, Aarons RJ, Minotti S, Doroudchi MM, Nalbantoglu J, Durham HD (2006) Induction of multiple heat shock proteins and neuroprotection in a primary culture model of familial amyotrophic lateral sclerosis. Neurobiol 24:213–225

Bayir H, Kapralov AA, Jiang J, Huang Z, Tyurina YY, Tyurin VA, Zhao Q, Belikova NA, Vlasova II, Maeda A, Zhu J, Na HM, Mastroberardino PG, Sparvero LJ, Amoscato AA, Chu CT, Greenamyre JT, Kagan VE (2009) Peroxidase mechanism of lipid-dependent cross-linking of synuclein with cytochrome C: protection against apoptosis versus delayed oxidative stress in Parkinson disease. J Biol Chem 284:15951–15969

Beal MF (1998) Mitochondrial dysfunction in neurodegenerative diseases. Biochim Biophys Acta 1366:211–223

Beaulieu JM, Julien JP (2003) Peripherin-mediated death of motor neurons rescued by overexpression of neurofilament NF-H proteins. J Neurochem 85:248–256

Behrens A (2003) Physiological and pathological functions of the prion protein homologue Dpl. Br Med Bull 66:35–42

Beyer K (2007) Mechanistic aspects of Parkinson's disease: alpha-synuclein and the biomembrane. Cell Biochem Biophys 47:285–299

Bi H, Sze CI (2002) *N*-methyl-D-aspartate receptor subunit NR2A and NR2B messenger RNA levels are altered in the hippocampus and entorhinal cortex in Alzheimer's disease. J Neurol Sci 200:11–18

Bitomsky N, Hofmann TG (2009) Apoptosis and autophagy: regulation of apoptosis by DNA damage signalling – Roles of p53, p73 and HIPK2. FEBS J 276:6074–6083

Bogdanov M, Brown RH, Matson W, Smart R, Hayden D, O'Donnell H, Flint Beal M, Cudkowicz M (2000) Increased oxidative damage to DNA in ALS patients. Free Radic Biol Med 29: 652–658

Bolognin S, Messori L, Zatta P (2009) Metal ion physiopathology in neurodegenerative disorders. Neuromolecular Med 28 Nov 2009 [Epub ahead of print]

Bonilla E (2000) Huntington disease. A review. Invest Clin 41:117–141

Bruegge K, Jelkmann W, Metzen E (2007) Hydroxylation of hypoxia-inducible transcription factors and chemical compounds targeting the HIF-alpha hydroxylases. Curr Med Chem 14:1853–1862

Bueler H (2009) Impaired mitochondrial dynamics and function in the pathogenesis of Parkinson's disease. Exp Neurol 218:235–246

Burke RE (2004) Recent advances in research on Parkinson disease: synuclein and parkin. Neurologist 10:75–81

Burns JM, Donnelly JE, Anderson HS, Mayo MS, Spencer-Gardner L, Thomas G, Cronk BB, Haddad Z, Klima D, Hansen D, Brooks WM (2007) Peripheral insulin and brain structure in early Alzheimer disease. Neurology 69:1094–1104

Busiguina S, Fernandes AM, Barrios V, Clark R, Tolbert DL, Berciano J, Torres-Aleman I (2000) Neurodegeneration is associated to changes in serum insulin-like growth factors. Neurobiol Dis 7:657–665

Caccamo D, Curro M, Condello S, Ferlazzo N, Ientile R (2009) Critical role of transglutaminase and other stress proteins during neurodegenerative processes. Amino Acids [Epub ahead print]

Caraci F, Battaglia G, Busceti C, Biagioni F, Mastroiacovo F, Bosco P, Drago F, Nicoletti F, Sortino MA, Copani A (2008) TGF-beta 1 protects against Abeta-neurotoxicity via the phosphatidylinositol-3-kinase pathway. Neurobiol Dis 30:234–242

Carvalho C, Correia SC, Santos RX, Cardoso S, Moreira PI, Clark TA, Zhu X, Smith MA, Perry G (2009) Role of mitochondrial-mediated signaling pathways in Alzheimer disease and hypoxia. J Bioenerg Biomembr 15 Oct [Epub ahead of print]

Cedazo-Minguez A, Popescu BO, Ankarcrona M, Nishimura T, Cowburn RF (2002) The presenilin 1 deltaE9 mutation gives enhanced basal phospholipase C activity and a resultant increase in intracellular calcium concentrations. J Biol Chem 277:36646–36655

Cepeda C, Ariano MA, Calvert CR, Flores-Hernandez J, Chandler SH, Leavitt BR, Hayden MR, Levine MS (2001) NMDA receptor function in mouse models of Huntington disease. J Neurosci Res 66:525–539

Chang Y, Kong Q, Shan X, Tian G, Ilieva H, Cleveland DW, Rothstein JD, Borchelt DR, Wong PC, Lin CL (2008) Messenger RNA oxidation occurs early in disease pathogenesis and promotes motor neuron degeneration in ALS. PLoS One 3:e2849

Chaudhuri TK, Paul S (2006) Protein-misfolding diseases and chaperone-based therapeutic approaches. FEBS J 273:1331–1449

Chen M, Ona VO, Li M, Ferrante RJ, Fink KB, Zhu S et al (2000) Minocycline inhibits caspase-1 and caspase-3 expression and delays mortality in a transgenic mouse model of Huntington disease. Nat Med 6:797–801

Chen YG (2005) Specific tau phosphorylation sites in hippocampus correlate with impairment of step-down inhibitory avoidance task in rats. Behav Brain Res 158:277–284

Chen LW, Yung KK, Chan YS, Shum DK, Bolam JP (2008a) The proNGF-p75NTR-sortilin signalling complex as new target for the therapeutic treatment of Parkinson's disease. CNS Neurol Disord Drug Targets 7:512–523

Chen JM, Gao C, Shi Q, Shan B, Lei TJ, Dong CF, An R, Wang GR, Zhang BY, Han J, Dong XP (2008b) Different expression patterns of CK2 subunits in the brains of experimental animals and patients with transmissible spongiform encephalopathies. Arch Virol 153:1013–1020

Chen L, Wei Y, Wang X, He R (2009) D-Ribosylated Tau forms globular aggregates with high cytotoxicity. Cell Mol Life Sci 66:2559–2571

Chen L, Wei Y, Wang X, He R (2010) Ribosylation rapidly induces alpha-synuclein to form highly cytotoxic molten globules of advanced glycation end products. PLoS One 5(2):e9052

Chesebro B, Trifilo M, Race R, Meade-White K, Teng C, LaCasse R, Raymond L, Favara C, Baron G, Priola S, Caughey B, Masliah E, Oldstone M (2005) Anchorless prion protein results in infectious amyloid disease without clinical scrapie. Science 308:1435–1439

Chiarini A, Dal Pra I, Whitfield JF, Armato U (2006) The killing of neurons by beta-amyloid peptides, prions, and pro-inflammatory cytokines. Ital J Anat Embryol 111:221–246

Churcher I (2006) Tau therapeutic strategies for the treatment of Alzheimer's disease. Curr Top Med Chem 6:579–595

Cobb NJ, Surewicz WK (2009) Prion diseases and their biochemical mechanisms. Biochemistry 48:2574–2585

Conforti P, Ramos C, Apostol BL, Simmons DA, Nguyen HP, Riess O, Thompson LM, Zuccato C, Cattaneo E (2008) Blood level of brain-derived neurotrophic factor mRNA is progressively reduced in rodent models of Huntington's disease: restoration by the neuroprotective compound CEP-1347. Mol Cell Neurosci 39:1–7

Conquer JA, Tierney MC, Zecevic J, Bettger WJ, Fisher RH (2000) Fatty acid analysis of blood plasma of patients with Alzheimer's disease, other types of dementia, and cognitive impairment. Lipids 35:1305–1312

Corcia P, Mayeux-Portas V, Khoris J, de Toffol B, Autret A, Muh JP, Andres C (2002) Abnormal SMN1 gene copy number is a susceptibility factor for amyotrophic lateral sclerosis. Ann Neurol 51:243–246

Corosaro A, Thellung S, Villa V, Princile DR, Palude D, Arena S, Millo E, Schettini D, Damonte G, Aceto A, Schettini G, Florio T (2003) Prion protein fragment 106–126 induces a p38 MAP kinase-dependent apoptosis in SH-SY5Y neuroblastoma cells independently from the amyloid fibril formation. Ann N Y Acad Sci 1010:610–622

Cotman CW (2005) The role of neurotrophins in brain aging: a perspective in honor of Regino Perez-Polo. Neurochem Res 30:877–881

Coyle JT, Schwarcz R (1976) Lesion of striatal neurones with kainic acid provides a model for Huntington's chorea. Nature 263:244–246

Craft S, Watson GS (2004) Insulin and neurodegenerative disease: shared and specific mechanisms. Lancet Neurol 3:169–178

Crews L, Tsigelny I, Hashimoto M, Masliah E (2009) Role of synucleins in Alzheimer's disease. Neurotox Res 16:306–317

Cui JG, Hill JM, Zhao Y, Lukiw WJ (2007) Expression of inflammatory genes in the primary visual cortex of late-stage Alzheimer's disease. Neuroreport 18:115–119

Cutler RG, Pedersen WA, Camandola S, Rothstein JD, Mattson MP (2002) Evidence that accumulation of ceramides and cholesterol esters mediates oxidative stress-induced death of motor neurons in amyotrophic lateral sclerosis. Ann Neurol 52:448–457

Cutler RG, Kelly J, Storie K, Pedersen WA, Tammara A, Hatanpaa K, Troncoso JC, Mattson MP (2004) Involvement of oxidative stress-induced abnormalities in ceramide and cholesterol metabolism in brain aging and Alzheimer's disease. Proc Natl Acad Sci USA 17:2070–2075

Davenport CM, Sevastou IG, Hooper C, Pocock JM (2009) Inhibiting p53 pathways in microglia attenuates microglial-evoked neurotoxicity following exposure to Alzheimer peptides. J Neurochem 6 Nov 2009 [Epub ahead of print]

DeArmond SJ, Prusiner SB (2003) Perspectives on prion biology, prion disease pathogenesis, and pharmacologic approaches to treatment. Clin Lab Med 23:1–41

De Felice FG, Vieira MN, Bomfim TR, Decker H, Velasco PT, Lambert MP, Viola KL, Zhao WQ, Ferreira ST, Klein WL (2009) Protection of synapses against Alzheimer's-linked toxins: insulin signaling prevents the pathogenic binding of Abeta oligomers. Proc Natl Acad Sci USA 106:1971–1976

Desplats PA, Denny CA, Kass KE, Gilmartin T, Head SR, Sutcliffe JG, Seyfried TN, Thomas EA (2007) Glycolipid and ganglioside metabolism imbalances in Huntington's disease. Neurobiol Dis 27:265–277

Desplats PA, Lambert JR, Thomas EA (2008) Functional roles for the striatal-enriched transcription factor, Bcl11b, in the control of striatal gene expression and transcriptional dysregulation in Huntington's disease. Neurobiol Dis 31:298–308

Dheen ST, Kaur C, Ling EA (2007) Microglial activation and its implications in the brain diseases. Curr Med Chem 14:1189–1197
Drachman DB, Rothstein JD (2000) Inhibition of cyclooxygenase-2 protects motor neurons in an organotypic model of amyotrophic lateral sclerosis. Ann Neurol 48:792–795
Drachman DB, Frank K, Dykes-Hoberg M, Teismann P, Almer G, Przedborski S, Rothstein JD (2002) Cyclooxygenase 2 inhibition protects motor neurons and prolongs survival in a transgenic mouse model of ALS. Ann Neurol 52:771–778
Dunah AW, Jeong H, Griffin A, Kim YM, Standaert DG, Hersch SM, Mouradian MM, Young AB, Tanese N, Krainc D (2002) Sp1 and TAFII130 transcriptional activity disrupted in early Huntington's disease. Science 296:2238–2242
Dwyer BE, Takeda A, Zhu XW, Perry G, Smith MA (2005) Ferric cycle activity and Alzheimer disease. Curr Neurovasc Res 2:261–267
Eslamboli A (2005) Assessment of GDNF in primate models of Parkinson's disease: comparison with human studies. Rev Neurosci 16:303–310
Facheris M, Beretta S, Ferrarese C (2004) Peripheral markers of oxidative stress and excitotoxicity in neurodegenerative disorders: tools for diagnosis and therapy? J Alzheimers Dis 6: 177–184
Farooqui AA, Rapoport SI, Horrocks LA (1997) Membrane phospholipid alterations in Alzheimer's disease: deficiency of ethanolamine plasmalogens. Neurochem Res 22:523–527
Farooqui AA, Horrocks LA (1998) Lipid peroxides in the free radical pathophysiology of brain diseases. Cell Mol Neurobiol 18:599–608
Farooqui AA, Ong WY, Horrocks LA (2003) Plasmalogens, docosahexaenoic acid and neurological disorders. Adv Exp Med Biol 544:335–354
Farooqui AA, Ong WY, Horrocks LA (2004) Biochemical aspects of neurodegeneration in human brain: involvement of neural membrane phospholipids and phospholipases $A_2$. Neurochem Res 29:1961–1977
Farooqui AA, Ong WY, Horrocks LA (2006) Inhibitors of brain phospholipase $A_2$ activity: their neuropharmacological effects and therapeutic importance for the treatment of neurologic disorders. Pharmacol Rev 58:591–620
Farooqui AA, Horrocks LA (2006) Phospholipase $A_2$-generated lipid mediators in the brain: the good, the bad, and the ugly. Neuroscientist 12:245–260
Farooqui AA, Horrocks LA (2007) Glycerophospholipids in the brain: phospholipases $A_2$ in neurological disorders. Springer, New York, NY, pp 1–394
Farooqui AA, Horrocks LA, Farooqui T (2007a) Modulation of inflammation in brain: a matter of fat. J Neurochem 101:577–599
Farooqui AA, Horrocks LA, Farooqui T (2007b) Interactions between neural membrane glycerophospholipid and sphingolipid mediators: a recipe for neural cell survival or suicide. J Neurosci Res 85:1834–1850
Farooqui T, Farooqui AA (2009) Aging: an important factor for the pathogenesis of neurodegenerative diseases. Mech Aging Dev 130:203–215
Farooqui AA (2009a) Hot topics in neural membrane lipidology. Springer, New York, NY
Farooqui AA (2009b) Lipid mediators in the neural cell nucleus: their metabolism, signaling, and association with neurological disorders. Neuroscientist 15:392–407
Farooqui AA (2010a) Neurochemical aspects of neuroinflammation in brain. In: Farooqui AA, Farooqui T (eds) Molecular aspects of neurodegeneration and neuroprotection. Bentham Science Publishers Ltd, in press
Farooqui AA (2010b) Studies on plasmalogen-selective phospholipase $A_2$ in Brain. Mol Neurobiol 41:267–273
Farooqui AA, Ong WY, Farooqui T (2010c) Lipid Mediators in the nucleus: their potential contribution to Alzheimer's disease. Biochim Biophys Acta 1801:906–916
Ferrer I (2009) Altered mitochondria, energy metabolism, voltage-dependent anion channel, and lipid rafts converge to exhaust neurons in Alzheimer's disease. J Bioenerg Biomembr 2 Oct 2009 [Epub ahead of print]

Fombonne J, Rabizadeh S, Banwait S, Mehlen P, Bredesen DE (2009) Selective vulnerability in Alzheimer's disease: amyloid precursor protein and p75(NTR) interaction. Ann Neurol 65:294–303

Forman MS, Trojanowski JQ, Lee VM (2004) Neurodegenerative diseases: a decade of discoveries paves the way for therapeutic breakthroughs. Nat Med 10:1055–1063

Fraser DA, Pisalyaput K, Tenner AJ (2009) C1q enhances microglial clearance of apoptotic neurons and neuronal blebs, and modulates subsequent inflammatory cytokine production. J Neurochem 16 Nov 2009 [Epub ahead of print]

Freixes M, Rodríguez A, Dalfó E, Ferrer I (2006) Oxidation, glycoxidation, lipoxidation, nitration, and responses to oxidative stress in the cerebral cortex in Creutzfeldt-Jakob disease. Neurobiol Aging 27:1807–1815

Fujii T, Kunugi H (2009) p75NTR as a therapeutic target for neuropsychiatric diseases. Curr Mol Pharmacol 2:70–76

Fumagalli F, Molteni R, Calabrese F, Maj PF, Racagni G, Riva MA (2008) Neurotrophic factors in neurodegenerative disorders: potential for therapy. CNS Drugs 22:1005–1019

Gandhi PN, Chen SG, Wilson-Delfosse AL (2009) Leucine-rich repeat kinase 2 (LRRK2): a key player in the pathogenesis of Parkinson's disease. J Neurosci Res 87:1283–1295

Gerlach M, Double KL, Youdim MB, Riederer P (2006) Potential sources of increased iron in the substantia nigra of Parkinsonian patients. J Neural Transm Suppl 2006(70):133–142

Gharami K, Xie Y, An JJ, Tonegawa S, Xu B (2006) Brain-derived neurotrophic factor over-expression in the forebrain ameliorates Huntington's disease phenotypes in mice. J Neurochem 105:369–379

Ghosh A, Roy A, Liu X, Kordower JH, Mufson EJ, Hartley DM, Ghosh S, Mosley RL, Gendelman HE, Pahan K (2007) Selective inhibition of NF-kappaB activation prevents dopaminergic neuronal loss in a mouse model of Parkinson's disease. Proc Natl Acad Sci USA 104:18754–18759

Gil JM, Rego AC (2008) Mechanisms of neurodegeneration in Huntington's disease. Eur J Neurosci 27:2803–2820

Gines S, Bosch M, Marco S, Gavalda N, Diaz-Harnandez M, Lucas JJ, Canais JM, Alberch J (2006) Reduced expression of the TrkB receptor in Huntington's disease mouse models and in human brain. Eur J Neurosci 23:649–658

Giratt A, Rodrigo T, Martin ED, Gonzalez JR, Mila M, Cena V, Diessen M, Canals JM, Alberch J (2009) Brain-derived neurotrophic factor modulates the severity of cognitive alterations induced by mutant huntingtin: involvement of phospholipaseCgamma activity and glutamate receptor expression. Neuroscience 158:1234–1250

Golde TE (2009) The therapeutic importance of understanding mechanisms of neuronal cell death in neurodegenerative disease. Mol Neurodegener 4:8

Golovko MY, Rosenberger TA, Faergeman NJ, Feddersen S, Cole NB, Pribill I, Berger J, Nussbaum RL, Murphy EJ (2006) Acyl-CoA synthetase activity links wild-type but not mutant alpha-synuclein to brain arachidonate metabolism. Biochemistry 45:6956–6966

Gorman AM (2008) Neuronal cell death in neurodegenerative diseases: recurring themes around protein handling. J Cell Mol Med 12:2263–2280

Graeber MB, Moran LB (2002) Mechanisms of cell death in neurodegenerative diseases: fashion, fiction, and facts. Brain Pathol 12:385–390

Granic I, Dolga AM, Nijholt IM, van Dijk G, Eisel UL (2009) Inflammation and NF-kappaB in Alzheimer's disease and diabetes. J Alzheimers Dis 16:809–821

Grossman A, Zeiler B, Sapirstein V (2003) Prion protein interactions with nucleic acid: possible models for prion disease and prion function. Neurochem Res 28:955–963

Guan Z, Söderberg M, Sindelar P, Prusiner SB, Kristensson K, Dallner G (1996) Lipid composition in scrapie-infected mouse brain: prion infection increases the levels of dolichyl phosphate and ubiquinone. J Neurochem 66:277–285

Guan Z, Wang Y, Cairns NJ, Lantos PL, Dallner G, Sindelar PJ (1999) Decrease and structural modifications of phosphatidylethanolamine plasmalogen in the brain with Alzheimer disease. J Neuropathol Exp Neurol 58:740–747

Guidetti P, Charles V, Chen EY, Reddy PH, Kordower JH, Whetsell WO Jr et al (2001) Early degenerative changes in transgenic mice expressing mutant huntingtin involve dendritic abnormalities but no impairment of mitochondrial energy production. Exp Neurol 169:340–350

Gunther EC, Strittmater SM (2010) Beta-amyloid oligomers and cellular prion protein in Alzheimer's disease. J Mol Med [Epub ahead of print]

Haass C, Selkoe DJ (2007) Soluble protein oligomers in neurodegeneration: lessons from the Alzheimer's amyloid beta-peptide. Nat Rev Mol Cell Biol 8:101–112

Han X, Holtzman DM, McKeel DW Jr (2001) Plasmalogen deficiency in early Alzheimer's disease subjects and in animal models: molecular characterization using electrospray ionization mass spectrometry. J Neurochem 77:1168–1180

Han X, Holtzman DM, McKeel DW Jr., Kelley J, Morris JC (2002) Substantial sulfatide deficiency and ceramide elevation in very early Alzheimer's disease: potential role in disease pathogenesis. J Neurochem 82:809–818

Han X (2007) Potential mechanisms contributing to sulfatide depletion at the earliest clinically recognizable stage of Alzheimer's disease: a tale of shotgun lipidomics. J Neurochem 103 (Suppl 1):171–179

Hartley DM, Zhao C, Speier AC, Woodard GA, Li S, Li Z, Walz T (2008) Transglutaminase induces protofibril-like amyloid beta-protein assemblies that are protease-resistant and inhibit long-term potentiation. J Biol Chem 283:16790–16800

Hashimoto M, Rockenstein E, Crews L, Masliah E (2003) Role of protein aggregation in mitochondrial dysfunction and neurodegeneration in Alzheimer's and Parkinson's diseases. Neuromolecular Med 4:21–36

He X, Huang Y, Li B, Gong CX, Schuchman EH (2010) Deregulation of sphingolipid metabolism in Alzheimer's disease. Neurobiol Aging 31:398–408

Hegde ML, Vasudevaraju P, Rao KJ (2010) DNA induced folding/fibrillation of alpha-synuclein: new insights in Parkinson's disease. Front Biosci 15:418–436

Heneka MT, O'Banion MK (2007) Inflammatory processes in Alzheimer's disease. J Neuroimmunol 184:69–91

Hermel E, Gafni J, Propp SS, Leavitt BR, Wellington CL, Young JE, Hackam AS, Logvinova AV, Peel AL, Chen SF, Hook V, Singaraja R, Krajewski S, Goldsmith PC, Ellerby HM, Hayden MR, Bredesen DE, Ellerby LM (2004) Specific caspase interactions and amplification are involved in selective neuronal vulnerability in Huntington's disease. Cell Death Differ 11: 424–438

Heverin M, Meaney S, Lütjohann D, Diczfalusy U, Wahren J, Björkhem I (2005) Crossing the barrier: net flux of 27-hydroxycholesterol into the human brain. J Lipid Res 46:1047–1052

Huang Y, Tanimukai H, Liu F, Iqbal K, Grunake-Iqbal I, Gong CX (2004) Elevation of the level and activity of acid ceramidase in Alzheimer's disease brain. Eur J Neurosci 20:3489–3497

Hynd MR, Scott HL, Dodd PR (2004) Differential expression of *N*-methyl-D-aspartate receptor NR2 isoforms in Alzheimer's disease. J Neurochem 90:913–919

Ikeda J, Kohriyama T, Nakamura S (2000) Elevation of serum soluble E-selectin and antisulfoglucuronyl paragloboside antibodies in amyotrophic lateral sclerosis. Eur J Neurol 7:541–547

Jeffrey M, Goodsir CM, Bruce ME, McBride PA, Scott JR, Halliday WG (1992) Infection specific prion protein (PrP) accumulates on neuronal plasmalemma in scrapie infected mice. Neurosci Lett 147:106–109

Jankovic J, Chen S, Le WD (2005) The role of Nurr1 in the development of dopaminergic neurons and Parkinson's disease. Prog Neurobiol 77:128–138

Jellinger KA (2009) Recent advances in our understanding of neurodegeneration. J Neural Transm 116:1111–1162

Jenkinson AM, Collins AR, Duthie SJ, Wahle KW, Duthie GG (1999) The effect of increased intakes of polyunsaturated fatty acids and vitamin E on DNA damage in human lymphocytes. FASEB J 13:2138–2142

Jenner P, Olanow CW (2006) The pathogenesis of cell death in Parkinson's disease. Neurology 66(10 Suppl 4):S24–S36

Jiang YM, Yamamoto M, Kobayashi Y, Yoshihara T, Liang Y, Terao S, Takeuchi H, Ishigaki S, Katsuno M, Adachi H, Niwa J, Tanaka F, Doyu M, Yoshida M, Hashizume Y, Sobue G (2005) Gene expression profile of spinal motor neurons in sporadic amyotrophic lateral sclerosis. Ann Neurol 57:236–251

Johnson R, Buckley NJ (2009) Gene dysregulation in Huntington's disease: REST, microRNAs and beyond. Neuromolecular Med 11:183–199

Junn E, Ronchetti RD, Quezado MM, Kim SY, Mouradian MM (2003) Tissue transglutaminase-induced aggregation of alpha-synuclein: implications for Lewy body formation in Parkinson's disease and dementia with Lewy bodies. Proc Natl Acad Sci USA 100:2047–2952

Juranek I, Bezek S (2005) Controversy of free radical hypothesis: reactive oxygen species – cause or consequence of tissue injury? Gen Physiol Biophys 24:263–278

Kabashi E, Valdmanis PN, Dion P, Rouleau GA (2007) Oxidized/misfolded superoxide dismutase-1: the cause of all amyotrophic lateral sclerosis? Ann Neurol 62:553–559

Kanfer JN, Sorrentino G, Sitar DS (1998) Phospholipases as mediators of amyloid beta peptide neurotoxicity: an early event contributing to neurodegeneration characteristic of Alzheimer's disease. Neurosci Lett 257:93–96

Katsel P, Li C, Haroutunian V (2007) Gene expression alterations in the sphingolipid metabolism pathways during progression of dementia and Alzheimer's disease: a shift toward ceramide accumulation at the earliest recognizable stages of Alzheimer's disease? Neurochem Res 32:845–856

Kegel KB, Sapp E, Alexander J, Valencia A, Reeves P, Li X, Masso N, Sobin L, Aronin N, DiFiglia M (2009) Polyglutamine expansion in huntingtin alters its interaction with phospholipids. J Neurochem 110:1585–1597

Kim JI, Ju WK, Choi JH, Choi E, Carp RI, Wisniewski HM, Kim YS (1999) Expression of cytokine genes and increased nuclear factor-kappa B activity in the brains of scrapie-infected mice. Brain Res Mol Brain Res 73:17–27

Kim J, Kim TY, Hwang JJ, Lee JY, Shin JH, Gwag BJ, Koh JY (2009) Accumulation of labile zinc in neurons and astrocytes in the spinal cords of G93A SOD-1 transgenic mice. Neurobiol Dis 34:221–229

King M, Nafar F, Clarke J, Mearow K (2009) The small heat shock protein Hsp27 protects cortical neurons against the toxic effects of beta-amyloid peptide. J Neurosci Res 87:3161–3175

Kivenyi P, Kiaei M, Gardian G, Calingasan NY, Beal MF (2004) Additive neuroprotective effects of creatine and cyclooxygenase 2 inhibitors in a transgenic mouse model of amyotrophic lateral sclerosis. J Neurochem 88:576–582

Klein SM, Behrstock S, McHugh J, Hoffmann K, Wallace K, Suzuki M, Aebischer P, Svendsen CN (2005) GDNF delivery using human neural progenitor cells in a rat model of ALS. Human Gene Ther 16:509–521

Kong Q, Shan X, Chang Y, Tashiro H, Lin CL (2008) RNA oxidation: a contributing factor or an epiphenomenon in the process of neurodegeneration. Free Radic Res 42:773–777

Koudinov A, Kezlya E, Koudinova N, Berezov T (2009) Amyloid-beta, tau protein, and oxidative changes as a physiological compensatory mechanism to maintain CNS plasticity under Alzheimer's disease and other neurodegenerative conditions. J Alzheimer Dis 18:381–400

Kriem B, Sponne I, Fifre A, Malaplate-Armand C, Lozac'h-Pillot K, Koziel V, Yen-Potin FT, Bihain B, Oster T, Olivier JL, Pillot T (2005) Cytosolic phospholipase $A_2$ mediates neuronal apoptosis induced by soluble oligomers of the amyloid-beta peptide. FASEB J 19:85–87

Kucic S, Kiernan MC (2009) Pathophysiology of neurodegeneration in familial amyotrophic lateral sclerosis. Curr Mol Med 9:255–272

Kumar B, Nahreini P, Hanson AJ, Andreatta C, Prasad JE, Prasad KN (2005) Selenomethionine prevents degeneration induced by overexpression of wild-type human alpha-synuclein during differentiation of neuroblastoma cells. J Am Coll Nutr 24:516–523

Lacor PN, Buniel MC, Chang L, Fernandez SJ, Gong Y, Viola KL, Lambert MP, Velasco PT, Bigio EH, Finch CE, Krafft G, Klein WL (2004) Synaptic targeting by Alzheimer's-related amyloid beta oligomers. J Neurosci 24:10191–10200

Lahiri DK, Maloney B, Basha MR, Ge YW, Zawia NH (2007) How and when environmental agents and dietary factors affect the course of Alzheimer's disease: the "LEARn" model (latent early-life associated regulation) may explain the triggering of AD. Curr Alzheimer Res 4:219–228

Lauren J, Gimbel DA, Nygaard HB, Gilbert JW, Strittmatter SM (2009) Cellular prion protein mediates impairment of synaptic plasticity by amyloid-beta oligomers. Nature 457: 1128–1132

Lee DW, Rajagopalan S, Siddiq A, Gwiazda R, Yang L, Beal MF, Ratan RR, Andersen JK (2009a) Inhibition of prolyl hydroxylase protects against 1-methyl-4-phenyl-1,2,3, 6-tetrahydropyridine-induced neurotoxicity: model for the potential involvement of the hypoxia-inducible factor pathway in Parkinson disease. J Biol Chem 284:29065–29076

Lee HJ, Bazinet RP, Rapoport SI, Bhattacharjee AK (2009b) Brain arachidonic acid cascade enzymes are upregulated in a rat model of unilateral parkinson disease. Neurochem Res 8 Dec 2009 [Epub ahead of print]

Leoni V, Mariotti C, Tabrizi SJ, Valenza M, Wild EJ, Henley SM, Hobbs NZ, Mandelli ML, Grisoli M, Björkhem I, Cattaneo E, Di Donato S (2008) Plasma 24S-hydroxycholesterol and caudate MRI in pre-manifest and early Huntington's disease. Brain 131:2851–2859

Levy-Lahad E, Tsuang D, Bird TD (1998) Recent advances in the genetics of Alzheimer's disease. J Geriatr Psychitry Neurol 11:42–54

Li SH, Lam S, Cheng AL, Li XJ (2000) Intranuclear huntingtin increases the expression of caspase-1 and induces apoptosis. Hum Mol Genet 9:2859–2867

Li SH, Li XJ (2004) Huntingtin and its role in neuronal degeneration. Neuroscientist 10:467–475

Lipton SA, Gu Z, Nakamura T (2007) Inflammatory mediators leading to protein misfolding and uncompetitive/fast off-rate drug therapy for neurodegenerative disorders. Int Rev Neurobiol 82:1–27

Liu R, Li B, Flanagan SW, Oberley LW, Gozal D, Oiu M (2002) Increased mitochondrial antioxidative activity or decreased oxygen free radical propagation prevent mutant SOD1-mediated motor neuron cell death and increase amyotrophic lateral sclerosis-like transgenic mouse survival. J Neurochem 80:488–500

Liu QY, Sooknanan RR, Malek LT, Ribecco-Lutkiewicz M, Lei JX, Shen H, Lach B, Walker PR, Martin J, Sikorska M (2006) Novel subtractive transcription-based amplification of mRNA (STAR) method and its application in search of rare and differentially expressed genes in AD brains. BMC Genomics 7:286

Liu D, Bao F, Wen J, Liu J (2007) Mutation of superoxide dismutase elevates reactive species: comparison of nitration and oxidation of proteins in different brain regions of transgenic mice with amyotrophic lateral sclerosis. Neuroscience 146:255–264

Loberto N, Prioni S, Bettiga A, Chigomo V, Prinetti A, Sonnino S (2005) The membrane environment of endogenous cellular prion protein in primary rat cerebellar neurons. J Neurochem 95:771–783

Lu JH, Teh BK, Wang YN, Tan YS, Lai MC, Reid KB (2008) The classical and regulatory functions of C1q in immunity and autoimmunity. Cell Mol Immunol 5:9–21

Lukas TJ, Luo WW, Mao H, Cole N, Siddique T (2006) Informatics-assisted protein profiling in a transgenic mouse model of amyotrophic lateral sclerosis. Mol Cell Proteomics 5:1233–1244

Lunn JS, Hefferan MP, Marsala M, Feldman EL (2009) Stem cells: comprehensive treatments for amyotrophic lateral sclerosis in conjunction with growth factor delivery. Growth Factors 27:133–140

Maccioni RB, Munoz JP, Barbeito L (2001) The molecular bases of Alzheimer's disease and other neurodegenerative disorders. Arch Med Res 32:367–381

Marcon G, Tell G, Perrone L, Garbelli R, Quadrifoglio F, Tagliavini F, Giaccone G (2009) APE1/Ref-1 in Alzheimer's disease: an immunohistochemical study. Neurosci Lett 466: 124–127

Martins IJ, Berger T, Sharman MJ, Verdile G, Fuller SJ, Martins RN (2009) Cholesterol metabolism and transport in the pathogenesis of Alzheimer's disease. J Neurochem 111: 1275–1308

Martinez Z, Zhu M, Han S, Fink AL (2007) GM1 specifically interacts with alpha-synuclein and inhibits fibrillation. Biochemistry 46:1868–1877

Masliah E, Mallory M, Alford M, DeTeresa R, Hansen LA, McKeel DW Jr., Morris JC (2001) Altered expression of synaptic proteins occurs early during progression of Alzheimer's disease. Neurology 56:127–129

Mast N, Norcross R, Andersson U, Shou M, Nakayama L, Bjorkhem I, Pikuleva IA (2003) Broad substrate specificity of human cytochrome P450 46A1 which initiates cholesterol degradation in the brain. Biochemistry 42:14284–14292

Matsuda N, Tanaka K (2010) Does impairment of the ubiquitin-proteasome system or the autophagy-lysosome pathway predispose individuals to neurodegenerative disorders such as Parkinson's disease? J Alzheimers Dis 19:1–9

Mattson MP, Culnsee C, Yu ZF (2000) Apoptotic and antiapoptotic mechanisms in stroke. Cell Tissue Res 30:173–187

Mattson MP, Meffert MK (2006) Roles for NF-kappaB in nerve cell survival, plasticity, and disease. Cell Death Differ 13:852–860

Migheli A, Piva R, Atzori C, Troost D, Schiffer D (1997) c-Jun, JNK/SAPK kinases and transcription factor NF-kappa B are selectively activated in astrocytes, but not motor neurons, in amyotrophic lateral sclerosis. J Neuropathol Exp Neurol 56:1314–1322

Migliore L, Fontana I, Colognato R, Coppede F, Siciliano G, Murri L (2005) Searching for the role and the most suitable biomarkers of oxidative stress in Alzheimer's disease and in other neurodegenerative diseases. Neurobiol Aging 26:587–595

Miranda HV, Outerio TF (2009) The sour side of neurodegenerative disorders: the effects of protein glycation. J Pathol 31 Dec [Epub ahead of print]

Mizutani K, Oka N, Kusunoki S, Kaji R, Kanda M, Akiguchi I, Shibasaki H (2003) Amyotrophic lateral sclerosis with IgM antibody against gangliosides GM2 and GD2. Intern Med 42: 277–280

Mobley WC, Nerve RL, Prusiner SB, McKinley MP (1988) Nerve growth factor increases mRNA levels for the prion protein and the beta-amyloid protein precursor in developing hamster brain. Proc Natl Acad Sci USA 85:9811–9815

Moore DJ, West AB, Dawson VL, Dawson TM (2005) Molecular pathophysiology of Parkinson's disease. Ann Rev Neurosci 28:57–87

Muchowski PJ, Wacker JL (2005) Modulation of neurodegeneration by molecular chaperones. Nat Rev Neurosci 6:11–22

Mufson EJ, Counts SE, Fahnestock M, Ginsberg SD (2007) Cholinotrophic molecular substrates of mild cognitive impairment in the elderly. Curr Alzheimer Res 4:340–350

Muller WEG, Ushijima H, Schroder HC, Forrest JMS, Schatton WFH, Rytik PG, Heffner-Lauc M (1993) Cytoprotective effect of NMDA receptor antagonist on prion protein (PrionSc)-induced toxicity in rat cortical cell culture. Eur J Pharmacol 246:261–267

Murer MG, Yan Q, Raisman-Vozari R (2001) Brain-derived neurotrophic factor in the control human brain, and in Alzheimer's disease and Parkinson's disease. Prog Neurobiol 63: 71–124

Muto Y, Hayashi T, Higashi Y, Endo T, Yamamoto T, Sato K (1999) Age-related decrease in brain-derived neurotrophic factor gene expression in the brain of the zitter rat with genetic spongiform encephalopathy. Neurosci Lett 271:69–72

Nakamura T, Lipton SA (2008) Emerging roles of S-nitrosylation in protein misfolding and neurodegenerative diseases. Antioxid Redox Signal 10:87–101

Napolitano M, Zei D, Centonze D, Palermo R, Bernardi G, Vacca A, Calabresi P, Gulino A (2008) NF-kB/NOS cross-talk induced by mitochondrial complex II inhibition: implications for Huntington's disease. Neurosci Lett 434:241–246

Naslavsky N, Shmeeda H, Friedlander G, Yanai A, Futerman AH, Barenholz Y, Taraboulos A (1999) Sphingolipid depletion increases formation of the scrapie prion protein in neuroblastoma cells infected with prions. J Biol Chem 274:20763–20771

Nathlie P, Jean-Noel O (2008) Processing of amyloid precursor protein and amyloid peptide neurotoxicity. Curr Alzheimer Res 5:92–99
Nelson TJ, Alkon DL (2005) Insulin and cholesterol pathways in neuronal function, memory and neurodegeneration. Biochem Soc Trans 33:1033–1036
Nicotera P, Lipton SA (1999) Excitotoxins in neuronal apoptosis and necrosis. J Cereb Blood Flow Metab 19:583–591
Nicotera P, Leist M, Manzo L (1999) Neuronal cell death: a demise with different shapes. Trends Pharmacol Sci 20:46–51
Niebroj-Dobos I, Rafalowska J, Fisziariska A, Gadarnski R, Grieb P (2007) Myelin composition of spinal cord in a model of amyotrophic lateral sclerosis (ALS) in SOD1G93A transgenic rats. Folia Neuropathol 45:236–241
Nunomura A, Takeda A, Smith MA, Perry G (2007) Oxidative RNA damage and neurodegeneration. Curr Med Chem 14:2968–2975
Nunomura A, Hofer T, Moreira PT, Castellani RJ, Smith MA, Perry G (2009) RNA oxidation in Alzheimer disease and related neurodegenerative disorders. Acta Neuropathol 118: 151–166
Nygaard HB, Strittmatter SM (2009) Cellular prion protein mediates the toxicity of beta-amyloid oligomers: implications for Alzheimer disease. Arch Neurol 66:1325–1328
Octave JN (2005) Alzheimer disease: cellular and molecular aspects Bull. Mem Acad R Med Belg 160:445–449
Oddo S, Caccamo A, Kitazawa M, Tseng BP, LaFerla FM (2003) Amyloid deposition precedes tangle formation in a triple transgenic model of Alzheimer's disease. Neurobiol Aging 24:1063–10670
Ong WY, Kim J-H, He X, Chen P, Farooqui AA, Jenner AM (2010) Changes in brain cholesterol metabolome after kainate excitotoxicity. Mol Neurobiol 41:299–313
Orrenius S, Gogvadze V, Zhivotovsky B (2007) Mitochondrial oxidative stress: implications for cell death. Annu Rev Pharmacol Toxicol 2007(47):143–183
Park KJ, Lee SH, Lee CH, Jang JY, Chung J, Kwon MH, Kim YS (2009a) Upregulation of Beclin-1 expression and phosphorylation of Bcl-2 and p53 are involved in the JNK-mediated autophagic cell death. Biochem Biophys Res Commun 382:726–729
Park S, Kim HT, Yun S, Kim IS, Lee J, Lee IS, Park KI (2009b) Growth factor-expressing human neural progenitor cell grafts protect motor neurons but do not ameliorate motor performance and survival in ALS mice. Exp Mol Med 41:487–500
Parkyn CJ, Vermeulen EG, Mootoosamy RC, Sunyach C, Jacobsen C, Oxvig C, Moestrup S, Liu Q, Bu G, Jen A, Morris RJ (2008) LRP1 controls biosynthetic and endocytic trafficking of neuronal prion protein. J Cell Sci 121:773–783
Parson CG, Stoffler A, Danysz W (2007) Memantine: a NMDA receptor antagonist that improves memory by restoration of homeostasis in the glutamatergic system–too little activation is bad, too much is even worse. Neuropharmacol 53:699–723
Patel NK, Gill SS (2007) GDNF delivery for Parkinson's disease. Acta Neurochir Suppl 97: 135–154
Patil S, Melrose J, Chan C (2007) Involvement of astroglial ceramide in palmitic acid-induced Alzheimer-like changes in primary neurons. Eur J Neurosci 26:2131–2141
Perini G, Della-Bianca V, Politi V, Della Valle G, Dal-Pra I, Rossi F, Armato U (2002) Role of p75 neurotrophin receptor in the neurotoxicity by beta-amyloid peptides and synergistic effect of inflammatory cytokines. J Exp Med 195:907–918
Perlmann T, Wallen-Mackenzie A (2004) Nurr1, an orphan nuclear receptor with essential functions in developing dopamine cells. Cell Tissue Res 318:45–52
Perovic S, Pergande G, Ushijima H, Kelve M, Forrest J, Müller WE (1995) Flupirtine partially prevents neuronal injury induced by prion protein fragment and lead acetate. Neurodegeneration 4:369–374
Perluigi M, Fai Poon H, Hensley K, Pierce WM, Klein JB, Calabrese V, De Marco C, Butterfield DA (2005) Proteomic analysis of 4-hydroxy-2-nonenal-modified proteins in G93A-SOD1

transgenic mice–a model of familial amyotrophic lateral sclerosis. Free Radic Biol Med 38:960–968
Petersen RB, Nunomura A, Lee HG, Casadesus G, Perry G, Smith MA, Zhu X (2007) Signal transduction cascades associated with oxidative stress in Alzheimer's disease. J Alzheimer Dis 11:143–152
Pettegrew JW, Kopp SJ, Minshew NJ, Glonek T, Feliksik JM, Tow JP, Cohen MM (1987) 31P nuclear magnetic resonance studies of phosphoglyceride metabolism in developing and degenerating brain: preliminary observations. J Neuropathol Exp Neurol 46:419–430
Pettegrew JW, Panchalingam K, Hamilton RL, McClure RJ (2001) Brain membrane phospholipid alterations in Alzheimer's disease. Neurochem Res 26:771–782
Petot GJ, Friedland RP (2004) Lipids, diet and Alzheimer disease: an extended summary. J Neurol Sci 226:31–33
Pfrieger FW (2003) Outsourcing in the brain do neurons depend on cholesterol delivery by astrocytes? Bioessays 25:72–78
Poole AC, Thomas RE, Andrews LA, McBride HM, Whitworth AJ, Pallanck LJ (2008) The PINK1/Parkin pathway regulates mitochondrial morphology. Proc Natl Acad Sci USA 105:1638–1643
Poon WW, Blurton-Jones M, Tu CH, Feinberg LM, Chabrier MA, Harris JW, Jeon NL, Cotman CW (2009) Beta-amyloid impairs axonal BDNF retrograde trafficking. Neurobiol Aging 18 Jun 2009 [Epub ahead of print]
Popescu A, Lippa CF, Lee VM, Trojanowski JQ (2004) Lewy bodies in the amygdala: increase of alpha-synuclein aggregates in neurodegenerative diseases with tau-based inclusions. Arch Neurol 61:1915–1919
Priller C, Dewachter I, Vassallo N, Paluch S, Pace C, Kretzschmar HA, van Leuven F, Herms J (2007) Mutant presenilin 1 alters synaptic transmission in cultured hippocampal neurons. J Biol Chem 282:1119–1127
Prusiner SB (2001) Shattuck lecture–neurodegenerative diseases and prions. N Engl J Med 344:1516–1526
Quintanilla RA, Orellana DI, González-Billault C, Maccioni RB (2004) Interleukin-6 induces Alzheimer-type phosphorylation of tau protein by deregulating the cdk5/p35 pathway. Exp Cell Res 295:245–257
Rachidi W, Riondel J, McMalion HM, Favier A (2005) Prion protein and copper: a mysterious relationship. Pathol Biol (Paris) 53:224–256
Rajawat YS, Bossis I (2008) Autophagy in aging and in neurodegenerative disorders. Hormones (Athens) 7:46–61
Rantham Prabhakara JP, Feist G, Thomasson S, Thompson A, Schommer E, Ghribi O (2008) Differential effects of 24-hydroxycholesterol and 27-hydroxycholesterol on tyrosine hydroxylase and alpha-synuclein in human neuroblastoma SH-SY5Y cells. J Neurochem 107: 1722–1729
Rao AV, Balachandran B (2002) Role of oxidative stress and antioxidants in neurodegenerative diseases. Nutr Neurosci 5:291–309
Rao SD, Weiss JH (2004) Excitotoxic and oxidative cross-talk between motor neurons and glia in ALS pathogenesis. Trends Neurosci 27:17–23
Ratovitski T, Gucek M, Jiang H, Chighladze E, Waldron E, D'Ambola J, Hou Z, Liang Y, Poirier MA, Hirschhorn RR, Graham R, Hayden MR, Cole RN, Ross CA (2009) Mutant huntingtin N-terminal fragments of specific size mediate aggregation and toxicity in neuronal cells. J Biol Chem 284:10855–108567
Rivaroli A, Prioni S, Loberto N, Bettiga A, Chigomo V, Prinetti A, Sonnino S (2007) Reorganization of prion protein membrane environment during low potassium-induced apoptosis in primary rat cerebellar neurons. J Neurochem 103:1954–1967
Rivera EJ, Goldin A, Fulmer N, Tavares R, Wands JR, de la Monte SM (2005) Insulin and insulin-like growth factor expression and function deteriorate with progression of Alzheimer's disease: link to brain reductions in acetylcholine. J Alzheimers Dis 8:247–268

Roberson ED, Scearce-Levie K, Palop JJ, Yan F, Cheng IH, Wu T, Gerstein H, Yu GQ, Mucke L (2007) Reducing endogenous tau ameliorates amyloid beta-induced deficits in an Alzheimer's disease mouse model. Science 316(5825):750–754

Rochet JC, Outeiro TF, Conway KA, Ding TT, Volles MJ, Lashuel HA, Bieganski RM, Lindquist SL, Lansbury PT (2004) Interactions among alpha-synuclein, dopamine, and biomembranes: some clues for understanding neurodegeneration in Parkinson's disease. J Mol Neurosci 23: 23–34

Rodriguez A, Martin M, Albasanz JL, Barrachina M, Espinosa JC, Torres JM, Ferrer I (2004) Group I mGluR signaling in BSE-infected bovine-PrP transgenic mice. Neurosci Lett 410: 115–120

Rodriguez A, Martin M, Albasanz JL, Barrachina M, Espinosa JC, Torres JM, Ferrer I (2006) Adenosine A1 receptor protein levels and activity is increased in the cerebral cortex in Creutzfeldt-Jakob disease and in bovine spongiform encephalopathy-infected bovine-PrP mice. J Neuropathol Exp Neurol 65:964–975

Rudinskiy N, Kaneko YA, Beesen AA, Gokce O, Regulier E, Deglon N, Luthi-Carter R (2009) Diminished hippocalcin expression in Huntington's disease brain does not account for increased striatal neuron vulnerability as assessed in primary neurons. J Neurochem 111:460–472

Russell DW, Halford RW, Ramirez DM, Shah R, Kotti T (2009) Cholesterol 24-hydroxylase: an enzyme of cholesterol turnover in the brain. Annu Rev Biochem 78:1017–1040

Sarlette A, Krampfl K, Grothe C, Neuhoff N, Dengler R, Petri S (2008) Nuclear erythroid 2-related factor 2-antioxidative response element signaling pathway in motor cortex and spinal cord in amyotrophic lateral sclerosis. J Neuropathol Exp Neurol 67:1055–1062

Sasaki S, Warita H, Murakami T, Shibata N, Komori T, Abe K, Kobayashi M, Iwata M (2005) Ultrastructural study of aggregates in the spinal cord of transgenic mice with a G93A mutant SOD1 gene. Acta Neuropathol 109:247–255

Sastry PS, Subba Rao K (2000) Apoptosis and the nervous system. J Neurochem 74:1–20

Sathasivam S (2010) Motor neurone disease: clinical features, diagnosis, diagnostic pitfalls and prognostic markers. Singapore Med J 51:367–372

Satoi H, Tomimoto H, Ohtani R, Kitano T, Kondo T, Watanabe M, Oka N, Akiguchi I, Furuya S, Hirabayashi Y, Okazaki T (2005) Astroglial expression of ceramide in Alzheimer's disease brains: a role during neuronal apoptosis. Neuroscience 130:657–666

Scheper W, Hoozemans JJ (2009) Endoplasmic reticulum protein quality control in neurodegenerative disease: the good, the bad and the therapy. Curr Med Chem 16:615–626

Scherzinger E, Lurz R, Turmaine M, Mangiarini L, Hollenbach B, Hasenbank R, Bates GP, Davies SW, Lehrach H, Wanker EE (1997) Huntingtin-encoded polyglutamine expansions form amyloid-like protein aggregates in vitro and in vivo. Cell 90:549–558

Schubert D, Soucek T, Blouw B (2009) The induction of HIF-1 reduces astrocyte activation by amyloid beta peptide. Eur J Neurosci 29:1323–1334

Seet RC, Lee CY, Lim EC, Tan JJ, Quek AM, Chong WL, Looi WF, Huang SH, Wang H, Chan YH, Halliwell B (2009) Oxidative damage in Parkinson disease: measurement using accurate biomarkers. Free Radic Biol Med 4 Dec 2009 [Epub ahead of print]

Selkoe DJ (2001) Alzheimer's disease: genes, proteins, and therapy. Physiol Rev 81:741–766

Sesti F, Liu S, Cai SQ (2009) Oxidation of potassium channels by ROS: a general mechanism of aging and neurodegeneration? Trends Cell Biol 20 Oct 2009 [Epub ahead of print]

Sethi P, Lukiw WJ (2009) Micro-RNA abundance and stability in human brain: specific alterations in Alzheimer's disease temporal lobe neocortex. Neurosci Lett 459:100–104

Shan X, Tashiro H, Lin CL (2003) The identification and characterization of oxidized RNAs in Alzheimer's disease. J Neurosci 23:4913–4921

Shaw PJ, Ince PG (1997) Glutamate, excitotoxicity and amyotrophic lateral sclerosis. J Neurol 244(Suppl 2):S3–S14

Shibata N, Kobayashi M (2008) The role for oxidative stress in neurodegenerative diseases. Brain Nerve 60:157–170

Shibata N, Kakita A, Takahashi H, Ihara Y, Nobukuni K, Fujimura H, Sakoda S, Sasaki S, Iwata M, Morikawa S, Hirano A, Kobayashi M (2009) Activation of signal transducer and activator of transcription-3 in the spinal cord of sporadic amyotrophic lateral sclerosis patients. Neurodegener Dis 6:118–126

Siddiq A, Aminova LR, Ratan RR (2007) Hypoxia inducible factor prolyl 4-hydroxylase enzymes: center stage in the battle against hypoxia, metabolic compromise and oxidative stress. Neurochem Res 32:931–946

Skinner PJ, Abbassi H, Chesebro B, Race RE, Reilly C, Haase AT (2006) Gene expression alterations in brains of mice infected with three strains of scrapie. BMC Genomics 7:114

Simons K, Ikonen E (2000) How cells handle cholesterol? Science 290:1721–1726

Siman R, Salidas S (2004) $\gamma$-secretase subunit composition and distribution in the presenilin wild-type and mutant mouse brain. Neuroscience 129:615–628

Singh N, Das D, Singh A, Mohan ML (2009) Prion protein and metal interaction: physiological and pathological implications. Curr Issues Mol Biol 12:99–108

Soane L, Kahraman S, Kristian T, Fiskum G (2007) Mechanisms of impaired mitochondrial energy metabolism in acute and chronic neurodegenerative disorders. J Neurosci Res 85:3407–3415

Söderberg M, Edlund C, Kristensson K, Dallner G (1991) Fatty acid composition of brain phospholipids in aging and in Alzheimer's disease Lipids 26:421–425

Sometani A, Kataoka H, Nitta A, Fukumitsu H, Nomoto H, Furukawa S (2001) Transforming growth factor-beta1 enhances expression of brain-derived neurotrophic factor and its receptor, TrkB, in neurons cultured from rat cerebral cortex. J Neurosci Res 66:369–376

Sorensen G, Medina S, Parchaliuk D, Phillipson C, Robertson C, Booth SA (2008) Comprehensive transcriptional profiling of prion infection in mouse models reveals networks of responsive genes. BMC Genomics 9:114

Soto C, Estrada LD (2008) Protein misfolding and neurodegeneration. Arch Neurol 65: 184–189

Sotthibundhu A, Sykes AM, Fox B, Underwood CK, Thangnipon W, Coulson EJ (2008) Beta-amyloid(1–42) induces neuronal death through the p75 neurotrophin receptor. J Neurosci 28:3941–3946

Speare JO, Offerdahl DK, Hasenkrug A, Carmody AB, Baron GS (2010) GPI anchoring facilitates propagation and spread of misfolded Sup35 aggregates in mammalian cells. EMBO J 29: 782–794

Steen E, Terry BM, Rivera EJ, Cannon JL, Neely TR, Tavares R, Xu XJ, Wands JR, de la Monte SM (2005) Impaired insulin and insulin-like growth factor expression and signaling mechanisms in Alzheimer's disease–is this type 3 diabetes? J Alzheimers Dis 7:63–80

Stephenson D, Rash K, Smalstig B, Roberts E, Johnstone E, Sharp J, Panetta J, Little S, Kramer R, Clemens J (1999) Cytosolic phospholipase $A_2$ is induced in reactive glia following different forms of neurodegeneration. Glia 27:110–128

Stack EC, Del Signore SJ, Luthi-Carter R, Soh BY, Goldstein DR, Matson S, Goodrich S, Markey AL, Cormier K, Hagerty SW, Smith K, Ryu H, Ferrante RJ (2007) Modulation of nucleosome dynamics in Huntington's disease. Hum Mol Genet 16:1164–1175

Stewart LR, White AR, Jobling MF, Needham BE, Maher F, Thyer J, Beyreuther K, Masters CL, Collins SJ, Cappai R (2001) Involvement of the 5-lipoxygenase pathway in the neurotoxicity of the prion peptide PrP106–126. J Neurosci Res 65:565–572

Stine OC, Sh. Li, Pleasant N, Wagster MV, Hedreen JC, Ross CA (1995) Expression of the mutant allele of IT-15 (the HD gene) in striatum and cortex of Huntington's disease patients. Hum Mol Genet 4:15–18

Stokes CE, Hawthorne JN (1987) Reduced phosphoinositide concentrations in anterior temporal cortex of Alzheimer-diseased brains. J Neurochem 48:1018–1021

Strong MJ (2010) The evidence for altered RNA metabolism in amyotrophic lateral sclerosis (ALS). J Neurol Sci 288:1–12

Su B, Wang X, Nunomura A, Moreira PI, Lee HG, Perry G, Smith MA, Zhu X (2008) Oxidative stress signaling in Alzheimer's disease. Curr Alzheimer Res 5:525–532

Sugama S, Takenouchi T, Cho BP, Joh TH, Hashimoto M, Kitani H (2009) Possible roles of microglial cells for neurotoxicity in clinical neurodegenerative diseases and experimental animal models. Inflamm Allergy Drug Target 8:277–284

Sultana R, Perluigi M, Butterfield DA (2009) Oxidatively modified proteins in Alzheimer's disease (AD), mild cognitive impairment and animal models of AD: role of Abeta in pathogenesis. Acta Neuropathol 118:131–150

Sun L, Liu SY, Zhou XW, Wang XC, Liu R, Wang Q, Wang JZ (2003) Inhibition of protein phosphatase 2A- and protein phosphatase 1-induced tau hyperphosphorylation and impairment of spatial memory retention in rats. Neuroscience 118:1175–1182

Sun GY, Xu J, Jensen MD, Simonyi A (2004) Phospholipase $A_2$ in the central nervous system: implications for neurodegenerative diseases. J Lipid Res 45:205–213

Suopanki J, Gotz C, Lutsch G, Schiller J, Harjes P, Herrmann A, Wanker EE (2006) Interaction of huntingtin fragments with brain membranes–clues to early dysfunction in Huntington's disease. J Neurochem 96:870–884

Supattapone S (2004) Prion protein conversion in vitro. J Mol Med 82:348–356

Tanimukai S, Hasegawa H, Nakai M, Yagi K, Hirai M, Saito N, Taniguchi T, Terashima A, Yasuda M, Kawamata T, Tanaka C (2002) Nanomolar amyloid beta protein activates a specific PKC isoform mediating phosphorylation of MARCKS in Neuro2A cells. Neuroreport 13: 549–553

Taylor DM, Tradewell ML, Minotti S, Durham HD (2007) Characterizing the role of Hsp90 in production of heat shock proteins in motor neurons reveals a suppressive effect of wild-type Hsf1. Cell Stress Chaperones 12:151–162

Teismann P, Vila M, Choi DK, Tieu K, Wu DC, Jackson-Lewis V, Przedborski S (2003) COX-2 and neurodegeneration in Parkinson's disease. Ann N Y Acad Sci 991:272–277

Terry RD, Masliah E, Salmon DP, Butters N, DeTeresa R, Hill R, Hansen LA, Katzman R (1991) Physical basis of cognitive alterations in Alzheimer's disease: synapse loss is the major correlate of cognitive impairment. Ann Neurol 30:572–580

Tell G, Damante G, Caldwell D, Kelley MR (2005) The intracellular localization of APE1/Ref-1: more than a passive phenomenon? Antioxid Redox Signal 7:367–384

Thellung S, Villa V, Corsaro A, Pellistri F, Venezia V, Russo C, Aceto A, Robello M, Florio T (2007) ERK1/2 and p38 MAP kinases control prion protein fragment 90–231-induced astrocyte proliferation and microglia activation. Glia 55:1469–1485

Thomas EA (2006) Striatal specificity of gene expression dysregulation in Huntington's disease. J Neurosci Res 84:1151–1164

Thomas SN, Cripps D, Yang AJ (2009) Proteomic analysis of protein phosphorylation and ubiquitination in Alzheimer's disease. Methods Mol Biol 566:109–121

Thompson LM, Aiken CT, Kaltenbach LS, Agrawal N, Illes K, Khoshnan A, Martinez-Vincente M, Arrasate M, O-Symbol-Rourke JG, Khashwji H et al (2009) IKK phosphorylates Huntingtin and targets it for degradation by the proteasome and lysosome. J Cell Biol 21 Dec 2009 [Epub ahead of print]

Tully AM, Roche HM, Doyle R, Fallon C, Bruce I, Lawlor B, Coakley D, Gibney MJ (2003) Low serum cholesteryl ester-docosahexaenoic acid levels in Alzheimer's disease: a case-control study. Br J Nutr 89:483–489

Urano Y, Hayashi I, Isoo N, Reid PC, Shibasaki Y, Noguchi N, Tomita T, Iwatsubo T, Hamakubo T, Kodama T (2005) Association of active $\gamma$-secretase complex with lipid rafts. J Lipid Res 46:904–912

Uversky VN (2008) Alpha-synuclein misfolding and neurodegenerative diseases. Curr Protein Pept Sci 9:507–540

Van Raamsdonk JM, Warby SC, Hayden MR (2007) Selective degeneration in YAC mouse models of Huntington disease. Brain Res Bull 72:124–131

Vargas MR, Johnson DA, Sirkis DW, Messing A, Johnson JA (2008) Nrf2 activation in astrocytes protects against neurodegeneration in mouse models of familial amyotrophic lateral sclerosis. J Neurosci 28:13574–13581

Velázquez E, Santos A, Montes A, Blázquez E, Ruiz-Albusac JM (2006) 25-Hydroxycholesterol has a dual effect on the proliferation of cultured rat astrocytes. Neuropharmacology 51:229–237

Vigh L, Smith RG, Soós J, Engelhardt JI, Appel SH, Siklós L (2005) Sublethal dose of 4-hydroxynonenal reduces intracellular calcium in surviving motor neurons in vivo. Acta Neuropathol 109:567–575

Viles JH, Klewpatinond M, Nadal RC (2008) Copper and the structural biology of the prion protein. Biochem Soc Trans 36:1288–1292

Vucic S, Kiernan MC (2009) Pathophysiology of neurodegeneration in familial amyotrophic lateral sclerosis. Curr Mol Med 9:255–272

Wacker JL, Huang SY, Steele AD, Aron R, Lotz GP, Nguyen Q, Giorgini F, Roberson ED, Lindquist S, Masliah E, Muchowski PJ (2009) Loss of Hsp70 exacerbates pathogenesis but not levels of fibrillar aggregates in a mouse model of Huntington's disease. J Neurosci 29:9104–9114

Walling HW, Baldassare JJ, Westfall TC (1998) Molecular aspects of Huntington's disease. J Neurosci Res 54:301–308

Wang R, Wang S, Malter JS, Wang DS (2009) Effects of HNE-modification induced by Abeta on neprilysin expression and activity in SH-SY5Y cells. J Neurochem 108:1072–1082

Watt JC, Westaway D (2007) The prion protein family: diversity, rivalry, and dysfunction. Biochim Biophys Acta 1772:654–672

Wells K, Farooqui AA, Liss L, Horrocks LA (1995) Neural membrane phospholipids in Alzheimer disease. Neurochem Res 20:1329–1333

Wilde GJC, Pringle AK, Wright P, Iannotti F (1997) Differential vulnerability of the CA1 and CA3 subfields of the hippocampus to superoxide and hydroxyl radicals in vitro. J Neurochem 69:883–886

Wilhelmus MM, Grunberg SC, Bol JG, van Dam AM, Hoozemans JJ, Rozemuller AJ, Drukarch B (2009) Transglutaminases and transglutaminase-catalyzed cross-links colocalize with the pathological lesions in Alzheimer's disease brain. Brain Path 19:612–622

Winkler C, Georgievska B, Carlsson T, Kink D (2006) Continuous exposure to glial cell line-derived neurotrophic factor to mature dopaminergic transplants impairs the graft's ability to improve spontaneous motor behavior in parkinsonian rats. Neuroscience 141:521–531

Wishart TM, Parson SH, Gillingwater TH (2006) Synaptic vulnerability in neurodegenerative disease. J Neuropathol Exp Neurol 65:733–739

Wissing D, Mouritzen H, Egeblad M, Poirer GG, Jaattela M (1997) Involvement of caspase-dependent activation of cytosolic phospholipase $A_2$ in tumor necrosis factor-induced apoptosis. Proc Natl Acad Sci USA 94:5073–5077

Wong J, Quinn CM, Brown AJ (2006) SREBP-2 positively regulates transcription of the cholesterol efflux gene, ABCA1, by generating oxysterol ligands for LXR. Biochem J 400:485–491

Woodruff TM, Costini KJ, Taylor SM, Noakes PG (2008) Role of complement in motor neuron disease: animal models and therapeutic potential of complement inhibitors. Adv Exp Med Biol 632:143–156

Wyss-Coray T (2006) Tgf-Beta pathway as a potential target in neurodegeneration and Alzheimer's. Curr Alzheimer Res 3:191–195

Xiang W, Windl O, Westner IM, Neuman M, Zerr I, Lederer RM, Kretzschmar HA (2005) Cerebral gene expression profiles in sporadic Creutzfeldt-Jakob disease. Ann Neurol 55:242–251

Xiang W, Hummel M, Mitteregger G, Pace C, Windl O, Mansmann U, Kretzschmar HA (2007) Transcriptome analysis reveals altered cholesterol metabolism during the neurodegeneration in mouse scrapie model. J Neurochem 102:834–847

Yamazaki T, Suzuki M, Irie T, Watanabe T, Mikami H, Ono S (2008) Amyotrophic lateral sclerosis associated with IgG anti-GalNAc-GD1a antibodies. Clin Neurol Neurosurg 110:722–724

Yao X (2009) Effect of zinc exposure on HNE and GLT-1 in spinal cord culture. Neurotoxicology 30:121–126

Yasojima K, Tourtellotte WW, McGeer EG, McGeer PL (2001) Marked increase in cyclooxygenase-2 in ALS spinal cord: implications for therapy. Neurology 57:952–956

Yasuhara T, Shingo T, Date I (2007) Glial cell line-derived neurotrophic factor (GDNF) therapy for Parkinson's disease. Acta Med Okayama 61:51–56

Yokota T (2009) MicroRNA and central nervous system. Brain Nerve 61:167–176

Yoshinaga N, Yasuda Y, Murayama T, Nomura Y (2000) Possible involvement of cytosolic phospholipase $A_2$ in cell death induced by 1-methyl-4-phenylpyridinium ion, a dopaminergic neurotoxin, in GH3 cells. Brain Res 855:244–251

Yun SW, Gerlach M, Riederer P, Klein MA (2006) Oxidative stress in the brain at early preclinical stages of mouse scrapie. Exp Neurol 201:90–98

Zabel C, Mao L, Woodman B, Rohe M, Wacker MA, Kläre Y, Koppelstätter A, Nebrich G, Klein O, Grams S, Strand A, Luthi-Carter R, Hartl D, Klose J, Bates GP (2009) A large number of protein expression changes occur early in life and precede phenotype onset in a mouse model for huntington disease. Mol Cell Proteomics 8:720–734

Zhai J, Ström AL, Kilty R, Venkatakrishnan P, White J, Everson WV, Smart EJ, Zhu H (2009) Proteomic characterization of lipid raft proteins in amyotrophic lateral sclerosis mouse spinal cord. FEBS J 276:3308–3323

Zhang X, Zhou K, Wang R, Cui J, Lipton SA, Liao FF, Xu H, Zhang YW (2007) Hypoxia-inducible factor 1alpha (HIF-1alpha)-mediated hypoxia increases BACE1 expression and beta-amyloid generation. J Biol Chem 282:10873–10880

Zhang J, Xue R, Ong WY, Chen P (2009) Roles of cholesterol in vesicle fusion and motion. Biophys J 97:1371–1380

Zhu JH, Horbinski C, Guo F, Watkins S, Uchiyama Y, Chu CT (2007) Regulation of autophagy by extracellular signal-regulated protein kinases during 1-methyl-4-phenylpyridinium-induced cell death. Am J Pathol 170:75–86

Zlokovic BV (2008) The blood-brain barrier in health and chronic neurodegenerative disorders. Neuron 57:178–201

Zuccato C, Cattaneo E (2007) Role of brain-derived neurotrophic factor in Huntington's disease. Prog Neurobiol 81:294–330

Zuccato C, Belyaev N, Conforti P, Ooi L, Tartari M, Papadimou E, MacDonald M, Fossale E, Zeitlin S, Buckley N, Cattaneo E (2007) Widespread disruption of repressor element-1 silencing transcription factor/neuron-restrictive silencer factor occupancy at its target genes in Huntington's disease. J Neurosci 27:6972–6983

# Chapter 9
# Potential Therapeutic Strategies for Neurodegenerative Diseases

## 9.1 Introduction

Neurodegenerative diseases include Alzheimer disease (AD), Parkinson disease (PD), Huntington disease (HD), amyotrophic lateral sclerosis (ALS), and prion diseases. The primary causes of neurodegenerative diseases are not known. However, these diseases share excitotoxicity, oxidative stress, and neuroinflammation along with the accumulation of misfolded proteins, mitochondrial and proteasomal dysfunction, and loss of synapses as common mechanisms of neurodegeneration (Farooqui and Horrocks, 2007; Farooqui, 2009a). Brain is rich in unsaturated fatty acids that are prone to oxidation. Growing evidence suggests that excitotoxicity, oxidative stress, and inflammatory processes contribute to neural cell death through the involvement of $PLA_2$, cyclooxygenases-2, stress kinases, JNK, MAPK, p38, and redox-sensitive transcription factors such as NF-κB and AP-1 (Farooqui et al., 2007a; Farooqui, 2009a). These transcription factors differentially regulate the genes for enzymes associated with the production of proinflammatory mediators and protective antioxidant genes such as γ-glutamylcysteine synthetase, Mn-superoxide dismutase, and hemeoxygenase-1 (Rahman and MacNee, 2000). In addition, AD and PD are characterized by a cerebral cholinergic and dopaminergic deficit and cerebral blood flow is diminished. Cerebrovascular dysfunction contributes to the cognitive decline and dementia in AD and PD.

Key to treat neurodegenerative diseases is an understanding of the mechanisms that trigger neurodegeneration. Although different neurodegenerative diseases are accompanied by different causes, genetic mutations, and different patterns of neuronal death, as stated in Chapter 8, they often display a number of common features, including endoplasmic reticulum stress, mitochondrial dysfunction, impairment of the proteasome, induction of oxidative stress, protein aggregation, alterations in ion homeostasis, redox status, and loss of synapse (Farooqui, 2009a). These changes are interrelated, causing disruption of normal neuronal function and eventually inducing neural cell death. In addition, neurodegenerative process is complicated by the fact that degenerating neurons also mount prosurvival responses to protect themselves from above neurochemical and neuropathological disease-related changes. This results in a very complex situation in which

A.A. Farooqui, *Neurochemical Aspects of Neurotraumatic and Neurodegenerative Diseases*, DOI 10.1007/978-1-4419-6652-0_9,

degenerating neurons go through a struggle between prodeath factors and prosurvival responses.

## 9.2 Factors Influencing the Onset of Neurodegenerative Diseases

As stated in Chapter 8, the most important risk factors for neurodegenerative diseases are old age and a positive family history. Neurodegenerative diseases are multifactorial illnesses caused by complex interactions among genetic factors, environmental factors, aging, and lifestyle. Environmental and dietary risk factors, such as heavy metals, hormones, cholesterol, high-fat diet, high alcohol intake, diet deficient in ω-3 fatty acids, antioxidants and vitamins, and reduced levels of physical activity (exercise), promote the onset and progression of neurodegenerative diseases. Red wine and Mediterranean diet may lower the risk of AD and PD (Luchsinger et al., 2007). In contrast, vascular risk factors (too much cigarette smoking and midlife high blood pressure) and chronic diseases (e.g., obesity, diabetes, traumatic brain injury, and cerebrovascular lesions) promote the early onset of neurodegenerative diseases. The links among risk factors and the development of oxidative stress and neuroinflammation in neurodegenerative diseases involve numerous complex metabolic and interactions among signaling pathways, which contribute to not only vascular compromise but also cognitive dysfunctions. Once oxidative stress and neuroinflammatory cascade of events is initiated, the process of oxidative stress and neuroinflammation can become overactivated causing cellular damage and loss of neuronal function (Farooqui et al., 2007a; Farooqui, 2009a, b). It is likely that long-term multidomain interventions toward the optimal control of multiple vascular risk factors and the maintenance of socially integrated lifestyles and psychosocial factors (e.g., high education, active social engagement, physical exercise, and mentally stimulating activity) reduce the risk of neurodegenerative diseases (Fratiglioni and Qiu, 2009; Qiu et al., 2009). The healthy lifestyle must be maintained throughout life (from childhood to old age) rather than after disease manifestation and may be particularly relevant to other factors, such as genetic predisposition and environmental factors, which may increase risk of neurodegenerative diseases. Cumulative evidence suggests that adherence to a healthy lifestyle may directly protect against neurodegenerative diseases or may delay chronic diseases, such as vascular disease and diabetes (Fratiglioni and Qiu, 2009).

Some neurodegenerative diseases, such as AD, are characterized by deficiencies in *S*-adenosylmethionine, vitamin B12, and folate and reduced glucose metabolism in the brain. The deficiency in these nutrients may cause alterations in gene promoter through DNA methylation resulting in upregulation of AD-related genes (Lahiri et al., 2007). The onset of AD is often subtle and usually occurs in mid- to late life (Graeber and Moran, 2002; Farooqui, 2009b). To explain the effect of lifestyle, environmental, and genetic factors, "Latent Early-Life Associated Regulation" (LEARn) hypothesis and its model have been proposed (Lahiri et al., 2007). According to this hypothesis, exposure to environmental agents (heavy metals), intrinsic factors (generation of cytokines due to high-fat diet consumption), and deficiency of dietary factors (vitamin B12, folate, and cholesterol) perturbs gene regulation in a long-term fashion, starting at early developmental stages (infancy), but that these perturbations do not have pathological results until significantly later in life (midlife) (Lahiri et al., 2007). The LEARn model is based on the regulatory structure common to eukaryotic genes and the effect of methylation at certain specific sites within the promoter (regulatory) region of specific genes, such as genes for APP, SP1, and BACE1. These changes alter the affinity for an AD-associated gene's promoter to transcription factors such as MeCP2 (hypomethyl derepression) or SP1 (hypomethyl activation). Increased expression of these genes results in greater production of Aβ, which contributes to neurodegeneration either by directly inducing oxidative stress or by indirectly through hyperphosphorylation of microtubule-associated τ-protein. This hypothesis has been proposed for AD, but can be extended to other neurodegenerative diseases.

### *9.2.1 Genetic and Environmental Factors*

Genes modulate the onset and frequency of neurodegenerative diseases (Coppede et al., 2006). Candidate genes associated with the pathogenesis of familial and sporadic neurodegenerative diseases modulate functioning of cholinergic, dopaminergic, and glutamatergic neurons. Some of these genes interact with environmental factors and increase the risk of neurodegenerative diseases. For example, P450 2D6 gene may increase the risk of PD among persons exposed to pesticides. Familial PD is associated with mutations in Parkin, PINK-1, or DJ-1 genes. These mutations are related to increased oxidative stress. Pathogenesis of sporadic PD may also involve exposure to metal ions, infections, stress, poor nutrition that may cause an increase in oxidative stress (Nunomura et al., 2007).

Three causative genes namely amyloid precursor protein gene (APP gene), presenilin-1 and -2 genes (PS-1 and PS-2), and apolipoprotein E (APOE) are associated with the pathogenesis of AD (Selkoe, 2001; Siman and Salidas, 2004; Priller et al., 2007). They have been discussed in Chapter 8. The human ApoE gene has three alleles (epsilon2, epsilon3, epsilon4), which are products of the same gene. The epsilon3 allele accounts for the majority of the ApoE gene pool (approximately 70–80%), the epsilon4 allele accounts for 10–15% and the epsilon2 allele for 5–10%. Inheritance of the epsilon4 allele strongly increases the risk for developing AD at an earlier age. ApoE is involved in cholesterol transport, neuronal repair, dendritic growth, and anti-inflammatory activities. In addition, APOE4 gene encodes a protein with crucial roles in cholesterol metabolism. APOE4 contributes to the pathogenesis of AD not only by modulating the metabolism and aggregation of Aβ peptide but also by directly regulating brain lipid metabolism and synaptic functions through APOE receptors (Bu, 2009). Early-life events such as infections, stress, and poor nutrition (see below) may contribute to the pathogenesis of AD and PD. There are similarities between neurochemical changes in infectious diseases and AD. Both these conditions are characterized by an increased production of many immune mediators, cytokines, chemokines, and complement proteins by infectious disease and AD patients (Urosevic and Martin, 2008).

Recent studies in mouse models of HD indicate that enriching the environment of transgenic animals delays the onset and slows the progression of HD-associated motor and cognitive symptoms. Environmental enrichment (EE) induces various molecular and cellular changes in specific brain regions of wild-type animals, including altered gene expression profiles, enhanced neurogenesis, and synaptic plasticity (Spires and Hannan, 2005). EE elicits not only transcriptional and translational events but also mediate neurogenic and neuroprotective responses, including restoration of brain-derived neurotrophic factor (BDNF) striatal transport in the R6/1 HD mice and elevation in the levels of amyloid-degrading enzyme (neprilysin) in the APPswe/PS1DeltaE9 AD mice (Li and Tang, 2005).

### *9.2.2 Lifestyle and Neurodegenerative Diseases*

Accumulating evidence suggests that lifestyle factors, such as physical exercise, high-fat diet, diet deficient in antioxidant-rich foods, and socializing with friends and family (see below), influence brain plasticity. Regular physical exercise ameliorates age-related neuronal loss and produces positive effect on neurodegenerative diseases (Trejo et al., 2002). In the brain, exercise induces both acute and long-term beneficial alterations, such as increased levels of various neurotrophic factors and enhanced cognition. Although the molecular mechanisms involved in exercise-mediated changes in the brain are not yet well understood, it is suggested that physical exercise increases the expression of insulin-like growth factor I (IGF-1) and BDNF in the brain. These neurotrophins are important for synaptic plasticity and learning and memory (Carro et al., 2001; Vaynman et al., 2004; Vaynman and Gomez-Pinilla, 2006). TrkB-IgG blocks beneficial effects of exercise on cognitive function (Vaynman et al., 2004). Insulin-like growth factor (IGF) is a member of the insulin gene family with known neurotrophic properties. The actions of IGF are mediated via the IGF type 1 (IGF1) and type 2 (IGF2) receptors as well as through the insulin receptors, all of which are widely expressed throughout the brain.

BDNF-TrkB signaling involves mitogen-activated protein kinase (MAPK), the phospholipase Cγ (PLCγ), and the phosphatidylinositol 3-kinase (PtdIns3K) pathways. MAPK and PtdIns3K play crucial roles in both translation and trafficking of proteins induced by synaptic activity, whereas PLCγ modulates intracellular $Ca^{2+}$ that can drive transcription via cAMP and a protein kinase C. BDNF secreted from active synapses and neurons recruits TrkB from extrasynaptic sites into lipid rafts (Fig. 9.1). Postsynaptic rises in cAMP concentrations facilitate translocation of TrkB into the postsynaptic density. Neuronal activity facilitates BDNF-mediated TrkB endocytosis, a signaling event important for many long-term BDNF events (Ji et al., 2005). Molecular analysis indicates that exercise significantly upregulates

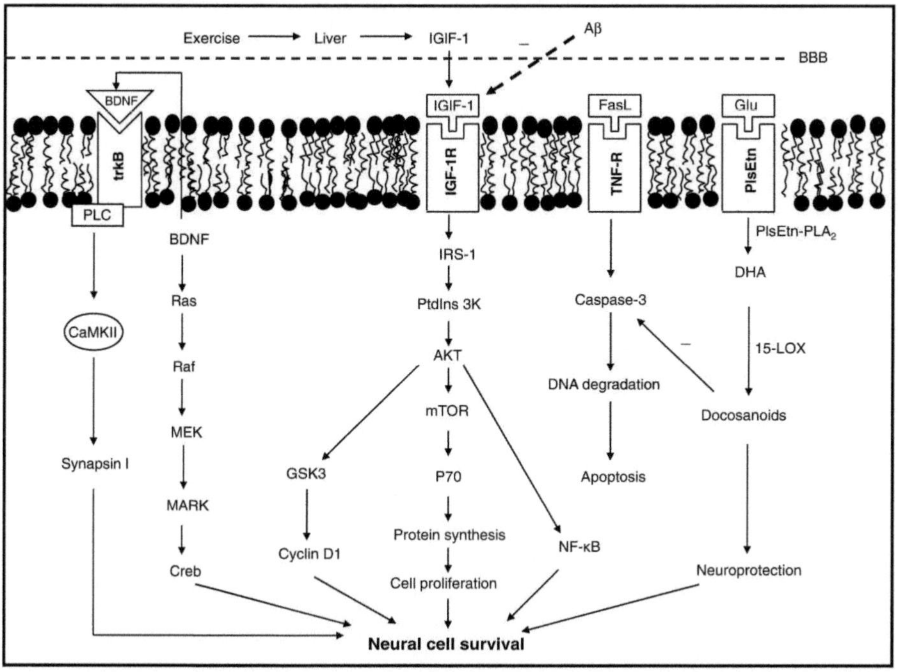


**Fig. 9.1** Hypothetical diagram showing interaction between IGF-1 and BDNF receptors in cell survival. Insulin-like growth factor-1 (IGF-1); Insulin-like growth factor-1 receptor (IGF-1R); trak B receptor (trkB); tumor necrosis factor receptor (TNF-R); phospholipase C (PLC); brain-derived growth factor (BDNF); insulin receptor substrate-1 (IRS-1); phosphatidylinositol 3-kinase (PtdIns 3 K); protein kinase B (Atk); mammalian Target of rapamycin (mTOR); mitogen-activated protein kinase (MARK); cAMP regulatory element binder (Creb); plasmalogen (PlsEtn); plasmalogen-selective phospholipase $A_2$ (PlsEtn-$PLA_2$); nuclear factor kappaB (NF-κB); and glycogen synthase kinase-3 (GS3K)

proteins downstream to BDNF activation that are important for synaptic function such as, synapsin I, and phosphorylated calcium/calmodulin protein kinase II and phosphorylated mitogen-activated protein kinase II (Ding et al., 2006). Exercise also increases the expression of several key intermediates of the PtdIns-3 K/Akt pathway, which is known for its role in enhancing neuronal survival (Chen and Russo-Neustdt, 2007). In addition, activation of cAMP/PKA and phosphorylation of synapsin I facilitate regenerative growth of neurons and promote neuronal survival. Blocking the IGF-I receptor retards the exercise-induced increases in signal transduction processes. These results provide information on the molecular mechanisms by which IGF-1 modulates the BDNF system to mediate exercise-induced synaptic and cognitive plasticity. BDNF not only facilitates long-term potentiation, an electrophysiological correlate of learning and memory, but also increases the activities of free radical scavenging enzymes and hence protect neurons against oxidative stress (Pelleymounter et al., 1996). Thus, interactions between IGF-1 and BDNF provide protection to neural cell in brain, where IGF-1 performs several functions,

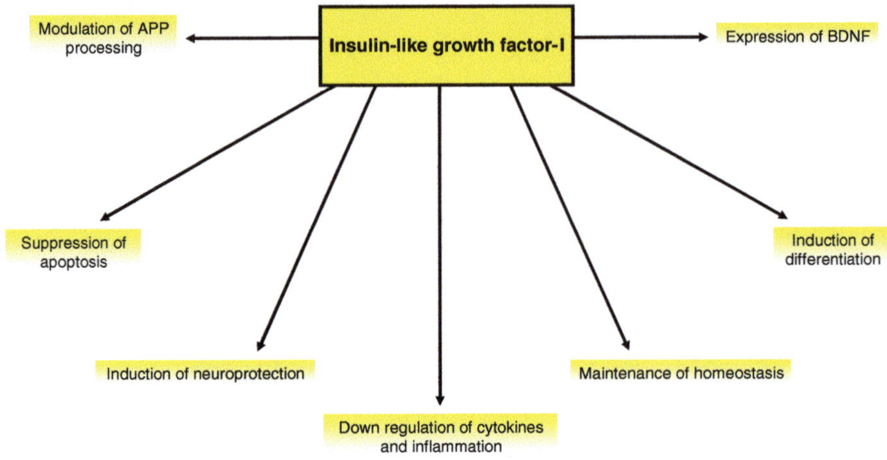


**Fig. 9.2** Roles of insulin-like growth factor-1 in the brain

including modulation of APP processing, expression of BDNF, suppression of apoptosis through downregulation of bax in neurons and bcl-X in astrocytes (Fig. 9.2) (Hoyer, 2004; Carro and Torres-Aleman, 2004).

In addition, exercise also upregulates the expression of the mitochondrial uncoupling protein 2, an energy-balancing factor concerned with ATP formation and free radical management (Vaynman et al., 2006), supporting the view that in brain tissue physical exercise promotes a fundamental mechanism by which key elements of energy metabolism may modulate the substrates of hippocampal synaptic plasticity.

### *9.2.3 Diet and Neurodegenerative Diseases*

Consumption of high-fat diet causes inflammation and oxidative stress in cardiovascular and cerebrovascular systems. The increase in plasma fatty acids due to high-fat diet mediates not only the activation of redox cycling of the copper–albumin complex and excessive lipid peroxidation, but accumulation of advanced glycation end products (AGEs) through the Maillard reaction and subsequent activation of the receptors for AGEs (RAGE). These receptors belong to multiligand receptor in the immunoglobulin superfamily, act as a cell surface binding site for Aβ, and mediate alternations in the phosphorylation state of mitogen-activated protein kinase (MAPKs) indicating that MAPKs are involved in neurodegenerative processes (Origlia et al., 2009). In particular, changes in the phosphorylation state of various MAPKs by aggregated proteins (Aβ, synuclein, huntingtin) lead to synaptic dysfunction and cognitive decline, as well as development of inflammatory responses in AD.

Browning of protein-enriched food and heating and reheating of sugar-enriched food and its consumption also generate AGEs in the body. So it may be good idea to stop eating not only processed foods but also foods that are superheated, broiled, or reheated multiple times. Formation and accumulation of AGEs occur during normal aging with lower rate, but progress with an accelerated rate of neurodegenerative diseases, heart disease, diabetes, and cancers following exposure to a high-fat diet (Ghosh et al., 2007). Identification of the cholesterol transporter apolipoprotein E4 as a major genetic risk factor for hypercholesterolemia, vascular dementia, and sporadic AD (Corder et al., 1994) reinforces the relationship between cholesterol and AD. Cholesterol-enriched diet and subsequent hypercholesterolemia not only alter the IGF-1 signaling pathway and decrease insulin degrading enzyme but also increase active p-Tyr276 GSK-3α levels leading to increase in levels of Aβ in rabbit hippocampus. These changes may be involved in the phosphorylation of CREB and the upregulation of the anti-apoptotic protein Bcl-2, events that may represent a defensive mechanism to prevent neurodegeneration (Sharma et al., 2008). High-fat diet also causes significant upregulation of gp91(phox) subunit of NADPH oxidase and downregulations of superoxide dismutase isoforms, glutathione peroxidase, and hemeoxygenase-2 in various body tissues (Roberts et al., 2006). These processes increase plasma levels of malondialdehyde and impair vasodilatory response to acetylcholine. These finding strongly support the presence of oxidative stress and endothelial dysfunction in rats consuming high-fat diet (Roberts et al., 2006). This is tempting to speculate that dietary modification may be important in managing neurodegenerative diseases.

Another lipid diet factor influencing the risk of neurodegenerative diseases is the intake of ω-3 fatty acids (docosahaexenoic acid, DHA and eicosapentaenoic acid, EPA). Epidemiological studies indicate that sufficient DHA intake reduces the risk of developing AD and other neurodegenerative diseases (Kalmijn et al., 1997; Morris et al., 2003; Schaefer et al., 2006; Farooqui, 2009b). In addition, dietary intake of fish oil may reduce cognitive decline (Farooqui, 2009b), and a recent trial (Freund-Levi et al., 2006) shows positive effects of DHA supplementation on cognition in patients with very mild AD. Similarly, investigations on three different transgenic models of AD indicate that animal models of AD are more vulnerable to DHA depletion than controls and that DHA exerts a beneficial effect against pathological signs of AD, including Aβ accumulation, cognitive impairment, synaptic marker loss, and hyperphosphorylation of τ (Lim et al., 2005; Calon and Cole, 2007).

Diet enriched in antioxidant and anti-inflammatory agents (curcumin, green tea, and ferulic acid) lowers the risk of developing neurodegenerative diseases (Farooqui and Farooqui, 2009). Dietary supplementation of colored fruit and vegetable extracts decreases the age-enhanced vulnerability to oxidative stress and inflammation. In addition, polyphenolic compounds found in red wine and fruits (such as blueberries) also exert their beneficial effects through signal transduction and neuronal communication, delaying dementia (Lau et al., 2007; Joseph et al., 2007). The consumption of extra-virgin olive oil, which contains micronutrients and polyphenolic antioxidants, including tyrosol [2-(4-hydroxyphenyl) ethanol], hydroxytyrosol, oleuropein, and oleocanthal, retards the development of neurodegenerative diseases (Lopez-Miranda et al., 2007). It is suggested that greater adherence to olive oil containing

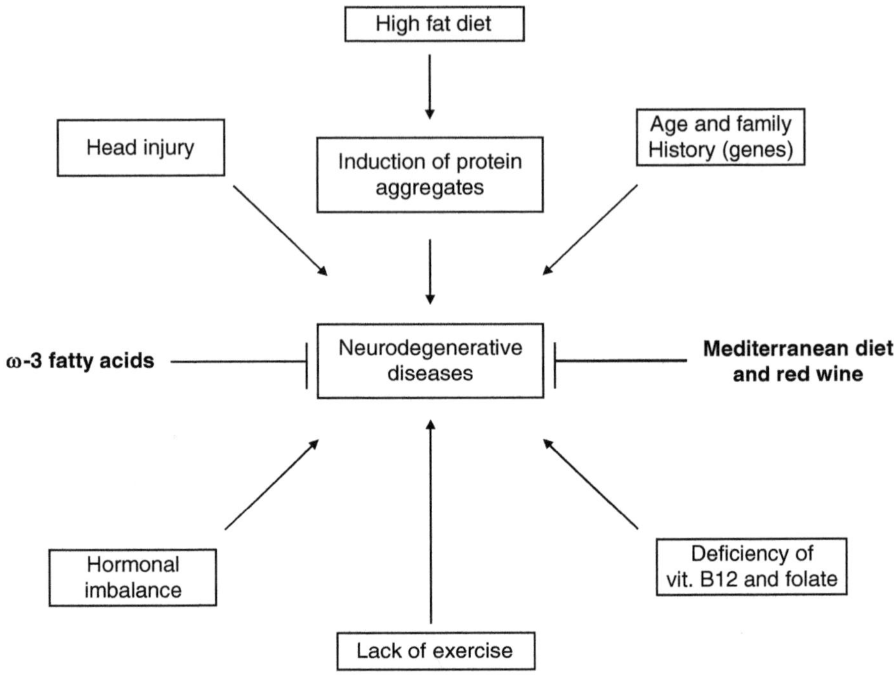


**Fig. 9.3** Factors promoting neurodegenerative diseases

Mediterranean diet results in a significant improvement in health status, as seen by a significant reduction in overall mortality (13%) in PD and AD patients (Fig. 9.3) (Sofi et al., 2008).

Excessive calorie intake increases the risk of chronic visceral and neurodegenerative diseases. Caloric restriction (CR) or intermittent fasting increases life span and protects brain against neurodegenerative diseases due to increase in cellular stress resistance (Lee et al., 2000; Mattson, 2008). CR lowers plasma insulin levels and mediates greater sensitivity to insulin; lowers body temperatures; reduces cholesterol, triglycerides, and blood pressure. It also elevates HDL and slows age-related decline in circulating levels of dehydroepiandrosterone sulfate. CR mediates the synthesis of cellular stress-response stimulating proteins (neurotrophic factors, neurotransmitter receptors, protein chaperones, and mitochondrial biosynthesis regulators) and enhances neuronal plasticity and resist oxidative and metabolic insults (Lee et al., 2000; Fontan-Lozano et al., 2008). CR also upregulates levels of IGF-1 resulting in prolonged life in mice. Modulation of aging by IGF-I may involve reduction in insulin signaling, enhancement of sensitivity to insulin, reduction in generation of ROS, improvement in antioxidant defenses resulting in reduced oxidative damage (Bartke et al., 2008). Based on above discussion, it can be proposed that CR as well as an active and stimulating lifestyle in late life as well as an optimal control of vascular and other chronic diseases both at middle age and late life may facilitate prevention or postponement of the onset of neurodegenerative diseases.

## 9.3 Therapeutic Approaches for AD

The earliest pathological events leading to AD include the accumulation of Aβ. The deposition is estimated to occur 10–15 years before the appearance of the first cognitive alterations in AD patients (Price and Morris, 1999). However, τ aggregation in tangles and in neurites does not begin to accelerate and build up in larger amounts in the neocortex until just prior to symptom onset. By the time the earliest clinical signs of AD appear, Aβ deposition may be close to reaching its peak and tangle formation and neuronal cell loss is substantial though still not at its maximal extent (Taraweneh and Holtzman, 2009). AD is then diagnosed late in life only after substantial neurodegeneration or synaptic damage has occurred. In addition, the molecular mechanism associated with the pathogenesis of AD remains unknown, so the pharmacotherapy of AD is mainly confined to symptomatic and disease-modifying treatments. The therapeutic approaches that are commonly available at the present time include cholinergic strategies, antioxidant and anti-inflammatory strategies, stabilization of mitochondrial dynamics, use of neurotrophic factors, neurosteroids, statins, memantine, NitroMemantine, ω-3 fatty acids, and gene therapy.

### *9.3.1 Cholinergic Strategies*

According to cholinergic hypothesis, AD is caused by reduction in the synthesis of acetylcholine cholinergic neurons of basal forebrain and loss of cholinergic neurotransmission in the cerebral cortex. The decrease in acetylcholine significantly contributes to the deterioration in cognitive function seen in AD patients (Birks, 2006). Although choline acetyltransferase (ChAT) is markedly reduced and cholinesterase activity is not affected in the cerebral cortex of AD brain, the location of cholinesterase is largely shifted to the neuritic plaques and neurofibrillary tangles. Cholinesterase hydrolyzes acetylcholine, a neurotransmitter that plays an important role in learning, remembering, and thinking. The inhibition of acetylcholinesterase by cholinesterase inhibitors reduces the breakdown of acetylcholine and increases availability acetylcholine at the synapse resulting in restoration of cognition and memory function. The earliest known cholinesterase (ChE) inhibitors include tacrine and physostigmine (Fig. 9.4). Although these drugs show modest improvement in the cognitive function in AD patients, clinical studies indicate that physostigmine has poor oral activity, brain penetration, and pharmacokinetic parameters while tacrine causes hepatotoxic liability. To overcome these disadvantages, new generation of AChE inhibitors such as donepezil, galantamine, and rivastigmine has been synthesized (Fig. 9.4) (Birks, 2006). These inhibitors not only reduce ChE

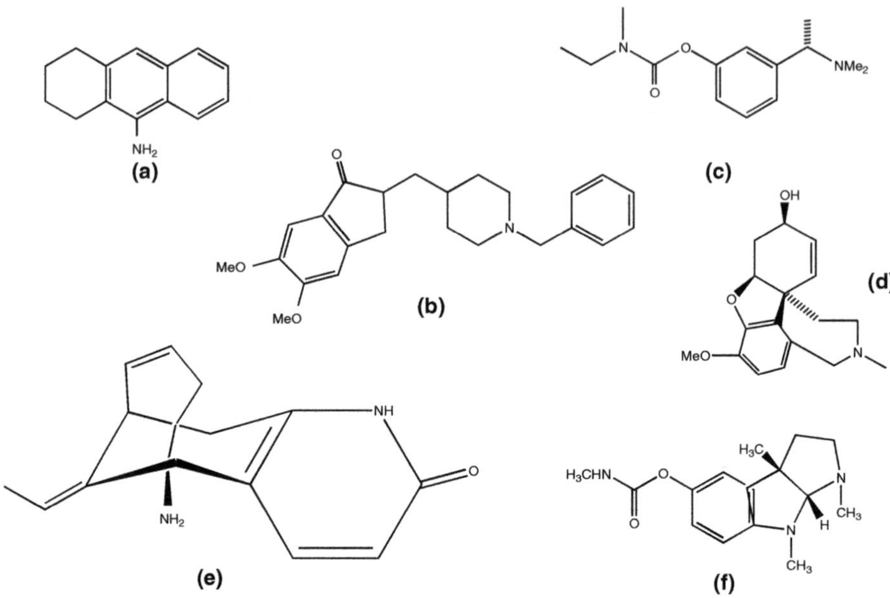


**Fig. 9.4** Chemical structures of tacrine (**a**); donepezil (**b**); rivastigmine (**c**); galantamine (**d**); huperzine (**e**); and physostigmine (**f**)

activity but also retard processing and deposition of Aβ (Munoz-Torrero, 2008). New AChE inhibitors (bis(7)-tacrine) are more potent than tacrine in inhibiting AChE activity and are designed to simultaneously alleviate cognitive deficits and inhibit Aβ peptide aggregation through binding to both catalytic and peripheral sites of the enzyme (Fig. 9.5). In addition, these inhibitors also increase the cerebral blood flow in AD patients both after acute and fairly short period of treatment (Nordberg, 1999). The M1 selective muscarinic agonists AF102B [Cevimeline], AF150(S) and AF267B not only increase αAPPs, reduce Aβ levels, τ hyperphosphorylation, and inhibit Aβ-induced neurotoxicity, in vitro, via M1 mAChR-modulation of kinases (e.g. PKC, MAPK and GSK3β), but also restore cognitive deficits, cholinergic markers, and retard τ hyperphosphorylation in AD models with a wide safety margin (Fisher, 2007). Triple transgenic AD mice, which show major AD pathologies and cognitive deficits, chronic AF267B treatment, rescue cognitive deficits and reduce Aβ42 and τ pathologies in the cortex and hippocampus (not amygdala), through the activation of M1 mAChR-activation and reduction in BACE1 steady-state levels and inhibition of GSK3β. These observations suggest that in comprehensive therapy many AD symptoms and hallmarks are possible by using AF102B [Cevimeline], AF150(S) and AF267B (Fisher, 2007). Collective evidence suggests that muscarinic agonists may thus influence the etiology of AD as well as provide symptomatic benefits. More studies are required on various animal models for better understanding of mechanism of action of M1 selective muscarinic agonists.

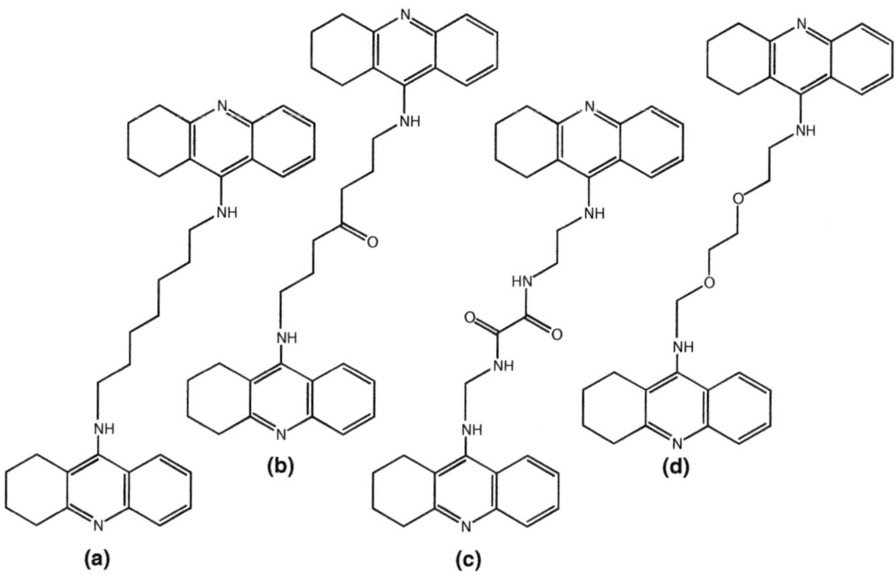


**Fig. 9.5** Chemical structures of tacrine-based dual binding site acetylcholinesterase (**a**); bis-tacrine-bearing ketone carbonyl group (**b**); bis-tacrine-bearing ketone oxalamide group (**c**); bis-tacrine-bearing ketone ethylenedioxy group (**d**)

Another reversible cholinergic inhibitor, huperzine A, a Chinese herb isolated from *Huperzia serrata* (Fig. 9.4), also improves cognitive deficits in a broad range of animal models and has been used for AD treatment in China for many years (Zhang et al., 2008). It is proposed that in addition to anticholinergic effect, huperzine A may also act by attenuating oxidative stress, regulating the expression of apoptotic proteins Bcl-2, Bax, P53, caspase-3, nerve growth factor, and its receptors, and interfering with APP processing and metabolism (Wang and Tang, 2005). It also ameliorates acute inflammation in transient focal cerebral ischemic rats and in an in vivo model of cerebral hypoperfusion. Thus, huperzine A suppresses the overexpression of inflammatory factor tumor necrosis factor-α (TNF-α) and over-phosphorylation of JNK and p38 mitogen-activated protein kinases (MAPKs) in a cell model of chronic hypoxia (Wang et al., 2010). Preincubation with mecamylamine, a nicotinic acetylcholine receptor (nAChR) antagonist, before hypoxia notably reverses the effects of huperzine A on TNF-α production and MAPKs phosphorylation (Wang et al., 2010). Studies on the effect of huperzine A in APP processing and Aβ generation in human neuroblastoma SK-N-SH cells overexpressing wild-type human APP695 indicate that huperzine A enhances αAPPs release in a dose-dependent manner (Peng et al., 2007). Studies on involvement of disintegrin and metalloprotease (ADAM10 and ADAM17) in huperzine A-mediated nonamyloidogenic APP metabolism indicate that huperzine A produces increase in the level of ADAM10, and the inhibitor of TNF-α converting enzyme

(TACE)/ADAM17 retards the huperzine A-mediated rise in αAPPs levels, supporting the view that huperzine A may modulate the non-amyloidogenic α-secretase pathway for APP processing in neuroblastoma SK-N-SH cells overexpressing wild-type human APP695 (Peng et al., 2007). αAPPs release is significantly blocked by muscarinic acetylcholine receptor antagonists (particularly by an M1 antagonist), protein kinase C (PKC) inhibitors, GF109203X and calphostin C, and the mitogen-activated protein kinase kinase (MEK) inhibitors, U0126 and PD98059. Furthermore, huperzine A markedly stimulates the phosphorylation of p44/p42 mitogen-activated protein (MAP) kinase, which is blocked by treatment with U0126 and PD98059. These results indicate that the activation of muscarinic acetylcholine receptors, PKC and MAP kinase, may be involved in huperzine A-mediated αAPPs secretion in neuroblastoma cells (Peng et al., 2007). Collective evidence suggests that huperzine A attenuates inflammation, improves spatial cognitive dysfunction, and modulates α-secretase-mediated non-amyloidogenic APP metabolism (Wang et al., 2010). In addition huperzine also reduces glutamate-induced cell death by interfering with glutamate receptor-gated ion channels in primary neuronal cultures (Zhou and Tang, 2002). It is stated that multiple neuroprotective effects of huperzine A may be responsible for additional beneficial effects in AD.

### *9.3.2 Antioxidant, Anti-inflammatory, and Antiexcitotoxic Strategies in AD*

Oxidative, inflammatory, and excitotoxic damage is a characteristic feature of many neurodegenerative diseases, including AD, but attempts to upregulate oxidative-, inflammatory-, and excitotoxic defenses by the therapeutic use of antioxidants-, anti-inflammatory drugs, and glutamate antagonists have either failed or provided very little benefits. The definition of oxidative, inflammatory, and excitotoxic damage should be extended to beyond the classical antioxidant, proinflammatory, and proexcitotoxic enzymes (SOD, catalase, NOS, COX-2, LOX, $PLA_2$, and calpains) and low molecular weight reductants (GSH, ascorbic acid, coenzyme Q10, and lipoic acid) (Perry and Smith, 2000; Farooqui, 2009a, b). In AD, a major role is played by long-term accumulation of Aβ, decrease in energy status, redox status, and alterations in ionic homeostasis of degenerating neurons. These long-term alterations in metabolism of neural cells are not compensated by the therapeutic use of antioxidants, anti-inflammatory drugs, and glutamate antagonists (Gilgun-Sherki et al., 2006; Wang et al., 2006; Tan et al., 2003; Farooqui, 2009a, b). This is tempting to speculate that other factors, such as stabilization of mitochondrial dynamics and enrichment of ω-3 fatty acids in diet may be added to therapeutics of AD.

Classical targets of antioxidant, anitiinflammatory agents, and antiexcitotoxic compounds include cyclooxygenase, NF-κB, and peroxisome proliferator-activated receptors (Townsend et al., 2004; Sastre et al., 2006a). Although the inhibition of these pathways may explain the effect of antioxidants, NSAID, and glutamate antagonists on AD progression, recent studies indicate that some NSAIDs, such as ibuprofen, indomethacin, and flurbiprofen, have a direct Aβ42 lowering properties though the inhibition of Rho and its downstream effector proteins (Rock 1 and 2) in cell cultures as well as transgenic models of AD-like amyloidosis (Townsend et al., 2004; Sastre et al., 2006b). Recent epidemiological studies have shown that long-term therapeutic use of non-steroidal anti-inflammatory drugs (NSAIDs) may reduce the risk of developing AD and delayed the onset of AD (Hirohata et al., 2008).

The molecular mechanism of neuroprotective effects associated with anti-inflammatory agents and antioxidant may not only depend on the general free radical trapping or antioxidant activity per se in neurons but also depend on the down-regulation of NF-κB activity (Shen et al., 2003), activation of Nrf2-ARE pathway, suppression of genes induced by proinflammatory cytokines and other mediators released by glial cells, and increase the production of ATP in degenerating neurons (Gilgun-Sherki et al., 2006; Wang et al., 2006). Collective evidence suggests that for antioxidants and anti-inflammatory agents to work properly their use should be started early in life so that they can block or delay onset of AD late in life.

### 9.3.3 Stabilization of Mitochondrial Dynamics and AD

Mitochondria play a critical role in initiating both apoptotic and necrotic cell death in neurodegenerative diseases. They not only generate ATP and maintain the ratios of ATP:ADP in favor of ATP hydrolysis into ADP + Pi but also produce ROS and act as cellular $Ca^{2+}$ sink. Under certain pathological conditions, such as extreme $Ca^{2+}$ load, elevated phosphate concentrations, and adenine nucleotide depletion, opening of the mitochondrial permeability transition pore (mPTP) results in the extrusion of mitochondrial $Ca^{2+}$ (Tsujimoto and Shimizu, 2003; Sullivan et al., 2005). The mPTP is a nonselective, high conductance channel composed of several proteins including adenine nucleotide translocase (ANT), mitochondrial inner membrane protein transporter, protein transporter at the outer mitochondrial membrane, the outer membrane voltage-dependent anion channel, and cyclophilin (CypD). The mitochondrial permeability transition (mPT) is defined as an increase in the permeability of the mitochondrial membranes to molecules of less than 1,500 Da in molecular weight. This process transforms mitochondria from organelles, which produce and sustain ATP to instruments of cell death through apoptotic and necrotic cell death (Farooqui, 2009a). Sanglifehrin A (mitochondria targeted antioxidant) (Fig. 9.6) and anti-apoptotic proteins (Bcl-2 and Bcl-$x_L$) block the mPT and can therefore inhibit mPT-dependent cell death. In addition, cyclosporin A (CsA), a potent immunosuppressive drug (Fig. 9.6), also blocks mitochondrial permeability transition (mPT) through its interactions with matrix cyclophilin D. Binding of cyclophilin D is increased in response to oxidative stress and some thiol reagents that sensitize the mPT to $Ca^{2+}$. Peripherally administered CsA attenuates mitochondrial dysfunction and neuronal damage in an experimental rodent model of TBI, in a dose-dependent manner (Sullivan et al., 2005). The underlying mechanism of

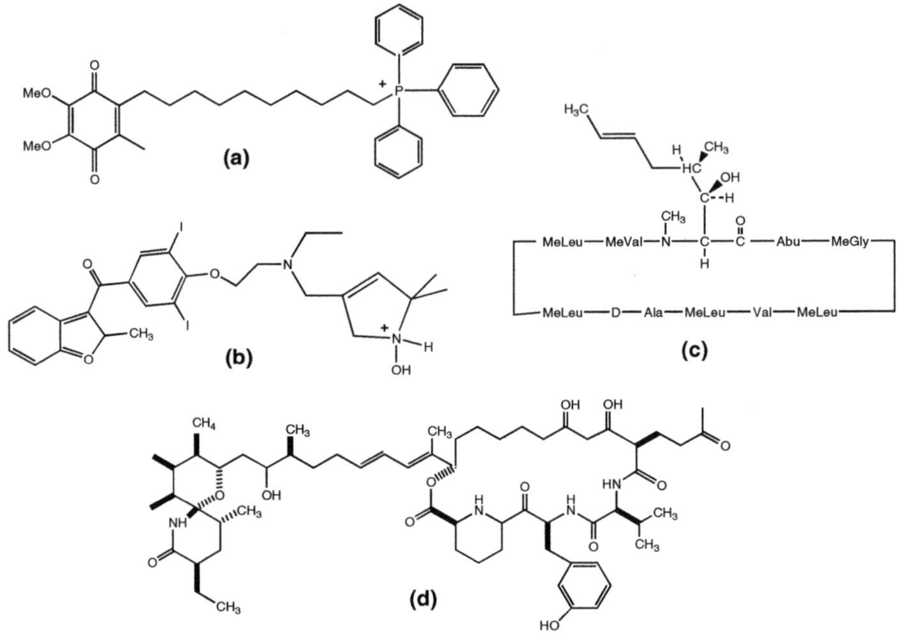


**Fig. 9.6** Chemical structures of the mitochondrial membrane stabilizers. MitoQ (ubiquinone linked to a triphenylphosphonium cation by an alkyl chain of unspecified length) (**a**); HO-3538 (superoxidase mimetic compound) (**b**); cyclosporin A (**c**); sanglifehrin A (**d**)

neuroprotection-mediated by CsA may involve interactions with the mPTP because FK506, which blocks mPT, but has some neuroprotective effects. CsA may also block mPT through the inhibition of calcineurin-mediated dephosphorylation of BAD (Waldmeier et al., 2003).

The accumulation of Aβ in mitochondria may contribute to mitochondrial dysfunction and oxidative stress in AD. Aβ interacts with ANT and cypD in the inner mitochondrial membrane, where it modulates transport of ATP and ADP through the regulation of mPT (Singh et al., 2009). Aβ also progressively accumulates within mitochondrial matrix, where it binds to a short-chain alcohol dehydrogenase called Aβ peptide binding alcohol dehydrogenase (ABAD). Interactions of ABAD with Aβ result in Aβ-mediated mitochondrial dysfunction as evidenced by increased ROS generation, mitochondrial membrane permeability formation, and caspase-3-like activity induction along with decreased activities of the Krebs cycle. These processes mediate neuronal perturbation, leading to impaired synaptic function and dysfunctional spatial learning/memory. An ABAD peptide specifically inhibits ABAD-Aβ interaction and suppresses Aβ-induced apoptosis and free radical generation in neurons. Transgenic mice overexpressing ABAD in an Aβ-rich environment manifest exaggerated neuronal oxidative stress and impaired memory. Collective evidence suggests that the blockade of mPTP may be a potential therapeutic strategy for AD (Chen and Yan, 2007a; Chen and Yan, 2007b).

### 9.3.4 Statins and AD Treatment

Statins inhibit 3-hydroxy-3-methylglutaryl coenzyme A (HMG-CoA) reductase, an enzyme that not only initiates the syntheses of cholesterol and isoprenoid lipids but also is a rate-limiting step for the synthesis of cholesterol. Statins (Fig. 9.7) significantly reduce risk for cardiovascular and cerebrovascular diseases (Endres, 2005; Vaughan, 2003). Beneficial effects of statins in cardiovascular and cerebrovascular systems are due to their antiexcitotoxic, antioxidant, vasculoprotective, and anti-inflammatory properties (Farooqui et al., 2007b; Farooqui, 2009b). In addition, treatment with statin also results in greater nitric oxide bioavailability, improvement in endothelial cell function, enhancement in cerebral blood flow, and decrease in platelet aggregation. Cholesterol is a major neural membrane component and is required for the formation of lipid rafts that are the platforms for signal transduction processes, including isoprenoid-dependent assembly and activation of raftophilic β- and γ-secretases that generate Aβ40 and 42 fragments from amyloid precursor protein (APP). In brain, statins activate β-secretase and reduce Aβ generation and accumulation in a transgenic mouse model of AD (Pedrini et al., 2005; Zimmermann et al., 2005). This process may involve statin-mediated inhibition of calpain, and increased production and secretion of soluble form of APP via α-secretase stimulation, involvement of PtdIns3K pathway, and inhibition of ROCK

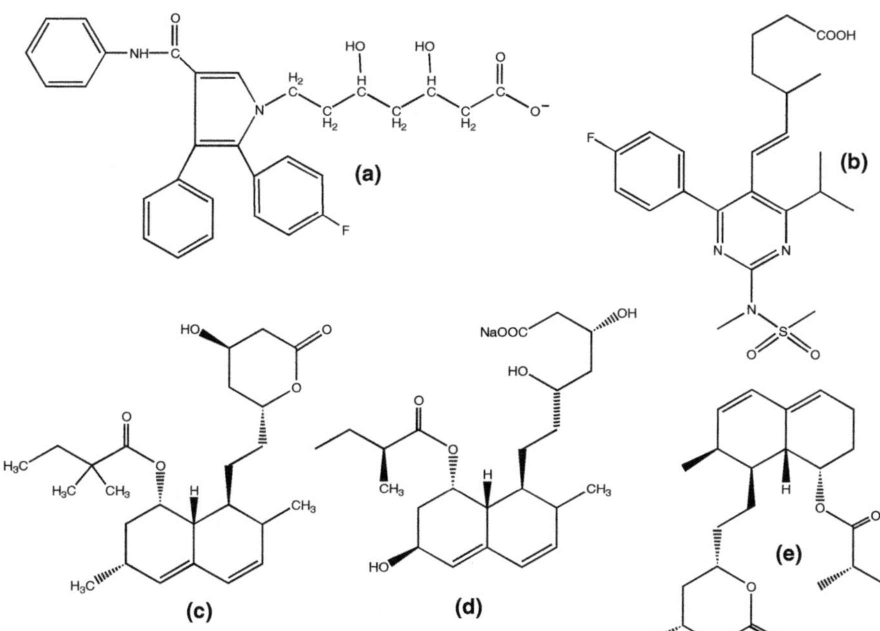


**Fig. 9.7** Chemical structures of statins. Lipitor (**a**); crestor (**b**); zocor (**c**); pravachol (**d**); and compactin (**e**)

signaling (Ma et al., 2009a, b). Statins also inhibit an Aβ-mediated inflammatory response through their ability to prevent the isoprenylation of members of the Rho family of small G proteins, resulting in the functional inactivation of these G proteins (Pedrini et al., 2005; Cordle et al., 2005). Treatment of microglia with statins results in perturbation of the cytoskeleton and morphological changes due to alteration in Rho family function. The neuroprotective effects of statins are blocked by mevalonate, a PtdIns3K inhibitor, and tyrphostin AG538, indicating the involvement of cholesterol and insulin/IGF-1 signaling in the neurotoxic response. Statins prevent calcium-dependent calpain activation, resulting in complete suppression of protein truncation events on multiple calpain substrates that are involved in neuronal death including CDK5 coactivator p35 cleavage to p25, GSK3, and β-catenin. This is followed by reduction and enhancement in nuclear translocation of p25 and β-catenin, respectively (Ma et al., 2009a, b). Statin (simvastatin) enhances learning and memory in transgenic and non-transgenic mice with AD-like pathology on a mixed genetic background (Mans et al., 2010). The molecular mechanism associated with enhancement of LTP and memory formation by statins remains elusive. A prolonged in vitro simvastatin treatment (2–4 hours) significantly increases the magnitude of LTP at CA3-CA1 synapses without altering basal synaptic transmission or the paired-pulse facilitation ratio in hippocampal slices from C57BL/6 mice. Increase in LTP is accompanied by the increased phosphorylation of Akt (protein kinase B) in the CA1 region following 2-h treatment with simvastatin. Inhibition of Akt phosphorylation suppresses the simvastatin-mediated enhancement of LTP suggesting the activation of Akt as a molecular pathway for augmentation of hippocampal LTP. It is suggested that simvastatin-mediated enhancement of hippocampal LTP may be a potential cellular mechanism, underlying the beneficial effects of simvastatin on cognitive function (Mans et al., 2010).

In cell cultures, statin-mediated reduction in Aβ production correlates with an inhibition of β-secretase dimerization into its more active form at several concentrations of statin (Parsons et al., 2006). These effects can be reversed by the administration of mevalonate indicating the involvement of pathways dependent on 3-hydroxy-3-methylglutaryl-CoA. At a low statin concentration, decrease in Aβ production and inhibition of β-secretase dimerization is mediated by inhibition of isoprenoid synthesis, but at high concentrations statins act by inhibiting β-secretase palmitoylation. Statins also modulate the phosphorylation of τ in humans. This may be another mechanism by which statins reduce the risk of AD (Riekse et al., 2006). Through their antioxidant and anti-inflammatory effects, statins not only block ROS-mediated brain damage but also inhibit the release of proinflammatory cytokines and nitric oxide synthesis (Sparks et al., 2005; Cordle and Landreth, 2005). Statin treatment also retards the rac1-dependent activation of NADPH oxidase and superoxide production. Collective evidence suggests that statins are novel and powerful drugs that modulate protein isoprenylation, small G protein function, antioxidant and anti-inflammatory activities of neural cells. Based on above findings, it is proposed that the long-term use of low doses of statins, starting as early as possible may slow the onset and progression of dementia and AD (Wolozin, 2002). Although biologically it seems feasible that statins may prevent dementia and slow the onset of AD due to their role in isoprenoids and cholesterol reduction, it is becoming increasingly evident that statins given in late life to individuals at risk of vascular disease have no effect in preventing AD or dementia (McGuinness et al., 2009). In addition, in some patients statin therapy is accompanied by some potential risks of psychiatric adverse drug reactions, including memory loss, depression, suicidality, aggression, sleep disorders, and antisocial behavior along with rare serious adverse drug reactions that affect mainly muscle, liver, and kidney (Farooqui et al., 2007a; Farooqui, 2009a, b). Several statin trials in AD patients are going on at the present time.

### 9.3.5 Memantine and AD Treatment

Memantine (1-amino-3,5-dimethyladamantane) (Fig. 9.8) a moderately low affinity and uncompetitive NMDA receptor antagonist, which is approved by the European Union and the US FDA for the treatment of moderate-to-severe AD. It enters NMDA receptor channel preferentially when channel is excessively open. The off-rate of memantine is relatively fast so that it does not substantially accumulate in the channel to interfere with normal synaptic transmission (Sonkusare et al., 2005; Lipton, 2006; Nakamura and Lipton, 2007). Thus low concentrations of memantine retard excessive glutamate stimulation, while still maintaining normal glutamate-mediated neurotransmission. The importance of maintaining normal synaptic NMDA signaling boosts intrinsic antioxidant defenses through the involvement of thioredoxin–peroxiredoxin system (Papadia et al., 2008). Under

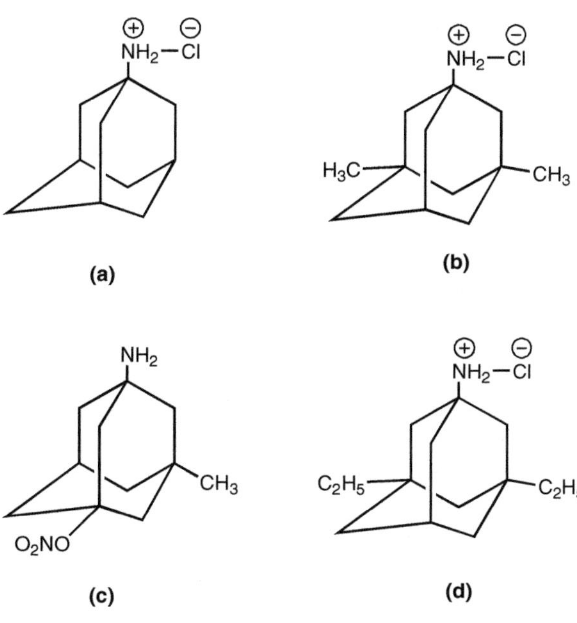


**Fig. 9.8** Chemical structures of memantine and its derivatives. Amantadine (1-aminoadamantane) (**a**); memantine (1-amino-3,5-dimethyladamantane) (**b**); NitroMemantine (**c**); and 1-amino-3, 5-diethyladmantane (**d**)

normal conditions, synaptic activity increases thioredoxin activity, facilitates the reduction of overoxidized peroxiredoxin, and promotes resistance to oxidative stress. Memantine has good tolerability, low side effect profiles, and a positive therapeutic impact in moderate-to-severe AD patients alone and in conjunction with donepezil (Schmitt et al., 2006, 2007). In addition, low levels of memantine also promote neuroplasticity and memory formation (Rogawski and Wenk, 2003; Reisberg et al., 2003). In addition, memantine also inhibits the internal ribosome entry site to block the expression of APP and τ and so ameliorates the symptoms of AD (Wu and Chen, 2009). Memantine is also under investigation as a potential treatment for other neurodegenerative disorders, such as HIV-associated dementia, neuropathic pain, glaucoma, depression, HD, ALS, and PD (Lipton, 2006).

Although the molecular mechanism associated with the beneficial effects of memantine is not fully understood in various animal and cell culture models of AD, memantine alleviates glutamatergic receptor overstimulation and promotes normal signaling among brain neurons. These processes may be associated with therapeutic benefits of memantine in the earliest stages of the disease (Rogawski and Wenk, 2003; Wenk et al., 2006; Planells-Cases et al., 2006). Aβ toxicity is induced by increased phosphorylation of τ-protein and activation of τ kinases, i.e., glycogen synthase kinase-3β and extracellular signal-related kinase ½ (Song et al., 2008). Additionally, Aβ-induced toxicity is accompanied by the cleavage of caspase-3 and decrease in phosphorylation of cyclic AMP response element-binding protein. Memantine treatment significantly protects cultured neurons against Aβ-induced toxicity by attenuating τ-phosphorylation and its associated signaling mechanisms. However, this drug does not alter either conformation or internalization of Aβ42 and is unable to attenuate Aβ-induced potentiation of extracellular glutamate levels.

In neuronal SK-N-SH cells, memantine significantly decreases the levels of the secreted form of sAPP, sAPPα, and Aβ40 compared to vehicle-treated cells. This change is initiated as early as 8 h and continues for up to 24 h of memantine treatment. Unlike sAPPα, a slight non-significant increase in total intracellular APP level is reported in 24 h treated memantine cells. It is proposed that memantine may be involved in the transport or trafficking of APP molecules away from the site of their proteolytic cleavage by the secretase enzymes (Ray et al., 2009). Although memantine may not stop or reverse AD, but through its moderating effect on glutamate metabolism, it may facilitate and restore normal neural cell signaling, a process closely associated with neuroprotection in AD patients. In addition to direct effect on NMDA channel, memantine not only increases the expression of BDNF in the limbic cortex in a concentration-dependent manner but also induces the BDNF receptor and trkB (Lu, 2003; Marvanova et al., 2001). Both processes may potentiate neuroprotection and memory enhancement in memantine-treated patients. Other neuropharmacological effects of memantine include non-competitive and voltage-independent inhibition of $5HT_3$ receptor current (Rammes et al., 2001) and inhibition of human neuronal nicotinic cholinergic receptor (Buisson and Bertrand, 1998), suggesting that detailed neurochemical and neuropharmacological investigations are required on the molecular mechanism of action of memantine in mammalian brain.

Second-generation memantines (NitroMemantine) have also been synthesized (Lipton, 2006; Nakamura and Lipton, 2007). According to Lipton, these drugs use memantine as a homing signal to target NO to hyperactivated NMDARs in order to avoid systemic side effects of NO such as hypotension (low blood pressure). The NitroMemantines have enhanced neuroprotective efficacy in vitro and in vivo animal models of neurological disorders. NitroMemantines not only lack the blood pressure lowering effect of nitroglycerin (Lipton, 2006) but also interact with sulfhydryl groups of cysteine residue in NMDA receptor channel causing S-nitrosylation and downregulation of (but not completely shut off) NMDA receptor activity (Lipton, 2006; Nakamura and Lipton, 2007). Thus, the nitrosylation of NMDA receptor at cysteine399 produces a conformational change in the NMDA receptor protein that facilitates tight binding of glutamate and zinc to the NMDA receptor channel. The tight binding of glutamate and zinc results in the desensitization of NMDA receptor causing its channel to close (Lipton et al., 2002). In addition to NMDA receptor, S-nitrosylation also enhances neuronal survival by inhibiting the activities of caspases and neuronal cell injury by regulating the (a) ubiquitin E3 ligase activity of parkin, (b) chaperone and isomerase activities of PDI, (c) nuclear translocation of GAPDH, and (d) activity of MMP-9 (Nakamura and Lipton, 2007). Several clinical trials are underway to test the efficacy for memantine and NitroMemantine for the treatment of AD and vascular dementia (Lipton, 2006).

### *9.3.6 Secretase Inhibitors and AD Treatment*

Proteolytic processing of amyloid precursor protein involves three proteases, namely α-secretase, β-secretase, and γ-secretase. First, β-secretase (β-site APP cleaving enzyme, BACE1) acts at the N-terminus of APP, followed by the action of one or more γ-secretase complexes (intramembrane aspartyl proteases) at the C-terminus of APP as part of the β-amyloidogenic pathway (Hussain et al., 1999; Lin et al., 2000). Processes that limit the accumulation of Aβ production and deposition by preventing formation, inhibiting aggregation, and/or enhancing clearance may offer effective treatments for AD. Since β- and γ-secretases mediate APP cleavage in the amyloidogenic pathway, inhibition of β- and γ-secretases may be an important therapeutic approach for treating AD and diminishing Aβ peptide formation in AD patients. In addition to Aβ generation, these enzymes are also involved in other vital physiological pathways. For example, involvement of γ-secretase in cell differentiation may preclude complete blockade of this enzyme for prolonged times in vivo. Furthermore, Notch receptor depends on α-secretase for its signaling function. Development of drugs that regulate the production of Aβ without affecting the Notch signaling is needed for the treatment of AD by γ-secretase inhibitors (Tomita, 2009).

β-Secretase has been identified as the rate-limiting enzyme for the production of Aβ. In addition to Aβ, this enzyme also participates in the proteolytic processing of neuregulin-1 (Willem et al., 2006), a ligand for the ErbB family of receptor tyrosine kinases. Neuregulin-1-mediated signaling pathway is associated with synapse formation, plasticity, neuronal migration, myelination of central and peripheral axons, and the regulation of neurotransmitter expression and function (Falls, 2003; Michailov et al., 2004). Most studies in knock-out mice indicate that inhibition of β-secretase may have minimal adverse effects. However, it is becoming increasingly evident that properties of the active site of this enzyme make it difficult to find small molecule inhibitors that bind with high affinity. Available inhibitors are large and peptidic in nature and, therefore, unsuitable as drug candidates. However, some good inhibitors of β- and γ-secretases have been recently synthesized (Hussain et al., 1999; Lin et al., 2000; Vassar et al., 1999; Ghosh et al., 2005). These inhibitors have low molecular weight with excellent cell permeability and possess enhanced pharmacokinetic profiles in cell culture and animal models of AD. Collective evidence suggests that there are many issues associated with the development of inhibitors of secretases for AD that must be addressed before these inhibitors can be used to test the "amyloid cascade hypothesis" in the clinic. The outcomes of trials of secretase inhibitors may open new doors and provide options for the treatment of AD.

Other approaches to downregulate Aβ production may be through the enhancement of Aβ degradation and may be through the modulation of α-secretase or "a disintegrin and metalloprotease (ADAM)'s" activity via protein kinase C (PKC), calcium ion, tyrosine kinase, MAP kinase, and hormonal signaling, which regulate catabolic processing of APP (Hooper and Turner, 2002; Kojro and Fahrenholz, 2005; Bandyopadhyay et al., 2007; Peng et al., 2007). α-Secretase attacks APP inside the Aβ sequence and therefore prevents formation of neurotoxic Aβ. Three membrane-anchored zinc-dependent metalloproteinases, ADAM10, ADAM17, and ADAM9 also, show α-secretase activity. Since the individual knock-out of these proteinases in above cases completely retards α-secretase processing of APP, it seems likely that different ADAM enzymes may compensate α-secretase activity mutually, and under different conditions may contribute to α-secretase-mediated cleavage of APP (Postina, 2008). Based on detailed investigations, it is suggested that α-secretase coactivators can be used for the treatment of AD in cell culture and animal models. The proteolytic cleavage of APP by the α-secretase within the Aβ sequence precludes formation of Aβ peptides and results in release of soluble APPsα, which has neuroprotective properties. α-Secretase co-activators include iron chelators. In addition, other agents, such as estrogen, testosterone, statins, huperzine, various neurotransmitters, and growth factors, also increase nonamyloidogenic cleavage of APP.

### 9.3.7 PPAR Agonists and AD Treatment

As stated in Chapter 7, peroxisome proliferator-activated receptors (PPARs) are a group of nuclear receptor proteins that function as transcription factors regulating the expression of genes associated with cell differentiation, inflammation, development, and metabolism through the modulation of carbohydrate and lipid metabolism (Heneka and Landreth, 2007). Three types of PPARsα, γ, and δ (β) have been identified in mammalian tissues. When activated, the receptors migrate to the nucleus, where they act as transcription factors and modulate a number of specific genes. PPARγ is a DNA-binding transcription factor whose transcriptional regulatory actions are activated after agonist binding (Kummer and Heneka, 2008).

Thiazolidinediones (Fig. 9.9), the drugs that have been developed to treat diabetes, are agonists for PPARγ. The endogenous agonists for PPARγ are polyunsaturated fatty acids and 15-deoxy-δ(12,14)- prostaglandin $J_2$. The activation of PPARγ not only reduces the differentiation of monocytes into activated macrophages but also inhibit the Aβ-mediated expression of the cytokine (interleukin-6 and tumor necrosis factor-α) in microglia and astrocytes. In addition, PPARγ agonists also reduce the expression of cyclooxygenase-2. PPARγ agonists are also capable of suppressing LPS-induced expression of iNOS and proinflammatory cytokines. Collective evidence suggests that PPARγ plays a critical role in regulating the inflammatory responses of microglia and monocytes to β-amyloid. Neuroinflammation is a major component of AD pathogenesis.

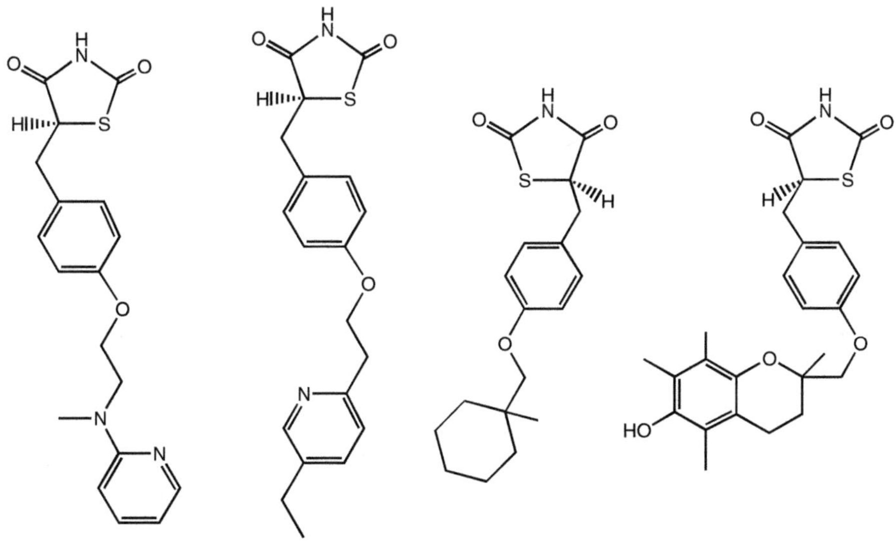


(a) (b) (c) (d)

**Fig. 9.9** Chemical structures of PPARγ agonists that have been used for the treatment of AD in animal and cell culture models. Rosiglitazone (**a**); pioglitazone (**b**); ciglitazone (**c**); and troglitazone (**d**)

In addition, PPARγ agonists prevent Aβ-induced neurodegeneration in hippocampal neurons, and PPARγ is activated by the nerve growth factor (NGF) survival pathway, suggesting a neuroprotective anti-inflammatory independent action (Fuenzalida et al., 2007). It is shown that PPARγ agonist rosiglitazone

(RGZ) protects hippocampal and dorsal root ganglion neurons against Aβ-induced mitochondrial damage and NGF deprivation-induced apoptosis, respectively, and also promotes PC12 cell survival. In neurons and in PC12 cells RGZ protective effects are associated with increased expression of the Bcl-2 anti-apoptotic protein (Fuenzalida et al., 2007). Cells overexpressing PPARγ contain a four- to fivefold increase in Bcl-2 protein content, whereas in dominant negative PPARγ-expressing cells, Bcl-2 is barely detected. Bcl-2 knockdown by small interfering RNA in cells overexpressing PPARγ results in increased sensitivity to Aβ and oxidative stress supporting the view that PPARγ protective effect involves Bcl-2 upregulation (Fuenzalida et al., 2007). In animal models of AD, PPARγ agonist treatment results in the decrease of amyloid plaque burden, reduction in neuroinflammation, and reversal of disease-related behavioral impairment (Jiang et al., 2008). The clinical trials of the PPARγ agonist rosiglitazone have indicated significant improvement in memory and cognition in AD patients (Landreth et al., 2008). Thus, PPARγ represents an important new therapeutic target in treating AD.

The effect of selective PPARδ agonist GW742 in 5xFAD mice, which harbor three mutations in amyloid precursor protein and two mutations in presenilin 1, indicating that GW742 significantly reduces amyloid plaque burden in the subiculum region of 3-month-old male and female 5xFAD mice (Kalinin et al., 2009). GW742 also significantly decreases astrocyte activation, suggesting anti-inflammatory effects on glial cells. The changes in plaque burden are accompanied by the upregulation in expression of the amyloid-degrading enzymes neprilysin and insulin degrading enzymes. These results suggest that like PPARγ agonists, PPARδ agonists also reduce amyloid burden through the clearance of Aβ.

### *9.3.8 Neurotrophins and AD Treatment*

Dysregulation of neurotrophin signaling has been implicated in several neurodegenerative diseases including AD. Thus, a marked reduction in neurotrophin levels (BDNF, TGF-β1, and proNGF) and their receptors have been reported to occur in AD patients brain compared to age-matched controls (Murer et al., 2001; Cotman, 2005). In AD, the accumulation of Aβ aggregates and increase in TNF-α and IL-1β signaling interferes with neurotrophin signaling by impairing the axonal transport of BDNF, TGF-β1, and proNGF in neurons of AD transgenic mice (Tg2576) (Poon et al., 2009). Although neurotrophins have proven to be elusive pharmacological targets, recent studies indicate that a novel multipotent neurotrophin antagonist, 3-[(5E)-4-oxo-5-[[5-(4-sulfamoylphenyl)-2-furyl]methylene]-2-thioxothiazolidin-3-yl]propanoic acid (Y1036) may provide beneficial effects in cell culture and animal models of neurodegenerative diseases (Eibl et al., 2010). Y1036 not only prevents BDNF or NGF from interacting with their obligate receptor(s) but also blocks both BDNF- and NGF-induced trk activation, downstream activation of the p44/42 mitogen-activated protein kinase pathway, and neurotrophin-mediated differentiation of dorsal root ganglion sensory neurons. It is speculated that identification and synthesis of a BDNF- and NGF-specific antagonist may be important for treating chronic diseases characterized by the dysregulation of multiple neurotrophins (Eibl et al., 2010).

Postmortem studies indicate that brain from AD patient shows a decrease in expression of cerebral insulin-like growth factor (IGF)-1 receptor (IGF-1R) and insulin receptor substrate (IRS) proteins (Freude et al., 2009a). It is also reported that not only IGF 1/IRS-2 signaling pathway plays an important role in regulation of α-secretase activity but circulating IGF-1 may modulate Aβ clearance from the brain by promoting Aβ transport over the blood–brain barrier (Freude et al., 2009b). In Tg2576 mice, IRS-2 deficiency not only completely reverses premature mortality but also delays Aβ accumulation in hippocampus. Studies on APP metabolism indicate that delay in Aβ accumulation is caused by the downregulation of APP processing. APP gene family includes the two paralogues APP-like protein (APLP) 1 and 2. The neurotoxic Aβ originates from APP by sequential cleavages via β- and γ-secretases. Insulin and insulin-like growth factor-1 (IGF-1) are involved in APP processing and modulating levels of Aβ in the brain. Thus, IGF-1 stimulates the shedding of APP, APLP1, and APLP2. IGF-1-induced shedding of both APP and APLP1 is dependent on PtdIns3-K, whereas sAPLP2 secretion is independent of this signaling pathway (Adlerz et al., 2009; Jacobsen et al., 2010). In human neuroblastoma SH-SY5Y cells, stimulation of APP and APLP1 processing involves multiple signaling pathways, whereas APLP2 processing is mainly dependent on PKC. Detailed investigations on shedding differences between APLP2 and APP indicate that APP is mainly cleaved by ADAM10, whereas APLP2 processing is mediated by TACE indicating that different α-secretases are involved in IGF-1-induced processing (Jacobsen et al., 2010).

### *9.3.9 ω-3 Fatty Acids and AD Treatment*

In human diet ω-6 fatty acids are represented by linoleic and arachidonic acids, while ω-3 or *n*-3 fatty acids are represented by linolenic, eicosapentaenoic, and docosahexaenoic acids (Fig. 9.10). In Western diet, the ratio of ω-6 to ω-3 fatty acids ranges from approximately 17–20:1 instead of the traditional range of 1–2:1. A high intake of ω-6 fatty acids shifts the physiological state of human body to one that is prothrombotic and proaggregatory as characterized by increase in blood viscosity, vasospasm, and vasoconstriction and reduction in bleeding time. In contrast, ω-3 fatty acids (Fig. 9.10) enriched diet produce anti-inflammatory, antithrombotic, antiarrhythmic, hypolipidemic, and vasodilatory effects (Farooqui, 2009b). Fish consumption and supplementation of ω-3 fatty acids in diet reduce the risk of having AD (Freund-Levi et al., 2006; Farooqui, 2009b). This suggestion is supported by epidemiological as well as experimental studies. Although the molecular mechanism associated with retardation of AD by ω-3 fatty acids is not fully understood, multiple mechanisms may be involved in beneficial effects of ω-3 fatty acids in AD. These mechanisms include antioxidant effects of ω-3 fatty acids, reduction in levels

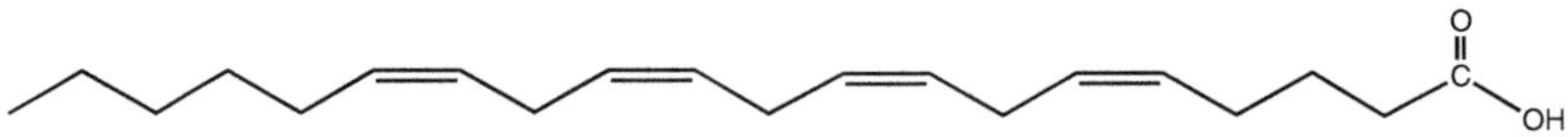


Arachidonic acid

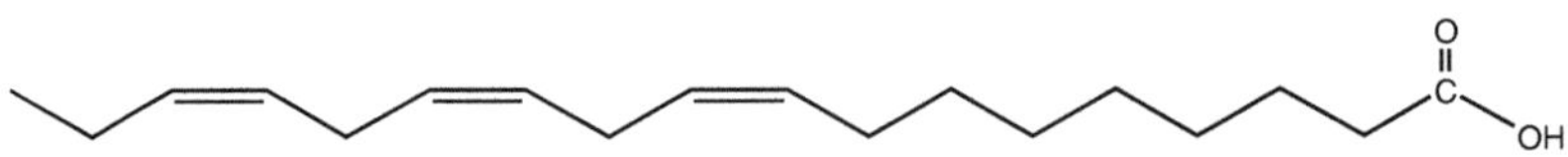


Alpha lino lenicacid

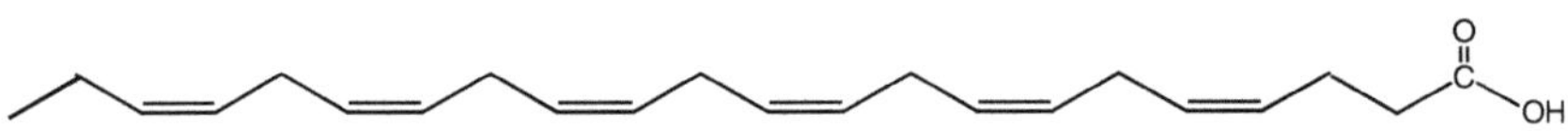


Docosahexaenoic acid

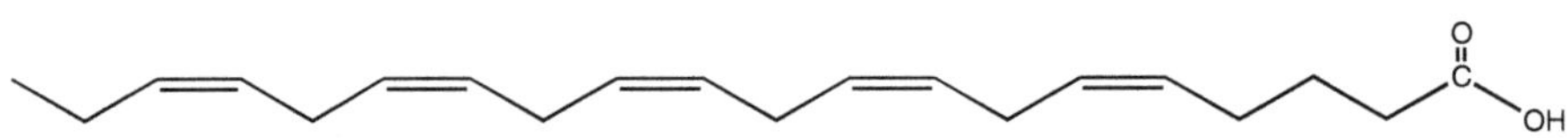


Eicosapentaenoic acid

**Fig. 9.10** Chemical structures of ω-6 and ω-3 fatty acids present in human diet

of arachidonic acid-derived metabolites, generation of neuroprotective metabolites of DHA, and increase in trophic factors or downstream trophic signal transduction processes (Cole et al., 2009).

In rat model of AD, dietary intake of DHA significantly reduces the levels of Aβ40, cholesterol, and saturated fatty acids (Hashimoto et al., 2006, 2008). Fluorescence and electron micrography studies on fibril formation indicate that DHA reduces the levels of oligomeric amyloid species in a concentration-dependent manner, supporting the view that dietary DHA-mediated suppression of in vivo Aβ aggregation occurs through the inhibitory effect of DHA on oligomeric amyloid species (Hashimoto et al., 2008). Chronic pre-administration of DHA also prevents β-amyloid-induced impairment of an avoidance ability-related memory function in a rat model of AD (Hashimoto et al., 2005) and protects mice from synaptic loss and dendritic pathology in another model of AD (Calon et al., 2004). DHA and its metabolite, neuroprotectin $D_1$ ($NPD_1$), not only decrease β-amyloid secretion from aging brain cells but also prevent apoptosis (Lukiw et al., 2005; Bazan, 2005, 2006). It is suggested that neuroprotective effect of DHA and $NPD_1$ may be due to the activation of PPARγ receptor and inhibition of proinflammatory cytokine and eicosanoid signaling, upregulation of γ-glutamylcysteinyl ligase and glutathione reductase activities, and inhibition of p65 subunit transcription factor NF-κB. DHA also inhibits c-jun N-terminal kinase and the phosphorylation of adaptor protein insulin receptor substrate-1 (IRS-1) and τ in cultured hippocampal neurons (Ma et al., 2009a, b). Collective evidence suggests that DHA deficiency in Western diet may lead to decrease in nerve growth factor, inflammatory signaling, apoptosis, and neuronal dysfunction in AD (Ikemoto et al., 2000; Lukiw et al., 2008). Agents and diet that stimulate $NPD_1$ synthesis may be useful for delaying and neurodegenerative diseases (Bazan, 2005; Serhan, 2005). Based on the above information, clinical trials of DHA are underway to prevent and treat AD.

### *9.3.10 Immunization Therapy in AD*

AD is a multifactorial disease that involves not only accumulation of Aβ but also increase in intensity of interactions among excitotoxicity, neuroinflammation, oxidative stress, mitochondrial dysfunction, and alterations in neurotransmission. Active and passive immunizations with Aβ produce Aβ antibodies, which successfully reduce the cerebral Aβ burden, Aβ-related astrocytosis, retardation of reaccumulation of Aβ, and restoration of Aβ-induced depletion of presynaptic SNAP-25 for at least 1 month and reduce inflammatory reactions for 1 week in AD murine models without producing inflammation, microhemorrhage, or systemic histotoxicity and improve cognitive functions in an AD mouse model (Chauhan and Siegel, 2004). These observations suggest that intracerebroventricular anti-Aβ may be a safe method for the rapid clearance of pre-existing Aβ and retarding reaccumulation of Aβ in human AD. However, in 2002, a Phase IIa clinical trial was interrupted due to the development of T-lymphocyte meningoencephalitis in approximately 6% of the AD patients. It was suggested that the immunogen (full-length Aβ42) may have led to an autoimmune response (Lemere et al., 2007). The failure of Aβ antibody immunization in AD patients has, however, not discouraged the interest for an inflammation-based therapy in AD patients but rather intensified the research on immunization in transgenic AD mice models. Shorter immunogen has been developed to target Aβ B-cell epitopes (within Aβ1–15). Shorter immunogen does not interact with Aβ-specific T-cell epitopes (Aβ16–42). This generates a safe and effective AD vaccine. Intranasal immunization with dendrimeric Aβ1–15 (16 copies of Aβ1–15 on a lysine core) or a tandem repeat of Aβ1–15 is joined by two lysines and conjugated to an RGD motif with a mutated form of an *Escherichia coli*-derived adjuvant-generated robust Aβ titers in both wild-type and APP Tg mice. Aβ antibodies recognize a B-cell epitope within Aβ1-7. Six months of intranasal immunization (from 6 to 12 months of age) of J20 mice with each immunogen reduce insoluble Aβ42 by 50%, decrease plaque burden and gliosis, and increase Aβ in plasma (Lemere et al., 2007; Lemere, 2009). Due to the adverse reactions to the active immunization in 2002 clinical trial, the irreversibility of immunization and the variable antibody response to vaccines in older individuals, passive immunization against the Aβ may be another attractive alternative immunotherapeutic strategy. The most advanced immunological approach is the use of bapineuzumab composed of humanized anti-Aβ monoclonal antibodies. It is now tested in two large late-stage trials (Frisardi et al., 2009; Kaufer and Gandy, 2009). It is anticipated that in future specifically selected anti-Aβ human monoclonal antibodies may reduce and inhibit the deposition of Aβ in brain while avoiding the cognitive decline that characterizes AD.

### *9.3.11 AL-108 or NAP Therapy in AD*

AL-108 is derived from an eight amino acid peptide (NAPVSIPQ: "NAP") synthesized from a naturally occurring neuroprotective brain protein known as activity-dependent neuroprotective protein (ADNP). This protein is secreted by astrocytes and has multiple cellular functions, including transcription factor, cytoplasmic, and extracellular activity. ADNP differentially interacts with chromatin to regulate essential genes. Complete ADNP deficiency is lethal. Partial deficiency of ADNP in $ADNP^{+/-}$ mice causes cognitive deficits, significant increases in phosphorylated τ, tangle-like structures, and neuronal loss compared with $ADNP^{+/+}$ mice (Vulih-Shultzman et al., 2007). Increased τ hyperphosphorylation in $ADNP^{+/-}$ mice may be responsible for memory impairments, a process that is a characteristic feature of neurodegenerative diseases associated with tauopathies and AD (Vulih-Shultzman et al., 2007).

Phase IIa clinical trials have recently shown that AL-108 has a positive impact on memory function in patients with amnestic mild cognitive impairment (aMCI) (Gozes et al., 2009). Although the molecular mechanism of its action is now known, but based on various in vivo and in vitro studies, it is suggested that this peptide interacts with tubulin and modulates microtubule assembly leading to enhanced cellular survival that is associated with fundamental cytoskeletal elements. In addition, AL-108 protects against toxicity-mediated by Aβ peptide, *N*-methyl-D-aspartate, electrical blockade, the envelope protein of the AIDS virus, dopamine, $H_2O_2$, nutrient starvation and zinc overload. It also provides neuroprotection in animal models of apolipoprotein E deficiency, cholinergic toxicity, closed head injury, stroke, middle aged anxiety, and cognitive dysfunction (Gozes et al., 2005). AL-108 also has a positive impact on memory function in patients with aMCI. This condition is a precursor to AD. AL-108 penetrates cells and crosses the blood–brain barrier after nasal or systemic administration (Gozes, 2007; Gozes et al., 2009). More studies are needed on therapeutic efficacy of AL-108 in larger double blind trials in AD patients.

## 9.4 Therapeutic Approaches for PD

Like AD, neurochemical changes in PD appear many years before the appearance of the first cognitive alterations in PD patients and PD is diagnosed late life only after substantial neurodegeneration in substantia nigra pars compacta has occurred. Although the molecular mechanisms associated with the pathogenesis of PD are unknown, several pathogenic factors, including oxidative stress, mitochondrial dysfunction, abnormal protein handling, inflammation, and excitotoxicity, have been identified. Intensity of cross talk among these factors may cause dopaminergic neuronal death in the substantia nigra. Manipulation of these factors may allow the development of disease-modifying treatment strategies to slow neuronal death. At present, the pharmacotherapy of PD is mainly confined to symptomatic and disease-modifying treatments. Although deep brain stimulation, ablative surgery, and fetal cell transplantation provide significant beneficial effects, these procedures do not change the onset and progression of PD.

Choice of pharmacotherapy includes consideration of short-term benefits as well as long-term consequences. Patients with mild PD often function adequately without symptomatic treatment. However, therapeutic approaches that are commonly used at the present time include dopaminergic strategies, antioxidant and anti-inflammatory strategies, stabilization of mitochondrial dynamics, use of statins, memantine, ω-3 fatty acids, and gene therapy (Farooqui, 2009a).

### *9.4.1 Dopaminergic Strategies in PD*

Current therapeutic approaches for PD are merely symptomatic, intended for the treatment of symptoms without any disease-modifying activity. Levodopa (3,4-dihydroxy-L-phenylalanine) is the most efficacious medication for the management of PD. Age-related changes in absorption, distribution, metabolism, and excretion of levodopa and its metabolite complicate the treatment of elderly PD patients. Long-term use of levodopa leads to dyskinesia. Trials with dopamine agonists indicate that the onset of dyskinesia can be delayed with the use of this therapy. Monoamine oxidase-B inhibitors, specifically selegiline, may also provide symptomatic improvement in PD patients (Romrell et al., 2003). Selective dopamine blockers, such as clozaril and quetiapine, have been shown to be effective for the treatment of psychosis associated with PD. Commonly used medications to manage PD symptoms also include monoamine oxidase type B (MAO-B) inhibitors, catechol-*O*-methyltransferase (COMT) inhibitors, and amantadine, a NMDA receptor antagonist (Jankovic and Stacy, 2007). Agents that block MAO-B, such as rasagiline and selegiline, increase concentrations of dopamine in the brain by blocking its reuptake from the synaptic cleft, a mechanism that can slow motor decline, increase "on" time and improve symptoms of PD. Adverse effects of these medications include confusion, hallucination, and orthostatic hypotension.

### *9.4.2 Antioxidant, Anti-inflammatory, and Antiexcitotoxic Strategies in PD*

PD is characterized by oxidative stress and neuroinflammation. Attempts to block oxidative stress and neuroinflammation by antioxidants in PD have been unsuccessful. This may be due to the fact that long-term oxidative stress and neuroinflammation in neurons cannot be compensated and fully or partially reversed by available antioxidants and anti-inflammatory agents because antioxidants and anti-inflammatory agents do not reach mitochondria, the primary source of ROS. In addition, preclinical studies in animal models have shown efficacy of mitochondrial-targeted antioxidants and the SS (Szeto-Schiller) peptides. The structural motif of SS peptides centers on alternating aromatic residues, and basic amino acids (aromatic-cationic peptides) can scavenge hydrogen peroxide and peroxynitrite, and inhibit lipid peroxidation (Sezeto, 2006). Another promising approach for enhancing antioxidant defenses is to transcriptionally upregulate the activity of the Nrf2/ARE pathway, which activates transcription of anti-inflammatory and antioxidant genes. A number of agents including sulforaphane, curcumin, and triterpenoids have been shown to activate Nrf2/ARE pathway and to produce neuroprotective effects (Beal, 2009).

The novel non-toxic and lipophilic brain-permeable iron chelators, VK-28 (5-[4-(2-hydroxyethyl) piperazine-1-ylmethyl]-quinoline-8-ol), and its multifunctional derivative, M-30 (5-[*N*-methyl-*N*-propargylaminomethyl]-8-hydroxyquinoline) (Fig. 9.11), as well as the main polyphenol constituent of green tea (–)-epigallocatechin-3-gallate (EGCG), which possesses iron metal chelating, radical scavenging, and neuroprotective properties, offer potential therapeutic benefits for PD (Avramovich-Tirosh et al., 2007). Pyrroloquinoline quinone (PQQ) is a free radical scavenger that has attacked considerable attention from both the nutritional and pharmacological viewpoints (Fig. 9.12). α-Synuclein, protein that accumulates in PD, has the propensity to oligomerize and form fibrils, and this tendency may play a crucial role in its toxicity. PQQ blocks the amyloid fibril formation and aggregation of α-synuclein in vitro in a PQQ concentration-dependent manner (Kobayyashi et al., 2006). Moreover, PQQ forms a conjugate with α-synuclein, and this PQQ-conjugated α-synuclein is also able to prevent α-synuclein amyloid fibril formation. It is suggested that PQQ may be a candidate for future anti-PD therapy in humans.

### 9.4.3 Stabilization of Mitochondrial Dynamics in PD

Mitochondrial dynamics (fission, fusion, migration) has been reported to play an important role in neurotransmission, synaptic stability, and maintenance and neuronal survival. PINK1 and Parkin are closely associated within the regulation of mitochondrial dynamics and function (Bueler, 2009). Mutations in DJ-1, Parkin, and PINK 1 render animals more susceptible to oxidative stress and mitochondrial toxins implicated in sporadic PD. DJ-1, Parkin, and PINK1 form a complex (termed PPD complex) to promote ubiquitination and degradation of Parkin substrates, including Parkin itself and synphilin-1. In addition, mutant proteins may retard the transport of nuclear-encoded mitochondrial proteins to mitochondria, interact with mitochondrial proteins and disrupt the electron transport chain, induce free radicals, cause

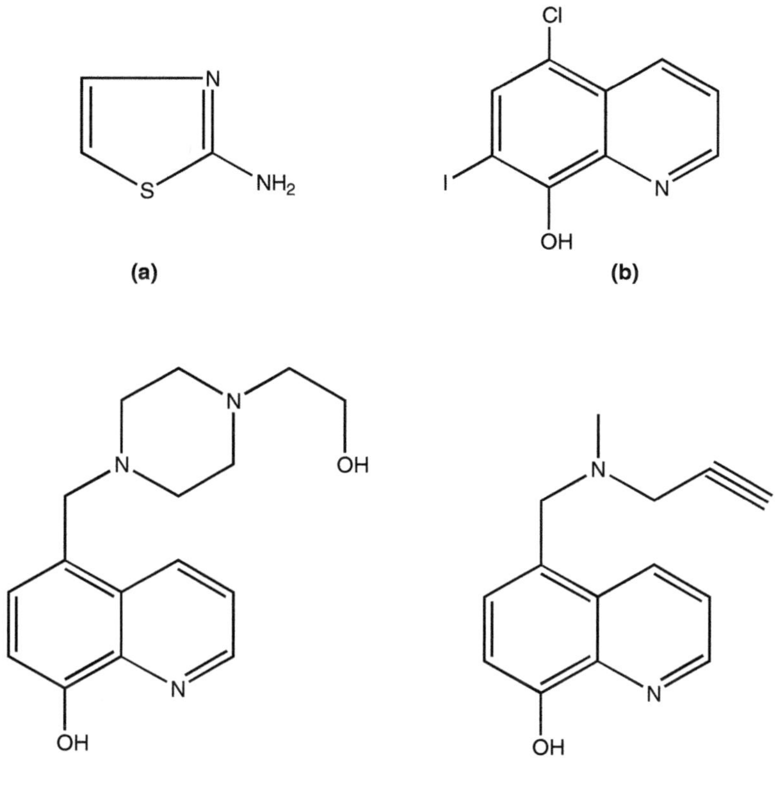

(a) (b) (c) (d)

**Fig. 9.11** Chemical structures of lipophilic brain permeable compounds that have been used for the treatment of PD in animal and cell culture models. Aminothiazole (**a**); clioquinol (**b**); VK28 (**c**); and M-30 (**d**)

mitochondrial dysfunction, and, ultimately, damage neurons (Reddy, 2008). Besides compromising cellular energy production, alterations in mitochondrial dynamics are involved in the induction of oxidative stress and apoptosis. It is speculated that drugs that modulate mitochondrial dynamics function and biogenesis may have important clinical applications in the future treatment of PD (Bueler, 2009). Mitochondrial enhancement and dynamics stabilizing strategies include trials of coenzyme Q10 (an essential cofactor in the mitochondrial respiratory chain) and creatine. These drugs are being tested in phase III clinical trials (Beal, 2009).

In transgenic *Drosophila* expressing human α-synuclein model of PD male transgenic flies fed with Regrapex-R (grape extract enriched in flavans, anthocyanins, emodin, and resveratrol) show a significant improvement in climbing ability compared to controls. Female transgenic flies show a significant extension in average life span (Long et al., 2009). This suggests that Regrapex-R is a potent free radical scavenger, a mitochondrial protector, and a candidate for further studies to assess its

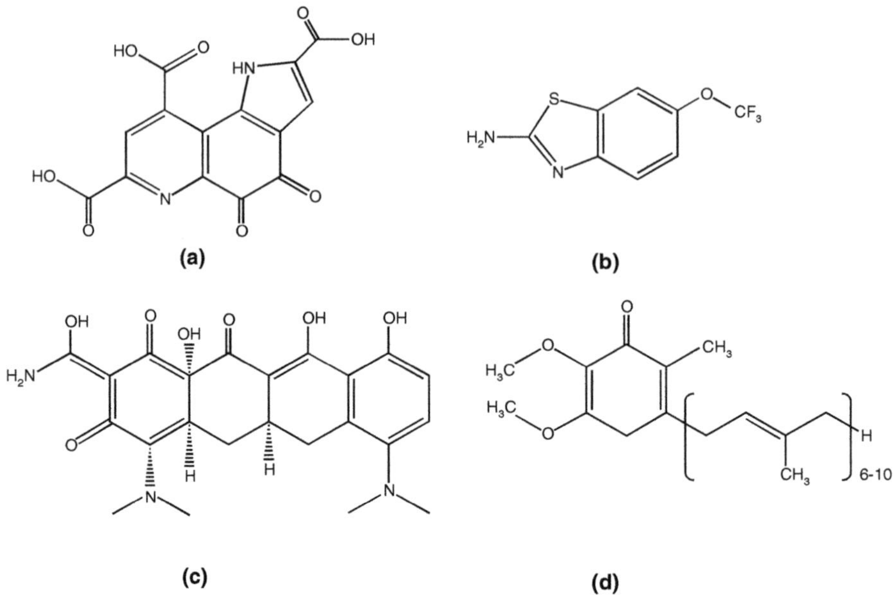


**Fig. 9.12** Chemical structures of drugs used for the treatment of PD, ALS, and HD in animal models. Pyrroloquinoline quinine (**a**); riluzole (**b**); minocycline (**c**); and coenzyme $Q_{10}$ (**d**)

ability to protect against neurodegenerative disease and potentially extend life span. Rotenone has been used to develop cell culture and animal models of PD. Thus, treatment of neuroblastomas SH-SY5Y cells with rotenone induces apoptotic cell death. Treatment with commercial extracts of *Anemopaegma mirandum* (Catuaba), a Brazilian tree and *Valeriana officinalis*, leads to preservation of mitochondrial membrane and protection from apoptotic indicating that *A. mirandum* extract has chemical component that stabilizes mitochondrial membrane integrity (Valverde et al., 2008; de Oliveria et al., 2009).

## *9.4.4 Statins and PD Treatment*

As mentioned above, L-DOPA treatment of PD often causes debilitating involuntary movements, termed L-DOPA-induced dyskinesia which in rodent is called as abnormal involuntary movements (AIMs). AIMs involve the activation of the Ras extracellular signal-regulated kinase 1/2 (ERK1/2) mitogen-activated protein kinase signaling pathway. Lovastatin, a specific inhibitor of HMG-CoA reductase, prevents Ras isoprenylation activity and subsequently phosphorylation of ERK1/2 (pERK1/2) (Schuster et al., 2008). It is suggested that lovastatin treatment before L-DOPA exposure reduces AIM incidence and severity of PD in the 6-hydroxydopamine (6-OHDA) rat model suggesting that lovastatin may represent a treatment option for managing L-DOPA-induced dyskinesia in PD (Schuster et al., 2008). In mice, orally administered simvastatin enters into the substantia nigra, inhibit nigral activation of ras (p21), attenuates nigral activation of NF-κB, blocks nigral expression of proinflammatory molecules, and suppresses nigral activation of glial cells (Ghosh et al., 2009). These observations parallel dopaminergic neuronal protection, normalization of striatal neurotransmitters, and improvement of motor functions in MPTP-intoxicated mice (Ghosh et al., 2009). In PD, statins not only improve blood flow and reduce coagulation but also modulate the immune system and reduce oxidative damage due to their antioxidant effects (Farooqui, 2009a). In transgenic mouse, lovastatin administration reduces α-synuclein aggregation and associated neuropathology of PD and supports the possibility that treatment with cholesterol-lowering drugs may be beneficial for patients with PD and/or DLB (Koob et al., 2010). Studies on the effect of statins and fibrates in a cohort of 419 patients with PD indicate that treatment with a statin or a fibrate delays the mean age of disease onset of PD by nearly 9 years, when compared with control (PD patients not taking statins or fibrate) (Mutez et al., 2009). However, it should be noted that inhibition of HMG-CoA reductase by statins may also result in the deficiency of CoQ10. This inhibition has been implicated in the pathophysiology of the myotoxicity associated with this pharmacotherapy by statins. CoQ10 and its analog, idebenone, have been widely used in the treatment of neurodegenerative and neuromuscular disorders. These compounds may potentially play a role in the treatment of PD (Mancuso et al., 2010). Another drawback of statin therapy is that statin-mediated reduction in cholesterol levels may deplete the cholesterol-rich membrane domains known as lipid rafts, which in turn could affect cellular signaling. Collective evidence suggests that long-term use of statins may produce protective as well as potential harmful effects in PD patients. Although the appropriateness of statin therapy in PD is not established at this time, beneficial effects of statins are more than harmful effects thereby statin therapy may be a novel approach for the treatment of PD.

### *9.4.5 Memantine and PD Treatment*

Although PD primarily involves abnormalities in motor function, in recent years increasing emphasis is being placed on cognitive and behavioral impairment in this disorder. Depression, dementia, and psychosis along with cognitive, autonomic, and sensory disturbances in PD have a major impact on quality of life for both patients and families. Studies on the treatment of PD with memantine have been controversial. Some studies demonstrate that prolonged memantine treatment in PD patients with dementia improves cognitive functions and preserves motor functional abilities as well as ameliorates neuropsychiatric symptoms, especially in patients with hyperhomocysteinemia (Aarsland et al., 2009; Leroi et al., 2009). In contrast, other groups have indicated that there is no significant change of PD and its psychotic symptoms severity (Menendez-Gonzalez et al., 2005; Ridha et al., 2005).

Memantine (Fig. 9.8) reduces the loss of dopamine neurons in the substantia nigra pars compacta in animal models of PD. Although molecular mechanism associated with neurotrophic and neuroprotective effects of memantine in dopaminergic neurons in cultures and animal models is not fully understood, recent studies indicate that memantine effect is mediated through astrocytes and not through microglia or neurons (Wu et al., 2009). Treatment of neuron-enriched cultures with memantine results in increased expression and secretion of glial cell line-derived neurotrophic factor (GDNF) along with histone hyperacetylation by inhibiting the cellular histone deacetylase activity. In addition, memantine also blocks the microglia activation (Wu et al., 2009). This process leads to reduction in proinflammatory factor production. Collective evidence suggests that the neuroprotective effects of memantine in cell culture are mediated in part through alternative novel mechanisms by reducing microglia-associated inflammation and by stimulating neurotrophic factor release from astroglia (Wu et al., 2009).

Amantadine has also been used for the treatment of PD-related dementia (Greulich and Fenger, 1995). Amantadine has been reported to reduce the duration of levodopa-induced dyskinesia and improves motor disability in PD. Although some beneficial effects are noted in patients after amantadine treatment, this drug is known to induce corneal edema that begins few months to several years after institutional therapy (Jeng et al., 2008).

### *9.4.6 PPAR Agonists and PD Treatment*

It is well known that peroxisome proliferator-activated receptor (PPAR) agonists (Fig. 9.9) modulate inflammatory responses in the brain (Chaturvedi and Beal, 2008). The PPARγ agonist, pioglitazone protects neurons in MPTP-intoxication mouse model of PD (Breidert et al., 2002). This drug acts against MPTP-induced neurotoxicity not only by inhibiting the conversion of MPTP to its active toxic metabolite $MPP^+$, via inhibition of monoamine oxidase-B (Quinn et al., 2008) and protecting tyrosine hydroxylase in substantia nigra, but also by reducing levels of iNOS in microglia and astrocytes, increasing the expression of the inhibitory protein-k-Bx and decreasing translocation of NF-kB in striatal neurons (Dehmer et al., 2004; Chaturvedi and Beal, 2008). Alterations in above signaling mediators are closely related to apoptosis, inflammation, oxidative damage, and proteosomal/mitochondrial dysfunction. Collective evidence suggests that PPARγ agonists activate PPARs, which bind to the DNA-binding site on promoters and regulate the expression of several target genes involved in the cell survival and neuroprotection. PPARγ agonists inhibit the catechol oxidation and oxidative damage by increasing the antioxidant enzyme expression and by retarding lipid peroxidation.

Some but not all non-steroidal anti-inflammatory drugs (NSAID), such as indomethacin, ibuprofen, naproxen, and fenoprofen, specifically bind and activate PPARβ and/or PPARγ (Sastre et al., 2006a, b; Heneka et al., 2007). These drugs have been recently tested for their efficacy to protect neurons in PD (Chen et al.,

2005; Esposito et al., 2007; Heneka and Landreth, 2007; Heneka et al., 2007). Several epidemiological studies have suggested an association between regular intake of PPAR activating NSAIDs and reduced prevalence of PD and AD (Chen et al., 2005; Esposito et al., 2007; Hencka et al., 2007). Regular intake of ibuprofen shows a 35% reduced risk of PD as compared to non-users (Esposito et al., 2007). Double-blind clinical trials are needed to test the efficacy of PPARγ agonists in PD patients.

### 9.4.7 Neurotrophins and PD Treatment

Among neurotrophic molecules, the potential of the glial cell line-derived neurotrophic factor (GDNF) and neurturin to protect the nigral dopaminergic neurons and/or rescue striatal dopamine levels has been extensively studied in animal models and PD patients (Hong et al., 2008; Ramaswamy et al., 2009; Herzog et al., 2009). Monthly administration of GDNF in bolus intracerebroventricular provides no clinical benefits probably because of the limited penetration of GDNF into the target brain areas (Patel and Gill, 2007). Thus, results from several controlled clinical studies delivering the GDNF directly into brain have not demonstrated efficacy and safety of this treatment. A major problem in clinical studies has been delivery. GDNF delivered by intracerebroventricular injection in patients has been ineffective, probably because GDNF did not reach the target. Administration of GDNF in the putamen and intraputamenal infusion were also ineffective, probably because of limited distribution within the putamen. A randomized clinical trial with gene therapy for neurturin is underway at the present time in an attempt to overcome these problems with targeting and distribution of neurturin (Peterson and Nutt, 2008). It is proposed that gene therapy may be a novel procedure for increasing local levels of GDNF by transplantation of GDNF-producing cells (carotid body cell aggregates or different genetically modified cells, including stem cells) and in vivo gene therapy utilizing recombinant adeno-associated viral vectors or lentivirus vectors for the treatment of PD (Lindvall and Wahlberg, 2008; Minguez-Castellanos and Escamilla-Sevilla, 2005; Mochizuki, 2009). These strategies are aimed not only to deliver GDNF to restore the lost brain functions but also in maintaining levels of neurotransmitters by boosting the function of remaining neurons in the PD brain. Postmortem analysis of patients who received fetal brain cell transplants indicates that implanted cells are prone to degeneration just like endogenous neurons in the same pathological area. This observation suggests that long-term efficacy of neural cell therapy in PD needs to overcome the degenerating environment in the brain. Thus, more studies are required to determine the efficacy of GDNF in PD patients.

### 9.4.8 ω-3 Polyunsaturated Fatty Acids and PD Treatment

ω-3 polyunsaturated fatty acids (Fig. 9.10) have neuroprotective effects in several neurodegenerative diseases including animal models of PD (Farooqui, 2009b).

Exposure of control and high ω-3 fatty acid fed mice to MPTP neurotoxicity indicates that ω-3 fatty acid fed mice are completely protected from MPTP neurotoxicity (Bousquet et al., 2008). Furthermore, dietary consumption of ω-3 fatty acid not only prevents the MPTP-mediated decrease in tyrosine hydroxylase-labeled nigral cells but also downregulates Nurr1 mRNA and dopamine transporter mRNA levels in the substantia nigra (Bousquet et al., 2008). Although ω-3 fatty acids dietary treatment has no effect on striatal dopaminergic terminals, high levels of ω-3 fatty acids in diet protect against the MPTP-mediated decrease in dopamine and its metabolite, dihydroxyphenylacetic acid, in the striatum. These observations suggest that consumption of high ω-3 fatty acid containing diet produces neuroprotective effects in an animal model of Parkinsonism (Bousquet et al., 2008). Similarly, chronic dietary supplementation fish oil, which contains ω-3 fatty acids, protects rat against 6-hydroxydopamine (6-OHDA) toxicity compared to control rats fed with commercially available diet (Delattre et al., 2009). Moreover, ω-3 fatty acid consuming rats show a marked reduction in rotational behavior caused by apomorphine, indicating retardation of dyskinesia behavior. Although in 6-OHDA model, ω-3 fatty acids neither alter tyrosine hydroxylase immunoreactivity in the substantia nigra pars compacta and in the ventral tegmental area nor deplete dopamine (DA) and its metabolites in the striatum, they markedly increase DA turnover suggesting that ω-3 fatty acid supplementation promotes DA turnover in the surviving neurons without modifying neuronal population (Delattre et al., 2009). Although the molecular mechanism associated with neuroprotective effects of ω-3 fatty acids is not fully understood, there are several possibilities. ω-3 fatty acids not only act as antioxidants but also bind with α-synuclein to interfere with its aggregation, a process closely associated with the pathogenesis of PD (Muntane et al., 2010). ω-3 fatty acids produce anti-inflammatory effects through the generation of neuroprotectins and resolvins (Bazan, 2006, 2007; Serhan, 2005; Farooqui, 2009a). Moreover, ω-3 fatty acids inhibit the synthesis and release of proinflammatory cytokines such as TNF-α and IL-1β and IL-2 that are released during induction and maintenance of inflammatory processes in the early course of PD (Farooqui, 2009b). In addition, ω-3 fatty acids have antidepressant effect in PD patients and improve their quality of life. DHA also reduces the severity or delay the development of levodopa-induced dyskinesias in MPTP-induced model of PD in monkeys (Samadi et al., 2006). Collective evidence suggests that DHA is not a drug, but a supplement that exerts beneficial effects through multiple mechanisms, including (a) regulation of the expression of potentially neuroprotective genes, (b) activation of anti-inflammatory pathways, and (c) modulation of neurotransmitters levels.

## 9.5 Therapeutic Approaches for ALS

The molecular mechanism associated with the pathogenesis of ALS still remains elusive, but oxidative stress, mitochondrial impairment, protein misfolding, cytoskeletal abnormalities and defective axonal transport, excitotoxicity, inflammation, growth factor deficiency, and apoptotic cell death have been closely associated with the pathogenesis of ALS. There is substantial evidence to support the hypothesis that oxidative stress is one mechanism by which degeneration of motor neurons can occur. This theory is supported by the discovery that mutation of the antioxidant enzyme, superoxide dismutase 1 (SOD1), causes disease in the familial ALS. However, the precise mechanism(s) by which mutant SOD1 leads to motor neuron degeneration have not been defined with certainty. Like other neurodegenerative diseases, treatments of ALS are symptomatic. Common therapeutic approaches include antiexcitotoxic, antioxidant and anti-inflammatory strategies, stabilization of mitochondrial dynamics, use of statins, memantine, ω-3 fatty acids, and gene therapy.

### *9.5.1 Riluzole and Memantine and ALS Treatment*

Riluzole, an antiglutamatergic agent with anticonvulsant properties (Fig. 9.12), is the only drug for the treatment of ALS approved by the food and drug administration. Riluzole acts through several mechanisms including the inhibition of glutamate release in the caudate nucleus (Cheramy et al., 1992), blockage of sodium channels in myelinated fibers (Benoit and Escande, 1991), and modulation of sodium current and late component of outward potassium current in cultured neurons (Zona et al., 1998). Furthermore, riluzole also activates a G protein-dependent processes that inhibit glutamate release (Doble et al., 1992). Two trials indicate that the drug is well-tolerated, efficacious and safe, and lengthens survival of patients with ALS by 3–6 months (Bensimon et al., 1994). The most common adverse reactions include asthenia, dizziness, gastrointestinal disorders, and increase in liver enzymic activities. These side effects occur at the 200 mg dose (Lacomblez et al., 1996).

As stated above, memantine is a non-competitive NMDA receptor antagonist. It protects neurons against NMDA- or glutamate-induced toxicity in vitro and in animal models of neurodegenerative diseases. Although in SOD1(G93A) mice, memantine neither delays the onset of rotarod deficits nor slows down the deterioration rate of motor performance at the initial phase of the disease, memantine treatment delays the progression of the disease to the paresis stage to a small extent relative to the saline-treated group. Most importantly, memantine significantly delays the hind limb paralysis and prolongs the survival of SOD1(G93A) mice compared to untreated SOD1(G93A) (Wang and Zhang, 2005). The delays in the hind limb paralysis in memantine-treated animals may be due to the inhibition of NMDA receptors in spinal motor neurons. Alternatively, memantine may attenuate the glutamate-induced damage of other spinal cord cells whose survival may prolong the longevity of motor neurons.

Recently, the kynurenine pathway (KP) has been implicated in the pathogenesis of ALS (Chen et al., 2009). The KP is a major route that metabolizes tryptophan and generates its neuroactive intermediates. The metabolic products of tryptophan include quinolinic acid (QUIN), a NMDA receptor agonist and kynurenic acid (KYNA), a neuroprotective NMDA receptor antagonist. In addition to the

NMDA receptor antagonism, an important feature of KYNA is the blockade of the α7-nicotinic acetylcholine receptor and its influence on the α-amino-3-hydroxy-5-methylisoxazole-4-proprionic acid receptor. Kynurenic acid has proven to be neuroprotective in several experimental settings. In contrast, QUIN is a potent neurotoxin that also has free radical-generating properties. In addition, these metabolites not only aid communication between the nervous and immune systems but also modulate cell proliferation. At the present time, KP inhibitors, teriflunomide (Sanofi-Aventis) and laquinimod (Teva Neuroscience), have entered clinical trials (Chen et al., 2009). In addition, the 8-hydroxyquinolinine metal attenuating compounds, clioquinol and PBT2, are similar to KYNA and QUIN. These metabolites can be used as therapeutic agents in future studies.

Cannabinoids also induce antiglutamatergic and anti-inflammatory effects through activation of the $CB_1$ and $CB_2$ receptors, respectively (Bilsland and Greensmith, 2008). Activation of $CB_1$ receptors may therefore retard glutamate release from presynaptic nerve terminals and reduce the postsynaptic calcium influx in response to glutamate receptor stimulation. Meanwhile, $CB_2$ receptors may influence inflammation, whereby receptor activation reduces microglial activation, resulting in a decrease in microglial secretion of neurotoxic mediators (Bilsland and Greensmith, 2008). In addition, cannabinoids may also produce antioxidant activity by a receptor-independent mechanism. It is proposed that the ability of cannabinoids to treat ALS should be carefully considered in ALS patients (Bilsland and Greensmith, 2008). Other antiexcitotoxic drugs that have been used for the treatment of ALS patient include branched chain amino acids (BCAA), dextromethorphan, L-threonine, topiramate, lamotrigine, ceftriaxone, and talampanel.

### *9.5.2 Antioxidant Strategies and ALS Treatment*

Oxidative stress is closely associated with the pathogenesis of ALS. Several antioxidants including vitamin E, acetylcysteine, methylcobalamin, and glutathione have been tested in clinical trials of ALS. Coenzyme Q10 (CoQ10) is an antioxidant and mitochondrial cofactor (Fig. 9.12). It not only prevents lipid peroxidation but also stabilizes the $Ca^{2+}$ channels. Its oral administration increases CoQ10 levels in mitochondria. CoQ10 produces promising effects in ALS transgenic mice and in clinical trials for neurodegenerative diseases other than ALS. Phase II clinical trial of CoQ10 in ALS patients indicate that this drug is ineffective in producing beneficial effects in ALS patients (Levy et al., 2006), and opinions on phase III trials have been controversial (Kaufmann et al., 2009).

### *9.5.3 Stabilization of Mitochondrial Dynamics and ALS Treatment*

ALS animal models and patients show abnormal mitochondrial morphology. Early stages of ALS involve mitochondrial cristae remodeling and matrix vesiculation in ventral horn neuron dendrites. Motor neurons cell bodies accumulate mitochondria derived from the distal axons projecting to skeletal muscle. Incipient disease in spinal cord involves enhancement in oxidative and nitrosative stress, indicated by protein carbonyls and nitration of cyclophilin D and adenine nucleotide translocator. Reducing the levels of cyclophilin D by genetic ablation significantly delays ALS onset and extends the life span of G93A-mSOD1 mice expressing high and low levels of mutant protein in a gender-dependent pattern (Martin et al., 2009; Petri et al., 2006; Zhou et al., 2010). In an another ALS model, G93A mouse, skeletal muscle fibers show localized loss of mitochondrial inner membrane potential in fiber segments near the neuromuscular junction. These defects occur in young G93A mice prior to disease onset, indicating that mitochondrial dynamics may also be disrupted in animal models of ALS. Studies on the effect of a novel peptide antioxidant (SS-31) that targets the inner mitochondrial membrane in the G93A mouse model of amyotrophic lateral sclerosis (ALS) indicate that SS-31 produces beneficial effects. Daily intraperitoneal injections of SS-31 starting at 30 days of age produce a significant improvement in survival and motor performance compared to vehicle-treated G93A mice (Petri et al., 2006). Furthermore, in comparison with vehicle-treated G93A mice, SS-31-treated mice exhibit not only a decrease in cell loss but a decrease in immunostaining for markers of oxidative stress in the lumbar spinal cord, supporting the view that treatment of ALS in animal model may still be possible through the inhibition of oxidative stress.

### *9.5.4 Neurotrophins and ALS Treatment*

Insulin-like growth factor-I (IGF-I) is a potent neurotrophic factor that has neuroprotective properties in the central and peripheral nervous systems. IGF-I has been reported to specifically enhance the extent and rate of motor neuron axonal outgrowth by acting through the IGF-I receptor and downstream signaling pathways. In contrast, BDNF enhances branching and arborization but not axon outgrowth of corticospinal motor neurons (Ozdinler and Macklis, 2006). In vitro and in vivo studies on the efficacy of IGF-I in ALS model systems indicate that IGF-I not only stimulates neurogenesis but also prolongs the life span by lowering the ALS progression in murine models of ALS. Systemic delivery of human recombinant IGF-I in clinical trials does not produce beneficial clinical effects in ALS patients. Lack of beneficial effects in ALS patients may be either due to the inactivation of IGF-I by IGF binding proteins (IGFBPs) or caused by limited delivery of IGF-I to motor neurons (Wilczak and de Keyser, 2005; Sakowski et al., 2009; Howe et al., 2009). Studies on transplantation of neural progenitor cells (NPs) engineered to express BDNF or IGF-1 or GDNF indicate that transplantation of GDNF- or IGF-1-expressing hNPs not only differentiates and attenuates the loss of motor neurons but also induces trophic changes in motor neurons of the spinal cord without adverse behavioral effects (Park et al., 2009). Although genetically modified hNPs survive, integrate, and release IGF-I or GDNF in the spinal cord, no improvement in motor performance and extension of life span is observed in all hNP transplantation groups compared to vehicle-injected controls (Park et al., 2009).

### 9.5.5 ω-3 Fatty Acids and ALS Treatment

Treatment of primary sensory neuronal cultures from aged animals with ARA, EPA, and DHA results in the stimulation of neurite outgrowth. ω-3 fatty acids, in particular DHA, show a remarkable effect on neurite outgrowth. The amplitude of ω-3 fatty acid effect is comparable to nerve growth factor and all-*trans*-retinoic acid adult and aged animals (Robson et al., 2008). ω-3 fatty acids also modulate various neurotransmitter receptors and are metabolized to docosanoids, which inhibit neuroinflammation (Farooqui et al., 2007b; Farooqui, 2009b). Based on these observations, it is suggested that DHA and EPA may be used for the treatment of ALS in animal models (Farooqui, 2009b). A combination of DHA, glycerophosphocholine, acetyl L-carnitine, and PtdSer may protect neural cells from neuronal cell death in ALS (Kidd, 2005). Clinical trials for the treatment of ALS with ω-3 fatty acids have not been performed in humans and are now urgently needed to judge the efficacy of ω-3 fatty acids.

### 9.5.6 Immunotherapy and ALS Treatment

Studies on antibody-mediated clearance of amyloid plaques in a transgenic mouse model of AD have encouraged investigators to try similar approach for the treatment of other neurodegenerative diseases, including ALS (White and Hawke, 2003). Studies on active and passive immunization of Mutant SOD1 indicate that vaccination can delay the onset and prolonging the life span of ALS mutant mice. Vaccination induces diverse inflammatory reactions, which are reported to modify both the onset and the progression of ALS (Urushitani, 2009). Passive immunization is also promising since mutant SOD1 can be targeted using a specific SOD1 monoclonal antibody. It is proposed that the development of the current immunization techniques is important in developing immunotherapy for ALS (Urushitani, 2009).

## 9.6 Therapeutic Approaches for HD

HD is a hereditary autosomal dominant neurodegenerative disease characterized by expended CAG repeats in the Huntingtin (Htt) gene. Mutated huntingtin (mHtt) has expended polyglutamine stretch and causes misfolding and conformational alterations of the disease-causing proteins, leading to pathogenic protein–protein interactions, including aggregate formation, and subsequently resulting in their deposition as inclusion bodies in affected neurons. The major symptoms of HD include chorea, progressive dementia, and psychiatric manifestations, such as depression, irritability, apathy, and psychosis. Native Htt is associated with synaptic vesicles and/or microtubules, where it is involved in vesicular transport and/or the binding to the cytoskeleton. In contrast, mHtt induces mitochondrial respiratory chain dysfunction and with other proteins promotes its own polymerization to form insoluble aggregates. These intraneuronal aggregates may induce transcriptional dysregulation, calcium dyshomeostasis, abnormal vesicle trafficking, defective mitochondrial bioenergetics and alterations in protein transport inside the nucleus and cytoplasm, and the vesicular transport (Reddy et al., 2009). There is no cure for HD, but current therapies of HD include suppression of mutant gene expression by RNAi, inhibition of protein misfolding/aggregation, activation of mitochondrial function, inhibition of neuronal cell death, and neuroprotection by neurotrophic factors.

### *9.6.1 Gene Silencing and HD Treatment*

Decreasing the levels of the mHtt may be the best strategy for the treatment of HD. Downregulation of abnormal gene expression has been demonstrated in a tetracycline-regulated mouse model (Yamamoto et al., 2000) as well as doxycycline-regulated SCA1 mouse models of HD (Zu et al., 2004). Nuclear inclusions, which appear after induction of the mHtt expression, disappear when the expression is blocked. In addition, behavioral abnormalities are considerably decreased. RNA interference (RNAi) and short hairpin have emerged as a potential therapeutic approach for neurodegenerative diseases, particularly those associated with autosomal dominant patterns of inheritance (Bonini and La Spada, 2005). Although this strategy may not be feasible and used in humans, it has been used to obtain important information in transgenic mice models of HD.

### *9.6.2 Enhancement of Protein Degradation and HD Treatment*

Enhancing the catabolism of mHtt is another way to decrease its toxicity. Studies in N171-82Q transgenic mice and in the *Drosophila* and the zebrafish models of HD indicate that the stimulation of mHtt degradation and induction of autophagy with the mTOR inhibitor rapamycin accelerates clearance of mHtt (Williams et al., 2008).

The activation of both processes involves mammalian target of rapamycin-dependent (e.g., by rapamycin analog CCI-779) or -independent (e.g., by lithium and calpain inhibitors) pathways. Combination treatment of these processes provides additive protection against polyQ-mediated-related neurodegeneration (Sarkar et al., 2008).

Amiloride and benzamil have also been used to reduce the polyQ aggregation and mHtt toxicity in HD models. Benzamil increases the life span of R6/2 mice by inhibiting polyQ aggregation, reducing motor deficits, and alleviating the inhibition of the ubiquitin-proteasome system (UPS) activity, leading to enhanced degradation of soluble htt-polyQ specifically in its pathological range (Wong et al., 2008). Another drug, Y-27632, a rho-associated kinases inhibitor increases UPS activity and reduces polyQ. It has been used in clinical trials of ischemia and hypertension (Lai and Frishman, 2005). This drug also increases the macroautophagy activity and its effect is mediated through the catabolic pathway, which reduces aggregation of mutant htt, ataxin-3, AR, and atrophin-1 in cell systems (Bauer et al., 2009). Collective evidence suggests that enhancement in the degradation of mHtt may prolongs the survival of neurons in animal models of HD.

### *9.6.3 Inhibition of Aggregation and HD Treatment*

Several small molecules, such as Congo Red, trehalose, and cystamine, prevent aggregation of huntingtin by blocking the formation of polyQ aggregates in animal models of HD. Thus, treatment of R6/2 mice with Congo Red or trehalose significantly increases the mice survival by 16.4 and 11.3%, respectively (Sanchez et al., 2003; Tanaka et al., 2004). Congo Red and trehalose activate macroautophagy in a rapamycin-independent manner (Sarkar et al., 2007). Cystamine reduces expanded polyQ aggregation by inhibiting TG that is thought to cross-link expanded polyQ proteins and facilitate their aggregation.

### *9.6.4 Creatine and Other Antioxidants and HD Treatment*

Based on progressive weight loss and metabolic defects in brain and muscle, a mechanistic link between cellular energetic defects and the pathogenesis of HD has been proposed long ago, and mitochondrial complex II inhibitors have been used to generate acute toxicity models that show many features of HD. Several drugs that improve energy metabolism defects or reduce oxidative stress mediated by polyQ aggregates have been successfully used to retard HD symptoms in mouse models. Dietary creatine supplementation significantly improves survival, slows the development of brain atrophy, and delays atrophy of striatal neurons, stabilizes the mitochondrial permeability transition, retards ATP depletion, and enhances the protein synthesis and the formation of huntingtin-positive aggregates in R6/2 mice. In addition, creatine treatment in R6/2 mice retards HD pathology, improves the phenotype, and increases the life span by 17.4% (Ferrante et al., 2000). Similarly in N171-82Q HD mice, creatine not only reduces brain atrophy and the formation of intranuclear inclusions but attenuates reductions in striatal *N*-acetylaspartate, delays the development of hyperglycemia, and increases the survival rate by 19.3% (Andreassen et al., 2001). Other drugs, such as dichloroacetate and triacetyluridine, produce beneficial effects in R6/2 mice (Saydoff et al., 2006).

Coenzyme Q10, an antioxidant and cofactor of the mitochondrial electron transport chain (Fig. 9.12) and inhibitor of mitochondrial permeability transition, is efficacious in R6/2 mouse lines. High-dose CoQ10 significantly extends survival in R6/2 mice. CoQ10 produces a marked improvement in motor performance and grip strength, with a reduction in weight loss, brain atrophy, and huntingtin inclusions in treated R6/2 mice (Smith et al., 2006). CoQ10 treatment increases its brain levels, while levels of 8-hydroxy-2-deoxyguanosine are decreased. Increase in CoQ10 not only extends the survival of R6/2 mice by up to 26.3% but greatly improves the motor performance and reduces weight loss and nuclear inclusions (Smith et al., 2006). Clioquinol treatment of transgenic Huntington's mice (R6/2) improves behavioral parameters, decreases huntingtin aggregation and accumulation, decreases striatal atrophy, improves rotarod performance, reduces weight loss, normalizes blood glucose and insulin levels, and facilitates extension of life span (Nguyen et al., 2005).

### 9.6.5 Minocycline and HD Treatment

Minocycline, a second-generation tetracycline (Fig. 9.12) that decreases activation and proliferation of microglia and macrophages, not only inhibits the release of apoptosis inducing factor, proapoptotic protein Smac/Diablo, and cytochrome *c* from mitochondria, but also downregulates the cleavage of proapoptotic factor Bid and activating caspases-1, -3, -8, and -9 (Wang et al., 2003). Intraperitoneal injections of minocycline in R6/2 mice extend its life span by 13.5% (Chen et al., 2000). In another study, minocycline and coenzyme Q10 improve survival of R6/2 mice 11.2 and 14.6%, respectively. Combined minocycline and CoQ10 therapy produces an enhanced beneficial effect, ameliorating behavioral and neuropathological alterations in the R6/2 mouse. Minocycline and CoQ10 treatment significantly extends survival and improves rotarod performance to a greater degree than either minocycline or CoQ10 alone. In addition, combined minocycline and CoQ10 treatment attenuated gross brain atrophy, striatal neuron atrophy, and huntingtin aggregation in the R6/2 mice relative to individual treatment (Stack et al., 2006). A small clinical trial minocycline has provided promising data in HD patients (Bonelli et al., 2004) and bigger and long-term clinical trials are needed to judge the clinical efficacy of minocycline in HD patients.

### 9.6.6 ω-3 Fatty Acids and HD Treatment

Randomized, placebo-controlled, double-blind studies show that EPA has beneficial effects in HD patients (Puri, 2005; Das and Vaddadi, 2004; Murck and Manku, 2007). Although the molecular mechanism associated with beneficial effects of EPA is not known, it is suggested that fatty acids may prevent or block polyQ aggregation, inhibit histone deacetylation, activate the ubiquitin-proteasome system, and restore mitochondrial dysfunction (Das and Vaddadi, 2004; Murck and Manku, 2007).

## 9.7 Therapeutic Approaches for Prion Diseases

Prion particle infection-mediated self-perpetuating conformational conversion of the cellular prion protein $PrP^{C}$ into the β-sheet-rich "scrapie" conformer $PrP^{Sc}$ is the central molecular event in the pathogenesis of a group of prion diseases (Aguzzi et al., 2008). Despite intense research, both the physiological function of $PrP^{C}$ and the molecular pathways leading to neurodegeneration in prion disease remain unknown. A large number of compounds have been identified as anti-prion agents and capable of reducing the $PrP^{Sc}$ levels in infected cells. These compounds include polyphenols (e.g., tannic acid and tea extracts), phenothiazines, antihistamines, statins, and antimalarial compounds. Among 17 compounds that have been evaluated in a solid-phase cell-free hamster PrP conversion assay, only one compound, polyphenols, inhibits the cell-free reaction with $IC_{50}$ of near 100 nM. Several of the new $PrP^{Sc}$ inhibitors cross the blood–brain barrier and thus have potential to be effective in prion diseases. However, none of these compounds have proven to be therapeutically effective against prion diseases (Sakaguchi, 2009).

### *9.7.1 Pentosan Polysulfate for the Treatment of Prion Diseases*

Pentosan polysulfate (PPS) is a large polyglycoside molecule (Fig. 9.13) with weak heparin-like activity. It has been shown to prolong the incubation period of the intracerebral infection when administered to the cerebral ventricles in a rodent scrapie model (Tsuboi et al., 2009; Doh-ura, 2009). PPS interacts with heparin binding sites on proteins and alters their physiological actions. PPS acts as a prophylactic agent against infection with prions both in vivo and in vitro (Dealler and Rainov, 2003). In addition, PPS has mild anticoagulant, anti-inflammatory, fibrinolytic, and hypolipidemic properties. In cell culture models, PPS also retards the production of further PrP(sc). These properties of PPS have encouraged its cerebroventricular administration in a young man with vCJD. Long-term continuous infusion of PPS for 18 months did not cause drug-related side effects. Follow-up CT scans show progressive brain atrophy during PPS administration (Todd et al., 2005). Although more human studies have been conducted in European countries and Japan, intraventricular PPS, no beneficial effects have been observed in patients with prion diseases (Doh-ura, 2009).

### *9.7.2 Quinacrine for the Treatment of Prion Diseases*

Quinacrine, an antimalarial drug (Fig. 9.13), inhibits $PrP^{Sc}$ formation. Quinacrine has been used for the treatment of Creutzfeldt–Jakob disease (CJD) (Korth et al., 2001; Love, 2001; May et al., 2003; Follette, 2003; Kobayashi et al., 2003; Doh-ura, 2004). The molecular mechanism involved in the inhibition of $PrP^{sc}$ formation

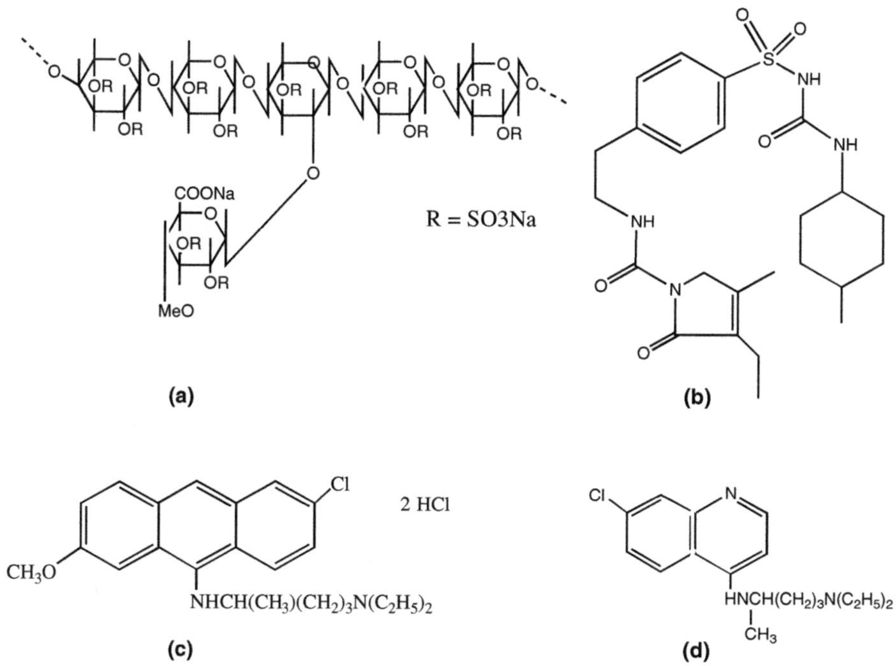


**Fig. 9.13** Chemical structures of drugs used for the treatment of prion disease. Pentosan polysulfate (**a**); glimepiride (**b**); quinacrine (**c**); and chloroquine (**d**)

by quinacrine remains unknown. However, it has been suggested that quinacrine blocks PrP (106–126)-mediated formation of channels (Farrelly et al., 2003). NMR spectroscopic studies indicate that the $PLA_2$ inhibitor, quinacrine, binds to human prion protein at Tyr225, Tyr226, and Gln227 residues of helix α3 (Vogtherr et al., 2003). Similarly, other antimalarial drugs, such as chloroquine and the phenothiazine derivatives (acepromazine, chlorpromazine, and promazine), also bind to prion protein between residues 121 and 230, suggesting that Tyr225, Tyr226, and Gln227 residues are necessary for the binding of antimalarial drugs and phenothiazine derivatives (Vogtherr et al., 2003) to $PrP^c$. It has also been reported that quinacrine acts as an antioxidant and reduces the toxicity of PrP106–126 (Turnbull et al., 2003). This once again suggests that the release of arachidonic acid and the oxidative stress generated by altered arachidonic acid metabolism may play an important role in the pathogenesis of prion diseases (Guentchev et al., 2000; Milhavet et al., 2000). In addition, quinacrine is a potent non-specific inhibitor of $PLA_2$ and also blocks PrP106–126-mediated stimulation of NMDA receptor (Stewart et al., 2001), indicating that multiple mechanisms may be involved in beneficial effects of quinacrine in prion disease patients.

### *9.7.3 Glimepiride for the Treatment of Prion Diseases*

Treatment with glimepiride, a third generation antidiabetic sulfonylurea approved for the treatment of diabetes mellitus (Fig. 9.13), releases of $PrP^C$ from the site where $PrP^C$ is converted into $PrP^{Sc}$. In vitro studies indicate that glimepiride retards $PrP^{Sc}$ formation in three prion infected neuronal cell lines (ScN2a, SMB and ScGT1 cells) (Bate et al., 2009). Glimepiride not only protects cortical and hippocampal neurons against the toxic effects of the prion-derived peptide PrP82–146 and significantly reduces the amount of PrP82–146 that binds to neurons but also blocks PrP82–146-mediated activation of $cPLA_2$ and the synthesis of prostaglandin $E_2$ that is associated with neuronal injury in prion diseases (Bate et al., 2009). In addition in cultured cells of neural and non-neural origin (adipocytes), glimepiride modulates the release and translocation of glycosylphosphatidylinositol (GPI)-anchored proteins (including PrPC) from plasma membrane lipid rafts to intracellular lipid droplets and this process is facilitated by a GPI-specific phospholipase C (GPI-PLC) and inhibited by a GPI-PLC inhibitor, *p*-chloromercuriphenylsulfonate. In addition, glimepiride upregulates PI3K/Akt-mediated signaling in heart and endothelial cells. Inhibition of PrPc to PrPSc conversion through the modulation of $PLA_2$ and PLC signaling suggests that glimepiride may be used for the treatment of prion diseases.

### *9.7.4 Vaccine for the Treatment of Prion Diseases*

Antibody-based immunotherapy represents an important approach for treating prion diseases, providing antibodies to the cellular prion protein $PrP^C$ can antagonize the conversion and deposition of $PrP^{Sc}$ in in vitro assays and in laboratory animals. However, induction of protective anti-prion immune responses in wild-type animals is difficult because of host tolerance to the endogenous $PrP^C$ (Heppner and Aguzzi, 2004).

In order to develop an anti-prion vaccine, a novel DNA fusion vaccine composed of mouse PrP and immune stimulatory helper T-cell epitopes of the tetanus toxin has been developed (Nitschke et al., 2007). This approach provokes a strong $PrP^C$-specific humoral and cellular immune response in PrP null mice, but only low antibody titers are found in vaccinated wild-type mice. Furthermore, prime-boost immunization with the DNA vaccine and recombinant PrP protein increased antibody titres in PrP null mice, but failed to protect wild-type mice from mouse scrapie (Nitschke et al., 2007).

Recent studies indicate that it is possible to overcome tolerance to $PrP^C$ and induce immune responses to bacterially expressed, recombinant PrP. Nevertheless, in vivo deleterious side effects of injected anti-PrP antibodies have been reported, mainly due to their Fc fragments and divalence. Removal of Fc fragments has no effect on prion replication inhibiting activity of Fabs in infected neuronal cells (Alexandrenne et al., 2009). It is suggested that for immunotherapy of prion diseases, the development and use of monovalent antibodies will be better than polyclonal antibodies (Alexandrenne et al., 2009). Towards this end, highly potent monoclonal antibodies (mAbs) have been raised in mice in which the prion protein gene has been deleted by gene targeting (Muller-Schiffmann and Korth, 2008). These mAbs show anti-prion activity not only in permanently scrapie-infected neuroblastoma (ScN2a) cells but also in vivo when injected intraperitoneally in mice. These studies also indicate that mAbs do not pass through blood–brain barrier (BBB). Thus, more studies are needed on immunotherapy of prion diseases.

## 9.8 Conclusion

Neurodegenerative diseases are a heterogeneous group of diseases, such as AD, PD, ALS, and HD. These diseases occur in sporadic and familial forms. Genetic and environmental factors along with lifestyle (diet and physical activity) play an important role in modulating the onset and pathogenesis of neurodegenerative diseases. In the past 30 years significant progress has been achieved on the molecular mechanisms associated with the pathogenesis of neurodegenerative diseases. Most neurodegenerative diseases are accompanied by oxidative and excitotoxic neuronal damage, neuroinflammation, mitochondrial dysfunction, and protein aggregation. Drugs that modulate above processes may combat onset and progression of neurodegenerative diseases. Such strategies include inhibitors of enzymes associated with signal transduction processes, anti-inflammatory drugs, and antioxidants that can positively affect clinical outcomes (Farooqui and Horrocks, 2007; Farooqui, 2009a, b; Jin et al. (2010)). The targeting of combinations of pathogenic events including clearance of disaggregated proteins together with neuroprotective and immune modulatory strategies may all be required to facilitate positive disease outcomes. Initial palliative treatments for neurodegenerative diseases using cholinergic and dopaminergic drugs, neurotrophins, enzyme inhibitors, antibiotics (memantine), statins, PPARγ inhibitors, non-steroidal inflammatory drugs, and anti-apoptotic agents have provided some beneficial effects and some drugs have gained FDA approval. The use of animal models for studying neurodegenerative diseases has achieved wider acceptance, and important insight into the potential causes and pathogenic variables associated with various neurodegenerative diseases continues to increase. Furthermore, development of immunotherapy has progressed and evolved considerably for the therapy of neurodegenerative diseases. Although proteomic, lipidomic, and genomic analyses have provided important information, there are still serious problems with their correct and straightforward interpretation. Moreover, better animal transgenic models of neurodegenerative diseases are required. A direct understanding of the molecular mechanism of protein aggregation and its effects on neuronal cell death and immunotherapy may open new therapeutic approaches for the treatment of neurodegenerative diseases.

## References

Aarsland D, Ballard C, Walker Z, Bostrom F, Alves G, Kossakowski K, Leroi I, Pozo-Rodriguez F, Minthon L, Londos E (2009) Memantine in patients with Parkinson's disease dementia or dementia with Lewy bodies: a double-blind, placebo-controlled, multicentre trial. Lancet Neurol 8:613–618

Adlerz L, Holback S, Multhaup G, Iverfeldt K (2009) IGF-1-induced processing of the amyloid precursor protein family is mediated by different signaling pathways. J Biol Chem 282: 10203–10209

Aguzzi A, Baumann F, Bremer J (2008) The prion's elusive reason for being. Ann Rev Neurosci 31:439–477

Alexandrenne C, Hanoux V, Dkhissi F, Boquet D, Couraud JY, Wijkhuisen A (2009) Curative properties of antibodies against prion protein: a comparative in vitro study of monovalent fragments and divalent antibodies. J Neuroimmunol 209:50–56

Andreassen OA, Dedeoglu A, Ferrante RJ, Jenkins BG, Ferrante KL, Thomas M, Friedlich A, Browne SE, Schilling G, Borchelt DR, Hersch SM, Ross CA, Beal MF (2001) Creatine increase survival and delays motor symptoms in a transgenic animal model of Huntington's disease. Neurobiol Dis 8:479–491

Avramovich-Tirosh Y, Reznichenko L, Mit T, Zheng H, Fridkin M, Weinreb O, Mandel S, Youdim MB (2007) Neurorescue activity, APP regulation and amyloid-beta peptide reduction by novel multi-functional brain permeable iron-chelating-antioxidants, M-30 and green tea polyphenol, EGCG. Curr Alzheimer Res 4:403–411

Bandyopadhyay S, Goldstein LE, Lahiri DK, Rogers JT (2007) Role of the APP non-amyloidogenic signaling pathway and targeting alpha-secretase as an alternative drug target for treatment of Alzheimer's disease. Curr Med Chem 14:2848–2864

Bartke A, Bonkowski M, Masternak M (2008) Thow diet interacts with longevity genes. Hormones 7:17–23

Bate C, Tayebi M, Diomede L, Salmona M, Williams A (2009) Glimepiride reduces the expression of PrP, prevents PrP formation and protects against prion mediated neurotoxicity in cell lines. PLoS One 4(12):e8221

Bauer PO, Wong HK, Oyama F, Goswami A, Okuno M, Kino Y, Miyazaki H, Nukina N (2009) Inhibition of rho kinases enhances the degradation of mutant huntingtin. J Biol Chem 284:13153–13164

Bazan NG (2005) Neuroprotectin $D_1$ ($NPD_1$): a DHA-derived mediator that protects brain and retina against cell injury-induced oxidative stress. Brain Pathol 15:159–166

Bazan NG (2006) The onset of brain injury and neurodegeneration triggers the synthesis of docosanoid neuroprotective signaling. Cell Mol Neurobiol 26:901–913

Bazan NG (2007) Omega-3 fatty acids, pro-inflammatory signaling and neuroprotection. Curr Opin Clin Nutr Metab Care 10:136–141

Beal MF (2009) Therapeutic approaches to mitochondrial dysfunction in Parkinson's disease. Parkinsonism Relat Disord 15(Suppl 3):S189–S194

Benoit E, Escande D (1991) Riluzole specifically blocks inactivated Na channels in myelinated nerve fibre. Pflugers Arch 419:603–609

Benesimon G, Lacomblez L, Meininger V (1994) A controlled trial of riluzole in amyotrophic lateral sclerosisALS/Riluzole Study Group. N Engl J Med 330:585–591

Bilsland LG, Greensmith L (2008) The endocannabinoid system in amyotrophic lateral sclerosis. Curr Pharm Des 14:2306–2316

Birks J (2006) Cholinesterase inhibitors for Alzheimer's disease. Cochrane Database Syst Rev (1):CD005593

Bonelli RM, Hodl AK, Hofmann P, Kapfhammer HP (2004) Neuroprotection in Huntington's disease: a 2-year study on minocycline. Int Clin Psychopharmacol 19:337–342

Bonini NM, La Spada AR (2005) Silencing polyglutamine degeneration with RNAi. Neuron 48:715–718

Bousquet M, Saint-Pierre M, Julien C, Salem N Jr., Cicchetti F, Calon F (2008) Beneficial effects of dietary omega-3 polyunsaturated fatty acid on toxin-induced neuronal degeneration in an animal model of Parkinson's disease. FASEB J 22:1213–1225

Breidert T, Callebert J, Heneka MT, Landreth G, Launay JM, Hirsch EC (2002) Protective action of the peroxisome proliferator-activated receptor-gamma agonist pioglitazone in a mouse model of Parkinson's disease. J Neurochem 82:615–624

Bu G (2009) Apolipoprotein E and its receptors in Alzheimer's disease: pathways, pathogenesis and therapy. Nat Rev Neurosci 10:333–344

Bueler H (2009) Impaired mitochondrial dynamics and function in the pathogenesis of Parkinson's disease. Exp Neurol 218:235–246

Buisson B, Bertrand D (1998) Open-channel blockers at the human alpha4beta2 neuronal nicotinic acetylcholine receptor. Mol Pharmacol 53:555–563

Calon F, Lim GP, Yang FS, Morihara T, Teter B, Ubeda O, Rostaing P, Triller A, Salem NJ, Ashe KH, Frautschy SA, Cole GM (2004) Docosahexaenoic acid protects from dendritic pathology in an Alzheimer's disease mouse model. Neuron 43:633–645

Calon F, Cole G (2007) Neuroprotective action of omega-3 polyunsaturated fatty acids against neurodegenerative diseases: evidence from animal studies. Prost Leukot Essent Fatty Acids 77:287–293

Carro E, Trejo JL, Busiguina S, Torres-Aleman I (2001) Circulating insulin-like growth factor I mediates the protective effects of physical exercise. J Neurosci 21:5678–5684

Carro E, Torres-Aleman I (2004) The role of insulin and insulin-like growth factor I in the molecular and cellular mechanisms underlying the pathology of Alzheimer's disease. Eur J Pharmacol 490:127–133

Chaturvedi RK, Beal MF (2008) PPAR: a therapeutic target in Parkinson's disease. J Neurochem 106:506–518

Chauhan NB, Siegel GJ (2004) Intracerebroventricular passive immunization in transgenic mouse models of Alzheimer's disease. Expert Rev Vaccines 3:717–725

Chen M, Ona VO, Li M, Ferrante RJ, Fink KB (2000) Minocycline inhibits caspase-1 and caspase-3 expression and delays mortality in a transgenic mouse model of Huntington disease. Nat Med 6:797–801

Chen H, Jacobs E, Schwarzschild MA, McCullough ML, Calle EE, Thun MJ, Ascherio A (2005) Nonsteroidal anti-inflammatory drug use and the risk for Parkinson's disease. Ann Neurol 58:963–967

Chen MJ, Russo-Neustdt AA (2007) Running exercise- and antidepressant-induced increases in growth and survival-associated signaling molecules are IGF-dependent. Growth Factor 25: 118–131

Chen JX, Yan SD (2007a) Amyloid-beta-induced mitochondrial dysfunction. J Alzheimer Dis 12:177–184

Chen JX, Yan SD (2007b) Pathogenic role of mitochondrial [correction of mitochondrial] amyloid-beta peptide. Expert Rev Neurother 7:1517–1525

Chen Y, Meiminger V, Guillemin GJ (2009) Recent advances in the treatment of amyotrophic lateral sclerosis emphasis on kynurenine pathway inhibitors. Cent Nerv Syst Agents Med Chem 9:32–39

Chéramy A, Barbeito L, Godeheu G, Glowinski J (1992) Riluzole inhibits the release of glutamate in the caudate nucleus of the cat in vivo. Neurosci Lett 147:209–212

Cole GM, Ma QL, Frautschy SA (2009) Omega-3 fatty acids and dementia. Prost Leukot Essent Fatty Acids 81:213–221

Coppede F, Mancuso M, Siciliano G, Migliore L, Murri L (2006) Genes and the environment in neurodegeneration. Biosci Rep 26:341–367

Corder EH, Saunders AM, Risch NJ, Strittmatter WJ, Schmechel DE, Gaskell PC Jr., Rimmler JB, Locke PA, Conneally PM, Schmader KE (1994) Protective effect of apolipoprotein E type 2 allele for late onset Alzheimer disease. Nat Genet 7:180–184

Cordle A, Landreth G (2005) 3-Hydroxy-3-methylglutaryl-coenzyme A reductase inhibitors attenuate β-amyloid-induced microglial inflammatory responses. J Neurosci 25:299–307

Cordle A, Koenigsknecht-Talboo J, Wilkinson B, Limpert A, Landreth G (2005) Mechanisms of statin-mediated inhibition of small G-protein function. J Biol Chem 280:34202–34209

Cotman CW (2005) The role of neurotrophins in brain aging: a perspective in honor of Regino Perez-Polo. Neurochem Res 30:877–881

Das VN, Vaddadi KS (2004) Essential fatty acids in Huntington's disease. Nutrition 20: 942–947

Dealler S, Rainov NG (2003) Pentosan polysulfate as a prophylactic and therapeutic agent against prion disease. IDrugs 6:470–478

Dehmer T, Heneka MT, Sastre M, Dichgans J, Schulz JB (2004) Protection by pioglitazone in the MPTP model of Parkinson's disease correlates with I kappa B alpha induction and block of NF kappa B and iNOS activation. J Neurochem 88:494–501

Delattre AM, Kiss A, Szawka RE, Anselmo-Franci JA, Bagatini PB, Xavier LL, Rigon P, Achaval M, Iagher F, de David C, Marroni NA, Ferraz AC (2009) Evaluation of chronic omega-3 fatty acids supplementation on behavioral and neurochemical alterations in 6-hydroxydopamine-lesion model of Parkinson's disease. Neurosci Res 24 Nov 2009 [Epub ahead of print]

de Oliveria DM, Barreto G, De Anrade DV, Dos El Bacha R, Goraldez LD (2009) Cytoprotective effect of Valeriana officinalis extract on an in vitro experimental model of Parkinson disease. Neurochem Res 34:215–220

Ding O, Vaynman S, Akhavan M, Ying Z, Gomez-Pinilla F (2006) Insulin-like growth factor I interfaces with brain-derived neurotrophic factor-mediated synaptic plasticity to modulate aspects of exercise-induced cognitive function. Neuroscience 140:823–833

Doble A, Hubert JP, Blanchard JC (1992) Pertussis toxin pretreatment abolishes the inhibitory effect of riluzole and carbachol on D-[3H]aspartate release from cultured cerebellar granule cells. Neurosci Lett 140:251–254

Doh-ura K (2004) Prion diseases: disease diversity and therapeutics. Rinsho Shinkeigaku 44: 855–856

Doh-ura K (2009) Innovation of therapeutics and prophylaxis for prion diseases. Rinsho Shinkeigaku 49:946–948

Eibl JK, Chapelsky SA, Ross GM (2010) Multipotent neurotrophin antagonist targets brain-derived neurotrophic factor and nerve growth factor. J Pharmacol Exp Ther 332:446–454

Endres M (2005) Statins and stroke. J Cereb Blood Flow Metab 25:1093–1110

Esposito E, Di Matteo V, Benigno A, Pierucci M, Crescimanno G, Di Giovanni G (2007) Non-steroidal anti-inflammatory drugs in Parkinson's disease. Exp Neurol 205:295–312

Falls DL (2003) Neuregulins: functions, forms, and signaling strategies. Exp Cell Res 284:14–30

Farooqui AA, Horrocks LA (2007) Glycerophospholipids in the brain: phospholipases $A_2$ in neurological disorders. Springer, New York, NY, pp 1–394

Farooqui AA, Horrocks LA, Farooqui T (2007a) Modulation of inflammation in brain: a matter of fat. J Neurochem 101:577–599

Farooqui AA, Ong WY, Horrocks LA, Chen P, Farooqui T (2007b) Comparison of biochemical effects of statins and fish oil in brain: the battle of the titans. Brain Res Rev 56:443–471

Farooqui AA (2009a) Hot topics in neural membrane lipidology. Springer, New York, NY

Farooqui AA (2009b) Beneficial effects of fish oil on human brain. Springer, New York, NY

Farooqui T, Farooqui AA (2009) Aging: an important factor for the pathogenesis of neurodegenerative diseases. Mech Ageing Dev 130:203–215

Ferrante RJ, Andreassen OA, Jenkins BG, Dedeoglu A, Kuemmerle S, Kubilus JK, Kaddurah-Daouk R, Hersch SM, Beal MF (2000) Neuroprotective effects of creatine in a transgenic mouse model of Huntington's disease. J Neurosci 20:4389–4397

Farrelly PV, Kenna BL, Laohachai KL, Bahadi R, Salmona M, Forloni G, Kourie JI (2003) Quinacrine blocks PrP (106–126)-formed channels. J Neurosci Res 74:934–941

Fisher A (2007) M1 muscarinic agonists target major hallmarks of Alzheimer's disease – an update. Curr Alzheimer Dis 4:577–580

Follette P (2003) New perspectives for prion therapeutics meeting Prion disease treatment's early promise unravels. Science 299:191–192

Fontan-Lozano A, Lopez-lluch G, Delgado-Garcia JM, Navas P, Carrion AM (2008) Molecular bases of caloric restriction regulation of neuronal synaptic plasticity. Mol Neurobiol 38: 167–177

Fratiglioni L, Qiu C (2009) Prevention of common neurodegenerative disorders in the elderly. Exp Gerontol 44:46–50

Freude S, Hettich MM, Schumann C, Stohr O, Koch L, Kohler C, Udelhoven M, Leeser U, Muller M, Kubota N, Kadowaki T, Krone W, Schroder H, Bruning JC, Schubert M (2009a) Neuronal IGF-1 resistance reduces Abeta accumulation and protects against premature death in a model of Alzheimer's disease. FASEB J 23:3315–3324

Freude S, Schilbach K, Schubert M (2009b) The role of IGF-1 receptor and insulin receptor signaling for the pathogenesis of Alzheimer's disease: from model organisms to human disease. Curr Alzheimer Res 6:213–223

Freund-Levi Y, Eriksdotter-Jonhagen M, Cederholm T, Basun H, Faxen-Irving G, Garlind A, Vedin I, Vessby B, Wahlund LO, Palmblad J (2006) Omega-3 fatty acid treatment in 174 patients with mild to moderate Alzheimer disease: OmegAD study: a randomized double-blind trial. Arch Neurol 63:1402–1408

Frisardi V, Solfrizzi V, Imbimbo BP, Capurso C, D'Introno A, Colacicco AM, Vendemiale G, Seripa D, Pilotto A, Capurso A, Panza F (2009) Towards disease-modifying treatment of alzheimer's disease: drugs targeting beta-amyloid. Curr Alzheimer Res 26 Nov 2009 [Epub ahead of print]

Fuenzalida K, Quintanila R, Ramos P, Pident D, Fuentealba RA, Martinez G, Inestrosa NC, Bronfman M (2007) Peroxisome proliferator-activated receptor gamma up-regulates the Bcl-2 anti-apoptotic protein in neurons and induces mitochondrial stabilization and protection against oxidative stress and apoptosis. J Biol Chem 282:37006–37015

Ghosh A, Kumaragurubaran N, Tang J (2005) Recent developments of structure-based β-secretase inhibitor's for Alzheimer's disease. Curr Top Med Chem 5:1609–1622

Ghosh S, Novak EM, Innis SM (2007) Cardiac proinflammatory pathways are altered with different dietary n-6 linoleic to n-3 α-linolenic acid ratios in normal, fat-fed pigs. Am J Physiol Heart Circ Physiol 293:H2919–H2927

Ghosh A, Roy A, Matras J, Brahmchari S, Gendelman HE, Pahan K (2009) Simvastatin inhibits the activation of p21ras and prevents the loss of dopaminergic neurons in a mouse model of Parkinson's disease. J Neurosci 29:13543–13556

Gilgun-Sherki Y, Melamed E, Offen D (2006) Anti-inflammatory drugs in the treatment of neurodegenerative diseases: current state. Curr Pharmaceut Des 12:3509–3519

Gozes I, Morimoto BH, Tiong J, Fox A, Sutherland K, Dangoor D, Holser-Cochav M, Vered K, Newton P, Aisen PS, Matsuoka Y, van Dyck CH, Thal L (2005) NAP: research and development of a peptide derived from activity-dependent neuroprotective protein (ADNP). CNS Drug Rev 11:353–368

Gozes I (2007) Activity-dependent neuroprotective protein: from gene to drug candidate. Pharm Therp 114:146–154

Gozes I, Stewart A, Morimoto B, Fox A, Sutherland K, Schmeche D (2009) Addressing Alzheimer's disease tangles: from NAP to AL-108. Curr Alzheimer Res 6:455–460

Graeber MB, Moran LB (2002) Mechanisms of cell death in neurodegenerative diseases: fashion, fiction, and facts. Brain Pathol 12:385–390

Greulich W, Fenger E (1995) J Neural Transm 46(Suppl):415–421

Joseph JA, Shukitt-Hale B, Lau FC (2007) Fruit polyphenols and their effects on neuronal signaling and behavior in senescence. Ann NY Acad Sci 1100:470–485

Guentchev M, Voigtlander T, Haberler C, Groschup MH, Budka H (2000) Evidence for oxidative stress in experimental prion disease. Neurobiol Dis 7:270–273

Hashimoto M, Hossain S, Agdul H, Shido O (2005) Docosahexaenoic acid-induced amelioration on impairment of memory learning in amyloid beta-infused rats relates to the decreases of amyloid beta and cholesterol levels in detergent-insoluble membrane fractions. Biochim Biophys Acta 1738:91–98

Hashimoto M, Hossain S, Shimada T, Shido O (2006) Docosahexaenoic acid-induced protective effect against impaired learning in amyloid β-infused rats is associated with increased synaptosomal membrane fluidity. Clin Exp Pharmacol Physiol 33:934–939

Hashimoto M, Shahdat HM, Yamashita S, Katakura M, Tanabe Y, Fujiwara H, Gamoh S, Miyazawa T, Arai H, Shimada T, Shido O (2008) Docosahexaenoic acid disrupts in vitro amyloid beta fibrillation and concomitantly inhibits amyloid levels in cerebral cortex of Alzheimer's disease model rats. J Neurochem 107:1634–1646

Heneka MT, Landreth GE (2007) PPARs in the brain. Biochim Biophys Acta 1771:1031–1045

Heneka MT, Landreth GE, Hull M (2007) Drug insight: effects mediated by peroxisome proliferator-activated receptor-gamma in CNS disorders. Nat Clin Pract Neurol 3:496–504

Heppner FL, Aguzzi A (2004) Recent developments in prion immunotherapy. Curr Opin Immunol 16:594–598

Herzog CD, Brown L, Gammon D, Kruegel B, Lin R, Wilson A, Bolton A, Printz M, Gasmi M, Bishop KM, Kordower JH, Bartus RT (2009) Expression, bioactivity, and safety 1 year after adeno-associated viral vector type 2-mediated delivery of neurturin to the monkey nigrostriatal system support cere-120 for Parkinson's disease. Neurosurgery 64:602–612

Hirohata M, Ono K, Yamada M (2008) Non-steroidal anti-inflammatory drugs as anti-amyloidogenic compounds. Curr Pharm Des 14:3280–3294

Hong M, Mukhida K, Mendez I (2008) GDNF therapy for Parkinson's disease. Expert Rev Neurother 8:1125–1139

Hooper NM, Turner AJ (2002) The search for alpha-secretase and its potential as a therapeutic approach to Alzheimer's disease. Curr Med Chem 9:1107–1119

Hoyer S (2004) Glucose metabolism and insulin receptor signal transduction in Alzheimer disease. Eur J Pharmacol 490:115–125

Howe CL, Vergstrom RA, Horazdovsky BF (2009) Subcutaneous IGF-1 is not beneficial in 2-year ALS trial. Neurology 73:1247

Hussain I, Powell D, Howlett DR, Tew DG, Meek TD, Chapman C, Gloger IS, Murphy KE, Southan CD, Ryan DM, Smith TS, Simmons DL, Walsh FS, Dingwall C, Christie G (1999) Identification of a novel aspartic protease (Asp2) as β-secretase. Mol Cell Neurosci 14:419–427

Ikemoto A, Nitta A, Furukawa S, Ohishi M, Nakamura A, Fujii Y, Okuyama H (2000) Dietary n-3 fatty acid deficiency decreases nerve growth factor content in rat hippocampus. Neurosci Lett 285:99–102

Jacobsen KT, Adlerz L, Multhaup G, Iverfeldt K (2010) Insulin-like growth factor-1 (IGF-1)-induced processing of amyloid-{beta} precursor protein (APP) and APP-like protein 2 is mediated by different metalloproteinases. J Biol Chem 5 Feb 2010 [Epub ahead of print]

Jankovic J, Stacy M (2007) Medical management of levodopa-associated motor complications in patients with Parkinson's disease. CNS Drugs 21:677–692

Jeng BH, Galor A, Lee MS, Meisler DM, Hellyfield JG (2008) Amantadine-associated corneal edema potentially irreversible even after cessation of the medication. Opthalmology 115: 1540–1544

Ji Y, Pang PT, Feng L, Lu B (2005) Cyclic AMP controls BDNF-induced TrkB phosphorylation and dendritic spine formation in mature hippocampal neurons. Nat Neurosci 8:164–172

Jiang Q, Heneka M, Landreth GE (2008) The role of peroxisome proliferator-activated receptor-gamma (PPARgamma) in Alzheimer's disease: therapeutic implications. CNS Drugs 22:1–14

Jin H, Randazzo J, Zhang P, Kador PF (2010) Multifunctional antioxidants for the treatment of age-related diseases. J Med Chem 53:1117–1127

Kalinin S, Richardson JC, Feinstein DL (2009) A PPARdelta agonist reduces amyloid burden and brain inflammation in a transgenic mouse model of Alzheimer's disease. Curr Alzheimer Res 6:431–437

Kalmijn S, Launer LJ, Ott A, Witteman JC, Hofman A, Breteler MM (1997) Dietary fat intake and the risk of incident dementia in the Rotterdam Study. Ann Neurol 42:776–782

Kaufer D, Gandy S (2009) APOE {epsilon}4 and bapineuzumab: infusing pharmacogenomics into Alzheimer disease therapeutics. Neurology 73:2052–2053

Kaufmann P, Thompson JL, Levy G, Buchsbaum R, Shefner J, Krivickas LS, Katz J, Rollins Y, Barohn RJ, Jackson CE, et al (2009) Phase II trial of CoQ10 for ALS finds insufficient evidence to justify phase III. Ann Neurol 66:235–244

Kidd PM (2005) Neurodegeneration from mitochondrial insufficiency: nutrients, stem cells, growth factors, and prospects for brain rebuilding using integrative management. Altern Med Rev 10:268–293

Kobayashi Y, Hirata K, Tanaka H, Yamada T (2003) Quinacrine administration to a patient with Creutzfeldt-Jakob disease who received a cadaveric dura mater graft – an EEG evaluation. Rinsho Shinkeigaku 43:403–408

Kobayashi M, Kim J, Kobayashi N, Han S, Nakamura C, Ikebukuro K, Soda K (2006) Pyrroloquinoline quinone (PQQ) prevents fibril formation of alpha-synuclein. Biochem Biophys Res Commun 349:1139–1144

Kojro E, Fahrenholz F (2005) The non-amyloidogenic pathway: structure and function of alpha-secretases. Subcell Biochem 38:105–127

Koob AO, Ubhi K, Paulsson JF, Kelly J, Rockenstein E, Mante M, Adame A, Masliah E (2010) Lovastatin ameliorates alpha-synuclein accumulation and oxidation in transgenic mouse models of alpha-synucleinopathies. Exp Neurol 221:267–274

Korth C, May BC, Cohen FE, Prusiner SB (2001) Acridine and phenothiazine derivatives as pharmacotherapeutics for prion disease. Proc Natl Acad Sci USA 98:9836–9841

Kummer MP, Heneka MT (2008) PPARs in Alzheimer's disease. PPAR Res 2008

Lacomblez L, Bensiman G, Leigh PN, Guillet P, Meininger V (1996) Dose-ranging study of riluzole in amyotrophic lateral sclerosis Amyotrophic Lateral Sclerosis/Riluzole Study Group II. Lancet 347:1425–1431

Lahiri DK, Maloney B, Basha MR, Ge YW, Zawia NH (2007) How and when environmental agents and dietary factors affect the course of Alzheimer's disease: the "LEARn" model (latent early-life associated regulation) may explain the triggering of AD. Curr Alzheimer Res 4:219–228

Lai A, Frishman WH (2005) Rho-kinase inhibition in the therapy of cardiovascular disease. Cardiol Rev 13:285–292

Landreth G, Jiang Q, Mandrekar S, Henecka M (2008) PPARgamma agonists as therapeutics for the treatment of Alzheimer's disease. Neurotherapeutics 5:481–489

Lau FC, Shukitt-Hale B, Joseph JA (2007) Nutritional intervention in brain aging: reducing the effects of inflammation and oxidative stress. Subcell Biochem 42:299–318

Lee J, Duan W, Long JM, Ingram DK, Mattson MP (2000) Dietary restriction increases the number of newly generated neural cells, and induces BDNF expression, in the dentate gyrus of rats. J Mol Neurosci 15:99–108

Lemere CA, Mair M, Peng Y, Jiang L, Seabrook T (2007) Novel Abeta immunogens: is shorter better? Curr Alzheimer Res 4:427–436

Lemere CA (2009) Developing novel immunogens for a safe and effective Alzheimer's disease vaccine. Prog Brain Res 175:83–93

Leroi I, Overshott R, Byrne EJ, Daniel E, Burns A (2009) Randomized controlled trial of memantine in dementia associated with Parkinson's disease. Mov Disord 24:1217–1221

Levy G, Kaufmann P, Buchsbaum R, Montes J, Barsdorf A, Arbing R, Battista V, Zhou X, Mitsumoto L, Levin B, Thompson JL (2006) A two-stage design for a phase II clinical trial of coenzyme Q10 in ALS. Neurology 66:660–663

Li L, Tang BL (2005) Environmental enrichment and neurodegenerative diseases. Biochem Biophys Res Commun 334:293–297

Lim GP, Calon F, Morihara T, Yang F, Teter B, Ubeda O, Salem N Jr, Frautschy SA, Cole GM (2005) A diet enriched with the omega-3 fatty acid docosahexaenoic acid reduces amyloid burden in an aged Alzheimer mouse model. J Neurosci 25:3032–3040

Lin X, Koelsch G, Wu S, Downs D, Dashti A, Tang J (2000) Human aspartic protease memapsin 2 cleaves the β-secretase site of β-amyloid precursor protein. Proc Natl Acad Sci USA 97: 1456–1460

Lindvall O, Wahlberg LU (2008) Encapsulated cell biodelivery of GDNF: a novel clinical strategy for neuroprotection and neuroregeneration in Parkinson's disease? Exp Neurol 209:82–88

Lipton SA, Choi YB, Takahashi H, Zhang D, Li W, Godzik A, Bankston LA (2002) Cysteine regulation of protein function – as exemplified by NMDA-receptor modulation. Trends Neurosci 25:474–480

Lipton SA (2006) Paradigm shift in neuroprotection by NMDA receptor blockade: memantine and beyond. Nat Rev Drug Discov 5:160–170

Long J, Gao H, Sun L, Liu J, Zhao-Wilson X (2009) Grape extract protects mitochondria from oxidative damage and improves locomotor dysfunction and extends lifespan in a Drosophila Parkinson's disease model. Rejuvenation Res 12:321–331

Lopez-Miranda J, Delgado-Lista J, Perez-Martinez P, Jimenez-Gómez Y, Fuentes F, Ruano J, Marin C (2007) Olive oil and the haemostatic system. Mol Nutr Food Res 51: 1249–1259

Love R (2001) Ethnic labels may not predict individuals' responses to drugs. Lancet 358(9292):1519

Lu B (2003) BDNF and activity-dependent synaptic modulation. Learn Mem 10:86–98

Luchsinger JA, Noble JM, Scarmeas N (2007) Diet and Alzheimer's disease. Curr Neurol Neurosci Rep 7:366–372

Lukiw WJ, Cui JG, Marcheselli VL, Bodker M, Botkjaer A, Gotlinger K, Serhan CN, Bazan NG (2005) A role for docosahexaenoic acid-derived neuroprotectin D1 in neural cell survival and Alzheimer disease. J Clin Invest 115:2774–2783

Lukiw WJ, Zhao Y, Cui JG (2008) An NF-kappaB-sensitive micro RNA-146a-mediated inflammatory circuit in Alzheimer disease and in stressed human brain cells. J Biol Chem 283:31315–31322

Ma T, Zhao Y, Kwak YD, Yang Z, Thompson R, Luo Z, Xu H, Liao FF (2009a) Statin's excitoprotection is mediated by sAPP and the subsequent attenuation of calpain-induced truncation events, likely via rho-ROCK signaling. J Neurosci 29:11226–11236

Ma QL, Yang F, Rosario ER, Ubeda QJ, Beech W, Gant DJ, Chen PP, Hudspeth B, Chen C, Zhao Y, Vinters HV, Frautchy SA, Cole GM (2009b) Beta-amyloid oligomers induce phosphorylation of tau and inactivation of insulin receptor substrate via c-Jun N-terminal kinase signaling: suppression by omega-3 fatty acids and curcumin. J Neurosci 29:9078–9089

Mancuso M, orsucci D, Volpi L, Calsolaro V, Siciliano G (2010) Coenzyme Q10 in neuromuscular and neurodegenerative disorders. Curr Drug Targets 11:111–121

Mans RA, Chowdhury N, Cao D, McMahon LL, Li L (2010) Simvastatin enhances hippocampal long-term potentiation in C57BL/6 mice. Neuroscience 166:435–444

Martin LJ, Gertz B, Pan Y, Price AC, Molkentin JD, Chang Q (2009) The mitochondrial permeability transition pore in motor neurons: involvement in the pathobiology of ALS mice. Exp Neurol 218:333–346

Marvanova M, Lakso M, Pirhonen J, Nawa H, Wong G, Castren E (2001) The neuroprotective agent memantine induces brain-derived neurotrophic factor and trkB receptor expression in rat brain. Mol Cell Neurosci 18:247–258

Mattson MP (2008) Awareness of hormesis will enhance future research in basic and applied neuroscience. Crit Rev Toxicol 38:633–639

May BCH, Fafarman AT, Hong SB, Rogers M, Deady LW, Prusiner SB, Cohen FE (2003) Potent inhibition of scrapie prion replication in cultured cells by bis-acridines. Proc Natl Acad Sci USA 100:3416–3421

McGuinness B, Craig D, Bullock R, Passmore P (2009) Statins for the prevention of dementia. Cochrane Database Syst Rev 2:CD003160

Menendez-Gonzalez M, Calatayud MT, Blazquez-Menes B (2005) Exacerbation of Lewy bodies dementia due to memantine. J Alzheimer Dis 8:289–291

Michailov GV, Sereda MW, Brinkmann BG, Fischer TM, Haug B, Birchmeier C, Role L, Lai C, Schwab MH, Nave KA (2004) Axonal neuregulin-1 regulates myelin sheath thickness. Science 304:700–703

Milhavet O, McMahon HE, Rachidi W, Nishida N, Katamine S, Mange A, Arlotto M, Casanova D, Riondel J, Favier A, Lehmann S (2000) Prion infection impairs the cellular response to oxidative stress. Proc Natl Acad Sci USA 97:13937–13942

Minguez-Castellanos A, Escamilla-Sevilla F (2005) Cell therapy and other neuroregenerative strategies in Parkinson's disease (I). Rev Neurol 41:604–614

Mochizuki H (2009) Current status of gene therapy for Parkinson disease. Brain Nerve 61:485–493

Morris MC, Evans DA, Bienias JL, Tangney CC, Bennett DA, Wilson RS, Aggarwal N, Schneider J (2003) Consumption of fish and n-3 fatty acids and risk of incident Alzheimer disease. Arch Neurol 60:940–946

Muller-Schiffmann A, Korth C (2008) Vaccine approaches to prevent and treat prion infection: progress and challenges. BioDrugs 22:45–52

Munoz-Torrero D (2008) Acetylcholinesterase inhibitors as disease-modifying therapies for Alzheimer's disease. Curr Med Chem 15:2433–2455

Muntané G, Janué A, Fernandez N, Odena MA, Oliveira E, Boluda S, Portero-Otin M, Naudí A, Boada J, Pamplona R, Ferrer I (2010) Modification of brain lipids but not phenotype in alpha-synucleinopathy transgenic mice by long-term dietary n-3 fatty acids. Neurochem Int 12 Nov 2009 [Epub ahead of print]

Murck H, Manku M (2007) Ethyl-EPA in Huntington disease: potentially relevant mechanism of action. Brain Res Bull 72:159–164

Murer MG, Yan Q, Raisman-Vozari R (2001) Brain-derived neurotrophic factor in the control human brain, and in Alzheimer's disease and Parkinson's disease. Prog Neurobiol 63: 71–124

Mutez E, Duhamel A, Defebvre L, Bordet R, Destée A, Kreisler A (2009) Lipid-lowering drugs are associated with delayed onset and slower course of Parkinson's disease. Pharmacol Res 60:41–45

Nakamura T, Lipton SA (2007) S-Nitrosylation and uncompetitive/fast off-rate (UFO) drug therapy in neurodegenerative disorders of protein misfolding. Cell Death Differ 14:1305–1314

Nguyen T, Hamby A, Massa SM (2005) Clioquinol down-regulates mutant huntingtin expression in vitro and mitigates pathology in a Huntington's disease mouse model. Proc Natl Acad Sci USA 102:11840–11845

Nitschke C, Fiechsig E, van den Brandt J, Lindner N, Luhrs T, Dittmer U, Klein MA (2007) Immunisation strategies against prion diseases: prime-boost immunisation with a PrP DNA vaccine containing foreign helper T-cell epitopes does not prevent mouse scrapie. Vet Microbiol 123:367–376

Nordberg A (1999) PET studies and cholinergic therapy in Alzheimer's disease. Rev Neurol (Paris) 155(Suppl 4):S53–S63

Nunomura A, Moreira PI, Lee HG, Zhu X, Castellani RJ, Smith MA, Perry G (2007) Neuronal death and survival under oxidative stress in Alzheimer and Parkinson diseases. CNS Neurol Disord Drug Targets 6:411–423

Origlia N, Arancio O, Domenici L, Yan SS (2009) MAPK, beta-amyloid and synaptic dysfunction: the role of RAGE. Expert Rev Neurother 9:1635–1645

Ozdinler PH, Macklis JD (2006) IGF-I specifically enhances axon outgrowth of corticospinal motor neurons. Nat Neurosci 9:1371–1381

Papadia S, Soriano FX, Léveillé F, Martel MA, Dakin KA, Hansen HH, Kaindl A, Sifringer M, Fowler J, Stefovska V, McKenzie G, Craigon M, Corriveau R, Ghazal P, Horsburgh K, Yankner BA, Wyllie DJ, Ikonomidou C, Hardingham GE (2008) Synaptic NMDA receptor activity boosts intrinsic antioxidant defenses. Nat Neurosci 11:476–487

Park S, Kim HT, Yun S, Kim IS, Lee J, Lee IS, Park KI (2009) Growth factor-expressing human neural progenitor cell grafts protect motor neurons but do not ameliorate motor performance and survival in ALS mice. Exp Mol Med 41:487–500

Parsons RB, Price GC, Farrant JK, Subramaniam D, Adeagbo-Sheikh J, Austen BM (2006) Statins inhibit the dimerization of beta-secretase via both isoprenoid- and cholesterol-mediated mechanisms. Biochem J 399:205–214

Patel NK, Gill SS (2007) GDNF delivery for Parkinson's disease. Acta Neurochir Suppl 97: 135–154

Pedrini S, Carter TL, Prendergast G, Petanceska S, Ehrlich ME, Gandy S (2005) Modulation of statin-activated shedding of Alzheimer APP ectodomain by ROCK. PLoS Med 2:e18

Pelleymounter MA, Cullen MJ, Baker MB, Gollub M, Wellman C (1996) The effects of intrahippocampal BDNF and NGF on spatial learning in aged Long Evans rats. Mol Chem Neuropathol 29:211–226

Peng Y, Lee DY, Jiang L, Ma Z, Schachter SC, Lemere CA (2007) Huperzine A regulates amyloid precursor protein processing via protein kinase C and mitogen-activated protein kinase pathways in neuroblastoma SK-N-SH cells over-expressing wild type human amyloid precursor protein 695. Neuroscience 150:386–395

Perry G, Smith MA (2000) Promise and pitfalls of therapeutic to modify oxidant balance in chronic diseases. IDrugs 3:371–372

Peterson AL, Nutt JG (2008) Treatment of Parkinson's disease with trophic factors. Neurotherapeutics 5:270–280

Petri S, Kiaei M, Damiano M, Hiller A, Wille E, Manfredi G, Calingasan NY, Szeto HH, Beal MF (2006) Cell-permeable peptide antioxidants as a novel therapeutic approach in a mouse model of amyotrophic lateral sclerosis. J Neurochem 98:1141–1148

Planells-Cases R, Lerma J, Ferrer-Montiel A (2006) Pharmacological intervention at ionotropic glutamate receptor complexes. Curr Pharm Des 12:3583–3596

Postina R (2008) A closer look at alpha-secretase. Curr Alzheimer Res 5:179–186

Price JL, Morris JC (1999) Tangles and plaques in nondemented aging and "preclinical" Alzheimer's disease. Ann Neurol 45:358–368

Poon WW, Blurton-Jones M, Tu CH, Feinberg LM, Chabrier MA, Harris JW, Jeon NL, Cotman CW (2009) Beta-Amyloid impairs axonal BDNF retrograde trafficking. Neurobiol Aging 18 Jun 2009 [Epub ahead of print]

Priller C, Dewachter I, Vassallo N, Paluch S, Pace C, Kretzschmar HA, van Leuven F, Herms J (2007) Mutant Clioquinol 1 alters synaptic transmission in cultured hippocampal neurons. J Biol Chem 282:1119–1127

Puri BK (2005) Treatment of Huntington's disease with eicosapentaenoic acid. In: Yehuda S, Mostofsky DI (eds) Nutrients, stress and medical disorders. Nutrition and health (Series). Humana Press Inc, Totowa, NJ, pp 279–286

Qiu C, Kivipetto M, von Strauss E (2009) Epidemiology of Alzheimer's disease: occurrence, determinants, and strategies toward intervention. Dialogues Clin Neurosci 11:111–128

Quinn LP, Crook B, Hows ME, Vidgeon-Hart M, Chapman H, Upton N, Medhurst AD, Virley DJ (2008) Br J Pharmacol 154:226–233

Rahman I, MacNee W (2000) Regulation of redox glutathione levels and gene transcription in lung inflammation: therapeutic approaches. Free Radic Biol Med 28:1405–1420

Ramaswamy S, Soderstrom KE, Kordower JH (2009) Trophic factors therapy in Parkinson's disease. Prog Brain Res 175:201–216

Rammes G, Rupprecht R, Ferrari U, Zieglgänsberger W, Parsons CG (2001) The *N*-methyl-D-aspartate receptor channel blockers memantine, MRZ 2/579 and other amino-alkyl-cyclohexanes antagonise 5-HT(3) receptor currents in cultured HEK-293 and N1E-115 cell systems in a non-competitive manner. Neurosci Lett 306:81–84

Ray B, Banerjee PK, Geig NH, Lahiri DK (2009) Memantine treatment decreases levels of secreted Alzheimer's amyloid precursor protein (APP) and amyloid beta (Abeta) peptide in the human neuroblastoma cells. Neurosci Lett [Epub ahead of print]

Reddy PH (2008) Mitochondrial medicine for aging and neurodegenerative diseases. Neuromolecular Med 10:291–315

Reddy PH, Mao P, Manczak M (2009) Mitochondrial structural and functional dynamics in Huntington's disease. Brain Res Rev 61:33–48

Reisberg B, Doody R, Stöffler A, Schmitt F, Ferris S, Möbius HJ Memantine Study Group (2003) Memantine in moderate-to-severe Alzheimer's disease. N Engl J Med 348:1333–1341

Riekse RG, Li G, Petrie EC, Leverenz JB, Vavrek D, Vuletic S, Albers JJ, Montine TJ, Lee VM, Lee M, Seubert P, Galasko D, Schellenberg GD, Hazzard WR, Peskind ER (2006) Effect of statins on Alzheimer's disease biomarkers in cerebrospinal fluid. J Alzheimer Dis 10:399–406

Ridha BH, Josephs KA, Rosser MN (2005) Delusions and hallucinations in dementia with Lewy bodies: worsening with memantine. Neurology 65:481–482

Roberts CK, Barnard RJ, Sindhu RK, Jurczak M, Ehdaie A, Vaziri ND (2006) Oxidative stress and dysregulation of NAD(P)H oxidase and antioxidant enzymes in diet-induced metabolic syndrome. Metabolism 55:928–934

Robson LG, Dyalls S, Sidloff D, Michael-Titus AT (2008) Omega-3 polyunsaturated fatty acids increase the neurite outgrowth of rat sensory neurones throughout development and in aged animals. Neurobiol Aging 10 Jly [Epub a head print]

Rogawski MA, Wenk GL (2003) The neuropharmacological basis for the use of memantine in the treatment of Alzheimer's disease. CNS Drug Rev 9:275–308

Romrell J, Fernandez HH, Okun MS (2003) Rationale for current therapies in Parkinson's disease. Exp Opin Pharmacother 4:1747–1761

Sakaguchi S (2009) Systematic review of the therapeutics for prion diseases. Brain Nerve 61: 929–938

Sakowski SA, Schuyler AD, Feldman EL (2009) Insulin-like growth factor-I for the treatment of amyotrophic lateral sclerosis. Amyyotroph Lateral Scler 10:63–73

Samadi P, Grégoire L, Rouillard C, Bédard PJ, Di Paolo T, Lévesque D (2006) Docosahexaenoic acid reduces levodopa-induced dyskinesias in 1-methyl-4-phenyl-1,2,3,6-tetrahydropyridine monkeys. Ann Neurol 59:282–288

Sanchez I, Mahlke C, Yuan J (2003) Pivotal role of oligomerization in expanded polyglutamine neurodegenerative disorders. Nature 421:373–379

Sarkar S, Perlstein EO, Imarisio S, Pineau S, Cordenier A, Maglathlin RL, Webster JA, Lewis TA, O'Kane CJ, Schreiber SL, Rubinsztein DC (2007) Small molecules enhance autophagy and reduce toxicity in Huntington's disease models. Nat Chem Biol 3:331–338

Sarkar S, Krishna G, Imarisio S, Saiki S, O'Kane CJ, Rubinsztein DC (2008) A rational mechanism for combination treatment of Huntington's disease using lithium and rapamycin. Hum Mol Genet 17:170–178

Sastre M, Dewachter I, Rossner S (2006a) Nonsteroidal anti-inflammatory drugs repress beta-secretase gene promoter activity by the activation of PPARgamma. Proc Natl Acad Sci USA 103:443–448

Sastre M, Klockgether T, Heneka MT (2006b) Contribution of inflammatory processes to Alzheimer's disease: molecular mechanisms. Int J Dev Neurosci 24:167–176

Saydoff JA, Garcia RA, Browne SE, Liu L, Sheng J, Brenneman D, Hu Z, Cardin S, Gonzalez A, von Borstel RW, Gregorio J, Burr H, Beal MF (2006) Oral uridine pro-drug PN401 is neuroprotective in the R6/2 and N171–82Q mouse models of Huntington's disease. Neurobiol Dis 24:455–465

Schaefer EJ, Bongard V, Beiser AS, Lamon-Fava S, Robins SJ, Au R, Tucker KL, Kyle DJ, Wilson PW, Wolf PA (2006) Plasma phosphatidylcholine docosahexaenoic acid content and risk of dementia and Alzheimer disease: the Framingham heart study. Arch Neurol 63:1545–1550

Schmitt FA, van Dyck CH, Wichems CH, Olin JT For the Memantine MEM-MD-02 Study Group (2006) Cognitive response to memantine in moderate to severe Alzheimer disease patients already receiving donepezil: an exploratory reanalysis. Alzheimer Dis Assoc Disord 20:255–262

Schmitt F, Ryan M, Cooper G (2007) A brief review of the pharmacologic and therapeutic aspects of memantine in Alzheimer's disease. Expert Opin Drug Metab Toxicol 3:135–141

Schuster S, Nadjar A, Guo JT, Li Q, Ittrich C, Hengerer B, Bezard E (2008) The 3-hydroxy-3-methylglutaryl-CoA reductase inhibitor lovastatin reduces severity of L-DOPA-induced abnormal involuntary movements in experimental Parkinson's disease. J Neurosci 28: 4311–4316

Selkoe DJ (2001) Alzheimer's disease: genes, proteins, and therapy. Physiol Rev 81:741–766

Serhan CN (2005) Novel ω-3-derived local mediators in anti-inflammation and resolution. Pharmacol Ther 105:7–21

Sezeto HH (2006) Mitochondria-targeted peptide antioxidants: novel neuroprotective agents. AAPS J 8:E521–E531

Sharma S, Prasanthi RPJ, Schommer E, Feist G, Ghribi O (2008) Hypercholesterolemia-induced Abeta accumulation in rabbit brain is associated with alteration in IGF-1 signaling. Neurobiol Dis 32(3):426–432

Shen WH, Zhang CY, Zhang GY (2003) Antioxidants attenuate reperfusion injury after global brain ischemia through inhibiting nuclear factor-kappa B activity in rats. Acta Pharmacol Sin 24:1125–1130

Siman R, Salidas S (2004) γ-secretase subunit composition and distribution in the presenilin wild-type and mutant mouse brain. Neuroscience 129:615–628

Singh P, Suman S, Chandna S, Das TK (2009) Possible role of amyloid-beta, adenine nucleotide translocase and cyclophilin-D interaction in mitochondrial dysfunction of Alzheimer's disease. Bioinformation 3:440–445

Smith KM, Matson S, Matson WR, Cormier K, Del Signore SJ, Hagerty SW, Stack EC, Ryu H, Ferrante RJ (2006) Dose ranging and efficacy study of high-dose coenzyme Q10 formulations in Huntington's disease mice. Biochim Biophys Acta 1762:616–626

Sofi F, Cesari F, Abbate R, Gensini GF, Casini A (2008) Adherence to Mediterranean diet and health status: meta-analysis. Br Med J 337:a1334

Song SM, Rauw G, Baker GB, Kar S (2008) Memantine protects rat cortical cultured neurons against beta-amyloid-induced toxicity by attenuating tau phosphorylation. Eur J Neurosci 28:1989–2002

Sonkusare SK, Kaul CL, Ramarao P (2005) Dementia of Alzheimer's disease and other neurodegenerative disorders–memantine, a new hope. Pharmacol Res 51:1–17

Sparks DL, Sabbagh MN, Connor DJ, Lopez J, Launer LJ, Petanceska S, Browne P, Wassar D, Johnson-Traver S, Lochhead J, Ziolkowski C (2005) Atorvastatin therapy lowers circulating cholesterol but not free radical activity in advance of identifiable clinical benefit in the treatment of mild-to-moderate AD. Curr Alzheimer Res 2:343–353

Spires TL, Hannan A (2005) Nature, nurture and neurology: gene-environment interactions in neurodegenerative disease. FEBS Anniversary Prize Lecture delivered on 27 June 2004 at the 29th FEBS Congress in Warsaw. FASEB J 272:2347–2361

Stack EC, Smith KM, Ryu H, Cormier K, Chen M, Hagerty SW, Del Signore SJ, Cudkowicz ME, Friedlander RM, Ferrante RJ (2006) Combination therapy using minocycline and coenzyme Q10 in R6/2 transgenic Huntington's disease mice. Biochim Biophys Acta 1762:373–380

Stewart LR, White AR, Jobling MF, Needham BE, Maher F, Thyer J, Beyreuther K, Masters CL, Collins SJ, Cappai R (2001) Involvement of the 5-lipoxygenase pathway in the neurotoxicity of the prion peptide PrP106–126. J Neurosci Res 65:565–572

Sullivan PG, Rabchevsky AG, Waldmeier PC, Spiringer JE (2005) J. Mitochondrial permeability transition in CNS trauma: cause or effect of neuronal cell death? Neurosci Res 79:231–239

Tan DX, Manchester LC, Sainz R, Mayo JC, Alvares FL, Reiter RJ (2003) Antioxidant strategies in protection against neurodegenerative disorders. Expert Opin Ther Patents 13:1513–1543

Tanaka M, Machida Y, Niu S, Ikeda T, Jana NR, Doi H, Kurosawa M, Nekooki M, Nukina N (2004) Trehalose alleviates polyglutamine-mediated pathology in a mouse model of Huntington disease. Nat Med 10:148–154

Tarawneh R, Holtzman DM (2009) Critical issues for successful immunotherapy in Alzheimer's disease: development of biomarkers and methods for early detection and intervention. CNS Neurol Disord Drug Targets 8:144–159

Todd NV, Morrow J, Doh-Ura K, Dealler S, O'Hare S, Farling P, Duddy M, Rainov NG (2005) Cerebroventricular infusion of pentosan polysulphate in human variant Creutzfeldt-Jakob disease. J Infect 50:394–396

Tomita T (2009) Alzheimer's disease treatment by inhibition/modulation of the gamma-secretase activity. Rinsho Shinkeigaku 49:845–847

Townsend L, Ongini E, Wenk G (2004) Novel therapeutic opportunities for Alzheimer's disease: focus on nonsteroidal anti-inflammatory drugs. FASEB J 19:1592–1601

Trejo JL, Carro E, Nunez A, Torres-Aleman I (2002) Sedentary life impairs self-reparative processes in the brain: the role of serum insulin-like growth factor-I. Rev Neurosci 13:365–374

Tsuboi Y, Doh-ura K, Yamada T (2009) Continuous intraventricular infusion of pentosan polysulfate: clinical trial against prion diseases. Neuropathology 29(5):632–636

Tsujimoto Y, Shimizu S (2003) Role of the mitochondrial membrane permeability transition in cell death. Apoptosis 12:835–840

Turnbull S, Tabner BJ, Brown DR, Allsop D (2003) Quinacrine acts as an antioxidant and reduces the toxicity of the prion peptide PrP106–126. NeuroReport 14:1743–1745

Urosevic N, Martin RN (2008) Infection and Alzheimer's disease: the APOE epsilon4 connection and lipid metabolism. J Alzheimer Dis 13:421–435

Urushitani M (2009) Future perspectives of immunotherapy against ALS. Rinsho Shinkeigaku 49:818–820

Valverde G, De Andrade D, Madureira de Oliveria D, Barreto G, Bertolino LA, Saraceno E, Capani F, Giraldez LD (2008) Effects of the extract of Anemopaegma mirandum (Catuaba) on Rotenone-induced apoptosis in human neuroblastomas SH-SY5Y cells. Brain Res 1198: 188–196

Vassar R, Bennett BD, Babu-Khan S, Kahn S, Mendiaz EA, Denis P, Teplow DB, Ross S, Amarante P, Loeloff R, et al (1999) β-Secretase cleavage of Alzheimer's amyloid precursor protein by transmembrane aspartic protease BACE. Science 286:735–741

Vaughan CJ (2003) Prevention of stroke and dementia with statins: effects beyond lipid lowering. Am J Cardiol 91:23B–29B

Vaynman S, Ying Z, Gomez-Pinilla F (2004) Hippocampal BDNF mediates the efficacy of exercise on synaptic plasticity and cognition. Eur J Neurosci 20:2580–2590

Vaynman S, Gomez-Pinilla F (2006) Revenge of the "sit": how lifestyle impacts neuronal and cognitive health through molecular systems that interface energy metabolism with neuronal plasticity. J Neurosci Res 84:699–715

Vaynman S, Ying Z, Wu A, Gomez-Pinilla F (2006) Coupling energy metabolism with a mechanism to support brain-derived neurotrophic factor-mediated synaptic plasticity. Neuroscience 139:1221–1234

Vogtherr M, Grimme S, Elshorst B, Jacobs DM, Fiebig K, Griesinger C, Zahn R (2003) Antimalarial drug quinacrine binds to C-terminal helix of cellular prion protein. J Med Chem 46:3563–3564

Vulih-Shutzman I, Pinhasov A, Mandel S, Grigoriadis N, Touloumi O, Pittel Z, Gozes I (2007) Activity-dependent neuroprotective protein snippet NAP reduces tau hyperphosphorylation and enhances learning in a novel transgenic mouse model. J Pharmacol Exp Ther 323: 438–449

Waldmeier PC, Zimmermann K, Qian T, Tintelnot-Blomley M, Lemasters JJ (2003) Cyclophilin D as a drug target. Curr Med Chem 10:1485–1506

Wang X, Zhu S, Drozda M, Zhang W, Stavrovskaya IG, Cattaneo E, Ferrante RJ, Kristal BS, Friedlander RM (2003) Minocycline inhibits caspase-independent and -dependent mitochondrial cell death pathways in models of Huntington's disease. Proc Natl Acad Sci USA 100:10483–10487

Wang R, Tang XC (2005) Neuroprotective effects of huperzine A. A natural cholinesterase inhibitor for the treatment of Alzheimer's disease. Neurosignals 14:71–82

Wang R, Zhang D (2005) Memantine prolongs survival in an amyotrophic lateral sclerosis mouse model. Eur J Neurosci 22:2376–2380

Wang JY, Wen LL, Huang YN, Chen YT, Ku MC (2006) Dual effects of antioxidants in neurodegeneration: direct neuroprotection against oxidative stress and indirect protection via suppression of glia-mediated inflammation. Curr Pharmaceut Des 12:3521–3533

Wang J, Zhang HY, Tang XC (2010) Huperzine A improves chronic inflammation and cognitive decline in rats with cerebral hypoperfusion. J Neurosci Res 88:807–815

Wenk GL, Parsons CG, Danysz W (2006) Potential role of *N*-methyl-D-aspartate receptors as executors of neurodegeneration resulting from diverse insults: focus on memantine. Behav Pharmacol 17:411–424

White AR, Hawke SH (2003) Immunotherapy as a therapeutic treatment for neurodegenerative disorders. J Neurochem 87:801–808

Wilczak N, de Keyser J (2005) Insulin-like growth factor system in amyotrophic lateral sclerosis. Endocr Dev 9:160–169

Willem M, Garratt AN, Novak B, Citron M, Kaufmann S, Rittger A, DeStrooper B, Saftig P, Birchmeier C, Haass C (2006) Control of peripheral nerve myelination by the beta-secretase BACE1. Science 314:664–666

Williams A, Sarkar S, Cuddon P, Ttofi EK, Saiki S, Siddiqi FH, Jahreiss L, Fleming A, Pask D, Goldsmith P, O'Kane CJ, Floto RA, Rubinsztein DC (2008) Novel targets for Huntington's disease in an mTOR-independent autophagy pathway. Nat Chem Biol 4:295–305

Wolozin B (2002) Cholesterol and Alzheimer's disease. Biochem Soc Trans 30:525–529

Wong HK, Bauer PO, Kurosawa M, Goswami A, Washizu C, Machida Y, Tosaki A, Yamada M, Knöpfel T, Nakamura T, Nukina N (2008) Blocking acidsensing ion channel 1 alleviates Huntington's disease pathology via an ubiquitin-proteasome system-dependent mechanism. Hum Mol Genet 17:3223–3235

Wu TY, Chen CP (2009) Dual action of memantine in Alzheimer disease: a hypothesis. Taiwan J Obstet Gynecol 48:273–277

Wu HM, Tzeng NS, Qian L, Wei SJ, Hu X, Chen SH, Rawls SM, Flood P, Hong JS, Lu RB (2009) Novel neuroprotective mechanisms of memantine: increase in neurotrophic factor release from astroglia and anti-inflammation by preventing microglial activation. Neuropsychopharmacology 34:2344–2357

Yamamoto A, Lucas JJ, Hen R (2000) Reversal of neuropathology and motor dysfunction in a conditional model of Huntington's disease. Cell 101:57–66

Zhang HY, Yan H, Tang XC (2008) Non-cholinergic effects of huperzine A: beyond inhibition of acetylcholinesterase. Cell Mol Neurobiol 28:173–183

Zhou J, Tang XC (2002) Huperzine A attenuates apoptosis and mitochondria-dependent caspase-3 in rat cortical neurons. FEBS Lett 526:21–25

Zhou J, Yi J, Fu R, Liu E, Siddique T, Ríos E, Deng HX (2010) Hyperactive intracellular calcium signaling associated with localized mitochondrial defects in skeletal muscle of an animal model of amyotrophic lateral sclerosis. J Biol Chem 285:705–712

Zimmermann M, Gardoni F, Di Luca M (2005) Molecular rationale for the pharmacological treatment of Alzheimer's disease. Drugs Aging 22(Suppl 1):27–37

Zona C, Siniscalchi A, Mercuri NB, Bernardi G (1998) Riluzole interacts with voltage-activated sodium and potassium currents in cultured rat cortical neurons. Neuroscience 85:931–938

Zu T, Duvick LA, Kaytor MD, Berlinger MS, Zoghbi HY, Clark HB, Orr HT (2004) Recovery from polyglutamine-induced neurodegeneration in conditional SCA1 transgenic mice. J Neurosci 24:8853–8861

# Chapter 10
# Perspective and Direction for Future Developments on Neurotraumatic and Neurodegenerative Diseases

## 10.1 Introduction

Neurotraumatic and neurodegenerative diseases are mediated by synergistic action of excitotoxicity, oxidative stress, neuroinflammation, and misfolding and deposition of specific proteins in the brain tissue (Fig. 10.1) (Farooqui, 2009a). Among neurotraumatic diseases, the onset of stroke may be modulated by age, genes, diet, and lifestyle. Spinal cord injury (SCI) and traumatic brain injury (TBI) are caused by mechanical insults to the spinal cord and brain tissues (Farooqui and Horrocks, 2007), whereas stroke is a metabolic insult induced by severe reduction or blockade in cerebral blood flow. In contrast, neurodegenerative diseases occur in familial and sporadic forms. Familial mutations play an important role in protein misfolding and aggregation, but the majority of cases of neurodegenerative diseases are sporadic, indicating that other factors namely age, diet, lifestyle may also contribute to the pathogenesis of neurodegenerative diseases (Farooqui and Horrocks, 2007). In addition, post-transcriptional modifications of proteins, particularly phosphorylation and glycation, also play an important role in modification of amyloid-β (Aβ), tau (τ), prions, and transthyretin, and patients with neurodegenerative diseases contain high levels of advance glycation end products (AGEs) (Takeuchi et al., 2004; Chen et al., 2009, 2010). As stated earlier, AGE through their receptor, RAGE, may cause an increase in oxidative stress and inflammation through the formation of ROS and the induction of NF-κB (Schmitt, 2006; Miranda and Outerio, 2009; Yan et al., 2009).

Neuronal death in most neurotraumatic and neurodegenerative diseases is accompanied by the upregulation of interplay among excitotoxicity, oxidative stress, and neuroinflammation (Farooqui and Horrocks, 2007; Farooqui et al., 2008). As stated in Chapter 1, neurotraumatic diseases are associated with massive release of glutamate and overstimulation of glutamate receptors (excitotoxicity), ROS production by mitochondrial dysfunction, oxidation of arachidonic acid and activation of NADPH oxidase, and neuroinflammation caused by the generation of eicosanoids and platelet-activating factor. Neurodegeneration in neurotraumatic diseases occurs rapidly (in a matter of hours to days) because of sudden lack of oxygen, rapid decrease in ATP, disturbance in transmembrane potential, and sudden collapse of ion gradients at very early stage (Farooqui and Horrocks, 2007). In addition,

A.A. Farooqui, *Neurochemical Aspects of Neurotraumatic and Neurodegenerative Diseases*, DOI 10.1007/978-1-4419-6652-0_10,

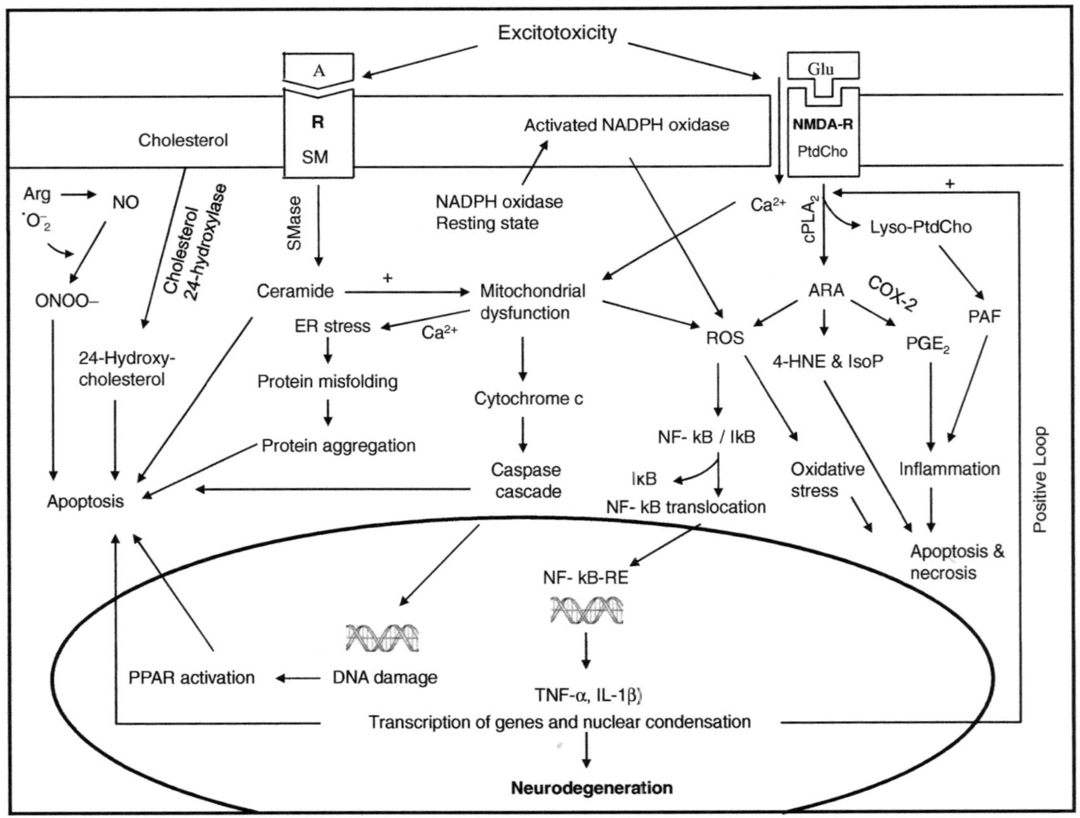


**Fig. 10.1** Diagram showing excitotoxic-, oxidative stress-, and inflammation-mediated injury in neurodegenerative diseases. Glutamate (Glu); NMDA receptor (NMDA-R); sphingomyelin (SM); sphingomyelinase (SMase); cytosolic phospholipase $A_2$ ($cPLA_2$); phosphatidylcholine (PtdCho); arachidonic acid (ARA); lyso-phosphatidylcholine (Lyso-PtdCho); reactive oxygen species (ROS); cyclooxygenase-2 (COX-2); 4-hydroxynonenal (4-HNE); prostaglandin $E_2$ ($PGE_2$); platelet-activating factor (PAF); isoprostane (IsoP). L-Arginine (Arg); nitric oxide (NO); peroxynitrite ($ONOO^-$); nuclear transcription-κB (NF-κB); tumor nectosis factor-α (TNF-a) and interleukin-1β (IL-1β) and poly(ADP-ribose) polymerase (PARP)

in neurotraumatic diseases, acute neuroinflammation develops rapidly because of rapid generation and accumulation of eicosanoids, platelet-activating factor, and the release of proinflammatory cytokines. In contrast, in neurodegenerative diseases, oxygen, nutrients, and reduced levels of ATP continue to be available to the nerve cells and ionic homeostasis is maintained to a limited extent. The interplay among excitotoxicity, oxidative stress, and neuroinflammation occurs at a slow rate, leading to a neurodegenerative process that takes several years to develop (Farooqui and Horrocks, 2007; Farooqui et al., 2008; Farooqui, 2009a). Furthermore, in neurodegenerative diseases due to abnormalities in immune system, chronic inflammation lingers for years, causing continued insult to the brain tissue and ultimately reaching the threshold of detection many years after the onset of the neurodegenerative diseases (Wood et al., 1998; Farooqui et al., 2007a). Among neurodegenerative diseases, at least in AD, alterations in neural membrane glycerophospholipids precede the clinical manifestations of the disease (dementia) (Pettegrew et al., 1995). Alterations in glycerophospholipid metabolism, increased generation of lipid mediators, and abnormal protein aggregation initiate vicious cycles of aberrant neuronal activity and compensatory alterations in neurotransmitter receptor signaling leading to loss of synapse, disintegration of neural networks, and, ultimately, failure of neurological functions. Despite of above differences, there are many similarities in molecular mechanisms of neuronal cell death observed in ischemic stroke and AD. In addition, mitochondrial and endoplasmic reticulum damage in neurotraumatic and neurodegenerative diseases shares common pathological findings such as activation of apoptotic cell death (Hayashi et al., 2006). In neurotraumatic and neurodegenerative diseases, neurons die through apoptotic cell death. This kind of cell death involves upstream effectors, such as Par-4, p53, and pro-apoptotic Bcl-2 family members, which mediate mitochondrial dysfunction and subsequent release of pro-apoptotic proteins, such as cytochrome *c* or apoptosis-inducing factor (AIF), and subsequent caspase-dependent and caspase-independent pathways, which finally lead to the degradation of cytoskeletal proteins and nuclear DNA (Mattson et al., 2001; Culmsee and Landshamer, 2006; Farooqui, 2009a). The regulation of apoptotic cascades is very complex and involves transcriptional control as well as post-transcriptional protein modifications, such as protease-mediated cleavage, ubiquitination or poly(ADP-ribosylation). In addition, the regulation of protein phosphorylation by kinases and phosphatases has emerged as a prerequisite mechanism in the control of the apoptotic cell death program in neurotraumatic and neurodegenerative diseases (Culmsee and Landshamer, 2006; Farooqui, 2009a).

## 10.2 Factors Contributing to Increased Frequency of Neurotraumatic and Neurodegenerative Diseases

The prevalence of neurotraumatic and neurodegenerative diseases has increased considerably in twenty-first century and is still increasing with a significant rate. Although traumatic injuries to brain and spinal cord result in SCI and TBI, reasons for the increased occurrence and commencement of stroke and neurodegenerative diseases remain elusive. Several factors may facilitate the onset of stroke and neurodegenerative diseases in human population (Farooqui et al., 2007b; Farooqui, 2009b). These factors include levels of ω-6 fatty acids in diet (Farooqui and Farooqui, 2009; Farooqui, 2009b) and consumption of processed food, which contain toxins, such as monosodium glutamate, aspartame, and other neurotoxins, although, above toxins may not cause stroke or neurodegenerative diseases, they may promote and facilitate the intensification of neuropathology of these diseases.

### *10.2.1 Diet and Frequency of Occurrence of Neurotraumatic and Neurodegenerative Diseases*

Ancestral humans obtained about 35% of their dietary energy from fats, 35% from carbohydrates, and 30% from protein. Saturated fats contributed approximately 7.5% total energy and harmful trans-fatty acids contributed in negligible amounts (Eaton, 2006). The introduction of food staples and food processing procedures in recent years has fundamentally changed seven crucial nutritional characteristics of ancestral human diets: (a) glycemic load, (b) fatty acid composition, (c) macronutrient composition, (d) micronutrient density, (e) acid–base balance, (f) sodium–potassium ratio, and (g) fiber content (Cordain et al., 2005). The present-day Western diet has a ratio of ω-6 to ω-3 fatty acids of about 15–20:1. The Paleolithic diet on which human beings evolved, and lived for most of their existence, had a ratio of 1:1 (Simopoulos, 2002, 2006; Cordain et al., 2005). Changes in eating habits, natural versus processed food, and agriculture development within the past 100 to 200 years have resulted in a marked increase in the ω-6 to ω-3 ratio (15–20:1). In addition, the present Western diet has decreased levels of antioxidants and micronutrients. Diet enriched in ω-6 fatty acids not only excessively generates ROS but also produces proinflammatory affects. It promotes the pathogenesis of many chronic diseases including cardiovascular diseases, autoimmune diseases, neurotraumatic and neurodegenerative diseases. In contrast, a diet enriched in ω-3 fatty acids produces cardioprotective, immunosuppressive, and neuroprotective effects (Simopoulos et al., 2006; Farooqui and Horrocks, 2007). A lower AA:DHA ratio suppresses neurodegenerative diseases (Farooqui, 2009b).

Neural membrane glycerophospholipids are synthesized from three dietary components: polyunsaturated fatty acids, uridine monophosphate (UMP), and choline (Farooqui and Horrocks, 2007). Administration of above nutrients increases the level of glycerophospholipids, specific pre- or postsynaptic proteins, and the number of dendritic spines – a requirement for new synapse formation (Wurtman et al., 2009; Kamphnis and Wurtman, 2009). These effects are markedly enhanced when animals receive all three compounds together. This multi-nutrient approach in animals has also been shown to decrease Aβ plaque burden, improve learning and memory through increased cholinergic neurotransmission, and have a neuroprotective effect in several mouse models of AD (Wurtman et al., 2009;

Kamphnis and Wurtman, 2009). It remains to be seen whether these potential therapeutic effects of a multi-nutrient can also be replicated in clinical settings or not and more studies are required on this aspects of nutrition in human subjects.

The consumption of high-fructose corn syrup (HFCS) has increased >1,000% between 1970 and 1990 (Bray et al., 2004). This is a drastic change in intake of any other food or food group. HFCS now represents >40% of caloric sweeteners added to foods and beverages and is the sole caloric sweetener in soft drinks in the USA. The digestion, absorption, and metabolism of fructose differ markedly from glucose. Brain does not metabolize fructose. In liver, utilization and metabolism of fructose favors de novo lipogenesis and generates uric acid. In addition, unlike glucose, fructose does not stimulate insulin secretion or enhance leptin production (Bray et al., 2004). Because insulin and leptin act as key afferent signals in the regulation of food intake and body weight, the consumption of fructose contributes to increased energy intake, weight gain, and hypertension (metabolic syndrome) (Bray et al., 2004). Rats fed with a high-fat, high-glucose diet supplemented with high-fructose corn syrup show alterations not only in energy and lipid metabolism but also in elevation in fasting glucose and also increase in cholesterol and triglyceride (Stranahan et al., 2008). These characteristics are similar to diabetes. Rats fed diet enriched in high-fructose corn syrup for 8 months show impairment in spatial learning ability, reduction in hippocampal dendritic spine density, and decrease in long-term potentiation at Schaffer collateral–CA1 synapses (Stranahan et al., 2008). Based on these findings, it is suggested that high-calorie diet reduces hippocampal synaptic plasticity and impairs cognitive function, possibly through BDNF-mediated effects on dendritic spines (Molteni et al., 2002; Stranahan et al., 2008). It is also reported that BDNF modulates synaptic plasticity not only by functioning as metabolic modulators but also by responding to peripheral signals, such as food intake (Gomez-Pinilla, 2008). There is a growing interest in possible links among impaired insulin signaling obesity, type 2 diabetes mellitus, and the pathogenesis of AD. Insulin requires nitric oxide to stimulate glucose uptake, it is likely that fructose-mediated hyperuricemia may aid to the pathogenesis of AD. In spite of above view, investigators have not been able to observe many well-known features of AD and more studies are needed to reach meaningful conclusion (Revill et al., 2006).

### *10.2.2 Detection of Neurotraumatic and Neurodegenerative Diseases*

The key for successful treatment of neurodegenerative diseases is early detection. In recent years, neuroimaging, a noninvasive functional imaging technique, has been used to assist with the diagnosis of neurodegenerative diseases, dementia, and determining chemical biomarkers for clinical progression in patients with neurodegenerative diseases and normal age-matched controls (Caroli and Frisoni, 2009; Vemuri et al., 2009). Similarly, clinical structural imaging (CT or MRI) is used to identify space-occupying lesions and stroke. In the earliest clinical stages of neurodegenerative diseases when symptoms are mild, clinical diagnosis is difficult. Since the pathology of neurodegenerative disease precedes its symptoms, biomarkers can serve as early diagnostic indicators and used to monitor preclinical pathologic changes (Hampel et al., 2008; de Leon et al., 2004). Neurodegeneration-mediated changes in the brain may be investigated using neuroimaging techniques. The in vivo techniques are useful for monitoring major changes. In addition, neuroimaging can also be used for monitoring progressing brain abnormalities (Langstrom et al., 2007). However, quantification of minor abnormalities requires postmortem brain tissue. These in vitro methods are complementary to the in vivo techniques and contribute to the knowledge on pathophysiology and etiology of the neurodegenerative diseases (Langstrom et al., 2007). Positron emission tomography (PET) and single photon emission computed tomography (SPECT) are novel neuroimaging techniques that not only provide early detection but can also be used for studying in vivo neurochemical, hemodynamic, or metabolic consequences of the degeneration of neurons in neurodegenerative diseases (Thobois et al., 2001). Investigators are making attempts to synthesize and develop radiotracers for in vivo imaging Aβ plaques in the aging human brain and AD patients. Quantitative evaluation of Aβ plaques in the brain may allow the evaluation of the efficacy of anti-amyloid therapies in AD patients and aged-matched human subjects. Studies on radiolabeled amyloid imaging agents using [$^{18}$F]FDDNP, [$^{11}$C]PIB, [$^{11}$C]SB-13, and [$^{123}$I]IMPY indicate that detecting Aβ plaques in the living human brain with amyloid imaging agents may be feasible (Ono, 2007). In addition, studies on inverse correlations between Aβ load measured by Pittsburgh Compound-B (PiB) positron emission tomography (PET) and cerebral metabolism using [$^{18}$F]fluoro-2-deoxy-D-glucose (FDG) in AD patients suggest that local Aβ-induced metabolic insult may initiate the pathogenesis of AD (Cohen et al., 2009; Small et al., 2008). Thus, PET scanning can be used to differentiate glucose metabolism in AD patients from patients with frontotemporal dementia. This can help to guide clinicians in symptomatic treatment strategies in these neurological conditions. In addition, magnetic resonance imaging (MRI) can also be used to detect excessive iron in brains of multiple sclerosis, AD, and PD (Bass et al., 2006). Collective evidence suggests that functional imaging techniques may not only provide insight into the pathophysiology of neurodegenerative diseases but also provide information on mechanism(s) of their progression. They also provide information in assessing the efficacy of putative neuroprotective and restorative therapy (Small et al., 2008; Cohen et al., 2009). The availability and frequent use of neuroimaging techniques have made it easy to diagnose more cases of neurodegenerative diseases.

## 10.3 Proteomics and Lipidomics in Neurotraumatic and Neurodegenerative Diseases

With the empowerment of proteomics and lipidomics of tissue and biological fluid samples, investigators are able to detect low levels of ideal biomarkers for neurotraumatic and neurodegenerative diseases. Proteomics and lipidomics have made it easy to quantify and measure reproducible biomarkers (hyperphosphorylation of tau, alterations in Aβ42 levels, variation in levels of $F_2$-isoprostanes, prostaglandins, leukotrienes, lipoxins, hydroxyeicosatetraenoic acids, nitrotyrosine, carbonyls in proteins, oxidized DNA bases, and 4-HNE, pattern and rate of atrophy along with functional and cognitive decline) that show little variation in the general population and unaffected by comorbid factors (Henley et al., 2005; Migliore et al., 2005). An ideal biomarker for the detection of neurotraumatic and neurodegenerative diseases not only should be reliable and distinguishable between biological fluid from normal and neurotraumatic and neurodegenerative diseases but also should be specific for each disease. It should be reproducible and easy to quantify (Henley et al., 2005). Establishment of automatic systems including databases and accurate analyses of above mediators will facilitate the identification of key biomarkers associated with neurotraumatic and neurodegenerative diseases (Lu et al., 2006). Although it is unlikely that any one biomarker may be able to fulfill all characteristics of an ideal biomarker, it is likely that determination of more than one biomarkers may not only promote early diagnosis of patients with neurotraumatic and neurodegenerative diseases but also facilitate monitoring and evaluation of effect of therapeutic agents in stroke and neurodegenerative disease patients. Studies on the relationship between baseline MRI and CSF biomarkers and subsequent changes in continuous measures of cognitive and functional abilities in cognitively normal (CN) subjects and patients with amnestic mild cognitive impairment (aMCI) and Alzheimer disease (AD) are beginning to appear in literature. It is reported that MRI and CSF provide complimentary predictive information about time to conversion from aMCI to AD and combination of these procedures may provide better prediction of AD than either source alone (Farooqui, 2009a). To date, most established CSF biomarkers (Aβ, τ-protein, and hyperphosphorylated τ) and structural and functional changes observed during neuroimaging have not achieved widespread clinical application. Although the development and validation of precise, reliable, and robust biomarker in CSF is a step in the right direction, identification and development of biomarker in blood, plasma, or serum will be an ideal step for the diagnosis of large populations with the risk of neurotraumatic and neurodegenerative diseases (Schneider et al., 2009).

## 10.4 Vaccines for the Treatment of Neurotraumatic and Neurodegenerative Diseases

In recent years, vaccines have been developed for treating AD, PD, HD, epilepsy, multiple sclerosis (MS), spinal cord injury (SCI), and stroke. Although studies on the treatment of neurodegenerative diseases by various vaccines have failed due to side effects, investigators continue to make advances at the immunology of neurodegenerative diseases. DNA vaccines have emerged as novel therapeutic agents because of the simplicity of their production and application (Nile et al., 2007). Myelin components, such as neurite outgrowth inhibitory protein (NOGO), myelin-associated glycoprotein (MAG), and oligodendrocyte-myelin glycoprotein (OMGP), promote demyelinating autoimmunity and prevent axonal regeneration. The development of DNA vaccines encoding NOGO, MAG, and OMGP and their fragments make them suitable vehicles for treatment of SCI. Recombinant DNA vaccine activates the immune system but does not induce experimental autoimmune encephalomyelitis (EAE) in Lewis rats (Xu et al., 2004; Nile et al., 2007). Recombinant DNA vaccine also promotes axonal regeneration in a spinal cord injury model. Thus, the use of DNA vaccine that encodes multiple specific domains of major inhibitory proteins and/or their receptors (NgR1 and NgR2) provides another promising approach to overcome the inhibitory barriers during CNS regeneration. Advances in understanding the immunologic mechanisms underlying the neuroprotective immunity to optimize the design of DNA vaccines for their use in clinical setting will facilitate therapy for neurodegenerative diseases.

## 10.5 Reasons for the Failure of Treatment in Neurotraumatic and Neurodegenerative Diseases

One major goal of current research in neurotraumatic and neurodegenerative diseases is the discovery of novel drugs not only to improve symptomatic management but also to block or retard the primary pathogenic mechanism(s). Therapeutic agent should be able to provide effective symptom control throughout the course of the disease without the development of side effects. Results of several clinical trials using combination of drugs, such as neurotransmitter replacement combined with a drug to protect against the toxic effect of accumulating aggregated proteins in neurotraumatic and neurodegenerative diseases from last decade, have been negative or unsatisfactory (Farooqui, 2009).

There are several reasons for the failure of treatment in neurotraumatic and neurodegenerative diseases, including not only the understanding of molecular mechanism of neurotraumatic and neurodegenerative diseases but also half-life, blood–brain barrier permeability safety, tolerability, effect on cognitive function, and molecular mechanism associated with the therapeutic actions of drugs. Delivery of optimal dose of drugs into the brain is one of the most challenging problem faced in the treatment of neurotraumatic and neurodegenerative diseases. Most neurotraumatic and neurodegenerative diseases are accompanied by excitotoxicity, oxidative stress, and neuroinflammation. Very little is known about the relationship between start of excitotoxicity, oxidative stress, and neuroinflammation and onset of neurotraumatic and neurodegenerative diseases. For excitotoxic, antioxidant, and anti-inflammatory therapy to work, drugs should be taken early in life because long-term excitotoxic, oxidative stress, and inflammatory imbalance and damage in neurons of specific area in brain cannot be compensated and fully or partially reversed by drugs that are given after the onset of neurodegenerative process. Neurons are more susceptible to ROS-mediated oxidative injury than glial cells. The chronic activation of microglia and astrocytes when a neurotraumatic or neurodegenerative disease starts may also cause damage to the brain–blood barrier. For the excitotoxic, oxidative, and neuroinflammatory effect to occur, the direct contact between activated microglia and degenerating neurons is not necessary because immune and inflammatory mediators (nitric oxide, proinflammatory cytokines, chemokines, and complement proteins) secreted and released by activated glial cells diffuse and reach neurons to act as endogenous neurotoxins to facilitate neurodegeneration (Block and Hong, 2005; Farooqui and Horrocks, 2007). Therefore, the use of a cocktail of antiexcitotoxic, antioxidant, and anitiinflammatory compounds cocktail has been recommended for correcting the fundamental oxidant/antioxidant and proinflammatory/anti-inflammatory imbalance in patients suffering from neurotraumatic and neurodegenerative diseases at the earliest stages (Gilgun-Sherki et al., 2006; Wang et al., 2006). However, once the onset of a neurotraumatic or neurodegenerative disease has occurred, reversal through the use of cocktail of antiexcitotoxic, antioxidant, and anti-inflammatory agents may be mechanistically improbable.

The efficacy of a drug or cocktail of drugs for treating neurotraumatic and neurodegenerative diseases may depend not only on half-lives and ability of drugs to cross the blood–brain barrier but also on potential ability of drug components to effectively reach various subcellular particles and their synergistic actions (Gilgun-Sherki et al., 2006; Tan et al., 2003). Thus, for the treatment of neurotraumatic or neurodegenerative diseases, a localized and controlled delivery of drugs at the site, where neurons are dying, is preferred because it reduces drug toxicity and increases treatment efficiency. Drugs aimed to treat neurotransmitter deficit have failed because neuronal signaling pathway network has been disrupted by the accumulated abnormal protein aggregates, creating a downstream block that cannot be overcome by upstream modulation of receptor function by neurotransmitter stimulating drugs (Palop et al., 2006). This view is supported by studies on animal models of AD and HD. It is reported that abnormal protein aggregates disrupt glutamatergic neurotransmission and calcium signaling (Handley et al., 2006; Xie, 2004). Another factor is the heterogeneity of neurotraumatic and neurodegenerative diseases. Many if not all neurotraumatic and neurodegenerative diseases are heterogeneous with respect to their clinical, biochemical, and genetic features. It is likely that these variants of these diseases may have different courses and responses to therapeutic agents.

## 10.6 Future Studies on the Treatment of Neurotraumatic and Neurodegenerative Diseases

At present, investigators are making attempts to eliminate the abnormal protein assemblies with the hope that enhancing the elimination of abnormal proteins and their assemblies may improve neuronal survival. Molecular chaperones are the first line of defense against misfolded, aggregation-prone proteins. Hsp70 and Hsp40 have been shown to exert therapeutic effects against various experimental models of the polyQ diseases (Waza et al., 2006). The discovery of small chemical activators of heat shock transcription factor 1 (HSF1), such as geldanamycin and its derivative, 17-allylamino-17-demethoxygeldanamycin (17-AAG), which induce multiple endogenous molecular chaperones, is a positive step. 17-AAG not only induces Hsp70 and Hsp40 in vivo but also enhances the degradation of mutant proteins (Waza et al., 2006; Naigai et al., 2010). The ability of 17-AAG to preferentially degrade mutant protein is directly applicable to spinal and bulbar muscular atrophy and animal models of neurodegenerative diseases. 17-AAG has shown to be effective not only in polyQ disease models but also in models of other neurodegenerative disease (Waza et al., 2006; Naigai et al., 2010). Moreover, knocking down of HSF1 abolishes the induction of molecular chaperones and the therapeutic effect of 17-AAG, indicating that its therapeutic effects depend on HSF1 activation (Waza et al., 2006; Naigai et al., 2010). Thus, the development of more blood–brain barrier-permeable molecular chaperone inducers will facilitate new treatment for a wide range of neurodegenerative diseases.

Although it is not known when misfolding and abnormal accumulation of proteins in neurodegenerative diseases start and when treatment for eliminating these proteins should be started, but there is some information that indicates that deposition of abnormal proteins occurs in the fourth decade of life. For example, Aβ levels in the plasma begin to rise after 40 years of age. Furthermore, there is a hypothesis that Aβ accumulation starts around the time of menopause (Finch et al., 1999). If the functional decline in neurodegenerative diseases is primarily caused by the slow neurodegeneration, it may take years to detect benefits of these disease-modifying treatments. However, if the pathogenic proteins actively interfere with signaling network and synaptic dysfunction then removal of abnormal proteins and their assemblies may produce apparent effects within weeks or months (Palop et al., 2006). Availability of this information not only can facilitate better planning of clinical trial periods (shorter or longer) but can also promote the evaluation of many drugs with shorter or longer half-lives. For drug evaluating companies, these steps will increase the pace of drug validation at a low budget. Consideration of above factors in development of drugs with multiple actions (Jin et al., 2010; Palop et al., 2006) along with agents that increase the production of ATP in degenerating neurons can improve the therapeutic outcome for the treatment of neurotraumatic and neurodegenerative diseases. A clearer appreciation of the potential therapeutic ability of antiexcitotoxic, anti-inflammatory, and antioxidant cocktail can emerge only when in vivo importance of interactions among excitotoxicity, neuroinflammation, and oxidative stress is realized and fully understood at the molecular level (Farooqui and Horrocks, 2007; Farooqui et al., 2007a, b).

It is well known that mammalian brain has ability to undergo experience-mediated adaptations. This property is reflected in the ability of neural cells to continuously modify the neural circuitry not only to feelings and behavior but also to interact effectively with their environment and to cope better with neural injuries. This process is called as neural plasticity. Four core factors modulate neural plasticity. They include reduced schedules of brain activity, noisy processing, weakened neuromodulatory control, and negative learning. The locus of this plasticity occurs at the level of synapses, the specialized junctions where one neuron receives chemical signals from another (Fleming and England, 2010). Synaptic connections become stronger or weaker in response to specific patterns of activity. This activity modulates changes not only in the release of neurotransmitters at presynaptic neurons but also in the receptors localized on postsynaptic neurons. It is proposed that accumulation of abnormal aggregated proteins in neurodegenerative diseases may impair the integrity or function of presynaptic terminals and postsynaptic specializations through interplay among excitotoxicity, neuroinflammation, and oxidative stress (Farooqui and Horrocks, 2007; Palop et al., 2006). Although cell death in neurotraumatic and neurodegenerative diseases involves interplay among excitotoxicity, neuroinflammation, and oxidative stress, the intensity of this interplay is faster in neurotraumatic diseases than neurodegenerative diseases.

We are in the midst of a national crisis. As stated above, the number of patients with neurotraumatic and neurodegenerative diseases is increasing with constant rate. As baby boomer generation grows older, enormous impact of neurotraumatic and neurodegenerative diseases will be felt by the American society (Brookmeyer et al., 1998; Cogan and Mitchell, 2003; Hodes, 2006; Trojanowski, 2008). In 2005, the number of patients with neurodegenerative diseases in the world was about 25 million, with more than 4 million new cases occurring each year. It is stated that the number of people affected will double every 20 years to 80 million by 2040, if a cure of neurodegenerative diseases is not discovered. Among neurodegenerative diseases, more than 30–35% of cases are due to AD. Today, approximately 5 million people in the USA suffer from AD, representing one in eight people over the age of 65. The projected cost to Medicare for treating AD patients is estimated to be about 1 trillion dollars by 2050. This number does not include other neurotraumatic and neurodegenerative diseases. Such a budget not only will burst NIH budget but will seriously affect US economy. Thus, developing strategies for the treatment of neurotraumatic and neurodegenerative diseases and use of substances that protect and promote a healthy nervous system is extremely important (Hodes, 2006; Trojanowski, 2008).

## 10.7 Conclusion

Neurodegenerative diseases are multifactorial disease of unknown causes. Since the number of patients with neurodegenerative diseases is increasing with a significant rate, finding therapeutic ways to prevent and lower the risk of neurodegenerative diseases is a crucial matter. Although some information is available on risk factors (dietary habits, genetics and heredity, age, and lifestyle, exposure to neurotoxins) for developing neurodegenerative diseases, information on optimal preventive strategies as well as drugs development is still in developing state. Hypothesis that a common mechanism involving misfolded proteins triggers a toxic cascade that leads to neuronal degeneration is very interesting. This hypothesis is the basis of the therapeutic potential of heat shock proteins, which prevent protein misfolding and aggregation. The principal routes of intracellular protein metabolism are the ubiquitin-proteasome system and the autophagy-lysosome pathway. These routes collaborate to degrade wasted proteins and their interplay is involved in coping with the neurological diseases, in which molecular chaperones play collective role by assisting the protein targeting to the proteasome or autophagy. Establishing the molecular mechanism associated with protein misfolding of β-amyloid, τ-protein, huntingtin, and α-synuclein is an important problem. Although the molecular mechanisms of different pathologies with regard to the disease development remain illusive, gene expression, protein–protein interactions, neuroplasticity, and synaptic dysfunction are closely associated with neurodegenerative process. Drugs that block misfolding and facilitate removal of misfolded protein from neurons may prevent or delay the pathogenesis of above chronic diseases. Overexpression of heat shock proteins reduces the number and size of inclusions and accumulation of disease-causing proteins. Hsp90 inhibitors also exert therapeutic effects through selective proteasome degradation of its client proteins.

Use of neuroimaging procedures (PET and SPECT) to diagnose, detect, and allow in vivo quantification of radiolabeled lipid mediator concentration in the subpicomolar range will be helpful in detection of neurodegenerative process at asymptomatic stages when there is no indication on CT and MRI. Collectively, neuroimaging may shed some light on the polymorphism and facilitate the identification of variants of neurodegenerative disorders. Identification of biomarkers for neurodegenerative diseases may not only lead to early diagnosis and follow-up of the progression of neurodegenerative diseases but also allow monitoring of therapeutic responses.

## References


Bass SD, Chen WK, Mulkem RV, Bakshi R (2006) Magnetic resonance imaging of iron deposition in neurological disorders. Top Magn Reson Imaging 17:31–40

Block ML, Hong J-S (2005) Microglia and inflammation-mediated neurodegeneration: multiple triggers with a common mechanism. Prog Neurobiol 76:77–98

Bray GA, Nielsen SJ, Popkin BM (2004) Consumption of high-fructose corn syrup in beverages may play a role in the epidemic of obesity. Am J Clin Nutr 79:537–543

Brookmeyer R, Gray S, Kawas C (1998) Projections of Alzheimer's disease in the United State and public health impact of delaying disease onset. Am J Public Health 88:1337–1342

Caroli A, Frisoni GB (2009) Quantitative evaluation of Alzheimer's disease. Expert Rev Med Devices 6:569–588

Chen L, Wei Y, Wang X, He R (2009) D-Ribosylated Tau forms globular aggregates with high cytotoxicity. Cell Mol Life Sci 66:2559–2571

Chen L, Wei Y, Wang X, He R (2010) Ribosylation rapidly induces alpha-synuclein to form highly cytotoxic molten globules of advanced glycation end products. PLoS One 5(2):e9052

Cogan JF, Mitchell OS (2003) Perspective from the President's commission on Social Security Reform. J Econ Perspect 7:149–172

Cohen AD, Price JC, Weissfeld LA, James J, Rosario BL, Bi W, Nebes RD, Saxton JA, Snitz BE, Aizenstein HA, Wolk DA, Dekosky ST, Mathis CA, Klunk WE (2009) Basal cerebral metabolism may modulate the cognitive effects of Abeta in mild cognitive impairment: an example of brain reserve. J Neurosci 29:14770–14778

Cordain L, Eaton SB, Sebastian A, Mann N, Lindeberg S, Watkins BA, O'Keefe JH, Brand-Miller J (2005) Origins and evolution of the Western diet: health implications for the 21st century. Am J Clin Nutr 81:341–354

Culmsee C, Landshamer S (2006) Molecular insights into mechanisms of the cell death program: role in the progression of neurodegenerative disorders. Curr Alzheimer Res 3:269–283

de Leon MJ, DeSante S, Zinkowski R, Kerkman D, DeBermadis J, Li J, Lair L, Reiberg B, Tsui W, Rusinnek H (2004) MRI and CSF studies in the early diagnosis of Alzheimer's disease. J Intern Med 256:205–223

Eaton SB (2006) The ancestral human diet: what was it and should it be a paradigm for contemporary nutrition? Proc Nutr Soc 65:1–6

Farooqui AA, Horrocks LA (2007) Glycerophospholipids in brain. Springer, New York, NY

Farooqui AA, Horrocks LA, Farooqui T (2007a) Modulation of inflammation in brain: a matter of fat. J Neurochem 101:577–599

Farooqui AA, Ong WY, Horrocks LA, Chen P, Farooqui T (2007b) Comparison of biochemical effects of statins and fish oil in brain: the battle of the titans. Brain Res Rev 56:443–471

Farooqui AA, Ong WY, Horrocks LA (2008) Neurochemical aspects of excitotoxicity. Springer, New York, NY

Farooqui T, Farooqui AA (2009) Aging: an important factor for the pathogenesis of neurodegenerative diseases. Mech Aging Dev 130:203–215

Farooqui AA (2009a) Hot topics in neural membrane lipidology. Springer, New York, NY

Farooqui AA (2009b) Beneficial effects of fish oil on human brain. Springer, New York, NY

Finch CE, Sapolsky RM (1999) The evolution of Alzheimer disease, the reproductive schedule, and apoE isoforms. Neurobiol Aging 20:407–428

Fleming JJ, England PM (2010) AMPA receptors and synaptic plasticity: a chemist's perspective. Nat Chem Biol 6:89–97

Gilgun-Sherki Y, Melamed E, Offen D (2006) Anti-inflammatory drugs in the treatment of neurodegenerative diseases: current state. Curr Pharmaceut Des 12:3509–3519

Gomez-Pinilla F (2008) Brain foods: the effects of nutrients on brain function. Nat Neurosci Rev 9:568–578

Hampel H, Burger K, Teipel SJ, Bokde AL, Zetterberg H, Blennow K (2008) Core candidate neurochemical and imaging biomarkers of Alzheimer's disease. Alzheimers Dement 4:38–48

Handley OJ, Naji JJ, Dunnett SB, Rosser AE (2006) Pharmaceutical, cellular and genetic therapies for Huntington's disease. Clin Sci (Lond) 110:73–88

Hayashi T, Shoji M, Abe K (2006) Molecular mechanisms of ischemic neuronal cell death–with relevance to Alzheimer's disease. Curr Alzheimer Disease 3:351–358

Henley SM, Bates GP, Tabrizi SJ (2005) Biomarkers for neurodegenerative diseases. Curr Opin Neurol 18:698–705

Hodes RJ (2006) Public funding for Alzheimer's disease research in United States. Nat Med 12:770–773

Jin H, Randazzo J, Zhang P, Kador PF (2010) Multifunctional antioxidants for the treatment of age-related diseases. J Med Chem 53:1117–1127

Kamphuis PJ, Wurtman RJ (2009) Nutrition and Alzheimer's disease: pre-clinical concepts. Eur J Neurol 16(Suppl 1):12–18

Langstrom B, Andren PE, Lindhe O, Svedberg M, Hall H (2007) In vitro imaging techniques in neurodegenerative diseases. Mol Imaging Biol 9:161–175

Lu Y, Hong S, Gotlinger K, Serhan CN (2006) Lipid mediator informatics and proteomics in inflammation-resolution. Sci World J 6:589–614

Mattson MP, Duan W, Pedersen WA, Culmsee C (2001) Neurodegenerative disorders and ischemic brain diseases. Apoptosis 6:69–81

Migliore L, Fontana I, Colognato R, Coppede F, Siciliano G, Murri L (2005) Searching for the role and the most suitable biomarkers of oxidative stress in Alzheimer's disease and in other neurodegenerative diseases. Neurobiol Aging 26:587–595

Miranda HV, Outerio TF (2009) The sour side of neurodegenerative disorders: the effects of protein glycation. J Pathol Dec 31 [Epub ahead of print]

Molteni R, Bernard RJ, Ying Z, Roberts CK, Gomez-Pinilla F (2002) A high-fat, refined sugar diet reduces hippocampal brain-derived neurotrophic factor, neuronal plasticity, and learning. Neuroscience 112:803–814

Naigai Y, Fujikake N, Popiel HA, Wada K (2010) Induction of molecular chaperones as a therapeutic strategy for the polyglutamine diseases. Curr Pharm Biotechnol Fab 16 [Epub ahead of print]

Nile DY, Xu G, Ahmed S, Xiao ZC (2007) DNA vaccine and the CNS axonal regeneration. Curr Pharm Des 13:2500–2506

Ono M (2007) Molecular imaging of beta-amyloid plaques in the brain. Brain Nerve 59:233–240

Palop JJ, Chin J, Mucke L (2006) A network dysfunction perspective on neurodegenerative diseases. Nature 443:768–773

Pettegrew JW, Klunk WE, Kanal E, Panchalingam K, McClure RJ (1995) Changes in brain membrane phospholipid and high-energy phosphate metabolism precede dementia. Neurobiol Aging 16:973–975

Revill P, Moral MA, Prous JR (2006) Impaired insulin signaling and the pathogenesis of Alzheimer's disease. Drug Today 42:785–790

Schmitt HP (2006) epsilon-Glycation, APP and Abeta in ageing and Alzheimer disease: a hypothesis. Med Hypotheses 66:698–906

Schneider P, Hampel H, Buerger K (2009) Biological marker candidates of Alzheimer's disease in blood, plasma, and serum. CNS Neurosci Ther 15:358–374

Simopoulos AP (2002) The importance of the ratio of omega-6/omega-3 essential fatty acids. Biomed Pharmacother 56:365–379

Simopoulos AP (2006) Evolutionary aspects of diet, the omega-6/omega-3 ratio and genetic variation: nutritional implications for chronic diseases. Biomed Pharmacother 60:502–507

Small GW, Bookheimer SY, Thompson PM, Cole GM, Huang SC, Kepe V, Barrio JR (2008) Current and future uses of neuroimaging for cognitively impaired patients. Lancet Neurol 7:161–172

Stranahan RS, Normal ED, Lee K, Cutler RG, Telljohan RS, Egan JM, Mattson MP (2008) Diet-induced insulin resistance impairs hippocampal synaptic plasticity and cognition in middle-aged rats. Hippocampus 18:1085–1088

Takeuchi M, Yamagishi S, Iwaki M, Nakamura K, Imaizumi T (2004) Advanced glycation end product (age) inhibitors and their therapeutic implications in diseases. Int J Clin Pharmacol 24:95–101

Tan DX, Manchester LC, Sainz R, Mayo JC, Alvares FL, Reiter RJ (2003) Antioxidant strategies in protection against neurodegenerative disorders. Expert Opin Ther Patents 13:1513–1543

Thobois S, Guillouet S, Bronssolle E (2001) Contributions of PET and SPECT to the understanding of the pathophysiology of Parkinson's disease. Neurophysiol Clin 31:321–340

Trojanowski JQ (2008) PENN neurodegenerative disease research – in the spirit of Benjamin Franklin. Neurosignals 16:5–10

Vemuri P, Wiste HJ, Weigand SD, Shaw LM, Trojanowski JQ, Weiner MW, Knopman DS, Petersen RC, Jack CR Jr Alzheimer's Disease Neuroimaging Initiative (2009) MRI and CSF biomarkers in normal, MCI, and AD subjects: predicting future clinical change. Neurology 73:294–301

Wang JY, Wen LL, Huang YN, Chen YT, Ku MC (2006) Dual effects of antioxidants in neurodegeneration: direct neuroprotection against oxidative stress and indirect protection via suppression of glia-mediated inflammation. Curr Pharmaceut Des 12:3521–3533

Waza M, Adachi H, Katsuno M, Minamiyama M, Tanaka F, Doyu M, Sobue G (2006) Modulation of Hsp90 function in neurodegenerative disorders: a molecular-targeted therapy against disease-causing protein. J Mol Med 84:635–646

Wood PL (1998) Neuroinflammation: mechanisms and management. Humana Press, Totowa, NJ

Wurtman RJ, Cancev W, Sakamoto T, Ulus IH (2009) Use of phosphatide precursors to promote synaptogenesis. Annu Rev Nutri 29:59–87

Xie CW (2004) Calcium-regulated signaling pathways: role in amyloid β-induced synaptic dysfunction. Neuromolecular Med 6:53–64

Xu G, Nie DY, Chen JT, Wang CY, Yu FG, Sun L, Luo XG, Ahmed S, Xiao ZC (2004) Recombinant DNA vaccine encoding multiple domains related to inhibition of neurite outgrowth: a potential strategy for axonal regeneration. J Neurochem 91:1018–1023

Yan SD, Bierhaus A, Nawroth PP, Stern DM (2009) RAGE and Alzheimer's disease: a progression factor for amyloid-beta-induced cellular perturbation? J Alzheimer Res 16:833–843

# Index




A.A. Farooqui, *Neurochemical Aspects of Neurotraumatic and Neurodegenerative Diseases*, DOI 10.1007/978-1-4419-6652-0,

MIX
Papier aus verantwortungsvollen Quellen
Paper from responsible sources
FSC® C105338

If you have any concerns about our products,
you can contact us on
**ProductSafety@springernature.com**

In case Publisher is established outside the EU,
the EU authorized representative is:
**Springer Nature Customer Service Center GmbH**
**Europaplatz 3, 69115 Heidelberg, Germany**

Printed by Libri Plureos GmbH
in Hamburg, Germany